High Mountain Conservation in a Changing World

ADVANCES IN GLOBAL CHANGE RESEARCH

VOLUME 62

More information about this series at http://www.springer.com/series/5588

Jordi Catalan · Josep M. Ninot
M. Mercè Aniz
Editors

High Mountain Conservation in a Changing World

Editors
Jordi Catalan
Biogeodynamics and Biodiversity Group
CREAF—CSIC, Campus UAB
Cerdanyola
Spain

M. Mercè Aniz
Aigüestortes i Estany de Sant Maurici National Park
Boi
Spain

Josep M. Ninot
University of Barcelona
Barcelona
Spain

ISSN 1574-0919 ISSN 2215-1621 (electronic)
Advances in Global Change Research
ISBN 978-3-319-55981-0 ISBN 978-3-319-55982-7 (eBook)
DOI 10.1007/978-3-319-55982-7

Library of Congress Control Number: 2017937113

Cover Illustration: *Reflection at Night.* Author: Miquel Jover Benjumea.

Printed on acid-free paper

This Springer imprint is published by Springer Nature
The registered company is Springer International Publishing AG
The registered company address is: Gewerbestrasse 11, 6330 Cham, Switzerland

Preface

Protected natural areas deserve the combination of nature uniqueness and conservation willing. The foundation stone for National Parks is the awareness of that unique character—actual or presumed, most times based on non-tangible properties or singularities —grown in some social sectors, followed by the aim of preserving particular biota, singular ecosystems or entire landscapes. This was the case for the Aigüestortes i Estany de Sant Maurici National Park. Like other protected areas in alpine mountains, its apparent wilderness was a key factor in declaring it National Park, in 1955. The rough relief carved in the Maladeta granodiorite batholith, including rocky crests and summits, steep slopes and glacial valley bottoms, had for long grabbed attention. The abundance of alpine lakes and other water-related systems must have been particularly attractive, in comparison with other Iberian mountains, and even with other Pyrenean areas.

The interest of these mountains has not been evenly valued by people during the last centuries. Indeed, an old recurrent feeling of the local inhabitants has been to consider them as 'damned mountains', a kind of God's punishment, given the poor use and strong limitations inherent to most of their extension. Contrarily, those Pyrenean mountains evoked in some cultivated people myths and legends or at least fantastic wilderness. This is best exemplified by the much acclaimed Romantic Catalan poet Jacint Verdaguer, who struck by the powerful landscape of the Pyrenean high mountains, wrote in the poem *Canigó* (1886):

Quins crits més horrorosos degué llançar la terra
infantant en ses joves anyades eixa serra!
que jorns de pernabatre!, que nits de gemegar!
per traure a la llum pura del sol eixes muntanyes,
del centre de sos cràters, del fons de ses entranyes,
com ones de la mar!
...

Passaren anys, passaren centúries de centúries
abans que s'abrigassen de terra i de boscúries
aqueixes ossamentes dels primitius gegants,
abans que tingués molsa la penya, flors les prades,
abans que les arbredes tinguessen aucellades,
les aucellades cants

What terrible wails the Earth must have issued
Giving birth to these mountains in her youth!
What restless days! What groaning nights!
To eject such mountains into the pure light of the Sun
From the heart of her craters, from the depths of her entrails,
Like waves sweeping across the sea!
...

Years passed; centuries of centuries passed
Before soil and wood began to cloak
These bones of the ancient giants,
Before the outcrops bore moss; the meadows, flowers,
Before the woodlands were filled with birds;
The birds with song.

The 60th birthday of the Aigüestortes i Estany de Sant Maurici National Park has been a suitable occasion to discuss the conservation of high mountain areas; to review the knowledge acquired on their natural and cultural inheritance, to evaluate their protecting role, and even to preview their fate under changing scenarios. These issues define the scope of this book, launched at the workshop "The High Mountain in a Changing World: Challenges for Conservation" held in Espot, Central Pyrenees, in November 2015. Like in other National Parks, its scientific knowledge has notably grown during the last decades, running through long, diversified ways. This progress may be tracked by browsing across the ten volumes produced in the proceedings *Jornades sobre Recerca al Parc Nacional d'Aigüestortes i Estany de Sant Maurici*, based on the Park's research workshops held once every 3 years. These volumes lead from the description or cataloguing of the most apparent nature components to the understanding of the functioning of ecosystems in a changing world, and to the realisation of the extent of the anthropic envelope—including cultural inheritance, protecting policies and changing environment. In fact, the weight of social matters on nature conservation has become more and more apparent. Thus, nature conservation remains science-based, but it is definitely a social affair.

Deepening and widening of local knowledge have progressively put it into a much broader frame, both spatial (i.e. other European high mountains, Boreal biomes) and temporary—from the far to the near geologic times, and to future. Apart from enabling the comparative analysis between high mountain reserves, this has also evidenced how very ancient anthropic activities are still compromising their presumed wilderness; and also how distinct elements of global change (i.e. long-distance pollution, biological invasion, climate warming) compromise their conservation purpose. The strength of high mountain reserves remains therefore in the study and monitoring of nature responses to ancient impacts and global change while keeping the anthropic influence at the lowest level possible. These new insights have to be the substrate on which to build a new paradigm on high mountain conservation. The previously mentioned workshop and this book gather recognised authorities in particular and emerging scientific fields, although also able to cross-discuss and evaluate the present opportunity of consistently studying, evaluating and foreseeing high mountain conservation. Most of the book core information is from the Pyrenees, but there are other chapters focussing in the Alps and Sierra Nevada. Overall, the discussion that emerges extends worldwide where the high mountain is a valuable concept.

The two first chapters (Part I) set the conceptual scope of the book, review historic and current conservation issues, and try to connect with the following thematic chapters. Part II focus on the historical perspective of high mountain systems, centred in the most influential facts and processes occurred through the last millennia. This standpoint involves the awareness of the role played by major climatic changes and, particularly, humans on biota, soils and landscapes. This anthropic heritage poses particular challenges to the wilderness in protected areas, mostly concerning the introduction and re-introduction of vertebrates related to fishing or hunting, but also affecting soil and water biochemistry and processes (Part III). Acknowledging how natural areas are subjected to global change has led to value the interest in monitoring at long-term horizons. Chapters in Part IV review distinct biotic and abiotic components from the recent past to future. Examples of response, shift or adaptation, illustrate possible scenarios for entire high mountain areas.

This book and the preceding workshop are mainly due to the work, expertise and willing of the chapters' authors, who have grown upon the scientific atmosphere created in the previous decades by so many colleagues. Nevertheless, we have also to acknowledge the institutions that planned, managed or funded the event and the present book, which includes the Spanish Organismo Autónomo Parques Nacionales, the technicians of the National Park, the local government of Espot, Diputació Lleida through the Institut d'Estudis Ilerdencs, and 'la Caixa'. Jordi Vicente (National Park) and Meritxell Batalla (CREAF) provided valuable help in the organisation and manuscript editing, respectively. Several colleagues read and provided comments for different parts of the book, which have contributed to improving them, yet the responsibility remains with the authors. We are glad to thank Anna Agustí Panareda, Anna Avila, Enric Ballesteros, Rick Battarbee, Anton Brancelj, Begoña García, Josep M. Gasol, Gareth Griffith, Erik Jeppesen, Joachim W. Kadereit, Felix Knauer, Jiri Kopacek, Christoph Marty, Yolanda Melero, Guillermo de Mendoza, Laszlo Nagy, Sergi Pla-Rabès, Roland Psenner, Joan Real, Daniel Sol, Jaume Terradas, Jean-Paul Theurillat, Luis Villar and Kevin Walsh.

Barcelona, Spain
December 2016

Jordi Catalan
Josep M. Ninot
M. Mercè Aniz

Contents

Contributors

Marie-Claude Bal Geolab UMR 6042 CNRS, Université de Limoges, Limoges, France

Francisco J. Bonet Departamento de Ecología, Facultad de Ciencias & Instituto Interuniversitario Sistema Tierra en Andalucía (IISTA), Universidad de Granada, Granada, Spain

Teresa Buchaca Integrative Freshwater Ecology Group, Center For Advanced Studies of Blanes (CEAB-CSIC), Blanes, Girona, Catalonia, Spain

Danilo Buñay Integrative Freshwater Ecology Group, Center For Advanced Studies of Blanes (CEAB-CSIC), Blanes, Girona, Catalonia, Spain

J. Julio Camarero Instituto Pirenaico de Ecología (IPE-CSIC), Zaragoza, Spain

Lluís Camarero Centre d'Estudis Avançats de Blanes, CSIC, Blanes, Girona, Spain

Pere Casals Forest Sciences Centre of Catalonia, CEMFOR-CTFC, Solsona, Catalonia, Spain

Emilio O. Casamayor Integrative Freshwater Ecology Group, Center for Advanced Studies of Blanes (CEAB), Spanish Council for Research (CSIC), Girona, Spain

Jordi Catalan CREAF - CSIC, Cerdanyola del Vallès, Catalonia, Spain

Ignacio Clemente Conte Institució Milà I Fontanals, Spanish National Research Council (CSIC), Barcelona, Spain

Raquel Cunill Facultat de Filosofia i Lletres, Department of Geography, Edifici B, Universitat Autònoma de Barcelona, Cerdanyola del Vallès, Bellaterra, Spain

Sara Díaz Bonilla Department of Prehistory, Autonomous University of Barcelona, Cerdanyola del Vallès, Bellaterra, Spain

Juan Carlos García-Codron Department of Geography, Urban Studies and Land Planning, Avenida de Los Castros s/n, Universidad de Cantabria, Santander, Spain

Jordi Garcia-Pausas Forest Sciences Centre of Catalonia, CEMFOR-CTFC, Solsona, Catalonia, Spain

Ermengol Gassiot Ballbè Department of Prehistory, Autonomous University of Barcelona, Cerdanyola del Vallès, Bellaterra, Spain

Paola Laiolo Research Unit of Biodiversity, Spanish National Research Council, Principado de Asturias, Oviedo University, Mieres, Spain

Francisco Lloret CREAF-UAB, Cerdanyola del Vallès, Spain

Juan Ignacio López-Moreno Pyrenean Institute of Ecology, Consejo Superior de Investigaciones Científicas, Zaragoza, Spain

Antoni Margalida Faculty of Life Sciences and Engineering, Department of Animal Science, University of Lleida, Lleida, Spain; Division of Conservation Biology, Institute of Ecology and Evolution, University of Bern, Bern, Switzerland

Niccolò Mazzucco Préhistoire et Technologie UMR 7055, Université Paris Ouest Nanterre La Défense, Nanterre, France

M. Mercè Aniz Parc Nacional d'Aigüestortes i Estany de Sant Maurici, Boí, Catalonia, Spain

Alexandre Miró Integrative Freshwater Ecology Group, Center For Advanced Studies of Blanes (CEAB-CSIC), Blanes, Girona, Catalonia, Spain

Francesc Montané Forest Sciences Centre of Catalonia, CEMFOR-CTFC, Solsona, Catalonia, Spain; School of Natural Resources and the Environment, University of Arizona, Tucson, AZ, USA

Enrique Morán-Tejeda Department of Geography, Universitat de Les Illes Balears, Palma (Illes Balears), Spain

Josep M. Ninot Departament de Biologia Evolutiva, Ecologia i Ciències Ambientals & Institut de Recerca de la Biodiversitat (IRBio), Universitat de Barcelona, Barcelona, Catalonia, Spain

Laura Obea Gómez Department of Prehistory, Autonomous University of Barcelona, Cerdanyola del Vallès, Bellaterra, Spain

José Ramón Obeso Research Unit of Biodiversity, Spanish National Research Council, Principado de Asturias, Oviedo University, Mieres, Spain

Santiago Palazón Fauna and Flora Service, Ministry of Territory and Sustainability, Government of Catalonia, Barcelona, Spain

Albert Pèlachs Facultat de Filosofia i Lletres, Department of Geography, Edifici B, Universitat Autònoma de Barcelona, Cerdanyola del Vallès, Bellaterra, Spain

Antonio J. Pérez-Luque Departamento de Ecología, Facultad de Ciencias & Instituto Interuniversitario Sistema Tierra en Andalucía (IISTA), Universidad de Granada, Granada, Spain

Ramon Pérez-Obiol Botany Unit, Facultat de Biociències, Department of Animal Biology, Plant Biology, and Ecology, Edifici C, Universitat Autònoma de Barcelona, Cerdanyola del Vallès, Bellaterra, Spain

Manuel Quesada Carrasco Department of Prehistory, Autonomous University of Barcelona, Cerdanyola del Vallès, Bellaterra, Spain

Ana I. Rios Forest Sciences Centre of Catalonia, CEMFOR-CTFC, Solsona, Catalonia, Spain

Christian Rixen WSL Institute for Snow and Avalanche Research SLF, Davos Dorf, Switzerland

David Rodríguez Antón Department of Prehistory, Autonomous University of Barcelona, Cerdanyola del Vallès, Bellaterra, Spain

Joan Romanyà Department of Biology, Health and Environment, Universitat de Barcelona, Barcelona, Catalonia, Spain

Pere Rovira Forest Sciences Centre of Catalonia, CEMFOR-CTFC, Solsona, Catalonia, Spain

Ibor Sabás Integrative Freshwater Ecology Group, Center For Advanced Studies of Blanes (CEAB-CSIC), Blanes, Girona, Catalonia, Spain

Alba Sanmiguel-Vallelado Pyrenean Institute of Ecology, Consejo Superior de Investigaciones Científicas, Zaragoza, Spain

Thomas Schmitt Senckenberg German Entomological Institute, Müncheberg, Germany; Faculty of Natural Sciences I, Department of Zoology, Institute of Biology, Martin Luther University Halle-Wittenberg, Halle (Saale), Germany

Joan Manuel Soriano Facultat de Filosofia i Lletres, Department of Geography, Edifici B, Universitat Autònoma de Barcelona, Cerdanyola del Vallès, Bellaterra, Spain

Marc Taull Forest Sciences Centre of Catalonia, CEMFOR-CTFC, Solsona, Catalonia, Spain

Rocco Tiberti Dipartimento di Scienze della Terra e dell'Ambiente (DSTA), University of Pavia, Pavia, Italy; Alpine Wildlife Research Centre, Gran Paradiso National Park, Valsavarenche, Aosta, Italy

Marc Ventura Integrative Freshwater Ecology Group, Center For Advanced Studies of Blanes (CEAB-CSIC), Blanes, Girona, Catalonia, Spain

Sonja Wipf WSL Institute for Snow and Avalanche Research SLF, Davos Dorf, Switzerland

Regino Zamora Departamento de Ecología, Facultad de Ciencias & Instituto Interuniversitario Sistema Tierra en Andalucía (IISTA), Universidad de Granada, Granada, Spain

Part I
Current Challenges of High Mountain Conservation

Chapter 1
The High Mountain Conservation in a Changing World

Jordi Catalan, Josep M. Ninot and M. Mercè Aniz

Abstract The high mountains have retained a noticeable degree of wilderness even in the most populated regions of the planet. This is the reason why many nature reserves have been established in these landscapes. Currently, climate change and long-range transport of contaminants are affecting those protected areas, and thus conservation priorities may be challenged by these new pressures. In fact, many high mountains hold a legacy of on-site past human activities (e.g., pasturing, forestry, mining), which in some areas may partially persist, even increase, whereas in others are substituted by new uses (e.g., tourism, mountain sport). Therefore, high mountain nature reserves face a challenging future. The conservation goals have to be revised. Former alternative paradigms respectively based on the preservation of wilderness or a traditional cultural landscape will be insufficient. Indeed, global change provides new goals for the high mountain conservation areas as suitable places where to study the nature's response in the absence of, or combined with, other local pressures. Different branches of sciences may contribute to inform about the changes; however, conservation is ultimately a societal endeavour and thus their goals must be linked to the social demand for a fair society in a sustainable planet. As an added-value to this task, the high mountains hold a large amount of symbolism.

Keywords Nature reserves · Climate change · Landscape shifts · Alpine biota · Long-range atmospheric contamination · Land use change · Mountain resources

J. Catalan (✉)
CREAF - CSIC, Campus UAB, Edifici C, 08193 Cerdanyola del Vallès, Catalonia, Spain
e-mail: j.catalan@creaf.uab.cat

J.M. Ninot
Departament de Biologia Evolutiva, Ecologia i Ciències Ambientals,
Institut de Recerca de la Biodiversitat (IRBio), Universitat de Barcelona,
Diagonal 643, 08028 Barcelona, Catalonia, Spain

M.M. Aniz
Parc Nacional d'Aigüestortes i Estany de Sant Maurici,
Ca de Simamet, C. de les Graieres 2, 25558 Boí, Catalonia, Spain

J. Catalan et al. (eds.), *High Mountain Conservation in a Changing World*,
Advances in Global Change Research 62, DOI 10.1007/978-3-319-55982-7_1

1.1 Introduction

1.1.1 Conservation in a Changing World

The increasing recognition of global change may place traditional conservation goals in a deadlock. Mountains have been flagship lands of conservation around the world. Although plenty of natural resources, mountains have generally been less amenable to the settling of large human populations. Their relative societal marginal role, together with their intrinsic natural values, has facilitated the creation of nature reserves with the general goal of preserving species and natural landscapes. Although climatic variability has long been recognized, the declaration of nature reserves was based on the assumption that things would not change significantly for many generations and thus the preserved lands will remain pristine or recover a state close to that if they had been highly modified by humans. Obviously, there were differences among continents. In the less populated regions, preserving wilderness played a central role (e.g., western North America) whereas in countries with a long tradition of mountain use the conservation goal often included human activities that were considered traditional in those landscapes (Radkau 2008; Beniston and Stoffel 2014). Yet the central conservation goals may have differed among countries or shifted with time since nature reserves started to be formulated in the late 19th century, the alternative views were not challenged by anything that could not be apparently managed at a local, regional or national scale.

Now the situation has drastically changed. National Parks and other nature reserves will have to review their foundational goals. Beyond managing their internal problems and the pressures from their immediate natural and societal surroundings, they will have to deal with atmospheric drivers (Fig. 1.1) that may shift their natural systems to situations completely unexpected in the recent past (DeFries et al. 2012). These changing conditions will overlap with parallel changes in the local and regional socio-economic context that will also be reacting to the new situation (Pearson 2016). The larger the expected problem, the sooner the reaction should be planned (Margalef 1976). Given the global change scenarios, it is opportune to analyze the role of high-mountain nature reserves in a context of persistent change, evaluate whether there are non-sustainable goals and discuss alternative aims more in accordance to the new situation.

1.1.2 Vulnerability, Exposure, and Sensitivity

The eventual vulnerability of an ecosystem to some perturbation depends on both the degree of exposure, and the sensitivity to it (Lloret 2017). High mountains are particularly exposed to some of the atmospheric global change components (Steffen et al. 2015): high mountains are prompt to shifts in climatic extremes (Rangwala et al. 2012); the atmosphere is thinner, so UV radiation is higher, particularly in the

Fig. 1.1 Meteorological and atmospheric deposition monitoring station in Lake Contraix (Aigüestortes i Estany de Sant Maurici National Park (PNAESM), Pyrenees). High mountain nature reserves are particularly suitable for developing long-term studies to investigate the development and consequences of global change on ecosystems. To cope with the local characteristics of the atmospheric forcing, meteorological and deposition stations have to be deployed and maintained with a perspective of at least decades, to provide fundamental information to any other ecosystem study in the mountain catchments (Camarero 2017b). Photography: Lluis Camarero

UVB range (Blumthaler et al. 1997), which may be enhanced by stratospheric ozone reduction (Blumthaler and Ambach 1990); cold conditions may facilitate the condensation of semi-volatile compounds (Grimalt et al. 2001); and they are barriers to air-masses thus they are exposed to long-range transport of substances (Catalan et al. 2013), microorganisms (Barberan et al. 2014) and diaspores (Flo and Hagvar 2013). Although harsh conditions in the high mountains have been a constant for the organisms living there, this does not mean that they are not sensitive to fluctuations. In many instances, a large part of the high mountain organisms may be living at the edge of their respective possibilities; the tree-line illustrates this issue (Korner and Paulsen 2004).

The way in which humans have been occupying and using the mountains have been changing throughout history and locations (Walsh 2014). In many of the ranges around the world, large areas have been modified for pasturing and forestry purposes (Miehe et al. 2014); often reaching a landscape configuration that may appear natural to non-expert analysis. This traditional land use has not been sustained through centuries everywhere, since wax and wane have modified the pressures depending on both societal and climatic factors (Bocquet 1997). During the last decades, high mountains are experiencing a large cultural and socio-economical shift in many regions of the world (Ooi et al. 2015). Transport facilities have increased visitors, and tourism has become a significant economic element in the high mountain and around nature reserves (Oian 2013). In parallel, most traditional land uses—once diversified and recurrent—are vanishing as local pressures on mountain ecosystems. Mountain conservation involves a daunting task evaluating exposure and sensitivity to a wealth of exceptionally dynamic pressures. An accurate evaluation is fundamental for defining new conservation goals since a crude estimation indicates that traditional ones may not be possible any longer.

1.2 Mountain Exposure to Global Changes

1.2.1 Climate Change

It has been suggested that mountains experience stronger cold and warm climatic fluctuations than average lands (Dedieu et al. 2014; Beniston 2006). Despite this may be a matter of debate, there is no question that the recording and perception of warming during the last decades have been observed at many different ranges (Diaz and Bradley 1997). The particularity with mountains is that the altitudinal gradient induces contrasting climates in a short distance. With climate change, the regional means will change, but the altitudinal and other topographically-induced variability may also be modified. All in all, mountains will warm throughout the world (Nogues-Bravo et al. 2007).

In regions with relatively dry average climates, such as the Mediterranean, high-mountains constitute sources of water to lowlands (Boithias et al. 2014). Ascending from plains, one goes across water-driven vegetation to mostly temperature-driven ecosystems. Not only in these mountain ranges but in general, the way in which precipitation will regionally shift appears critical to project potential vegetation changes, including details of the topographical climatic variation in the high-mountain (McCullough et al. 2016). The Mid-Holocene has provided clear evidence that a complete shift in dominant landscape vegetation is possible at sub-millennial time scales (Carrion 2002), with transitional changes at local scales probably occurring over decadal periods. Mountain nature reserves may be submitted to progressive declines in annual precipitation and also to increasing frequency of drought events (Beniston et al. 2007). The character of the shift, either smooth or abrupt, may result in entirely different interactions with other processes, such as pests (Hodar et al. 2003), invasive species (Thomas 2010), pollution (Bogdal et al. 2010), etc.

Even in ranges not expected to experience changes in annual precipitation, warming will change the hydric balance. The seasonal pattern of warming would play a critical role in determining the kind of new situations that ecosystem experience. For instance, summer and autumn warming may lead to seasonal water limitation; spring warming to early snowmelt (Moran-Tejeda et al. 2014) and upward shift of the snow line in spring; winter warming to episodic events of melting (Gobiet et al. 2014) and shorter snow duration (Hantel and Hirtl-Wielke 2007); summer snowline uplift to a reduced nival belt (Gottfried et al. 2011), and so on. In general, the shift from snow to rain may have amplifying consequences for the hydrological cycle and the natural processes depending on it (Morán-Tejeda et al. 2017). Floods related to rain-on-snow events may increase (Beniston and Stoffel 2016), at least during a transition period as warming proceed. All things considered, climate change may increase synchrony between ecosystems compared to pre-industrial dynamics, which appears to be the case for forest over large areas (Shestakova et al. 2016).

1.2.2 Atmospheric Contaminants

During the last decades, awareness about the accumulation of some persistent organic pollutants (POPs) in high-mountain organisms has been increasing, mainly from fish (Schmid et al. 2007; Grimalt et al. 2001) and pine needles studies (Grimalt and van Drooge 2006; Davidson et al. 2004). The initially uncertainity of measuring high concentrations of some pollutants, far away from the areas where they are produced or used, has given way to understanding the mechanisms related to the semi-volatile character of these compounds (Catalan 2015). Although details may differ from site to site, the preferential accumulation in cold areas, such as the high mountains, is related to the air-water partition sensitivity to temperature for these compounds (Wania and Westgate 2008). They are extremely hydrophobic, so if in solution they quickly adhere to any organic material and thus organisms (Catalan et al. 2004). The toxicological consequences of the POPs bioaccumulation in mountain organisms are scarcely known; there are only a few pioneering studies (Jarque et al. 2015; Quiros et al. 2007).

In fact, environmental dynamics of synthetic substances and their ecological consequences is one of the major unknowns among global change components. The assumed general behaviour may be plenty of challenging particularities for each compound (Bartrons et al. 2012). Conservation biology should pay more attention to the investigation of the potential problem. It may be affecting wildlife in a way still difficult to evaluate, but that could be non-negligible according to the indications provided by some studies in wild predators (Elliott et al. 2012) and domestic herbivores (Shunthirasingham et al. 2013). An added interest to the topic is the interaction of this dynamics with climate warming (Noyes et al. 2009). On the one hand, higher temperatures will decrease the tendency to condensate of these compounds in some mountain areas. On the other hand, high temperatures will increase their release from soils were they might have accumulated. In any case, there will be a redistribution of substances trapped in natural reservoirs [e.g., glaciers, Schmid et al. (2011)] and long-distance air transport will be probably enhanced. New synthetic organic substances are discovered every day (Muir and

Howard 2006). Even if the substances are not particularly persistent in the environment as much as POPs (e.g. PCBs), the use of some of them in large amounts (e.g. in agriculture) may result in a steady transport to the mountains and the exposition of the organisms there to high concentrations (Weber et al. 2010).

Metals are other atmospheric pollutants that have accumulated in the mountains (Camarero 2017b). In fact, they have a longer history than POPs. Evidence of early pollution in Europe dating from the Roman period and also from ancient times in other parts of the world becomes stronger the more studies on natural registers exist (Catalan 2015). Mountains have also been sites of historical interest for mining provided that ores of different minerals are common. Depending on the economic context even the exploitation of small, difficult to access mines have taken place (e.g. Trou des Romains, Val Sapin, Italian Alps). The regional context of this mining throughout history is well recorded in lacustrine sediments and peats (Camarero et al. 1998). The ecological consequences, if any, of these atmospheric pollutants have not been evaluated yet. In some areas, the accumulation of some trace metals in soils is very high so, under lower current deposition, they have become sources of pollutants to sediments, plants, and animals, rather than sinks (Bacardit et al. 2012). There is a legacy of pollutants in soils that may maintain high pollutant fluxes for some decades.

Ozone depicts a unique case in the global change context. On the one hand, the current decline in stratospheric ozone causes an increase in the UV radiation reaching the ground. In high mountains, the effect is enhanced compared to low lands. First, because the atmosphere is thinner and thus both total and relative UV is higher (Blumthaler et al. 1997); second because many high mountains are above a cloud belt that protects valley lands from high radiation (Blumthaler et al. 1994). However, the direct consequences for the organisms' life of the increase in UV and, particularly, in the more harmful UVB, might not be huge, since mountain organisms have evolved in high UV environments, thus have developed many protective and repairing mechanisms. Nevertheless, UV exposure is repeatedly pointed as a potential factor of some species decline (Mitchell et al. 2015). On the other hand, tropospheric ozone is increasing. At ground level, ozone protective role against UV matters little and what becomes important is its harmful highly oxidative effect (Wittig et al. 2009). Observatories in high mountains have indicated a sustained increase in tropospheric ozone in many areas of the world. It can be considered a global hazard. Indeed, there are mountains particularly exposed to air masses that bring high ozone concentrations from source areas (Elvira et al. 2016). Ozone is harmful both to plants and humans. The effects of ozone on trees may be confounded at first instance with drought effects. It would be reasonable to develop specific surveillance protocols for high mountains, particularly in nature reserves. Thus environmental assessment will provide a double benefit. With climate change, the stratosphere-to-troposphere ozone flux will be modified with latitudinal differences in UV radiation effects and stratospheric ozone (Hegglin and Shepherd 2009). Some technically sophisticated observatories already deal with estimations of trends and global averages (Li et al. 2007; Cristofanelli and Bonasoni 2009; Cristofanelli et al. 2010). Simpler systems may serve as sentinels for conservation purposes, and for investigating marked orographical differences within relatively small areas (Burley et al. 2015).

1.2.3 Long-Distance Atmospheric Fertilization

Part of the decline in stratospheric ozone is due to the emission of nitrogen oxides to the atmosphere. Nitrogen emissions have many other effects. The perturbation of the nitrogen cycle is probably also beyond the limits of a sustainable Earth system (Steffen et al. 2015). Crystalline rocks dominate in many high-mountain landscapes. Soils and waters on these bedrocks show little acid neutralizing capacity. Therefore, mountains in areas receiving acid deposition due to emissions of sulfur and nitrogen oxides to the atmosphere suffer acidification (Psenner and Catalan 1994). Visible consequences (e.g., fish kills, tree defoliation) had prompt to an international reaction of successful results in controlling sulfur emissions and severe acidification (Catalan et al. 2013). However, nitrogen emissions are still high and difficult to reduce because of their multiple sources—industrial, agricultural and urban. In addition to being acidifying agents, N compounds are also fertilizers for plants (Stevens et al. 2015). Increasing atmospheric CO_2, nitrogen deposition and temperature all point to an acceleration of primary productivity. In recent years, P atmospheric deposition has also been identified as being enhanced by human industrialization (Penuelas et al. 2013a). Mountain soils and waters are exposed to all these fertilizing agents, which are changing the biogeochemical cycles in the high mountains in a way not sufficiently understood (Camarero and Catalan 2012).

1.2.4 Biotic Dispersal Enhancement

Increased transport of goods and people is a feature of the post-industrial society. This has facilitated the dispersal of all kinds of organisms to longer distances and with higher frequency. Mountain natural reserves are increasingly exposed to exotic and invasive species and new diseases and pests (Pauchard et al. 2009). Amphibians have been particularly suffering this pressure. Chytridiomycosis has been spreading quickly around the world and devastating some of their populations (Wake and Vredenburg 2008). The interaction between pests and climate change is a major source of uncertainty in conservation. Tree populations at their ecological limits may be more prompt to infection, which may accelerate and otherwise slow substitution of species (Camarero 2017a).

In some ranges, fish stocking is the primary pressure on amphibians (Miró and Ventura 2013). Human enhancement of fish dispersal in mountain lakes, which are naturally fishless, is a complex behavioral and economic phenomenon where ignorance and interest play significant roles and poses major difficulties to conservation (Ventura et al. 2017). In the current shifting environmental conditions, which are challenging the foundational goals of a natural reserve, the role that leisure may play becomes a matter of discussion. With rising inhabitants, the number of visitors to natural reserves will continue increasing, both from nearby and long-distance origins and with different motivation, including from purely touristic attraction to scientific interest. Planning and protocols to prevent facilitated dispersal of undesired organisms should develop and spread at a quicker pace than the problems are diffusing.

Enhanced dispersal also includes the controlled introduction of new species, or re-introduction of formerly extinct species, most typically of vertebrates. Reintroducing some of these emblematic species from more or less remote populations is now feasible (Fig. 1.2), and adds new socio-economical interest to the areas beyond conservation purposes and, in parallel, the need for new managing measures (Palazón 2017). Consequently, some key species in mountain ranges are becoming more and more managed.

Fig. 1.2 A female brown bear with a cub of the year in the Pyrenees (September 5, 2016). Beyond the ecological consequences, the reintroduction of large carnivorous in areas where they have been extinct or nearly so indicates a change of the societal attitude towards nature of valuable symbolism. The last female bear from Central Pyrenees died in 2004 and the last male in 2010 (Palazón 2017). The current bear population was reintroduced from a set of individuals captured in the Balkans (Slovenia) and released in the Pyrenees in 1996–1997. There was only one male born from a released Slovenian male and the last Pyrenees female. In 2015, the Central Pyrenees subpopulation had more than 30 bears identified. Photography: Departament de Territori i Sostenibilitat, Generalitat de Catalunya

1.3 Mountain Exposure to Regional Changes

Economic globalization has brought many changes to the local and regional socio-economical context of mountains. Low cost of transport has modified commercial yields in forestry, mining and mountain pasturing. On the other hand, mountain activities for leisure are exponentially increasing. Conservation faces global change hazards exposed to regional socio-economic changes, which ultimately will also be conditioned by climate change.

1.3.1 Pasture Shifting Systems

Mountain pasturing dates from several millennia ago in some ranges (Schmidt et al. 2002; Pèlachs et al. 2017). Depending on the socio-economic context of the mountains the intensity and type of pasturing have been fluctuating across time. Mountain archaeology has provided evidence of the early Neolithic use of rock shelters and caves used as protections for sheep and goats in European mountains (Gassiot Ballbè et al. 2017). No signs of ecological impact have been associated with them yet. The opening of extensive pasturing areas at altitude during the Bronze Age is clearly documented by archaeological sites and palaeoecological records from lakes, bogs, and stone fences and shelters (Fig. 1.3). This use of high-mountain meadows during millennia cannot have been ecologically innocuous; indeed, it may even have constituted a selection factor favoring particular plant treats, species and assemblages. This is a scarcely studied subject, as mountain pasturing has usually been seen by ecologists as a traditional activity rooted in medieval times. Also, the biogeochemical cycles in mountain meadows have been probably accelerated by large herds pasturing during a portion of the year, whose winter survival is assured by migration to much lower altitudes or far-away locations. This has led to much higher herbivorism at the high mountain than in non-herding situations, where much smaller wild herbivore populations survived resisting winter conditions with short altitudinal migrations. With the development of high mountain pastures, the C/N ratio distinctly declined in sediments of the lakes in the catchment, probably reflecting changes in soil conditions (Schmidt et al. 2002).

The way to manage pasturing is different from range to range (even from valley to valley), and it has been shifting through time according to societal changes (Berrocal et al. 2014). A general tendency seems to have been increasing the number of animals per field shepherd by joining several small herds into a large one during summer pasturing; even that the ownership was maintained by many individuals. This seems to have been a tendency as societies have become more complex from the Bronze to Middle Ages and modern times. Many protected areas include large pasturing meadows, whose management has been a matter of debate. Arguments in favor of maintaining traditional practices are not probably aware of the real extent of this adjective. Ironically, relatively recent shifting from sheep to

Fig. 1.3 Despoblat de Casesnoves archaeological site. Medieval human occupation at 2225 m a.s.l. that lasted until the 13th century (PNAESM, Pyrenees). Although the high mountain landscape retains a lot of the wilderness character, the human footprint can be recognized in many ranges of the world. Archaeological evidence indicates an early Holocene occupation (Gassiot Ballbè et al. 2017) and a development of a cultural landscape at least since the Bronze Age. Natural and cultural heritages may coexist in high mountain nature reserves. However, management decisions should not respond to pressures based on weak arguments about the merits of traditional uses. Nature reserves should promote the proper understanding of the ecological incidence of humans on high mountain ecosystems and landscape across time, but not necessarily reproduce them. Photography: Grup d'Arqueologia de l'Alta Muntanya, Archive of the Aigüestortes i Estany de Sant Maurici National Park

cows has raised new concerns in some areas. The two species influence the meadows in an entirely different way, thus changing the vegetation and soil processes. Soils in the mountains are usually thin, and meadows provide the areas with the largest capacity for carbon storage (Garcia-Pausas et al. 2017). All in all, nowadays the so called traditional uses have seen sharp shifts, in the form of abandonment of extensive, low-income activities. This has led to encroachment of less productive pasture into shrub land and to secondary woodland (Lasanta-Martinez et al. 2005). To allow forest recovering *vs.* maintaining sheep or cow pasturing may become an issue for some conservation reserves that are fairly conditioned by a cultural view of the mountain, and do not consider enhancing wilderness as a primary goal (Fig. 1.4). On the other hand, the scenario to manage may change in a faster way than the time required for implementing protection measures (Garrard et al. 2016). Indeed, rapid economic development related to the climatic shifts may be faster changing the landscape than the vegetation natural response. Projections of climate change effects on nature reserves should include potential new economic scenarios in the region.

1.3.2 Conservation Versus Extraction

Through history, the high mountains have been regarded and used by humans as a source of goods (timber, rangeland, hydropower, fishing, etc.), with the particularity of being a space marginal to the permanent human habitat. Thus, being strongly subject to seasonality, human settlements have been placed for most time at low altitudes [but see Pèlachs et al. (2017)]. In contrast with other exploited ecosystems, high mountain apparently maintained a high degree of wilderness, to some extent

Fig. 1.4 Sheep flock in the Vall d'Àssua (Pyrenees). Enhancement of high altitude pasturing meadows lowering the treeline may have lasted for several millennia in many mountain ranges of the world. This practice had to have ecological and also evolutionary implications, which are still scarcely understood. The current socio-economic context of the mountains is changing the land use. This driver of ecological shifts is still stronger than climate change in many regions. Conservation requires distinguishing between the two drivers and eventually understanding their interaction (Pèlachs et al. 2017). Photography: Jordi Peró, Archive of the Aigüestortes i Estany de Sant Maurici National Park

related to the imposed winter cessation of activities and to the noticeable proportion of land scarcely or not at all exploitable (rocky and scree areas, alpine heathland, fell-fields, etc.). However, growing evidence is emerging on the profound shifts caused by human activity at several levels of the high mountain wilderness, such as depletion or vanishing of large mammal's populations, alien fish introduction, uneven grazing by domestic herbivores, overuse of running waters, or forest exploitation.

Up to some decades ago, exploitation was thought as limited to shape semi-natural ecosystems, hampered by the limitations inherent to the environment. Short seasonal forest growing led to longer logging periodicity in subalpine woodlands than in lower altitude forests, and grazing intensity necessarily remained lower than the resprouting capacity of grasslands. However, as high mountain ecosystems become more finely analyzed, more footprints of exploitation practices emerge to identify these activities as key factors in the current ecosystem functioning. The present structure and functioning of woodlands are in many cases a delayed response to ancient logging. For instance, this included recent tree mortality in the Pyrenean *Abies alba* forests, a phenomenon predisposed by historical logging that enhanced dense tree populations and induced by recent climatic changes (Camarero 2017a). Similarly, shifts in alpine grazing have changed grassland structure and composition (Komac et al. 2014).

Forestry is the most controversial issue concerning resource extraction in the mountains; particularly if a view of maintaining a cultural landscape predominates above wilderness enhancement (Agnoletti 2007). Where nature reserves include large forested areas there is little argument for selective extraction, since they can maintain natural dynamics—including catastrophic events. However, in the tiny reserves of many European ranges forest management may be an important issue. Arguments against forest aging and consequently selective cutting may appear even with the conservationist support. Global change tendencies and the need for pest control may increase the supporters of this latter view.

Mountains are full of natural resources susceptible to economic exploitation. As a result of the orographic processes many ranges are rich in metals. This was very early appreciated, and even in dry high mountains (e.g. Sierra Nevada, Iberian Peninsula), where high-altitude agriculture or herding was not particularly suitable, mining was an old practice (Martin Civantos 2014). The techniques for mineral extraction in former times were not particularly concerned with the environment. As a consequence, a legacy of pollution is maintained in the soils affected by atmospheric transport from the mining sites in the region. Environmental history studying palaeoenvironmental registers is progressively unveiling the distribution and patterns of this old extractive activity (Catalan 2015). In some valleys, rich in metals close to the surface, an associated metallurgic industry has impacted forest to provide wood fuel (Pèlachs et al. 2009). This activity also has a changing dynamics of centuries; current landscapes bearing apparently well-develop forests may hide a history of several centuries of exploitation. Historically, mining industry/business was firmly driven by economic constraints, compared to pasturing or agriculture that could be closely related to the local domestic activities. Therefore, except in mountain where there is still an economical yield, mining becomes a historical issue rather than a conservation problem. The historical heritage that mining may have left on the mountain is an on-going research subject. Technological improvements have usually lead to increased extraction and atmospheric pollution when resources of economic value exists (Uglietti et al. 2015). Unfortunately, environmentally careless mining keeps going in some ranges of the world nowadays (Pond et al. 2008; Wickham et al. 2013).

The increase in human populations and water demand, both for cities and agriculture irrigation, has raised the value of mountains as water source. Large regions, even whole countries, may depend on the precipitation in faraway mountains [e.g., Himalayan region (Xu et al. 2009)]. Glaciers may provide a year-long regular water supply to areas of marked seasonality in precipitation. Some large cities in the Andes depend on reservoirs fed by such glaciers, which shrinking and final disappearance may cause large societal problems (Carey et al. 2014). Probably, little can be done on glacier preservation at this stage of climatic change (WGMS 2015; Zemp et al. 2015). Thus plans for long distance transport to these cities are required, which may change hydrological regimes in other areas, and challenge conservation of mountain ecosystems.

In some mountains, there has been a traditional management of water for irrigation purposes. An extraordinary example is the Sierra Nevada (Iberian Peninsula) acequia system built during the medieval Muslim period (Martin Civantos 2014). Thousands of kilometers of small channels bring water from the mountain (sierra) to the irrigation fields (vega), starting at about 2000 m a.s.l. and modifying the hydrological regime of the lower part of the mountains. Other channels at higher altitude (i.e., careos) drive water towards groundwater to increase spring supply. This change in the hydrology has to have altered natural vegetation. A large proportion of the modified area is now part of a National Park and nature reserve surrounding it.

Hydrological extraction has been more recent in the mountains located farther from areas requiring high amounts of water supply, mostly starting at early- or mid-20th century and associated with the development of hydropower stations. In this case, lakes and streams have been particularly exposed to alterations (Catalan et al. 1997). Surface connectivity between lakes has been enhanced by undergrown galleries that may even connect lakes in different watersheds. In some areas, little control of water level regulation or failure of old valves have produced huge water level oscillations in lakes, resulting in severe impacts. The temporal overlapping between this kind of industrial exploitation of mountain resources and the declaration of some mountain parks have led to odd situations in which both coexist in the same area (e.g., Aigüestortes i Estany de Sant Maurici National Park, Pyrenees). Beyond obvious impacts on the most affected lakes and streams by seasonal desiccation or strong water oscillations, flow reduction and smoothening of the seasonal fluctuations tend to produce a banalisation of the aquatic biota. Landscape visual impacts are usually scarcely compatible with conservation reserves. Yet the ultimate goal should be the eradication of this extraction activity from natural reserves, actions against light contamination and general visual impacts of buildings and services are easily achievable actions. Climate change may exacerbate debates between conservation and water extraction even in the mountains with water surplus (Beniston 2012).

1.3.3 Tourism and Sport Pressures

In many mountain valleys around the world, tourism is growing as an economic component replacing productive or extractive activities (Nyaupane et al. 2014). Tourism development has led to separation into different categories such as ecotourism, sustainable tourism, wildlife tourism, nature-based tourism, heritage tourism and cultural tourism (Rotherham 2013). Each of these categories has their own idiosyncrasy, which should be taken into account in the high mountain conservation planning.

Nature reserves become a reclaim for the tourism sector. Declaration of new protection areas sometimes includes among their benefits that they will favor this economic field. Handling visitor's affluence may become a primary issue for some natural reserves of relatively reduced size or without possibility to diversify visitors towards different zones. Access to mountain reserves usually follows some main routes that become both a constraint and an opportunity for managing the number of visitors. To traditional mountaineers, urban inhabitants on holidays and outdoor long-distance sports practitioners provide new challenges to conservation. An accepted self-responsibility for the risk that mountains constitute, which mountaineers and people working in the mountains had, is being replaced by a safety delegation to the community. This may increase the gap between visitors and locals. To an increasing degree, rural landscapes are being transformed into sites for leisure. Even though tourism is welcomed as a rescue plan of rural mountain economies in decline, it may at the same time be associated with unwanted changes. In the cases in which the nature of these landscapes is labeled as wilderness by conservationists and tourist industry alike, inhabitants of local communities may perceive that the social and cultural aspects of the landscape they strongly identify with are being disregarded (Oian 2013). This applies, for instance, to hunting and angling in nature reserves and neighbouring areas for animal population control or leisure but not for local consumption.

Climate change is modifying conditions at high altitudes, changing the spatial distribution of risk and accessibility. Global factors cannot be mitigated locally but improved management practices that aid local conservation and development in this high mountain ecosystem are required (Garrard et al. 2016). At lower altitudes within large ranges, a shift from traditional practices to a more diversified blend of agropastoralism, tourism services, and cash-crop production may become adaptive for local economies (Konchar et al. 2015). Conservation must find its place within this changing dynamics. On the other hand, in mountain ranges supporting a winter snow industry, changes in snow deposition and snowpack persistence may result in new demands upon preserved areas, enhancing social debates about conservation and economic sustainability at local and regional scales (Beniston 2012). Sensitivity analyses of snow patterns to projected climate change may help to anticipate conflicts; not only altitude matters for snow distribution (Uhlmann et al. 2009).

1.4 High Mountain Idiosyncratic Sensitivity

1.4.1 Temperature Versus Water

The elevation increase in mountains provides an associated decline in temperature, atmospheric pressure and land area availability. Therefore, in a short distance environmental conditions change markedly providing the cues for a highly diverse landscape and richness in organisms. This marked gradient also determines the high sensitivity of mountains to climate change. The great diversity in forest formations—and vegetation in general—is at expenses of a lower available surface for each of them compared to plains, where similar conditions extend over large areas. Therefore, climate change may imply a significant modification of suitable areas for a particular type of forest, scrub or meadow (Dullinger et al. 2012).

The orographic barriers tend to increase precipitation at the slope facing the ascending air masses. Mountains in general are, therefore, richer in water resources than surrounding low lands and usually become a net source of water for the latter. If air masses rich in water are mostly coming from the same direction, the vegetation contrast between mountain slopes may be remarkable. Consequently, mountain vegetation is particularly sensitive to changes in direction and average moisture content of air masses. In contrast to temperature, there is not a global pattern of precipitation change with altitude (Körner 2007). In the temperate zones of the planet, precipitation increases with elevation, either as rainfall or snow. But in other parts of the globe the tendency can be the opposed, or the maximum can be at intermediate levels. The mean altitude of the surrounding ranges is another factor determining the characteristics of the altitudinal precipitation pattern. If climate warming forces in a similar direction most of the mountains on the planet, changes in air mass direction and moisture would show more distinctive regional (even local) characteristics (Engler et al. 2011). Whereas long-term conservation plans can be based on general warming projections, tendencies on precipitation would be better assessed locally, and monitored at different sites in nature reserves with valleys facing different directions (Beniston and Stoffel 2014).

A critical issue in areas with current positive water balance, where temperature drives vegetation distribution, is whether changes in climate will bring to a situation of water deficit during vegetation growth periods. The short-term and long-term response of mountain tree populations to episodic droughts are still scarcely studied (Cocozza et al. 2016). Nature reserves entirely or partially within this situation would face the main challenges in the near future. Drought episodes may compensate for any fertilizing (CO_2 and N deposition increase) or warming growth effect that may exist upon tree species usually controlled by temperature rather than water availability (Camarero et al. 2014). Vegetation on mountains with current water limitation may be more resilient to extreme drought events (Herrero and Zamora 2014), albeit that recurrent episodes may produce a progressive loss of resilience (Lloret et al. 2004). Long-term palaeoecological evidence warns about non-linear responses to water availability (Anderson 2012).

Most temperate mountain ranges are characterized by an altitudinal partition of the slopes during spring into a snow-free belt and a white upper belt (Fig. 1.5). In fact, there is a changing role of temperature and precipitation on snowpack. At lower altitudes, temperature influence predominates and precipitation is a better predictor of snowpack variability above certain altitudinal threshold (Moran-Tejeda et al. 2013). As climate warms, the threshold will move upwards. Warming may also affect heavy snowfall frequency differently with altitude, increasing the contrast between the upper snow belt and the lower altitudes with a shorter snow period (Ignacio Lopez-Moreno et al. 2011).

Fig. 1.5 Landscape view of Val de Saboredo (Aran, Pyrenees). The typical spring division of the high mountain into *a white snow belt* and a lower one where fields flourish will probably come earlier and extend its duration with climate warming (Morán-Tejeda et al. 2017). This shift increases potential risks (avalanches, floods) beyond the non-expert perception. Nature reserves should be prepared for handling these increasing likely situations. Photography: Francesc Xavier Bové Carbó, Archive of the Aigüestortes i Estany de Sant Maurici National Park

1.4.2 Persistence Versus Migration

The high mountain biota includes cases of log-term persistence within a given range of some species. They went through contrasting climatic conditions such as those corresponding to the Quaternary succession of cold and temperate periods. In this sense, some ancient plant and animal endemics have apparently remained from Pliocene onwards roughly within their present range (Schmitt 2017). Interestingly, most of these species, called paleoendemic, correspond to more or less isolated tips in phylogenetic trees and nowadays are found in conservative habitats (Garcia et al. 2012) and are focus of conservation biology research (Segarra-Moragues and Catalan 2010). Aside from these survivors, however, most probably there were close relatives that vanished through changing ecological conditions. In parallel, more dynamic ecosystems (e.g., grasslands) must have enhanced radiating speciation, which is well exemplified in rich taxonomic complexes exhibiting narrow endemicity at the levels of species or lower (García and Gómez 2007). Therefore, a high mountain with fragmented landscapes in which different habitats are densely arranged has been a noticeable arena for various biological groups where fine-scale isolation has favored speciation. All in all, the altitudinal gradient imposes contrasting environmental conditions and any overall shift in climate results in pressure upon the current altitudinal species distribution (Fig. 1.6).

In simple terms, the response to an environmental change beyond the ideal conditions for a species consists of two options (Berg et al. 2010). Either persisting by acclimation (and eventually adaptation) to the new situation (Lapenis et al. 2005; Reich et al. 2016) or migrating following the direction of suitable conditions for the species (Hickling et al. 2006). The actual response depends on both the characteristics of the species and the pace at which environmental changes occur (Theurillat and Guisan 2001). In extreme cases, a third option may occur, the sudden collapse and local extinction of the population (Penuelas et al. 2013b).

With warming in the mountains, we can expect an upwards shift of the populations. This may apply from flying invertebrates (Konvicka et al. 2003) to trees (Seppa et al. 2002). However, the time response may be markedly different (Dullinger et al. 2004), with a variety of factors playing a role. Not only matters the capacity for displacement and generation time, but also the interaction with other species (Laiolo and Obeso 2017). Dynamics at ecotones between mountain forest belts is particularly difficult to predict (Dullinger et al. 2005) and simulate (Wiegand et al. 2006); the two, or more, tree species implied may not be responding in the same way at the climate change (Rabasa et al. 2013). Therefore, we can expect not only the displacement of the vegetation in the mountains, but also variations in the relative thickness of the belts, or even the number of belts. At present, except for the top mountain (Rixen and Wipf 2017), vegetation response to climate change may be still obscured for concurrent land use shifts. The response of invertebrates may be less ambiguous at the current stages of mountain warming (Wilson et al. 2005). The different velocity of reaction may be causing a rearrangement of species interactions, which increases the difficulty in predicting the ecological outcomes of

Fig. 1.6 *Parnassius apollo* (**a**) and *Limenitis reducta* (**b**) from Aigüestortes i Estany de Sant Maurici National Park (Pyrenees). *Parnassius apollo* has experienced strong regressions in its lower altitudinal limit in France and likely in many southern places of the Iberian Peninsula. The regression is related to a shorter snow cover duration and thus thermal insulation of the immature instars. On the contrary, the Mediterranean species *L. reducta* may be expanding its altitudinal range in the mountains (C. Stefanescu, com. per.). Butterflies and other invertebrates provide early evidence of nature's response to climate change in the mountains. Although climate warming will change the distribution of the biota across altitude in most of the mountains of the world, not all the species will respond in the same way and at the same pace (Laiolo and Obeso 2017). They possess different capacities to withstand changes and to migrate. Short living and motile organisms are responding in a faster way, and we can expect a long period of continuous rearrangement of the interactions among them. Nature reserve aims have to switch from a purely conservative strategy towards a stewardship of the changes that may happen. Photography: Marta Avizanda, Archive of the Aigüestortes i Estany de Sant Maurici National Park

climatic change (Tylianakis et al. 2008). However, one should also consider that mountain landscapes have gone through marked climatic fluctuations even in recent time (e.g., Little Ice Age), so fixed stability in the species interactions and assemblages may be the exception rather than the rule. Species with two altitudinal fronts in their distribution in the mountain may show different predominant processes in each of them. Dispersal may dominate the 'leading edge,' whereas in the 'rear edge' acclimation, adaptation, and genetic drift may predominate (Hampe and Petit 2005). Nature reserves may provide key sites for studying these processes.

1.4.3 Regional Fingerprints

There are many features in common around the mountains of the world. On the other hand, each range is different. Even at a relatively short distance, the environmental and socio-economical context of the mountains may differ and with them so do the risk and sensitivity to perturbations. A paradigmatic case is the contrast between the surrounding ranges at the north and south halves of the Mediterranean Sea. Roughly, temperature controls northern vegetation through limiting growth period, whereas water deficit shortens biologic activity at the southern areas. The idiosyncratic aspects may extend to concepts apparently clearly established. The available land area at high altitudes is lower than at montane stages, for instance. However, what happens at medium altitudes may largely differ between ranges. One can find all sort of altitude-areal distributions affecting the upper limit of montane vegetation belts, with contrasting implications in case of upward migration of the montane species (Elsen and Tingley 2015).

1.5 Conservation Synergies and Challenges

The combination of exposure and sensitivity determines the eventual vulnerability to environmental changes of the landscape and ecosystems of National Parks and nature reserves in high mountains. Due to the variety of large-scale and local changes to which most reserve areas are currently exposed, the achievement of conservation synergies may be the general goal for natural reserves. The challenge is how to handle the different trade-offs that the regional context may define.

1.5.1 Conservation Versus Stewardship (Franciscans Vs. Benedictines)

The declaration of nature reserves and, notably, National Parks was rooted in a concept of stable nature in which the (only) perturbing element was human activity. It was assumed that controlling the latter influence, conservation of landscape, ecosystems, and emblematic organisms would proceed on their own. Today, direct human pressure around the reserve areas is still a major problem to deal with for conservation in many places. A radical conservationist position (Franciscan) might be useful in front of local and regional socio-economic influences. But it is certainly useless when the change is driven at much larger scale, exceeding any local or regional countermeasure. Conservation becomes a matter of stewardship of the changes (Benedictines position) in agreement with goals at planetary scale (Steffen et al. 2011). The first step is acknowledging the situation. Climate change is indeed occurring and, whatever the final mitigation of the problem, significant local changes in species distribution and ecosystem dynamics will occur.

A new primary goal of conservation is to handle a smoother transition as possible to new states. This requires the projection of the potential changes and understanding the dynamics leading to them. As we are immersed in the dynamics, this exercise has to be permanently recursive. Migration and invasion would become current issues in nature reserves, and decisions on how to handle them would be better if planned in advance. The principles of nature reserve stewardship have to be developed keeping in mind the multifaceted essence of global change and the mountain idiosyncrasy to enhance some of them. Three basic strategies have been suggested to make the best use of current understanding in an environment of inevitable uncertainty and likely sudden change: reducing the magnitude of, and exposure and sensitivity to, known stresses; focusing on proactive policies that shape change; and avoiding or escaping unsustainable social-ecological traps (Chapin et al. 2010). Likely, all conservation measures are vulnerable to projected changes, but also they should involve sources of adaptive capacity and resilience that can sustain active stewardship of nature reserves.

Climate shift in mountains may result in an extinction debt to pay decades ahead (Hanski 2013). Decisions would have to be taken to what extent it merits to fight against species disappearance that sooner or later may occur (Dullinger et al. 2012). Ecological knowledge here becomes critical, and there is an increasing demand for assessments considering the details of the species distribution. Nature reserves can be just a small portion of the territory occupied by a particular species, but within which actions can be undertaken. As long as a species predicted to become extinct still persists, there is time for conservation measures such as habitat restoration and landscape management (Kuussaari et al. 2009). Standardised long-term monitoring (Zamora et al. 2017), more high-quality empirical studies on key taxa (Fig. 1.7) and ecosystems, and further development of analytical methods will help to quantify the extinction debt better and to more successfully protect mountain biodiversity.

The water balance will certainly play a significant role in mountain reserves' fate. Warming will increase growth temperature and thus water demand; drought risk will increase even in areas with average positive water balance. Does it make sense to undertake mitigation engineering measures? Climate has oscillated enormously during the last million years. Many mountains have been glaciated and deglaciated several times. Fragmentation and refugia during harsh periods explain the current distribution of many species and biogeographical paradoxes among sister species (Schmitt 2017). Do we have the knowledge and tools to identify potential refugia within nature reserves (Gavin et al. 2014)? This could be a main task in conservation research for the near future (Birks and Willis 2008).

Fig. 1.7 Bearded vulture (*Gypaetus barbatus*) in Aigüestortes i Estany de Sant Maurici National Park. There are species that play a key role in ecosystem dynamics and, at the same time, become flagship organisms for conservation, either in general or for a particular nature reserve. They attain a high symbolic value both for the scientific community and the population in general. Correct management of these species requires accurate knowledge of their behaviour that needs to be based on long-term research plans (Margalida 2017). Nature reserves are the appropriate scenario for sustaining this type of studies. Eventually, this will facilitate more objective management decisions. Photography: Mario Lancha. Archive of the Aigüestortes i Estany de Sant Maurici National Park

On the other hand, molecular studies may identify priority areas for conservation of the genetic resources of endangered species (Petit et al. 1998).

Conservation has a strong imprinting of the habitat concept. It has been useful for mapping, and a great effort has been put in habitat classification. Patterns of habitat invasions are consistent between regions and some of the high mountain habitats are among the less susceptible to invasion [e.g., heathlands and high-mountain grasslands (Chytry et al. 2008)]. However, about the habitat concept and the interaction between species in general, there is a marked influence of the Franciscan conservation view. Habitats are sometimes erroneously thought as places to be filled with particular species. In fact, it may be argued the opposed, the species make the habitat (Rosenzweig 1995). One may recognize different habitats because they hold different species but would not make any distinction if the same species, or a subgroup of them, were filling the space. The distribution of species will certainly change at a different pace, thus rearranging the current interactions at local and regional levels. Conservation will have to deal with that. Coexisting species will not migrate at the same pace with climate change, they even may move in opposite directions if some migrate with the gradient and other counter it (Laiolo and Obeso 2017).

There is a wealth of information about ecological changes with climate fluctuations and human land use in the mountains that still requires a deep analysis beyond patterns' description. Progressively, environmental history is providing a better temporal and spatial resolution about the changes occurred in the mountains during the Holocene (Pèlachs et al. 2017). Yet we may be switching towards a situation without analogues in the past, during this transition we may face situations that similarly occurred during the last 10,000 years. Introducing a long-term historical perspective (centuries to several millennia) in the tool box of mountain conservation management appears as a sensible measure to achieve a correct compromise between observation and action.

1.5.2 Loss of Uniqueness

Natural reserves are proud of being representative of the natural values of a particular territory and landscape, but also they are selected because they hold elements of singularity. Scenic values have not to be dismissed (Fig. 1.8). Sometimes the singularity may be more aesthetical than substantial (e.g., certain erosive forms). The very term "National Park" tracks back to some of the initial triggers for the creation of nature reserves during a historical period of effervescent national feelings (Radkau 2008). Identity was reinforced protecting areas of "unique landscape." These scenic values are still perceived as a primary value for visitors to mountain nature reserves (Schirpke et al. 2013). Abandonment of traditional land use may lead to more homogenous landscapes, which may be perceived negatively by visitors. Thus regional mountain planning and political decision makers should make compatible demands from the wilderness, and cultural landscape conservation visions. This is not an easy task. It requires the understanding of nature and socio-economic dynamics at multiple scales. In many ways, the global challenges have its equivalent at regional and local scales.

High-mountains, particularly those in ranges of intermediate size, become territories with a collection of communities and species that mostly differ from those dominant in low lands and middle mountains. This applies in particular when the elevation is in a dry region so that large differences in water availability are added to the altitudinal thermal gradient. In areas without water shortage, warming will produce an uplift of the vegetation belts with the risk of losing the high-mountain character to some extent (Rixen and Wipf 2017). An extreme situation will be the collapse of the alpine flora in a few submits and its impoverishment in many high-mountain temperate areas. This may lead to challenging decisions to be taken in nature reserves. Studies on the genetic structure of small plant populations have provided unexpected results about the lack of the correspondence between isolation

Fig. 1.8 Sant Esperit waterfall in the Aigüestortes i Estany de Sant Maurici National Park at 1950 (**a**) and 2010 (**b**). Scenic values are part of the mountain nature reserves. Historically, National Parks emerged in a period of national enthusiasm as a way to remark the singularities of a country's landscape. Although this view may have evolved, the symbolic elements of the landscape pervade from local to global scales, becoming icons of locations, values and natural processes. Photography: Ricard Novell archive (**a**) and Archive of the Aigüestortes i Estany de Sant Maurici National Park (**b**)

and genetic richness (Blanco-Pastor et al. 2013). Ad hoc studies may be helpful, before expending efforts in a wrong direction.

1.5.3 Functional Versus Phylogenetic Conservation

Criticism to estimated excessive effort on conservation actions towards preserving genetic variability and details in phylogeography issues advocates that conservation should aim to maintain functional diversity rather than to pay attention to taxonomic and phylogenetic details. This appears to be a false debate. It primarily depends on where one stops defining functional traits. The more detail we gain in functional attributes, the more we approach function to phylogenetic structures (Flynn et al. 2011). Trait convergence among scarcely related phylogenetic groups is a fact; however, convergence in some characters does not invalidate differences in others. So, in the end, any species is phylogenetically and ecologically unique.

Preserving species preserves functions, with the added value that no judgments are required on which functions have to be prioritized. On the other hand, partial functional redundancy is a fact in nature. Probably simply by natural selection, essential functions have become more redundant than the ones less critical for the organization of ecosystem functioning regarding cycling of matter and energy flow.

1.5.4 Size Matters

It is unclear how much of the present distribution of species in high mountains is related to human influence. Treeline has been modified at least since about 3000 years ago in many European mountain valleys (Pèlachs et al. 2017). Fire was the tool to open the landscape (Tinner et al. 2005). The artificial maintenance of pasturing meadows at high altitude should have modified the species assemblage and favored some of them, according to the enhanced grazing and fertilization. It would be interesting to check whether a selection of species or morphotypes can be related to these increased pressures. Similarly, the human pressure on some trees may not have been genetically innocuous. Fir (*Abies alba*) is currently mostly restricted to north-facing slopes in the Pyrenean valleys. However, there is growing evidence that in the past, along the second half of the Holocene, it also extended through south-facing slopes over a larger area. Apparently, there are not climatic reasons for the observed change. Rather, there is growing palaeoecological evidence that humans have preferred to leave fir in the shaded areas and facilitate other species (e.g., pines (*Pinus sylvestris*, *P. uncinata*), or beech (*Fagus sylvatica*), among others) in the sunny slopes and plains. This practice may have lasted during centuries, perhaps actually introducing a shift towards these conditions in the Pyrenean fir metapopulation. Studies across the range of *Abies alba* indicate an intense human pressure on this tree in Southern Europe (Ruosch et al. 2016). In any case, with decreased forestry activity within and outside the nature reserves, current tree distribution in many mountains would be changing even without climate change. Many nature reserves may be too small to cope appropriately with these changes and forest mass conservation in general. Reserve size matters, at least concerning trees and large mammals, thus conservation policy outside natural reserves becomes a vital element for them. Natural reserves may provide space for pilot studies and observations to inform decisions over large mountain-areas regulation.

1.5.5 Local Contribution to Global Ecological Services

The task that mountain nature reserves may play in mitigating global change may appear as insignificant. Although carbon stock in soils and forests may be high per surface unit (Garcia-Pausas et al. 2017), the overall figure may become irrelevant when considering large areas. It may happen similarly with biodiversity preservation and any other of the components of current global change. Although ecological services to mitigate global change issues can be modest, the symbolism of mountain nature reserves may be valuable. Actions in mountain reserves may be mirrored in other areas covering larger extensions. It is not only a matter of education but also of symbolism. Mountains have had a symbolic appeal since ancient times, and they are still being so. Current conservation goals may emotionally link people to those traditional values [e.g., sacred lakes and forests (Brandt et al. 2013)].

1.5.6 Conservation Beyond Conservation

The conservation paradigm was built in a context in which a large part of the population had actual contact with natural environments through a rural life or recent memory of them. Even in those circumstances, the understanding and emotional connection to nature varied among cultures (Crowley 2013). Nowadays, more than half of the population of the planet has moved to the cities and a large part of the population in developed countries has only incidental contact with nature. Even among those that expend a lot of time outdoors, most are of urban origin, many are sports practitioners or specialized collectors with a limited understanding (interest) for how nature works as a whole. Conservation has had always a double appeal: emotional and rational. The hankering of preserving a past legacy is now challenged by global change but also by the loss of tight connections of humankind with nature.

The conservation paradigm has to face the new situation and accumulate as many as possible rational and emotional arguments. A leading token in that sense is the global stewardship humans are responsible for (Steffen et al. 2011). In this situation, new conservation values emerge; for instance, microbial conservation (Casamayor 2017) and biogeochemical ecosystem services (Garcia-Pausas et al. 2017). Saving big trees and bears is fine, but consciousness about our intervention in the dynamics of the whole planet system and the need to respond in an appropriate way to this challenge becomes a priority in the conservation paradigm. In contrast to trees and bears, microbes and biogeochemical pathways are scarcely apparent. Communication and visualization techniques in nature reserves have a great role to play in this issue, including the use of new technologies.

The more conservation suffers from social pressure, the more it is involved in the economic arena. This has pros and cons, but the tendency of a growing economics around conservation calls for more expert attention (Sala et al. 2013). The environment has been progressively accepted as an economic sector as pollution problems have been growing. Conservation has not fully been included yet in the same dynamics. It is still viewed in aesthetical terms rather than as a critical part of the socio-economic affairs. This flawy attitude is even shared by some professionals of environmental issues. They see conservation as a matter of preserving beautiful butterflies, instead of recognizing that the conservation paradigm provides the roots in which their activity is grounded.

The planet requires areas (land and sea) in which nature free dynamics can develop. These areas have to be sufficiently large to be more than an open zoo. On the other hand, conservation has to facilitate the contact of the citizens with the natural processes beyond TV shows. Therefore, rather than firm boundaries, a complex system of progressive zonation with different degrees of human presence and activities is desirable. The high-mountain landscape is naturally prompt to this situation. This is the reason why many nature reserves are located there. The harsher conditions with increasing altitude provide a natural softening of direct human incidence. However, preserving the highest lands cannot be at the expense of inhibition about lower areas. Nature conservation is crying for new developments in science and technology on natural and social fields according to the current challenging changing times.

Acknowledgements Although the authors are exclusively responsible for the content of this chapter, they have been largely inspired by the presentations and discussions in the international workshop on "The High Mountain in a Changing World: Challenges for conservation", which took place in Espot (4–6 November, 2015) as part of the activities for the 60th anniversary of the Aigüestortes i Estany de Sant Maurici National Park (Catalan Pyrenees, Spain). The authors acknowledge Jordi Vicente for the photograph selection and thank the authors for permission to include them in the chapter. The research projects CUL-PA (ref. 998/2013) and ARBALMONT (ref. 634S/2012) from the Spanish National Parks research program are also acknowledged.

References

Agnoletti M (2007) The degradation of traditional landscape in a mountain area of Tuscany during the 19th and 20th centuries: implications for biodiversity and sustainable management. For Ecol Manag 249:5–17

Anderson L (2012) Rocky Mountain hydroclimate: Holocene variability and the role of insolation, ENSO, and the North American Monsoon. Glob Planet Change 92–93:198–208

Bacardit M, Krachler M, Camarero L (2012) Whole-catchment inventories of trace metals in soils and sediments in mountain lake catchments in the Central Pyrenees: apportioning the anthropogenic and natural contributions. Geochim Cosmochim Acta 82:52–67

Barberan A, Henley J, Fierer N, Casamayor EO (2014) Structure, inter-annual recurrence, and global-scale connectivity of airborne microbial communities. Sci Total Environ 487:187–195

Bartrons M, Grimalt JO, de Mendoza G, Catalan J (2012) Pollutant dehalogenation capability may depend on the trophic evolutionary history of the organism: PBDEs in freshwater food webs. PLoS ONE 7:e41829

Beniston M (2006) Mountain weather and climate: a general overview and a focus on climatic change in the Alps. Hydrobiologia 562:3–16

Beniston M (2012) Impacts of climatic change on water and associated economic activities in the Swiss Alps. J Hydrol 412:291–296

Beniston M, Stoffel M (2014) Assessing the impacts of climatic change on mountain water resources. Sci Total Environ 493:1129–1137

Beniston M, Stoffel M (2016) Rain-on-snow events, floods and climate change in the Alps: events may increase with warming up to 4 °C and decrease thereafter. Sci Total Environ 571:228–236

Beniston M, Stephenson DB, Christensen OB, Ferro CAT, Frei C, Goyette S, Halsnaes K, Holt T, Jylha K, Koffi B, Palutikof J, Schoell R, Semmler T, Woth K (2007) Future extreme events in European climate: an exploration of regional climate model projections. Clim Change 81:71–95
Berg MP, Kiers ET, Driessen G, van der Heijden M, Kooi BW, Kuenen F, Liefting M, Verhoef HA, Ellers J (2010) Adapt or disperse: understanding species persistence in a changing world. Glob Change Biol 16:587–598
Berrocal MC, Lopez MS, Gonzalez AU, Lopez-Saez JA (2014) Landscape construction and long-term economic practices: an example from the Spanish Mediterranean Uplands through rock art archaeology. J Archaeol Method Th 21:589–615
Birks HJB, Willis KJ (2008) Alpines, trees, and refugia in Europe. Plant Ecol Divers 1:147–160
Blanco-Pastor JL, Fernandez-Mazuecos M, Vargas P (2013) Past and future demographic dynamics of alpine species: limited genetic consequences despite dramatic range contraction in a plant from the Spanish Sierra Nevada. Mol Ecol 22:4177–4195
Blumthaler M, Ambach W (1990) Indication of increasing solar ultraviolet-B radiation flux in alpine regions. Science 248:206–208
Blumthaler M, Ambach W, Salzgeber M (1994) Effects of cloudiness on global and diffuse UV irradiance in a high-mountain area. Theor Appl Clim 50:23–30
Blumthaler M, Ambach W, Ellinger R (1997) Increase in solar UV radiation with altitude. J Photoch Photobio B 39:130–134
Bocquet A (1997) Archéologie et peuplement des Alpes françaises du Nord au néolithique et aux âges des métaux. L'Anthropologie 101:291–393
Bogdal C, Nikolic D, Luthi MP, Schenker U, Scheringer M, Hungerbuhler K (2010) Release of legacy pollutants from melting glaciers: model evidence and conceptual understanding. Environ Sci Technol 44:4063–4069
Boithias L, Acuna V, Vergonos L, Ziv G, Marce R, Sabater S (2014) Assessment of the water supply: demand ratios in a Mediterranean basin under different global change scenarios and mitigation alternatives. Sci Total Environ 470:567–577
Brandt JS, Wood EM, Pidgeon AM, Han LX, Fang ZD, Radeloff VC (2013) Sacred forests are keystone structures for forest bird conservation in southwest China's Himalayan Mountains. Biol Conserv 166:34–42
Burley JD, Theiss S, Bytnerowicz A, Gertler A, Schilling S, Zielinska B (2015) Surface ozone in the Lake Tahoe Basin. Atmos Environ 109:351–369
Camarero JJ (2017a) The multiple factors explaining decline in mountain forests: historical logging and warming-related drought stress cause silver fir dieback in the Aragón Pyrenees. In: Catalan J, Ninot JM, Aniz MM (eds) Challenges for high mountain conservation in a changing world. Springer, pp 131–154
Camarero JJ, Gazol A, Galván JD, Sangüesa-Barreda G, Gutiérrez E (2014) Disparate effects of global-change drivers on mountain conifer forests: warming-induced growth enhancement in young trees vs. CO_2 fertilization in old trees from wet sites. Glob Change Biol 21:738–749
Camarero L (2017b) Atmospheric chemical loadings in the high mountain: current forcing and legacy pollution. In: Catalan J, Ninot JM, Aniz MM (eds) Challenges for high mountain conservation in a changing world. Springer, pp 325–341
Camarero L, Catalan J (2012) Atmospheric phosphorus deposition may cause lakes to revert from phosphorus limitation back to nitrogen limitation. Nat Commun 3:1118
Camarero L, Masque P, Devos W, Ani-Ragolta I, Catalan J, Moor HC, Pla S, Sanchez-Cabeza JA (1998) Historical variations in lead fluxes in the Pyrenees (northeast Spain) from a dated lake sediment core. Water Air Soil Pollut 105:439–449
Carey M, Baraer M, Mark BG, French A, Bury J, Young KR, McKenzie JM (2014) Toward hydro-social modeling: merging human variables and the social sciences with climate-glacier runoff models (Santa River, Peru). J Hydrol 518:60–70
Carrion JS (2002) Patterns and processes of Late Quaternary environmental change in a montane region of southwestern Europe. Quat Sci Rev 21:2047–2066

Casamayor EO (2017) Towards a microbial conservation perspective in high-mountain lakes. In: Catalan J, Ninot JM, Aniz MM (eds) Challenges for high mountain conservation in a changing world. Springer, pp 157–180

Catalan J (2015) Tracking long-range atmospheric transport of trace metals, polycyclic aromatic hydrocarbons, and organohalogen compounds using lake sediments of mountain regions. In: Blais JM, Rosen MR, Smol JP (eds) Environmental contaminants, vol 18. Developments in paleoenvironmental research. Springer, Netherlands, pp 263–322

Catalan J, Vilalta R, Weitzman B, Ventura M, Comas E, Pigem C, Aranda R, Ballesteros E, Camarero L, García Serrano J, Pla S, Sáez A, Aiguabella P (1997) The hydraulic industry in the Pyrenees: evaluation, correction and prevention of the environmental impact at the Aigüestortes i estany de Sant Maurici National Park (In Catalan). La Caixa, Barcelona

Catalan J, Ventura M, Vives I, Grimalt JO (2004) The roles of food and water in the bioaccumulation of organochlorine compounds in high mountain lake fish. Environ Sci Technol 38:4269–4275

Catalan J, Bartrons M, Camarero L, Grimalt JO (2013) Mountain waters as witnesses of global pollution. In: Pechan P, Vries GEd (eds) Living with water: targeting quality in a dynamic world. Springer, New York, pp 31–67

Cocozza C, de Miguel M, Psidova E, Ditmarova L, Marino S, Maiuro L, Alvino A, Czajkowski T, Bolte A, Tognetti R (2016) Variation in ecophysiological traits and drought tolerance of beech (*Fagus sylvatica* L.) Seedlings from different populations. Front Plant Sci 7:886

Cristofanelli P, Bonasoni P (2009) Background ozone in the southern Europe and Mediterranean area: influence of the transport processes. Environ Pollut 157:1399–1406

Cristofanelli P, Bracci A, Sprenger M, Marinoni A, Bonafe U, Calzolari F, Duchi R, Laj P, Pichon JM, Roccato F, Venzac H, Vuillermoz E, Bonasoni P (2010) Tropospheric ozone variations at the Nepal climate observatory-pyramid (Himalayas, 5079 m a.s.l.) and influence of deep stratospheric intrusion events. Atmos Chem Phys 10:6537–6549

Crowley T (2013) Climbing mountains, hugging trees: a cross-cultural examination of love for nature. Emot Space Soc 6:44–53

Chapin FS III, Carpenter SR, Kofinas GP, Folke C, Abel N, Clark WC, Olsson P, Smith DMS, Walker B, Young OR, Berkes F, Biggs R, Grove JM, Naylor RL, Pinkerton E, Steffen W, Swanson FJ (2010) Ecosystem stewardship: sustainability strategies for a rapidly changing planet. Trends Ecol Evol 25:241–249

Chytry M, Maskell LC, Pino J, Pysek P, Vila M, Font X, Smart SM (2008) Habitat invasions by alien plants: a quantitative comparison among Mediterranean, subcontinental and oceanic regions of Europe. J Appl Ecol 45:448–458

Davidson DA, Wilkinson AC, Kimpe LE, Blais JM (2004) Persistent organic pollutants in air and vegetation from the Canadian Rocky Mountains. Environ Toxicol Chem 23:540–549

Dedieu JP, Lessard-Fontaine A, Ravazzani G, Cremonese E, Shalpykova G, Beniston M (2014) Shifting mountain snow patterns in a changing climate from remote sensing retrieval. Sci Total Environ 493:1267–1279

DeFries RS, Ellis EC, Chapin FS, Matson PA, Turner BL, Agrawal A, Crutzen PJ, Field C, Gleick P, Kareiva PM, Lambin E, Liverman D, Ostrom E, Sanchez PA, Syvitski J (2012) Planetary opportunities: a social contract for global change science to contribute to a sustainable future. Bioscience 62:603–606

Diaz HF, Bradley RS (1997) Temperature variations during the last century at high elevation sites. Clim Change 36:253–279

Dullinger S, Dirnbock T, Grabherr G (2004) Modelling climate change-driven treeline shifts: relative effects of temperature increase, dispersal and invasibility. J Ecol 92:241–252

Dullinger S, Dirnbock T, Kock R, Hochbichler E, Englisch T, Sauberer N, Grabherr G (2005) Interactions among tree-line conifers: differential effects of pine on spruce and larch. J Ecol 93:948–957

Dullinger S, Gattringer A, Thuiller W, Moser D, Zimmermann NE, Guisan A, Willner W, Plutzar C, Leitner M, Mang T, Caccianiga M, Dirnboeck T, Ertl S, Fischer A, Lenoir J,

Svenning J-C, Psomas A, Schmatz DR, Silc U, Vittoz P, Huelber K (2012) Extinction debt of high-mountain plants under twenty-first-century climate change. Nat Clim Change 2:619–622
Elsen PR, Tingley MW (2015) Global mountain topography and the fate of montane species under climate change. Nat Clim Change 5:772–776
Elvira S, Gonzalez-Fernandez I, Alonso R, Sanz J, Bermejo-Bermejo V (2016) Ozone levels in the Spanish Sierra de Guadarrama mountain range are above the thresholds for plant protection: analysis at 2262, 1850, and 995 m a.s.l. Environ Monit Assess 188
Elliott JE, Levac J, Guigueno MF, Shaw DP, Wayland M, Morrissey CA, Muir DCG, Elliott KH (2012) Factors influencing legacy pollutant accumulation in alpine osprey: biology, topography, or melting glaciers? Environ Sci Technol 46:9681–9689
Engler R, Randin CF, Thuiller W, Dullinger S, Zimmermann NE, Araujo MB, Pearman PB, Le Lay G, Piedallu C, Albert CH, Choler P, Coldea G, De Lamo X, Dirnbock T, Gegout J-C, Gomez-Garcia D, Grytnes J-A, Heegaard E, Hoistad F, Nogues-Bravo D, Normand S, Puscas M, Sebastia M-T, Stanisci A, Theurillat J-P, Trivedi MR, Vittoz P, Guisan A (2011) 21st century climate change threatens mountain flora unequally across Europe. Glob Change Biol 17:2330–2341
Flo D, Hagvar S (2013) Aerial dispersal of invertebrates and mosses close to a receding alpine glacier in Southern Norway. Arct Antarct Alp Res 45:481–490
Flynn DFB, Mirotchnick N, Jain M, Palmer MI, Naeem S (2011) Functional and phylogenetic diversity as predictors of biodiversity-ecosystem-function relationships. Ecology 92:1573–1581
García M, Gómez D (2007) Flora del Pirineo Aragonés. Patrones espaciales de biodiversidad y su relevancia para la conservación. Pirineos 162:71–88
Garcia MB, Espadaler X, Olesen JM (2012) Extreme reproduction and survival of a true Cliffhanger: the endangered plant *Borderea chouardii* (Dioscoreaceae). PLoS ONE 7
Garcia-Pausas J, Romanyà J, Montané F, Rios AI, Taull M, Rovira P, Casals P (2017) Are soil carbon stocks in mountain grasslands compromised by land use changes? In: Catalan J, Ninot JM, Aniz MM (eds) Challenges for high mountain conservation in a changing world. Springer, pp 207–230
Garrard R, Kohler T, Price MF, Byers AC, Sherpa AR, Maharjan GR (2016) Land use and land cover change in Sagarmatha National Park, a world heritage site in the Himalayas of Eastern Nepal. Mt Res Dev 36:299–310
Gassiot Ballbè E, Mazzucco N, Clemente Conte I, Rodríguez Antón D, Obea Gómez L, Quesada Carrasco M, Díaz Bonilla S (2017) The beginning of high mountain occupations in the Pyrenees. Human settlements and mobility from 18,000 cal BC to 2000 cal BC. In: Catalan J, Ninot JM, Aniz MM (eds) Challenges for high mountain conservation in a changing world. Springer, pp 75–105
Gavin DG, Fitzpatrick MC, Gugger PF, Heath KD, Rodríguez-Sánchez F, Dobrowski SZ, Hampe A, Hu FS, Ashcroft MB, Bartlein PJ, Blois JL, Carstens BC, Davis EB, de Lafontaine G, Edwards ME, Fernandez M, Henne PD, Herring EM, Holden ZA, W-s Kong, Liu J, Magri D, Matzke NJ, McGlone MS, Saltré F, Stigall AL, Tsai Y-HE, Williams JW (2014) Climate refugia: joint inference from fossil records, species distribution models and phylogeography. New Phytol 204:37–54
Gobiet A, Kotlarski S, Beniston M, Heinrich G, Rajczak J, Stoffel M (2014) 21st century climate change in the European Alps—a review. Sci Total Environ 493:1138–1151
Gottfried M, Hantel M, Maurer C, Toechterle R, Pauli H, Grabherr G (2011) Coincidence of the alpine-nival ecotone with the summer snowline. Environ Res Lett 6:014013
Grimalt JO, van Drooge BL (2006) Polychlorinated biphenyls in mountain pine (*Pinus uncinata*) needles from Central Pyrenean high mountains (Catalonia, Spain). Ecotoxicol Environ Saf 63:61–67

Grimalt JO, Fernandez P, Berdie L, Vilanova RM, Catalan J, Psenner R, Hofer R, Appleby PG, Rosseland BO, Lien L, Massabuau JC, Battarbee RW (2001) Selective trapping of organochlorine compounds in mountain lakes of temperate areas. Environ Sci Technol 35:2690–2697

Hampe A, Petit RJ (2005) Conserving biodiversity under climate change: the rear edge matters. Ecol Lett 8:461–467

Hanski I (2013) Extinction debt at different spatial scales. Anim Conserv 16:12–13

Hantel M, Hirtl-Wielke LM (2007) Sensitivity of Alpine snow cover to European temperature. Int J Climatol 27:1265–1275

Hegglin MI, Shepherd TG (2009) Large climate-induced changes in ultraviolet index and stratosphere-to-troposphere ozone flux. Nat Geosci 2:687–691

Herrero A, Zamora R (2014) Plant responses to extreme climatic events: a field test of resilience capacity at the southern range edge. PLoS ONE 9:e87842

Hickling R, Roy DB, Hill JK, Fox R, Thomas CD (2006) The distributions of a wide range of taxonomic groups are expanding polewards. Glob Change Biol 12:450–455

Hodar JA, Castro J, Zamora R (2003) Pine processionary caterpillar *Thaumetopoea pityocampa* as a new threat for relict Mediterranean Scots pine forests under climatic warming. Biol Conserv 110:123–129

Ignacio Lopez-Moreno J, Goyette S, Vicente-Serrano SM, Beniston M (2011) Effects of climate change on the intensity and frequency of heavy snowfall events in the Pyrenees. Clim Change 105:489–508

Jarque S, Quirós L, Grimalt JO, Gallego E, Catalan J, Lackner R, Piña B (2015) Background fish feminization effects in European remote sites. Sci Rep 5:11292

Komac B, Domenech M, Fanlo R (2014) Effects of grazing on plant species diversity and pasture quality in subalpine grasslands in the eastern Pyrenees (Andorra): implications for conservation. J Nat Conserv 22:247–255

Konchar KM, Staver B, Salick J, Chapagain A, Joshi L, Karki S, Lo S, Paudel A, Subedi P, Ghimire SK (2015) Adapting in the shadow of Annapurna: a climate tipping point. J Ethnobiol 35:449–471

Konvicka M, Maradova M, Benes J, Fric Z, Kepka P (2003) Uphill shifts in distribution of butterflies in the Czech Republic: effects of changing climate detected on a regional scale. Glob Ecol Biogeogr 12:403–410

Körner C (2007) The use of 'altitude' in ecological research. Trends Ecol Evol 22:569–574

Korner C, Paulsen J (2004) A world-wide study of high altitude treeline temperatures. J Biogeogr 31:713–732

Kuussaari M, Bommarco R, Heikkinen RK, Helm A, Krauss J, Lindborg R, Ockinger E, Partel M, Pino J, Roda F, Stefanescu C, Teder T, Zobel M, Steffan-Dewenter I (2009) Extinction debt: a challenge for biodiversity conservation. Trends Ecol Evol 24:564–571

Laiolo P, Obeso JR (2017) Life-history responses to the altitudinal gradient. In: Catalan J, Ninot JM, Aniz MM (eds) Challenges for high mountain conservation in a changing world. Springer, pp 253–283

Lapenis A, Shvidenko A, Shepaschenko D, Nilsson S, Aiyyer A (2005) Acclimation of Russian forests to recent changes in climate. Glob Change Biol 11:2090–2102

Lasanta-Martinez T, Vicente-Serrano SM, Cuadrat-Prats JM (2005) Mountain Mediterranean landscape evolution caused by the abandonment of traditional primary activities: a study of the Spanish Central Pyrenees. Appl Geogr 25:47–65

Li J, Wang ZF, Akimoto H, Gao C, Pochanart P, Wang XQ (2007) Modeling study of ozone seasonal cycle in lower troposphere over east Asia. J Geophys Res-Atmos 112

Lloret F (2017) Trade-offs in high mountain conservation. In: Catalan J, Ninot JM, Aniz MM (eds) Challenges for high mountain conservation in a changing world. Springer, pp 37–59

Lloret F, Siscart D, Dalmases C (2004) Canopy recovery after drought dieback in holm-oak Mediterranean forests of Catalonia (NE Spain). Glob Change Biol 10:2092–2099

Margalef R (1976) Bases ecològiques per a una gestió de la natura. In: Folch R (ed) Natura, ús o abús? Editorial Barcino, Barcelona, pp 23–64

Margalida A (2017) Importance of long-term studies to conservation practice: the case of the bearded vulture in the Pyrenees. In: Catalan J, Ninot JM, Aniz MM (eds) Challenges for high mountain conservation in a changing world. Springer, pp 343–383

Martin Civantos JM (2014) Mountainous landscape domestication. Management of non-cultivated productive areas in Sierra Nevada (Granada-Almeria, Spain). Eur J Post-Class Archaeol 4:99–103

McCullough IM, Davis FW, Dingman JR, Flint LE, Flint AL, Serra-Diaz JM, Syphard AD, Moritz MA, Hannah L, Franklin J (2016) High and dry: high elevations disproportionately exposed to regional climate change in Mediterranean-climate landscapes. Landsc Ecol 31:1063–1075

Miehe G, Miehe S, Bohner J, Kaiser K, Hensen I, Madsen D, Liu JQ, Opgenoorth L (2014) How old is the human footprint in the world's largest alpine ecosystem? A review of multiproxy records from the Tibetan Plateau from the ecologists' viewpoint. Quat Sci Rev 86:190–209

Miró A, Ventura M (2013) Historical use, fishing management and lake characteristics explain the presence of non-native trout in Pyrenean lakes: implications for conservation. Biol Conserv 167:17–24

Mitchell D, Paniker L, Lin K, Fernandez A (2015) Interspecific variation in the repair of UV damaged DNA in the genus *Xiphophorus* as a factor in the decline of the Rio Grande Platyfish. Photochem Photobiol 91:486–492

Moran-Tejeda E, Ignacio Lopez-Moreno J, Beniston M (2013) The changing roles of temperature and precipitation on snowpack variability in Switzerland as a function of altitude. Geophys Res Lett 40:2131–2136

Moran-Tejeda E, Lorenzo-Lacruz J, Ignacio Lopez-Moreno J, Rahman K, Beniston M (2014) Streamflow timing of mountain rivers in Spain: recent changes and future projections. J Hydrol 517:1114–1127

Morán-Tejeda E, López-Moreno JI, Sanmiguel-Vallelado A (2017) Changes in climate, snow and water resources in the Spanish Pyrenees: observations and projections in a warming climate. In: Catalan J, Ninot JM, Aniz MM (eds) Challenges for high mountain conservation in a changing world. Springer, pp 305–323

Muir DCG, Howard PH (2006) Are there other persistent organic pollutants? A challenge for environmental chemists. Environ Sci Technol 40:7157–7166

Nogues-Bravo D, Araujo MB, Errea MP, Martinez-Rica JP (2007) Exposure of global mountain systems to climate warming during the 21st century. Glob Environ Change-Hum Policy Dimens 17:420–428

Noyes PD, McElwee MK, Miller HD, Clark BW, Van Tiem LA, Walcott KC, Erwin KN, Levin ED (2009) The toxicology of climate change: environmental contaminants in a warming world. Environ Int 35:971–986

Nyaupane GP, Lew AA, Tatsugawa K (2014) Perceptions of trekking tourism and social and environmental change in Nepal's Himalayas. Tour Geogr 16:415–437

Oian H (2013) Wilderness tourism and the moralities of commitment: hunting and angling as modes of engaging with the natures and animals of rural landscapes in Norway. J Rural Stud 32:177–185

Ooi N, Laing J, Mair J (2015) Sociocultural change facing ranchers in the Rocky Mountain West as a result of mountain resort tourism and amenity migration. J Rural Stud 41:59–71

Palazón S (2017) The importance of reintroducing large carnivores: the brown bear in the Pyrenees. In: Catalan J, Ninot JM, Aniz MM (eds) Challenges for high mountain conservation in a changing world. Springer, pp 231–249

Pauchard A, Kueffer C, Dietz H, Daehler CC, Alexander J, Edwards PJ, Ramon Arevalo J, Cavieres LA, Guisan A, Haider S, Jakobs G, McDougall K, Millar CI, Naylor BJ, Parks CG, Rew LJ, Seipel T (2009) Ain't no mountain high enough: plant invasions reaching new elevations. Front Ecol Environ 7:479–486

Pearson RG (2016) Reasons to conserve nature. Trends Ecol Evol 31:366–371

Pèlachs A, Nadal J, Manuel Soriano J, Molina D, Cunill R (2009) Changes in Pyrenean woodlands as a result of the intensity of human exploitation: 2,000 years of metallurgy in Vallferrera, northeast Iberian Peninsula. Veg Hist Archaeobot 18:403–416

Pèlachs A, Pérez-Obiol R, Soriano JM, Cunill R, Bal M-C, García-Codron JC (2017) The role of environmental geohistory in high mountain landscape conservation. In: Catalan J, Ninot JM, Aniz MM (eds) Challenges for high mountain conservation in a changing world. Springer, pp 107–129

Penuelas J, Poulter B, Sardans J, Ciais P, van der Velde M, Bopp L, Boucher O, Godderis Y, Hinsinger P, Llusia J, Nardin E, Vicca S, Obersteiner M, Janssens IA (2013a) Human-induced nitrogen-phosphorus imbalances alter natural and managed ecosystems across the globe. Nat Commun 4

Penuelas J, Sardans J, Estiarte M, Ogaya R, Carnicer J, Coll M, Barbeta A, Rivas-Ubach A, Llusia J, Garbulsky M, Filella I, Jump AS (2013b) Evidence of current impact of climate change on life: a walk from genes to the biosphere. Glob Change Biol 19:2303–2338

Petit RJ, El Mousadik A, Pons O (1998) Identifying populations for conservation on the basis of genetic markers. Conserv Biol 12:844–855

Pond GJ, Passmore ME, Borsuk FA, Reynolds L, Rose CJ (2008) Downstream effects of mountaintop coal mining: comparing biological conditions using family- and genus-level macroinvertebrate bioassessment tools. J North Am Benthol Soc 27:717–737

Psenner R, Catalan J (1994) Chemical composition of lakes in crystalline basins: a combination of atmospheric deposition, geologic background, biological activity and human action. In: Margalef R (ed) Limnology now: a paradigm of planetary problems. Elsevier Science B.V., pp 255–314

Quiros L, Jarque S, Lackner R, Fernandez P, Grimalt JO, Pina B (2007) Physiological response to persistent organic pollutants in fish from mountain lakes: analysis of Cyp1A gene expression in natural populations of Salmo trutta. Environ Sci Technol 41:5154–5160

Rabasa SG, Granda E, Benavides R, Kunstler G, Espelta JM, Ogaya R, Penuelas J, Scherer-Lorenzen M, Gil W, Grodzki W, Ambrozy S, Bergh J, Hodar JA, Zamora R, Valladares F (2013) Disparity in elevational shifts of European trees in response to recent climate warming. Glob Change Biol 19:2490–2499

Radkau J (2008) Nature and power. A global history of the environment. Cambridge University Press, Cambridge

Rangwala I, Barsugli J, Cozzetto K, Neff J, Prairie J (2012) Mid-21st century projections in temperature extremes in the southern Colorado Rocky Mountains from regional climate models. Clim Dyn 39:1823–1840

Reich PB, Sendall KM, Stefanski A, Wei XR, Rich RL, Montgomery RA (2016) Boreal and temperate trees show strong acclimation of respiration to warming. Nature 531:633–636

Rixen C, Wipf S (2017) Non-equilibrium in alpine plant assemblages: current shifts in Europe's summit floras). In: Catalan J, Ninot JM, Aniz MM (eds) Challenges for high mountain conservation in a changing world. Springer, pp 285–303

Rosenzweig ML (1995) Species diversity in space and time. Cambridge University Press, Cambridge

Rotherham ID (2013) Emerging concepts and case studies of eco-cultural tourism. Cult Tour 74–89

Ruosch M, Spahni R, Joos F, Henne PD, van der Knaap WO, Tinner W (2016) Past and future evolution of *Abies alba* forests in Europe—comparison of a dynamic vegetation model with palaeo data and observations. Glob Change Biol 22:727–740

Sala E, Costello C, Dougherty D, Heal G, Kelleher K, Murray JH, Rosenberg AA, Sumaila R (2013) A general business model for marine reserves. PLoS ONE 8:e58799

Schirpke U, Holzler S, Leitinger G, Bacher M, Tappeiner U, Tasser E (2013) Can we model the scenic beauty of an alpine landscape? Sustainability 5:1080–1094

Schmid P, Kohler M, Gujer E, Zennegg M, Lanfranchi M (2007) Persistent organic pollutants, brominated flame retardants and synthetic musks in fish from remote alpine lakes in Switzerland. Chemosphere 67:S16–S21

Schmid P, Bogdal C, Bluthgen N, Anselmetti FS, Zwyssig A, Hungerbuhler K (2011) The missing piece: sediment records in remote mountain lakes confirm glaciers being secondary sources of persistent organic pollutants. Environ Sci Technol 45:203–208

Schmidt R, Koinig KA, Thompson R, Kamenik C (2002) A multi proxy core study of the last 7000 years of climate and alpine land-use impacts on an Austrian mountain lake (Unterer Landschitzsee, Niedere Tauern). Palaeogeogr Palaeoclimatol Palaeoecol 187:101–120

Schmitt T (2017) Molecular biogeography of the high mountain systems of Europe: an overview. In: Catalan J, Ninot JM, Aniz MM (eds) Challenges for high mountain conservation in a changing world. Springer, pp 63–74

Segarra-Moragues JG, Catalan P (2010) The fewer and the better: prioritization of populations for conservation under limited resources, a genetic study with *Borderea pyrenaica* (Dioscoreaceae) in the Pyrenean National Park. Genetica 138:363–376

Seppa H, Nyman M, Korhola A, Weckstrom J (2002) Changes of treelines and alpine vegetation in relation to post-glacial climate dynamics in northern Fennoscandia based on pollen and chironomid records. J Quat Sci 17:287–301

Shestakova TA, Gutiérrez E, Kirdyanov AV, Camarero JJ, Génova M, Knorre AA, Linares JC, Resco de Dios V, Sánchez-Salguero R, Voltas J (2016) Forests synchronize their growth in contrasting Eurasian regions in response to climate warming. Proc Natl Acad Sci USA 113:662–667

Shunthirasingham C, Wania F, MacLeod M, Lei YD, Quinn CL, Zhang XM, Scheringer M, Wegmann F, Hungerbuhler K, Ivemeyer S, Heil F, Klocke P, Pacepavicius G, Alaee M (2013) Mountain cold-trapping increases transfer of persistent organic pollutants from atmosphere to cows' milk. Environ Sci Technol 47:9175–9181

Steffen W, Persson A, Deutsch L, Zalasiewicz J, Williams M, Richardson K, Crumley C, Crutzen P, Folke C, Gordon L, Molina M, Ramanathan V, Rockstrom J, Scheffer M, Schellnhuber HJ, Svedin U (2011) The anthropocene: from global change to planetary stewardship. Ambio 40:739–761

Steffen W, Richardson K, Rockström J, Cornell SE, Fetzer I, Bennett EM, Biggs R, Carpenter SR, de Vries W, de Wit CA, Folke C, Gerten D, Heinke J, Mace GM, Persson LM, Ramanathan V, Reyers B, Sörlin S (2015) Planetary boundaries: guiding human development on a changing planet. Science 347

Stevens CJ, Lind EM, Hautier Y, Harpole WS, Borer ET, Hobbie S, Seabloom EW, Ladwig L, Bakker JD, Chu CJ, Collins S, Davies KF, Firn J, Hillebrand H, La Pierre KJ, MacDougall A, Melbourne B, McCulley RL, Morgan J, Orrock JL, Prober SM, Risch AC, Schuetz M, Wragg PD (2015) Anthropogenic nitrogen deposition predicts local grassland primary production worldwide. Ecology 96:1459–1465

Theurillat JP, Guisan A (2001) Potential impact of climate change on vegetation in the European Alps: a review. Clim Change 50:77–109

Thomas CD (2010) Climate, climate change and range boundaries. Divers Distrib 16:488–495

Tinner W, Conedera M, Ammann B, Lotter AF (2005) Fire ecology north and south of the Alps since the last ice age. Holocene 15:1214–1226

Tylianakis JM, Didham RK, Bascompte J, Wardle DA (2008) Global change and species interactions in terrestrial ecosystems. Ecol Lett 11:1351–1363

Uglietti C, Gabrielli P, Cooke CA, Vallelonga P, Thompson LG (2015) Widespread pollution of the South American atmosphere predates the industrial revolution by 240 y. Proc Natl Acad Sci USA 112:2349–2354

Uhlmann B, Goyette S, Beniston M (2009) Sensitivity analysis of snow patterns in Swiss ski resorts to shifts in temperature, precipitation and humidity under conditions of climate change. Int J Climatol 29:1048–1055

Ventura M, Tiberti R, Buchaca T, Buñay D, Sabás I, Miró A (2017) Why should we preserve fishless high-mountain lakes? In: Catalan J, Ninot JM, Aniz MM (eds) Challenges for high mountain conservation in a changing world. Springer, pp 181–205

Wake DB, Vredenburg VT (2008) Are we in the midst of the sixth mass extinction? A view from the world of amphibians. Proc Natl Acad Sci USA 105:11466–11473

Walsh K (2014) The archaeology of Mediterranean landscapes. Cambridge University Press, Cambridge

Wania F, Westgate JN (2008) On the mechanism of mountain cold-trapping of organic chemicals. Environ Sci Technol 42:9092–9098

Weber J, Halsall CJ, Muir D, Teixeira C, Small J, Solomon K, Hermanson M, Hung H, Bidleman T (2010) Endosulfan, a global pesticide: a review of its fate in the environment and occurrence in the Arctic. Sci Total Environ 408:2966–2984

WGMS (2015) Global glacier change bulletin no. 1 (2012–2013). Publication based on database version: doi:10.5904/wgms-fog-2015-11. Zurich, Switzerland

Wickham J, Wood PB, Nicholson MC, Jenkins W, Druckenbrod D, Suter GW, Strager MP, Mazzarella C, Galloway W, Amos J (2013) The overlooked terrestrial impacts of mountaintop mining. Bioscience 63:335–348

Wiegand T, Camarero JJ, Ruger N, Gutierrez E (2006) Abrupt population changes in treeline ecotones along smooth gradients. J Ecol 94:880–892

Wilson RJ, Gutierrez D, Gutierrez J, Martinez D, Agudo R, Monserrat VJ (2005) Changes to the elevational limits and extent of species ranges associated with climate change. Ecol Lett 8:1138–1146

Wittig VE, Ainsworth EA, Naidu SL, Karnosky DF, Long SP (2009) Quantifying the impact of current and future tropospheric ozone on tree biomass, growth, physiology and biochemistry: a quantitative meta-analysis. Glob Change Biol 15:396–424

Xu JC, Grumbine RE, Shrestha A, Eriksson M, Yang XF, Wang Y, Wilkes A (2009) The melting Himalayas: cascading effects of climate change on water, biodiversity, and livelihoods. Conserv Biol 23:520–530

Zamora R, Pérez-Luque AJ, Bonet FJ (2017) Monitoring global change in high mountains. In: Catalan J, Ninot JM, Aniz MM (eds) Challenges for high mountain conservation in a changing world. Springer, pp 385–413

Zemp M, Frey H, Gärtner-Roer I, Nussbaumer SU, Hoelzle M, Paul F, Haeberli W, Denzinger F, Ahlstrøm AP, Anderson B, Bajracharya S, Baroni C, Braun LN, Cáceres BE, Casassa G, Cobos G, Dávila LR, Delgado Granados H, Demuth MN, Espizua L, Fischer A, Fujita K, Gadek B, Ghazanfar A, Hagen JO, Holmlund P, Karimi N, Li Z, Pelto M, Pitte P, Popovnin VV, Portocarrero CA, Prinz R, Sangewar CV, Severskiy I, Sigurðsson O, Soruco A, Usubaliev R, Vincent C (2015) Historically unprecedented global glacier decline in the early 21st century. J Glaciol 61:745–762

Chapter 2
Trade-offs in High Mountain Conservation

Francisco Lloret

Abstract High mountain ecosystems present features that determine their conservation: isolation, harsh environmental conditions and steep gradients. The vulnerability of ecological systems to disruptive agents can be addressed by considering exposure to these agents and the sensitivity of the system. Conservation management usually offsets trade-offs of resources allocated to minimise exposure with strategies designed to reduce sensitivity. Although exposure to human action may be reduced in high mountains by isolation, this effect is offset by disruptive agents operating at global or regional scales, such as pollution and climate change. In the long term, climate change can be expected to have a strong impact on alpine habitats, as the dispersal of their native species is severely constrained. Alternatively, high mountains may provide refuges for threatened species currently populating lower altitudes. When reducing exposure is not a feasible strategy, the alternative is to reduce sensitivity, which in high mountains would focus on improving connectivity, preserving habitat quality and controlling antagonistic interactions such as grazing. Lowering vulnerability to climate change requires interventions in various contributing drivers. Cost-effective models make help to optimise the outcome of different goals subject to trade-offs, and they can also be useful for allocating alternative actions over time. The application of ecological trade-off concepts helps to frame conservation from a functional perspective. This approach should also take into account the fact that the functional properties of ecological entities are multifactorial and interactive. This concept is recognised in ecosystem services that present negative correlations—trade-offs—as well as positive ones—synergies.

Keywords Climate change · Conservation · Ecosystem services · Ecosystem management · Exposure · High mountain · Sensitivity · Trade-off

---

F. Lloret (✉)
CREAF-UAB, Edifici C, Campus UAB, 08193 Cerdanyola del Vallès, Spain
e-mail: Francisco.Lloret@uab.es

J. Catalan et al. (eds.), *High Mountain Conservation in a Changing World*,
Advances in Global Change Research 62, DOI 10.1007/978-3-319-55982-7_2

2.1 Introduction

Conservation involves the allocation of limited resources to actions aimed at preserving the diversity of natural heritage and the properties and functions of ecosystems, particularly if these actions are expected to provide outcomes beneficial to significant portions of human society (i.e. ecosystem services).

The managed subjects (belonging to different organisation levels: species, population, ecosystem) compete for the resources allocated to conservation. Furthermore, conservation strategies may differ according to whether they focus on minimising exposure to human-driven impact or on reducing the sensitivity of the exposed biological system. These two strategies thus also compete for conservation resources. This situation is made even more complex by the fact that conservation goals compete, in their turn, with other objectives of human societies closely related to economic well-being. Competition for shared resources between different functions, as exemplified by trade-offs, constitutes a basic principle of economics that is shared by ecological systems. Here I use a framework based on the ecological concept of trade-offs to analyse conservation issues relevant to high mountain ecosystems.

2.2 Distinctive Features of Conservation in High Mountain Ecosystems

High mountain ecosystems present several distinctive ecological features that have important consequences for conservation (Beniston 2003) (Fig. 2.1). First, they experience a remarkable degree of isolation, due to their topographic location and the presence of summits which limit both biotic flux and human access. While genetic and demographic flux is restricted in mountain populations, the latter also

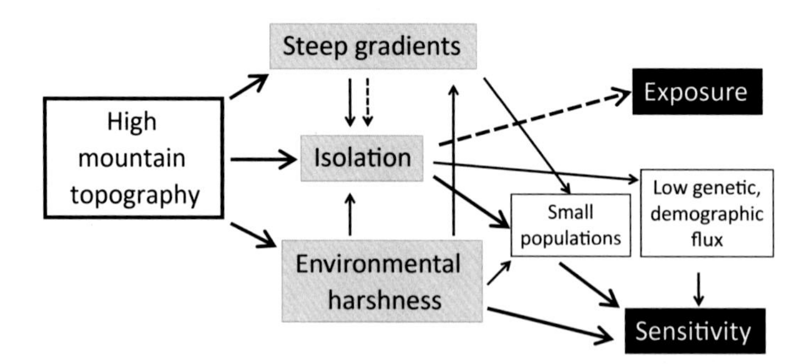

Fig. 2.1 Ecological characteristics of high mountain ecosystems (*grey boxes*) and their effects on components of vulnerability (exposure, sensitivity). *Solid* and *broken arrows* indicate negative and positive consequences for conservation, respectively (see text). *Arrow* body shows relevance of the relationship

benefit from greater protection from pathogens, pests and widespread disturbances than populations in well-connected areas. However, this isolation is not sufficient to protect them from global processes such as climate change and pollution, as these are transmitted through the atmosphere. Significantly, high mountains are usually situated in sparsely populated areas, due to their low accessibility, in contrast with lowlands and coastal areas, where the most human population is concentrated, all over the world. The human-driven impact on high mountain areas is, therefore, lower in comparison with these densely populated regions. Nevertheless, historically human presence has played an important role in configuring mountain ecosystems. For instance, in the Alps and Pyrenees, human activity has been regularly recorded since Neolithic times (Tinner et al. 2005; Gassiot Ballbè et al. 2017), and it has profoundly modified the landscape over the last centuries (Colombaroli et al. 2010; Pèlachs et al. 2017). Isolation and low population density also provide high mountains with emotional and aesthetic values that are often idealised. These habitats commonly play host to sanctuaries, or an entire mountain system can be seen as a sanctuary in itself. This perception coincides with the concept of preservation and may contribute to the conservation of natural systems. Interestingly, low accessibility may imply fewer resources for conservation. On the other hand, these remote areas may experience looser control by centres of decision over the conservation practices carried out there.

Second, high mountain habitats provide harsh living conditions. The altitudinal gradient implies a decrease in temperature and a prolonged duration of snow cover, which combine to shorten the periods of growth. Moreover, strong winds, low water availability at high altitudes and the scanty soil development associated with steep slopes and erosion result, overall, in a pronounced abiotic stress. In consequence, vegetation cover is reduced, leading to mutually reinforcing feedback (for instance, between vegetation cover, water retention and soil erosion). Also, high altitude favours the passage of pollutants from the troposphere to the ground (Camarero 2017b) and a loss of atmospheric protection against ionising radiation. Therefore, only relatively few species are able to persist in these extreme conditions. These species typically present low growth rates and life cycles adapted to the short duration of favourable conditions (Laiolo and Obeso 2017). The environmental conditions specific to high mountains, along with their geographic isolation, have forged an adaptive landscape that has shaped the characteristic functional and compositional traits of its biota. Another consequence is a noticeable fragility in these ecosystems, as the resident species are often pushed to their limits of ecophysiological tolerance. Nevertheless, selective pressure may have favoured adaption to these environments. Simultaneously, species tolerant of a broad range of conditions are often found here, far from the competition withstood by other species. Also, importantly, low growth rates and short periods of growth limit population recovery after disturbances or harsh environmental conditions, thus reducing resilience.

Another characteristic of high mountain ecosystems is that they tend to exhibit steep environmental gradients over relatively short distances. These gradients are largely determined by topography and aspect, which determine the radiation balance and the hydrological system, including water run-off and freshwater courses and reservoirs. Moreover, poorly structured soils make these gradients more dependent on the chemical and physical properties of bedrock, thereby enhancing the role of this source of environmental heterogeneity. Overall, high mountains tend to offer significant environmental heterogeneity (i.e. microhabitats) in combination with strong seasonal fluctuations. In fact, to some extent this heterogeneity associated with steep gradients counterbalances isolation, favouring the existence of altitudinal corridors and stepping stone routes that allow dispersal. The result is that we expect relatively low levels of biodiversity, although this is highly idiosyncratic and has a substantial spatial turnover.

All three ecological characteristics (isolation, harsh environment, steep environmental gradients) are the consequences of mountain topography and they interact mutually. Strong environmental changes over small distances, together with harsh abiotic conditions, may constrain a population's expansion by limiting its size, but they may also permit effective dispersal by saving relatively close barriers or allow populations to migrate across the altitudinal range in search of suitable conditions. Harsh conditions, in their turn, are a major component of pronounced gradients—in fact, mountains tend to correspond to the extremes of many abiotic gradients at the regional level—and they contribute to isolation (Fig. 2.1).

2.3 Conservation, Vulnerability and Trade-offs

One major goal of conservation is to deal with the vulnerability of natural systems (species, populations, habitats, ecosystems) in the face of risks associated with human activity, by maintaining or increasing the values related to the persistent functioning of such systems. The concept of vulnerability may be approached from different perspectives; in the ecological context, and particularly when assessing climate change vulnerability is defined as the degree to which a system is able to cope with adverse effects, being a function of (1) the exposure of the system in question to agents that can potentially diminish these values, (2) the system's sensitivity to subsequent changes, and (3) its eventual resilience or adaptability to the new context (Turner et al. 2003; Parry et al. 2007; Chapin et al. 2010). Let us concentrate on the two first components, exposure and sensitivity, which are affected by the immediate impact of environmental hazards: any situation involving an increase in exposure or sensitivity will result in greater vulnerability and should thus require conservation action. Similarly, conservation management could maintain a given degree of vulnerability by reducing exposure when sensitivity is increased (e.g. because populations become too small). Therefore, as a first approach vulnerability would result from the product of exposure and sensitivity. Given that vulnerability is defined in relation to a disruptive agent, if no such agent exists, exposure and vulnerability equal zero.

The ecological characteristics of high mountain ecosystems have different consequences on vulnerability (Fig. 2.1). Isolation reduces exposure to human intervention and the resulting loss of habitat and alteration of biogeochemical cycles, including those caused by local pollution. In contrast, environmental harshness explains the high sensitivity of these systems, particularly their difficulties in recovering from disturbances. In fact, this high sensitivity is reinforced by isolation, due to the limitations imposed on genetic and demographic flux through dispersal, and by the resulting small populations. Furthermore, steep gradients contribute to small population size as the habitat area is limited. In contrast, isolation can diminish exposure to deleterious biotic agents—pests, pathogens—at the landscape scale, although the high heterogeneity promoted by steep gradients may favour the dispersal of these agents. The spread of other disturbances, such as wildfires, can also be constrained by the low degree of connectivity, although rough topography may also enhance the propagation of fire uphill.

The concept of ecological trade-offs provides a useful framework for understanding and designing the allocation of resources devoted to conservation goals, such as the management of exposure and vulnerability. In organisms, the allocation of limited resources to different purposes or functions implies a negative relationship between these resources. However, the configuration of this relationship is not an easy task, first because it is essential to establish a common currency that accounts for the various functions (Reekie and Bazzaz 1987). A negative correlation between estimators of different functional properties is not in itself a proof of trade-off unless the mechanisms connecting functional properties can be properly determined and converted to this common currency. Another key issue is the fact that these functions are essential for the overall persistence of the system in question, which in its original ecological sense corresponds to organisms. Consequently, the product of the estimators of the different functions, after conversion to a common currency (e.g. biomass), should be a constant other than zero, as a zero value would mean that the system does not exist. One typical case of ecological trade-off describes the allocation of resources to seed production in plants (Harper et al. 1970). Assuming a constant amount of resources allocated to each seed set, seed size and seed number reveal different functions: small seed size and a large number of seeds would optimise dispersal, while large seed size—in detriment to number—would favour seedling survival. Obviously, the reality is much more complex since small seed size may contribute to other functions, such as minimising genotype losses by predation. Alternatively, a given function usually determines the involvement of different resources, generating a complex network of interacting functions and resources that are used to different degrees.

Conservation practice can be considered as analogous to the allocation of limited resources, and trade-offs would correspond to different management actions. While natural selection would be the main driver of resource allocation at the species level, in conservation this role would be performed by environmental managers involved in decision-making processes. Conservation has an economic component associated with the allocation of limited funding to different goals (i.e. economic trade-off). But this funding allocation is intrinsically associated with ecological properties, which are also subjected to trade-offs.

2.4 Conservation Management of Exposure and Sensitivity in High Mountains

According to the application of trade-off principles to conservation management, resources allocated to a reduction in the exposure of conserved systems would often be detrimental to those assigned to decreasing sensitivity (Fig. 2.2). The conservation of endangered species, such as grizzly and brown bears in mountain areas, illustrates the application of these principles (Martin et al. 2012; Braid and Nielsen 2015). The presence of these species in a landscape reflects a trade-off between food resources and human presence, which roughly correspond to the components of these species' vulnerability—sensitivity and exposure, respectively—in the territory. GIS techniques allow these properties of habitat quality and exposure to disturbance to be combined in spatially explicit contexts to determine areas of prioritised use at the regional level (e.g. road development or habitat restoration) (Braid and Nielsen 2015) (Fig. 2.3a). Such analyses can achieve a notable degree of detail when existing populations are recorded. They can, therefore, support conservation management by promoting the use of attractive sink-like habitats (with good food quality but also proximity to human structures) that connect segmented populations, provided disturbances of human origin are actively curtailed (i.e. reducing exposure). In contrast, in areas that are attractive sink-like habitats but are located far from pre-existing populations, the vulnerability of the brown bear may be reduced by discouraging bears to use these habitats (e.g. by allowing forest logging, building electrified barriers near potential food resources or minimising rubbish). Similarly, refuge habitats (with poor food quality but far from human exposure) can be managed to decrease the sensitivity of their populations by

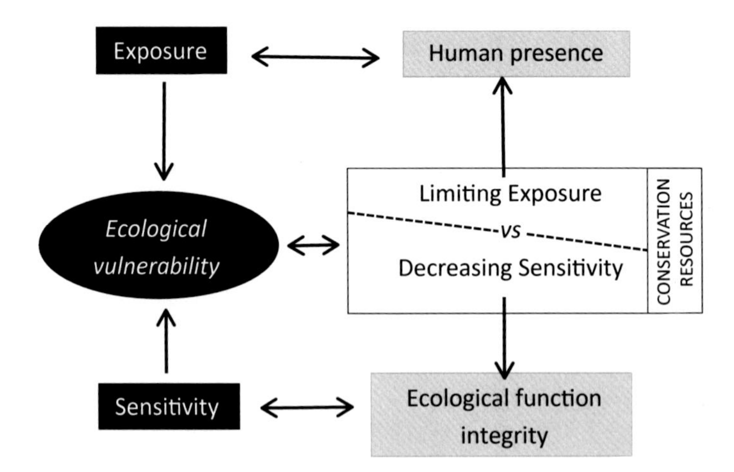

Fig. 2.2 Trade-off between conservation resources allocated to manage the two components of ecological vulnerability (exposure, sensitivity) by reducing human frequentation or enhancing ecological integrity

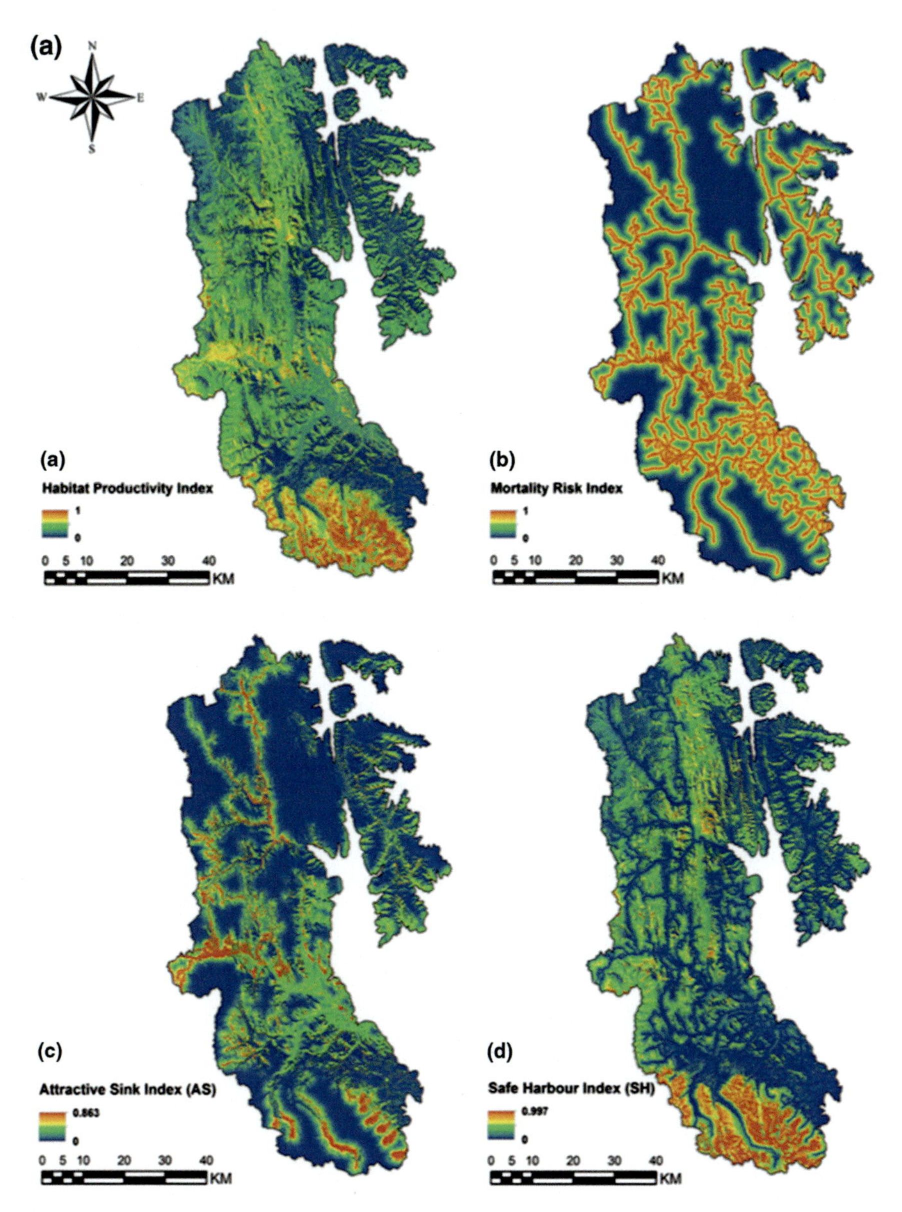

Fig. 2.3 Different cases of trade-off applied to conservation. **a** Maps of sensitivity ((*a*) habitat productivity) confronted to exposure ((*b*) road-based mortality risk) to identify sink (*c*) and refuge areas (*d*) for bear populations in Alberta, Canada (Braid and Nielsen 2015). **b** Solution for a reserve model considering the trade-off between owl populations (x-axis) and timber harvest (y-axis) (CM, current management scenario) in Oregon, USA (Nalle et al. 2004). **c** Trade-off among provisioning service (meat) and regulating services (carbon sequestration and water conservation) in alpine grasslands of Tibet, China (Pan et al. 2014). **d** Maps of modelled outputs of fire management considering a trade-off between fuel reduction by prescribed fires and limited resources: expected tree density (*A*) fire intensity (flame height) (*B*), and predictions after wildfire with and without previous fuel reduction treatments (Ager et al. 2013)

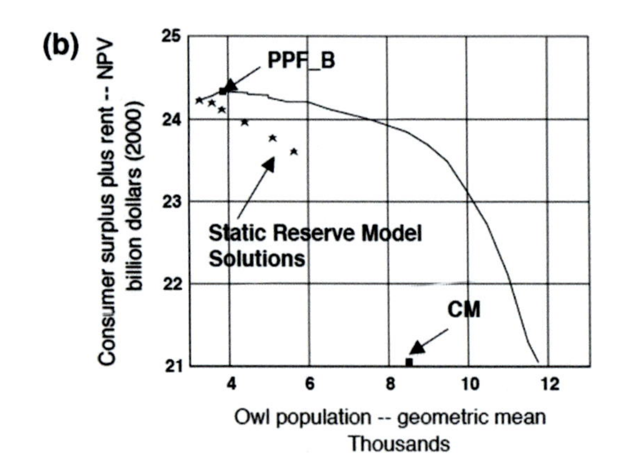

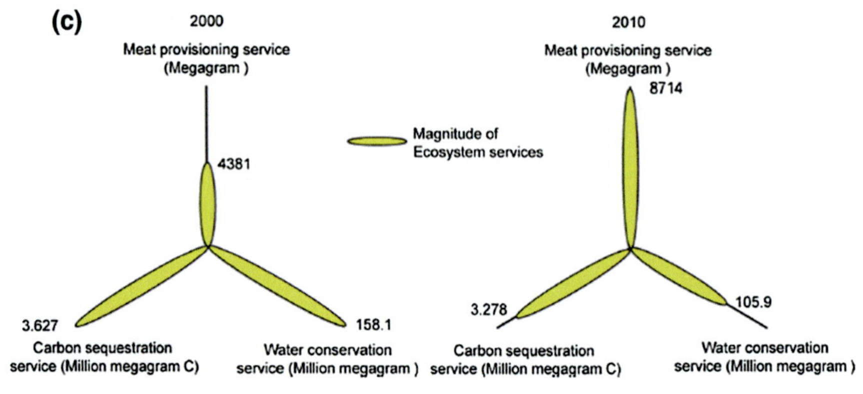

Fig. 2.3 (continued)

favouring the production of resources (e.g. by increasing forest hard mast species) (Martin et al. 2012). Because these actions involve spending money, a rationale based on using GIS, models and simulations to identify the distribution of trade-offs over space and time becomes a powerful tool for managers. These models may also incorporate the shifting balance of the trade-offs between services provided at the landscape level, such as food supply and shelter for vertebrates, over the course of seasons and life stages (May et al. 2010).

In general, the reduction of exposure in protected areas is a more common practice than the reduction of sensitivity. One major reason for this is the fact that it is easier and more operative to regulate on a territorial basis, for instance by controlling access, than to intervene in ecosystems and populations, where accurate knowledge of their functioning is often lacking. This strategy is particularly appropriate in high mountain areas due to their low accessibility for humans, which makes it possible to effectively control use by visitors and locals. Another example of exposure reduction is the control of exotic species by minimising their presence through limitations on their introduction and reinforcement of their eradication.

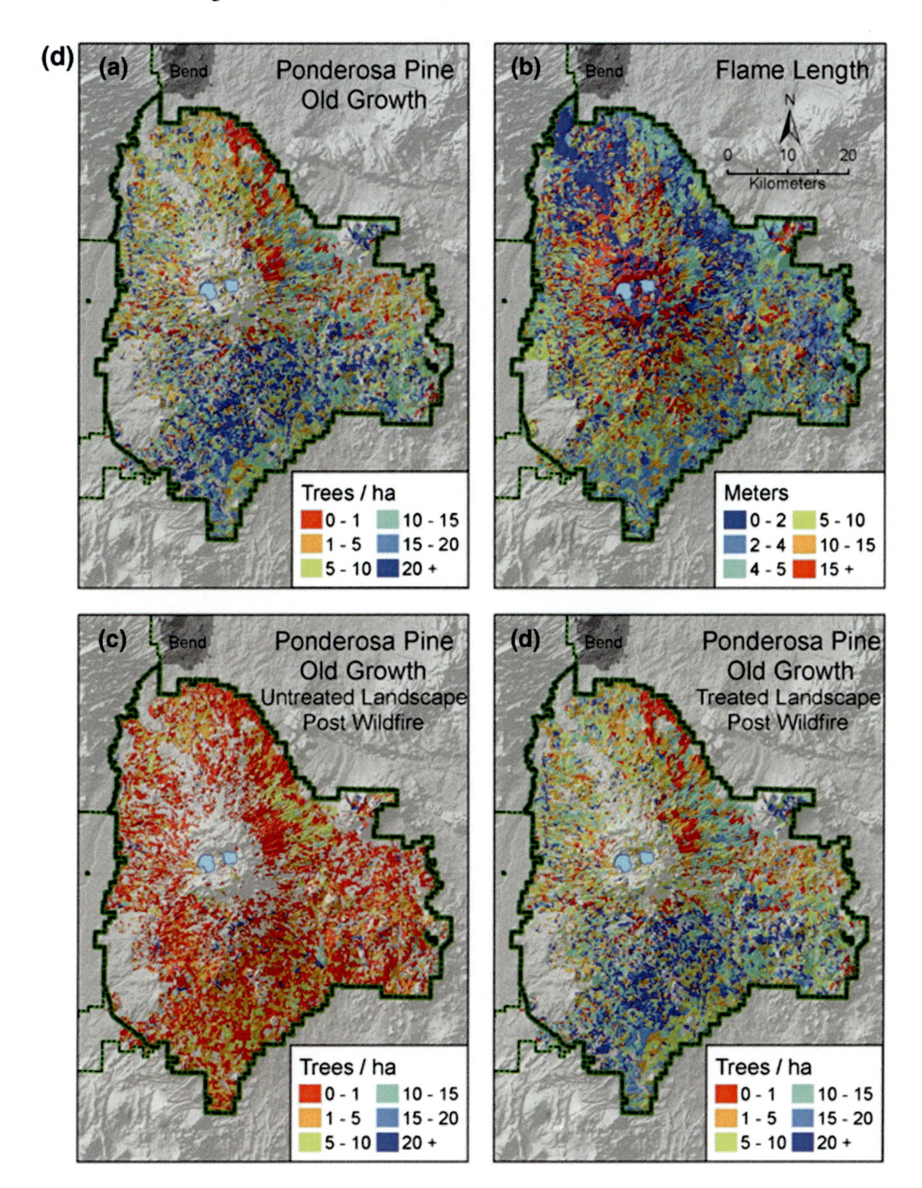

Fig. 2.3 (continued)

Although high mountain ecosystems are usually considered to be scarcely affected by biological invasions, this supposition probably underestimates this phenomenon, which will likely become increasingly relevant in mountain areas under climate change (Pauchard et al. 2009, 2016).

Conservation practice commonly adopts a precautionary principle, as illustrated by strategies based on the reduction of exposure to external agents (human-caused disturbances and pollution) that threaten natural systems, in concordance with the desire of societies to preserve their collective memory of natural heritage. This passive attitude to conservation is challenged by strategies focused on adaptive conservation that take the insufficiency of our current knowledge as the starting point for the development of more effective practices (Holling 1978; Armitage et al. 2008). Exposure reduction may obtain remarkable results at a local scale, particularly in the face of changes in land use, which represent a major threat to high mountain ecosystems (Theurillat and Guisan 2001; Spehn et al. 2006), but it clearly proves inefficient against exposure to agents that operate at global or regional levels, such as pollution and climate change. The capacity of local managers to reduce exposure to these agents is very limited. For instance, high mountain lakes are particularly exposed to airborne chemical loadings (see Camarero 2017b), and in this case reducing the sensitivity of these ecosystems involves the preservation of biodiversity and food webs (Ventura et al. 2017).

High mountain areas are particularly exposed to climate change due to their position at the extreme of regional climatic gradients. Accordingly, at the global scale, vulnerability to the vegetation shifts associated with climate change is considered particularly high in alpine biomes (Gonzalez et al. 2010). Isolation and habitat specialism contribute to this vulnerability (La Sorte and Jetz 2010). In Europe, alpine and Mediterranean mountain environments are projected to decline dramatically in comparison with other climatically defined environments (Metzger et al. 2008). Thus, we expect a significant loss of habitat for many plant species particularly as a result of decreased precipitation (Engler et al. 2011). This loss of habitat may not be exclusively due to a decline in a species' climatic suitability, but rather to the improvement in conditions for species such as trees, which can modify the habitat and competitively exclude current populations of high mountain specialists (Dirnböck et al. 2011). Nevertheless, at the species level, at least until now in Europe, mountain areas seem to exhibit a substantial inertia in the face of modifications to biodiversity caused by climate change (Theurillat and Guisan 2001). In fact, mountains may constitute a shelter for many species on account of their topographic characteristics, which provide altitudinal corridors (Loarie et al. 2009). Furthermore, high mountains may become a refuge for species threatened by global changes in their current distribution at lower altitudes (Sergio and Pedrini 2007), thereby emphasising the importance of preserving large-scale elevation gradients (Moritz et al. 2008).

Given local managers' inability to directly influence climatic trends, conservation trade-offs should focus on reducing sensitivity to climate change, in many cases by acting on co-drivers that produce deleterious synergies in combination with climate change (Hulme 2005; Mawdsley et al. 2009), or alternatively by enhancing mechanisms of stabilisation and resilience (Lloret et al. 2012). Nowadays, this strategy of reducing sensitivity to climate change has established a place on agendas for conservation. This issue is becoming particularly relevant and challenging in high mountain ecosystems, due to the harshness and distinctiveness of their habitats, but also due to the frequent involvement of small populations that have experienced directional selection for generations. Specifically, management focused on reducing sensitivity should consider connectivity and genetic flux (Moritz et al. 2008), preservation of microhabitat quality (Marini et al. 2011) and control of grazing (Nagy and Grabherr 2009). Similarly, reductions in the risk of disturbance may favour the preservation of habitats and small populations against the adverse effects of climate change (Millar et al. 2007). In fact, the management of sensitivity often comes to focus on population-level processes involving the enhancement of genetic variability (Maudet et al. 2002) and population size, as well as the control of antagonists (predators, pathogens, pests, parasites) (McKinney et al. 2009). Nevertheless, some of these actions may in themselves involve trade-offs: for instance, disturbance often contributes to species co-existence, giving rise to a complex picture that we shall discuss below.

Forest die-off clearly illustrates the difficulties in managing ecosystems, even locally, when the major threats are global. Forest die-off accompanied by tree mortality is increasingly being reported in many biomes across the world, including mountain areas of North and South America and Europe (Suarez et al. 2004; Bigler et al. 2006; van Mantgem et al. 2009; Allen et al. 2010; Smith et al. 2015) (see also Camarero 2017a). Many factors, such as the capacity of soils to store water, antagonistic biotic interactions and stand structure can significantly contribute to this phenomenon (Raffa et al. 2008; Galiano et al. 2010; Bell et al. 2014). In many cases climate, and more particularly drought and heat episodes, is closely associated with this decline (Suarez et al. 2004; Bigler et al. 2006; Allen et al. 2010, 2015; Anderegg et al. 2013; Williams et al. 2012; Smith et al. 2015). Importantly, the trend towards warming is increasingly accompanied by climatic variability, which results in pulses of extreme weather. This feature reveals a major component of the new abiotic environment of the next future (Easterling et al. 2000). Reducing exposure to this global threat is thus far beyond the scope of local managers. However, they probably can reduce forest's vulnerability to drought by acting on drivers that amplify tree mortality. In such forests, the vulnerability could be decreased by controlling antagonists (Sturrock et al. 2011) or by managing forest composition and structure (Grant et al. 2013). However, these practices, although common in forests managed for commercial purposes, could clash with the criteria for intervention in preserved areas. This conflict is particularly acute when adaptive management, which involves learning from experimental settings, is proposed as a rational alternative for improving the future health of forests (Millar et al. 2007). This controversy can only be solved by a straightforward definition and prioritisation of conservation goals by social agents. The transcendental values of forests as sanctuaries or social icons can support the effort to identify these goals and formulate specific local decisions. In any case, even in today's humid mountains in temperate regions, managers will probably have to come to terms with the management of water availability in the near future (Grant et al. 2013).

These reinforcing co-drivers may interact with climate and with each other in complex ways involving feedbacks (either positive or negative) and trade-offs (McDowell et al. 2011; Jactel et al. 2012). For instance, Scots pine is experiencing high mortality rates in some mountain areas in the Pyrenees due to a combination of increasing drought, poorly developed soils and mistletoe infestation

(Galiano et al. 2010). Stand density also appears to be a contributing factor since competition for scarce water is to be expected. Moreover, bark beetle proliferation, a common driver of conifer mortality in association with drought, has been detected in damaged Scots pine stands in the Alps (Dobbertin et al. 2007) and it is one of the main causes of forest dieback in Western North America (Hart et al. 2014). A parallel die-off is occurring in silver fir forest in the Pyrenees, associated with logging in the past (Camarero et al. 2011; Camarero 2017a). The primary or contributing role of pests and pathogens versus drought is also often hard to elucidate, as they can establish mutually reinforcing feedback (Hart et al. 2014; Oliva et al. 2014). Such multiple interactions between factors that contribute to the decline of forests are common in many mountain areas of the world and must be taken into account by conservation managers in the new climatic scenarios (Allen et al. 2015).

2.5 Managing Conflicting Goals

The allocation of resources to competing goals represents a clear example of the application of trade-offs to conservation issues. The paradigmatic case involves the economic benefits obtained from the harvesting of natural resources, which in the mountains usually correspond to timber, grass and fish, versus values associated with biodiversity, often exemplified by key or charismatic species, by species diversity or by ecosystem functioning. This approach, which can be spatially and temporally explicit, makes it possible to develop cost-effective models that optimise the outcome of different goals subject to trade-offs, after combining functions that share a common currency (e.g. economic value). These models have, for instance, been widely used to assess wildlife and timber production in mountain forest regions (Nalle et al. 2004) (Fig. 2.3b), and to assess the relationship between forage production and the abundance of particular plant species that denote environmental quality in mountain grasslands (Loucougaray et al. 2015). Another case is the trade-off between the financial income produced by the introduction of non-native fish to lakes and streams and the environmental costs (Ventura et al. 2017); in this case, the focal entity corresponds to the whole watershed ecosystem. These cost-effective models also allow us to simulate outcomes by applying alternative management actions at different times—i.e. spreading the investment of resources over time (Lampert et al. 2014). Nevertheless, one major challenge for such quantitative analyses is the parameterisation of the current common currency for the various alternative management actions.

Although functions subject to trade-off show a negative relationship of their estimators, not all negative correlations are the result of resource allocation for the overall maintenance of a system. Recent changes in land use in European mountain areas (Pèlachs et al. 2017), provide an interesting case for illustrating the complexity of this approach. In these areas, the human-induced transformation of the landscape has led to the loss of most woodland while agricultural and grazing areas have increased. Since the mid-twentieth century, however, significant depopulation resulting from profound socio-economic transformations is fomenting substantial afforestation (Roura-Pascual et al. 2005; MacNeill 2003; Lasanta-Martínez et al. 2005). This afforestation, in turn, results in loss of the grasslands and open habitats that play host to major elements of biodiversity. Managers can consider taking actions focused on enhancing some of these habitats: this situation could be interpreted as a trade-off. In these cases, the surface area occupied by different human activities—commercial exploitation, provisioning and regulation services, biodiversity conservation—represents a limiting resource that is likely to be promoted by management decisions. A negative correlation between different uses just reflects, however, that their sum, rather than their product, is constant (i.e. the whole territory), without any particular functional meaning. In contrast with an organism that requires different essential functions that compete for resources to persist, a territory will remain over time, regardless of whatever land cover it may support. For a proper application of the trade-off concept, the abundance of a given land use should be associated with any functional property of an upscaled, comprehensive ecological system. We can establish trade-offs between different land uses if they correspond to different conflicting benefits: for example, logging in forest areas—with direct financial revenues—as opposed to the preservation of open habitats for some species—a conservation goal. Similarly, different land uses may be associated with distinct components of the species' niche, such as foraging in open areas and refuge in forest lands (May et al. 2010; Martin et al. 2012). This concept is important because it highlights the fact that the conservation of ecological processes is not based solely on the patterns of abundance of categorical entities (species, land uses) but also on the functional properties with which they are associated. The application of the ecological trade-off concept to conservation issues helps to frame this functional perspective.

Alternatively, conservation can prioritise some categories (forests or open land, particular species) irrespective of their functional properties but according to social preferences, including rarity or aesthetic and iconic perceptions. This conservation approach is based more on heritage preservation than on functionality or market utility. Heritage diversity is high in mountain areas, given the particularities of these environments and the associated evolutionary processes that are enhanced by isolation. The heritage perspective of conservation can easily lead to efforts to increase species or habitats, especially if these have some distinctive value. So, in the face of the dilemmas arising from the allocation of different land uses in a territory, the obvious solution is to increase the total protected area, following a strategy of accumulating heritage. According to this approach, the goal will be to include the smallest surface of each land use that provides a plateau of diversity, according to the asymptotic relationship between diversity and area. Alternatively, if the territory's area is limited, one preliminary solution would be to find the optimal combination of the land use surfaces—according to the same asymptotic relationships for each land use—and then consider those elements (species or habitats) that are common to the different uses. Obviously, the procedure becomes much more complicated when the relationship between diversity and area is dependent on the spatial context, i.e. influenced by the proximity of other land uses (Bennett et al. 2006).

This approach, based on the complementarity of territories for determining total diversity, concurs with "gap analysis" techniques, which have been effectively developed in conservation practice (Flather et al. 1997). An important issue is that apart from the intrinsic value of biodiversity, the contribution of diversity to ecosystem functioning—productivity, water, C and nutrient cycling—should also be strongly emphasised, particularly because of its contribution to stability and resilience (Hooper et al. 2005).

2.6 Complex/Interacting Controls of Trade-offs

The functional properties of any ecological entity are multifactorial, and they often interact. At the ecosystem level, conservation benefits from a framework that recognises multiple services, which are equivalent to functional ecosystem properties that are relevant to humans. Ecosystem services are subjected to resource allocation, and consequently, the trade-off approach can be applied, as far as human societies invest distinctly in different ecosystem types or promote some functional properties of ecosystems. In fact, conflicts between provisioning and regulating services are common (Bennet et al. 2009), and they can be considered as trade-offs provided a common currency is regarded. Often the outcome of these services roughly corresponds to land use categories, considering the explicit spatial distribution of their properties in the territory, which in mountain regions correlates to topography and distance from settlements and roads (Grêt-Regamey et al. 2008; Paletto et al. 2015). Services that are typically provided by mountain ecosystems include revenues from forests, grasslands or watersheds, protection against natural hazards and outdoor recreation (Paletto et al. 2015; Vacchiano et al. 2015; Ventura et al. 2017) (Table 2.1).

Tourism and leisure constitute one of the most important economic activities in the mountain areas of developed countries. Despite its impact on habitats and water resources (Nagy and Grabherr 2009), this activity can potentially generate strong synergies with conservation goals. The economic value of these uses can be relatively easy to quantify in terms of consumption and investment (Paletto et al. 2015). The contribution of natural systems favouring such activity can also be estimated by various indirect methods for evaluating preferences (Grêt-Regamey et al. 2008), or from indicators of aesthetic value, (f.e. changes in the colour diversity of grasslands, Loucougaray et al. 2015). Nevertheless, the current state of uncertainty about the quantification of ecosystem services is particularly marked in high mountain regions (Grêt-Regamey et al. 2008).

The multiplicity of ecosystem services illustrates the potential existence of multiple trade-offs operating on a given ecological entity. Furthermore, a parameter used to estimate a given functional property may, in fact, respond to several functional processes. Multiple trade-offs may be explored by correlation matrix between services, highlighting the consistency of negative correlations. In contrast, positive correlations would indicate synergies between services (e.g. Raudsepp-Hearne et al. 2010).

Table 2.1 The relative contribution of high mountains to ecosystem services (agents providing the service are shown in *italics*), indicating potential trade-offs (negative correlations) and synergies (positive correlations)

	Contribution	Trade-offs	Synergies
Provisioning services			
Food	Moderate	Raw materials, biodiversity services	Freshwater, biodiversity services
Livestock, berries, mushrooms, hunting, wildlife, fish			
Raw materials	Moderate	Freshwater, carbon storage, soil protection, biodiversity services, cultural services	
Timber, fuelwood, hay, plant oils			
Freshwater	High	Tourism	Wastewater treatment, biodiversity services, cultural services
Snowpack, springs, rivers			
Medicinal resources	Moderate	Biodiversity services	Cultural services
Plants			
Regulating services			
Local climate and air quality	Moderate		Carbon storage, biodiversity services, cultural services
Forests			
Carbon storage and sequestration	Moderate	Habitat for species	Soil protection
Forests, soils			
Moderation of extreme events	High	Habitat for species	Provisioning services, soil protection, biodiversity services, cultural services
Avalanches, landslides, floods			
Wastewater treatment	Low		Soil protection, biodiversity services, cultural services
Wetlands			
Soil protection and erosion protection	High		Biodiversity services, cultural services
Soil integrity			
Pollination	Low		Biodiversity services
Insect biodiversity			
Biological control	Low	Species and genetics diversity	Biodiversity services, recreation, tourism

(continued)

Table 2.1 (continued)

	Contribution	Trade-offs	Synergies
Food webs			
Biodiversity services			
Habitat for species	High	Species and genetics diversity	Species and genetics diversity, cultural services
Refuges, corridors, buffers			
Maintenance of species and genetic diversity	High		
Species, genotypes, food webs, interaction networks			
Cultural services			
Recreation and health	High	Tourism	Tourism, aesthetic appreciation, spiritual experience
Hiking, picnicking, sailing			
Tourism	High	Aesthetic appreciation, spiritual experience	Aesthetic appreciation, spiritual experience
Skying, resorts, alpinism, trekking, hunting, fishing			
Aesthetic appreciation	High		Spiritual experience
Spiritual experience	High		

Note that some services may present both trade-offs and synergies with other services depending on the intensity and characteristics of the involved processes (f.i., overgrazing often causes soil erosion while moderate grazing can promote biodiversity). Trade-offs and synergies between services are only indicated in the service appearing first in the list

For instance, in alpine grasslands, livestock provisioning services represent a trade-off with regulatory services provided by NPP, which in turn determines carbon sequestration and water and soil conservation (Pan et al. 2014; Fig. 2.3c). While trade-offs between services lead to cost-effective analyses, synergies between them represent reinforcing mechanisms that make it possible to save resources or focus investment on obtaining multiple benefits.

Another source of complexity comes from the fact that the outcomes of ecosystem services may depend on previous management. For instance, land use transformation in Andean mountains over the past decades has led to the loss of ecosystem services, due to the conversion of cloud forests and paramo grasslands in the alpine and sub-alpine stages to agricultural use; later, pine plantations were developed on open alpine grasslands and agricultural land. While pine plantations produce an adverse impact through reducing the area previously occupied by native alpine grasslands, they can improve ecosystem services when they occupy land that had been previously degraded by agriculture (Balthazar et al. 2015).

Fire management is a complex management goal that involves several of these issues: synergies between management options, interacting agents and time lags. Although weather conditions and a low fuel load overall constrain wildfires in high mountain ecosystems, their incidence is far from negligible. They can lead to the lowering of the treeline (Nagy and Grabherr 2009) and constitute a natural disturbance in the conifer forests of the mountains of western North America (Sibold et al. 2006), where fire regime has been heavily altered by active fire suppression policies over the past century (Donovan and Brown 2007). In fact, in many of the world's mountain regions, the fire has been used as a major management tool to foment grasslands, eventually determining ecotones (Nagy and Grabherr 2009). In recent decades, the loss of local population in mountain areas, at least in developed countries, is leading to encroachment onto former grasslands, and the consequences on biodiversity and ecosystem functioning have yet to be fully explored (Roura-Pascual et al. 2005; Brandt et al. 2013; Formica et al. 2014).

Fire management, and particularly fire suppression, consumes a large proportion of the resources devoted to forest management in countries with a high climatic fire risk, dense populations in the wildland–urban interface or substantial forest revenues. Fire management is therefore suited to an analysis based on trade-offs. Both actions aiming to reduce ignition—i.e. public information, regulation of access to forests, lighting restrictions—and to suppress fires share the goal of minimising burned areas. A cost-efficiency analysis in both the ecological and social contexts may help to optimise the contribution of each different action to the common goal. Paradoxically, however, the reduction of burned surface area implies further development of vegetation and subsequently the accumulation of fuel for the future, which will likely produce more intense and extensive fires (Donovan and Brown 2007; Lloret et al. 2009; Loepfe et al. 2012, but see Odion et al. 2004). This situation illustrates the temporal dimension of conservation practices and how they modify the future environment. It also reveals the existence of feedbacks regulating ecological systems; in this case vegetation growth and wildfires are mutually regulated by negative feedback. Fire suppression policies may lead the system to a structure of fire sizes that tends to be less equitative, with many small fires—which are rapidly extinguished—and very few extremely large ones, although these are usually very intense (Lloret et al. 2009). In terms of ecological trade-offs, and in the context of prolonged periods, investment in a reduction to fire exposure—limitations on lightning and fire propagation caused by humans—corresponds with increasing future fire sensitivity—associated with more intense and severe fires—. Assuming that management resources are limited, if one major goal is the minimisation of megafires of extreme extension and intensity, a strict fire extinction policy is not in itself the best long-term solution, and it may have strong consequences for the long-term structural and functional properties of ecosystems (Donovan and Brown 2007). Given that fire extinction is mandatory in some areas close to populated areas or installations, and in areas with specific conservation values, explicit geographical models can be developed to establish areas in which fuel load accumulation resulting from fire suppression should be counterbalanced by mechanical fuel reduction or by restoring (i.e. prescribed) fire (Ager et al. 2013) (Fig. 2.3d). In many cases, the cost of mechanical fuel reduction is high and may imply a loss of commercial revenues—which can be compensated to some extent by biofuel production—or C stocks. In populated areas, these actions, therefore, tend to be concentrated in restricted, sensitive areas, often close to the wildland–urban interface (Driscoll et al. 2016). Consequently, the true trade-off regarding economic cost corresponds to fire suppression versus fuel reduction options, and the challenge is to optimise the respective actions through territory and time.

In addition to the fire-vegetation feedback, a fire-driven system is also controlled by the ambivalent effect of weather, since high temperatures and low humidity favour the ignition and propagation of fires. However, when these conditions continue over time they result in chronic drought—as in arid climates—and reduced fuel load (Loepfe et al. 2014). The increase in dry fuel after extreme drought episodes that exacerbate vegetation mortality can be considered transitory in the context of an overall fire regime, although it does represent a temporary window of opportunity for wildfires (Allen 2007). Thus, as in other ecosystems, the occurrence and severity of wildfires in mountain forests respond to a network of historical interactions involving climate, previous fires, management and other disturbances (e.g. insect outbreaks) (Bigler et al. 2005). Accordingly, a fire regime and its distribution in a landscape can be analysed by spatially explicit simulation models that include the characteristics of vegetation and management (Schumacher and Bugmann 2006; Loepfe et al. 2012). The empirical analysis of fire distribution at a regional scale reveals that in drier regions wildfires are controlled by fuel load availability, while in moist regions fire is determined by the occurrence of extreme dry periods (Loepfe et al. 2014). Therefore, the fuel vs. climate control of a fire regime can change over time, while the fuel load accumulates and climate changes (Kloster et al. 2012). In periods in which logging is intense or agricultural activity predominates, wildfires will mainly be determined by the local accumulation of fuel load; in contrast, when afforestation dominates a landscape, the limiting factors would be weather and drought (Pausas and Fernández-Muñoz 2012). In high mountains, these dynamics would often correspond to pastures and encroachment, respectively, but they have traditionally been disrupted by humans who have widely used wildfire to favour grazing (MacNeill 2003; Colombaroli et al. 2010). Thus, conservation in high mountains should come to terms with fire management, particularly because fire climatic risk is expected to increase in many regions (Moriondo et al. 2006). In addition to the financial component of the trade-off between fire suppression and fuel reduction, conservation in these areas should incorporate analysis of the trade-offs and synergies between biodiversity values and ecosystem services associated with a fire regime in terms of species composition, soil and vegetation, C stocks and erosion (Garcia-Pausas et al. 2017). For instance, while C stock and erosion losses respond similarly to wildfires, species diversity may be favoured by moderately frequent fires (Coop et al. 2010).

2.7 General Concluding Remarks

Conservation practice involves making decisions with only a limited knowledge of complex systems, uncertainty about the future and scarcity of resources. Management decisions often reflect opportunistic reactions to urgent problems or, alternatively, routines followed without any regular evaluation or updating. I advocate that these actions will gain in consistency and effectiveness if they are designed and put into practice within a rational framework based on resource allocation, in accordance with the functional outcomes of the alternative management options that affect an integrated ecological or social entity. The application of this approach to conservation in high mountain ecosystems should take into account their particular characteristics of isolation, environmental harshness and steep gradients. Generally speaking, the minimisation of exposure to detrimental anthropogenic agents can be enhanced by the isolation of these habitats, but the latter can become particularly sensitive due to their limited extension and their legacy of selection for specific, extreme conditions. This general approach, exemplified by the concept of ecological trade-offs and associated with basic economic principles, can be developed and modelled for any specific case, incorporating the complex interactions between ecological processes and social agents, as well as the temporal dimension that takes into account both the legacy of the past and future scenarios.

Acknowledgements I thank AGAUR, Generalitat de Catalunya for support for project 2014 SGR 453 and Spanish MEC for Project CGL2015-67419-R.

References

Ager A, Vaillant N, McMahan A (2013) A restoration of fire in managed forests: a model to prioritize landscapes and analyze tradeoffs. Ecosphere 4:1–19

Allen CD (2007) Interactions across spatial scales among forest dieback, fire, and erosion in northern New Mexico landscapes. Ecosystems 10:797–808

Allen CD et al (2010) A global overview of drought and heat-induced tree mortality reveals emerging climate change risks for forests. For Ecol Manage 259:660–684

Allen CD, Breshears DD, McDowell NG (2015) On underestimation of global vulnerability to tree mortality and forest die-off from hotter drought in the anthropocene. Ecosphere 6 art129

Anderegg LDL, Anderegg WRL, Abatzoglou J et al (2013) Drought characteristics' role in widespread aspen forest mortality across Colorado, USA. Glob Change Biol 19:1526–1537

Armitage D, Marschke M, Plummer R (2008) Adaptive co-management and the paradox of learning. Glob Environ Change 18:86–98

Balthazar V, Vanacker V, Molina A et al (2015) Impacts of forest cover change on ecosystem services in high Andean mountains. Ecol Indic 48:63–75

Bell DM, Bradford JB, Lauenroth WK (2014) Forest stand structure, productivity, and age mediate climatic effects on aspen decline. Ecology 95:2040–2046

Beniston M (2003) Climatic change in mountain regions. Clim Change 59:5–31

Bennet EM, Peterson GD, Gordon LJ (2009) Understanding relationships among multiple ecosystem services. Ecol Lett 12:1394–1404

Bennett AF, Radford JQ, Haslem A (2006) Properties of land mosaics: Implications for nature conservation in agricultural environments. Biol Conserv 133:250–264
Bigler C, Kulakowski D, Veblen TV (2005) Multiple disturbance interactions and drought influence fire severity in Rocky Mountain subalpine forests. Ecology 86:3018–3029
Bigler C, Bräker OU, Bugmann H et al (2006) Drought as an inciting mortality factor in Scots pine stands of the Valais, Switzerland. Ecosystems 9:330–343
Braid ACR, Nielsen SE (2015) Prioritizing sites for protection and restoration for grizzly bears (Ursus arctos) in southwestern Alberta, Canada. PLoS ONE 10:1–16
Brandt JS, Haynes MA, Kuemmerle T et al (2013) Regime shift on the roof of the world: alpine meadows converting to shrublands in the southern Himalayas. Biol Conserv 158:116–127
Camarero JJ (2017a) The multiple factors explaining decline in mountain forests: historical logging and warming-related drought stress cause silver fir dieback in the Aragón Pyrenees. In: Catalan J, Ninot JM, Aniz MM (eds) Challenges for high mountain conservation in a changing world. Springer, pp 131–154
Camarero L (2017b) Atmospheric chemical loadings in the high mountain: current forcing and legacy pollution. In: Catalan J, Ninot JM, Aniz MM (eds) Challenges for high mountain conservation in a changing world. Springer, pp 325–341
Camarero JJ, Bigler C, Linares JC et al (2011) Synergistic effects of past historical logging and drought on the decline of Pyrenean silver fir forests. For Ecol Manag 262:759–769
Chapin FS III, Carpenter SR, Kofinas GP et al (2010) Ecosystem stewardship: sustainability strategies for a rapidly changing planet. Trends Ecol Evol 25:241–249
Colombaroli D, Henne PD, Kaltenrieder P et al (2010) Species responses to fire, climate and human impact at tree line in the Alps as evidenced by palaeo-environmental records and a dynamic simulation model. J Ecol 98:1346–1357
Coop JD, Massatti RT, Schoettle AW (2010) Subalpine vegetation pattern three decades after stand-replacing fire: effects of landscape context and topography on plant community composition, tree regeneration, and diversity. J Veg Sci 21:472–487
Dirnböck T, Essl F, Rabitsch W (2011) Disproportional risk for habitat loss of high-altitude endemic species under climate change. Glob Change Biol 17:990–996
Dobbertin M, Wermelinger B, Bigler C et al (2007) Linking increasing drought stress to Scots pine mortality and bark beetle infestations. Sci World J 7:231–239
Donovan GH, Brown TC (2007) Be careful what you wish: the legacy of Smokey Bear. Front Ecol Environ 5:73–79
Driscoll DA, Bode M, Bradstock RA et al (2016) Resolving future fire management conflicts using multicriteria decision making. Conserv Biol 30:196–205
Easterling DR, Meehl GA, Parmesan C et al (2000) Climate extremes: observations, modeling, and impacts. Science 289:2068–2074
Engler R, Randin C, Thuiller W et al (2011) 21st century climate change threatens mountain flora unequally across Europe. Glob Chang Biol 17:2330–2341
Flather CH, Wilson KR, Dean DJ et al (1997) Identifying gaps in conservation networks: of indicators and uncertainty in geographic-based analyses. Ecol Appl 7:531–542
Formica A, Farrer EC, Ashton IW et al (2014) Shrub expansion over the past 62 years in Rocky Mountain alpine tundra: possible causes and consequences. Arctic Antarct Alp Res 46: 616–631
Galiano L, Martínez-Vilalta J Lloret F (2010) Drought-induced multifactor decline of Scots pine in the Pyrenees and potential vegetation change by the expansion of co-occurring oak species. Ecosystems 13:978–991
Garcia-Pausas J, Romanyà J, Montané F, Rios AI, Taull M, Rovira P, Casals P (2017) Are soil carbon stocks in mountain grasslands compromised by land use changes? In: Catalan J, Ninot JM, Aniz MM (eds) Challenges for high mountain conservation in a changing world. Springer, pp 207–230
Gassiot Ballbè E, Mazzucco N, Clemente Conte I, Rodríguez Antón D, Obea Gómez L, Quesada Carrasco M, Díaz Bonilla S (2017) The beginning of high mountain occupations in the Pyrenees. Human settlements and mobility from 18,000 cal BC to 2000 cal BC.

In: Catalan J, Ninot JM, Aniz MM (eds) Challenges for high mountain conservation in a changing world. Springer, pp 75–105
Gonzalez P, Neilson RP, Lenihan JM et al (2010) Global patterns in the vulnerability of ecosystems to vegetation shifts due to climate change. Glob Ecol Biogeog 19:755–768
Grant GE, Tague CL, Allen CD (2013) Watering the forest for the trees: an emerging priority for managing water in forest landscapes. Front Ecol Environ 11:314–321
Grêt-Regamey A, Walz A, Bebi P (2008) Valuing ecosystem services for sustainable landscape planning in Alpine Regions. Mt Res Dev 28:156–165
Harper JL, Lovell PH, Moore KG (1970) The shapes and sizes of seeds. Annu Rev Ecol Syst 1:327–356
Hart SJ, Veblen TT, Eisenhart KS et al (2014) Drought induces spruce beetle (Dendroctonus rufipennis) outbreaks across northwestern Colorado. Ecology 95:930–939
Holling CS (1978) Adaptive environmental assessment and management. Wiley, New York
Hooper DU, Chapin FS, Ewel JJ et al (2005) Effects of biodiversity on ecosystem functioning: a consensus of current knowledge. Ecol Monogr 75:3–35
Hulme PE (2005) Adapting to climate change: is there scope for ecological management in the face of a global threat? J Appl Ecol 42:784–794
Jactel H, Petit J, Desprez-Lostau ML et al (2012) Drought effects on damage by forest insects and pathogens: a meta-analysis. Glob Change Biol 18:267–276
Kloster S, Mahowald NM, Randerson JT et al (2012) The impacts of climate, land use, and demography on fires during the 21st century simulated by CLM-CN. Biogeosciences 9:509–525
La Sorte FA, Jetz W (2010) Projected range contractions of montane biodiversity under global warming. Proc Biol Sci 277:3401–3410
Laiolo P, Obeso JR (2017) Life-history responses to the altitudinal gradient. In: Catalan J, Ninot JM, Aniz MM (eds) Challenges for high mountain conservation in a changing world. Springer, pp 253–283
Lampert A, Hastings A, Grosholz ED et al (2014) Optimal approaches for balancing invasive species eradication and endangered species management. Science 344:1028–1031
Lasanta-Martínez T, Vicente-Serrano SM, Cuadrat-Prats JM (2005) Mountain Mediterranean landscape evolution caused by the abandonment of traditional primary activities: a study of the Spanish Central Pyrenees. Appl Geogr 25:47–65
Lloret F, Piñol J, Castellnou M (2009) Wildfires. In: Woodward J (ed) The physical geography of the Mediterranean. Oxford University Press, Oxford, pp 541–558
Lloret F, Escudero A, Iriondo JM et al (2012) Extreme climatic events and vegetation: the role of stabilizing processes. Glob Chang Biol 18:797–805
Loarie SR, Duffy PB, Hamilton H et al (2009) The velocity of climate change. Nature 462:1052–1055
Loepfe L, Martinez-Vilalta J, Pinol J (2012) Management alternatives to offset climate change effects on Mediterranean fire regimes in NE Spain. Clim Change 115:693–707
Loepfe L, Rodrigo A, Lloret F (2014) Two thresholds determine climatic control of forest fire size in Europe and northern Africa. Reg Environ Chang 14:1395–1404
Loucougaray G, Dobremez L, Gos P et al (2015) Assessing the effects of grassland management on forage production and environmental quality to identify paths to ecological intensification in mountain grasslands. Environ Manage 56:1039–1052
MacNeill JR (2003) The mountains of the Mediterranean world. Cambridge University Press, Cambridge
Marini L, Klimek S, Battisti A (2011) Mitigating the impacts of the decline of traditional farming on mountain landscapes and biodiversity: a case study in European Alps. Environ Sci Policy 14:258–267
Martin J, Revilla E, Quenette PY et al (2012) Brown bear habitat suitability in the Pyrenees: transferability across sites and linking scales to make the most of scarce data. J Appl Ecol 49:621–631
Maudet C, Miller C, Bassano B et al (2002) Microsatellite DNA and recent statistical methods in wildlife conservation: applications in Alpine ibex [Capra ibex (ibex)]. Mol Ecol 11:421–436

Mawdsley JR, O'Malley R, Ojima DS (2009) A review of climate-change adaptation strategies for wildlife management and biodiversity conservation. Conserv Biol 23:1080–1089
May R, van Dijk J, Landa A et al (2010) Spatio-temporal ranging behaviour and its relevance to foraging strategies in wide-ranging wolverines. Ecol Modell 221:936–943
McDowell NG, Beeerling DJ, Breshears DD et al (2011) The interdependence of mechanisms underlying climate-driven vegetation mortality. Trends Ecol Evol 26:523–532
McKinney ST, Fiedler CE, Tomback DA (2009) Invasive pathogen threatens bird-pine mutualism: implications for sustaining a high-elevation ecosystem. Ecol Appl 19:597–607
Metzger MJ, Bunce RGH, Leemans R et al (2008) Projected environmental shifts under climate change: European trends and regional impacts. Environ Conserv 35:64–75
Millar CI, Stephenson NL, Stephens SL (2007) Climate change and forests of the future: managing in the face of uncertainty. Ecol Appl 17:2145–2151
Moriondo M, Good P, Durao R et al (2006) Potential impact of climate change on fire risk in the Mediterranean area. Clim Res 31:85–95
Moritz C, Patton JL, Conroy CJ et al (2008) Impact of a century of climate change on small-mammal communities in Yosemite National Park, USA. Science 322:261–264
Nagy L, Grabherr G (2009) The biology of alpine habitats. Oxford University Press, Oxford
Nalle DJ, Montgomery CA, Arthur JL et al (2004) Modeling joint production of wildlife and timber. J Environ Econ Manage 48:997–1017
Odion DC, Frost EJ, Stritth JR et al (2004) Patterns of fire severity and forest conditions in the Western Klamath Mountains, California. Conserv Biol 18:927–936
Oliva J, Stenlid J, Martínez-Vilalta J (2014) The effect of fungal pathogens on the water and carbon economy of trees: implications for drought-induced mortality included in theoretical models for drought-induced mortality. New Phytol 203:1028–1035
Paletto A, Geitner C, Grilli G et al (2015) Mapping the value of ecosystem services: a case study from the Austrian Alps. Ann For Res 58:157–175
Pan Y, Wu J, Xu Z (2014) Analysis of the tradeoffs between provisioning and regulating services from the perspective of varied share of net primary production in an alpine grassland ecosystem. Ecol Complex 17:79–86
Parry M, Canziani O, Palutikof J, van del Linden P, Hanson C (eds) (2007) Climate change 2007: impacts, adaptation and vulnerability. Contribution of working group II to the fourth assessment report of the intergovernmental panel on climate change. Cambridge University Press, Cambridge
Pauchard A, Kueffer C, Hansjoerg D et al (2009) Ain't no mountain high enough: plant invasions reaching new elevations. Front Ecol Environ 7:479–486
Pauchard A, Milbau A, Albihn A et al (2016) Non-native and native organisms moving into high elevation and high latitude ecosystems in an era of climate change: new challenges for ecology and conservation. Biol Invasions 18:345–353
Pausas JG, Fernández-Muñoz S (2012) Fire regime changes in the Western Mediterranean Basin: from fuel-limited to drought-driven fire regime. Clim Change 110:215–226
Pèlachs A, Pérez-Obiol R, Soriano JM, Cunill R, Bal M-C, García-Codron JC (2017) The role of environmental geohistory in high mountain landscape conservation. In: Catalan J, Ninot JM, Aniz MM (eds) Challenges for high mountain conservation in a changing world. Springer, pp 107–129
Raffa KF, Aukema BH, Bentz BJ et al (2008) Cross-scale drivers of natural disturbances prone to Anthropogenic amplification: the dynamics of bark beetle eruptions. Bioscience 58:501–516
Raudsepp-Hearne C, Peterson GD, Bennett EM (2010) Ecosystem service bundles for analysing tradeoffs in diverse landscapes. Proc Natl Acad Sci USA 107:5242–5247
Reekie EG, Bazzaz FA (1987) Reproductive effort in plants. 1. Carbon allocation to reproduction. Am Nat 129:876–896
Roura-Pascual N, Pons P, Etienne M et al (2005) Transformation of a rural landscape in the Eastern Pyrenees between 1953 and 2000. Mt Res Dev 25:252–261

Schumacher S, Bugmann H (2006) The relative importance of climatic effects, wildfires and management for future forest landscape dynamics in the Swiss Alps. Glob Change Biol 12:1435–1450
Sergio F, Pedrini P (2007) Biodiversity gradients in the Alps: the overriding importance of elevation. Biodivers Conserv 16:3243–3254
Sibold JS, Veblen TT, González ME (2006) Spatial and temporal variation in historic fire regimes in subalpine forests across the Colorado Front Range in Rocky Mountain National Park, Colorado, USA. J Biogeogr 33:631–647
Smith JM, Paritsis J, Veblen TT, Chapman TB (2015) Permanent forest plots show accelerating tree mortality in subalpine forests of the Colorado Front Range from 1982 to 2013. For Ecol Manage 341:8–17
Spehn E, Lieberman M, Körner C (2006) Land use change and mountain biodiversity. Taylor & Francis, CRC Press, Boca Raton, London
Sturrock RN, Frankel SJ, Brown AV et al (2011) Climate changes and forest disease. Plant Pathol 60:133–149
Suarez ML, Ghermandi L, Kitzberger T (2004) Factors predisposing episodic drought-induced tree mortality in Nothofagus—site, climatic sensitivity and growth trends. J Ecol 92:954–966
Theurillat J, Guisan A (2001) Potential impact of climate change on vegetation in the European Alps: a review. Clim Change 50:77–109
Tinner W, Conedera M, Ammann B et al (2005) Fire ecology North and South of the Alps since the last Ice Age. Holocene 15:1214–1226
Turner BL, Kasperson RE, Matson PA et al (2003) A framework for vulnerability analysis in sustainability science. Proc Natl Acad Sci USA 100:8074–8079
Vacchiano G, Maggioni M, Perseghin G et al (2015) Effect of avalanche frequency on forest ecosystem services in a spruce-fir mountain forest. Cold Reg Sci Technol 115:9–21
van Mantgem PJ, Stephenson NL, Byrne JC et al (2009) Widespread increase of tree mortality rates in the Western United States. Science 323:521–524
Ventura M, Tiberti R, Buchaca T, Buñay D, Sabás I, Miró A (2017) Why should we preserve fishless high-mountain lakes? In: Catalan J, Ninot JM, Aniz MM (eds) Challenges for high mountain conservation in a changing world. Springer, pp 181–205
Williams AP, Allen CD, Macalady AK et al (2012) Temperature as a potent driver of regional forest drought stress and tree mortality. Nat Clim Change 3:292–297

Part II
Developing a Historical Perspective of the High Mountain Social-Ecological System

Chapter 3
Molecular Biogeography of the High Mountain Systems of Europe: An Overview

Thomas Schmitt

Abstract The biogeography of alpine and arctic–alpine species is complex, much more complex than thought until relatively recently. Alpine species survived glacial periods mostly within refugia in close proximity to the mountains where they are found today. One mountain range can be colonised from several glacial refugia, while one refugium can be the source of colonisation of more than one mountain range. The zonal distributions in the glacial cold steppes are only of importance for arctic–alpine species. Their arctic ranges normally derive from there, while the southern mountains were colonised from there or from near-mountain refugia as in the cases of the alpine species.

Keywords Alpine disjunct species · Arctic–alpine disjunctions · Refugia · Range dynamics · Nunataks · Biogeography · Phylogeography

3.1 Introduction

Europe is a continent characterised by a larger number of different high mountain systems, especially in its southern parts. These mountains vary considerably in their size and in their height. By far the highest and largest mountain system in Europe is the Alps, if the even higher Caucasus that forms part of the border with Asia is not considered. The Pyrenees and the Scandes also represent large systems of continuous high mountains. This applies less to the Carpathians, which have larger stretches along their length of 1300 km without true high mountain habitats (i.e. the alpine zone or oreal, naturally not covered by forests). Furthermore, many generally

T. Schmitt (✉)
Senckenberg German Entomological Institute, Eberswalder Straße 90, 15374 Müncheberg, Germany
e-mail: thomas.schmitt@senckenberg.de

T. Schmitt
Faculty of Natural Sciences I, Department of Zoology, Institute of Biology, Martin Luther University Halle-Wittenberg, 06099 Halle (Saale), Germany

J. Catalan et al. (eds.), *High Mountain Conservation in a Changing World*,
Advances in Global Change Research 62, DOI 10.1007/978-3-319-55982-7_3

small blocks of high mountain areas are scattered through the southern European peninsulas, like those of the Balkans, the Apennine Peninsula and Iberia. Finally, isolated and small high mountain systems are found elsewhere, such as the Massif Central, the Vosges, the Harz and some parts of the Sudety Mountains. Accordingly, Europe can be visualised as a highly diverse archipelago composed of mountain islands of very different sizes and very different degrees of isolation embedded in a matrix of lowland areas. This complexity of high mountains translates into a high diversity of their biogeographic structures (Fig. 3.1).

In this chapter, I therefore give a short overview of the biogeographic structures of the high mountains of Europe. On the one hand, I focus on the biogeographic structures within single mountain systems, on the other hand, I work out the different biogeographic links among different high mountain systems. A special focus of this chapter is on molecular analyses, these being highly suitable for unravelling biogeographic structures. Most of the examples presented refer to invertebrates and plants. Based on these genetic patterns, a comprehensive analysis of range dynamics in high mountain ecosystems in space and time is presented.

Before considering the more detailed biogeographic structures of high mountain species, we have to define two fundamental distribution types, the arctic–alpine and the alpine disjunct species. Arctic–alpine species are distributed in the alpine belt of high mountain systems in the South and in the Arctic realm and have long been interpreted as having resulted from the postglacial disjunction of an extended zonal distribution during the last ice age, with retreat to high altitudes in the South and high latitudes in the North (Scharff 1899; Holdhaus 1954). Alpine disjunct species are lacking in the tundra belt of the North, but are distributed in several of the southern high mountain systems in the subalpine and alpine belt. The distributional

Fig. 3.1 Centres of genetic endemism of alpine elements. The *bold line* indicates the Alps, the most important centre of alpine endemism in Europe. *Continuous lines* highlight further mountain ranges with high numbers of genetic endemism. *Broken lines* indicated mountain systems with frequent genetic endemism, while *dotted lines* mark those mountains with just few cases of genetic endemism. Map based on Google Maps

group of arctic–alpine species must not be mixed up with boreo-montane species, which have a somewhat similar distribution, but are confined to the montane and subalpine forests in the mountains and to the boreal forest belt in the North (Schmitt 2009); furthermore, these elements are also found in the forests of the lower European mountain ranges without an alpine belt. Similar differences exist between alpine disjunct and montane disjunct species. As boreo-montane and montane disjunct species are not true high mountain elements, they are not addressed in this chapter.

3.2 Different Genetic Lineages Within High Mountain Systems

Following the picture of an island archipelago of high mountain systems in the 'European Sea of Lowlands', the Alps are the largest 'island', or even a 'continent' surrounded by other mountain 'islands'. As such, the Alps harbour a large number of endemic high mountain species, some of them distributed throughout these mountains, others with rather narrow distributions in some parts of the Alps, with many of these distributions located in the south-western or south-eastern Alps (Varga and Schmitt 2008). Although some of these geographically restricted endemics are genetically impoverished (e.g. *Erebia sudetica inalpina*; Haubrich and Schmitt 2007), perhaps as a result of constantly low numbers of individuals and the hereby resulting genetic bottlenecks, others are genetically even more diverse than their lowland relatives (e.g. *Coenonympha darwiniana, C. macromma*; Schmitt and Besold 2010). This high genetic diversity might be the consequence of simple uphill—downhill shifts within one region as conditions changed from interglacial to glacial and vice versa without major genetic bottlenecks.

In most cases, more widely distributed Alpine species comprise several genetic lineages, which can be 'translated' into several centres of differentiation, i.e. refugia that later served as centres of dispersal. The classic pattern, repeated with little variation in numerous plant and animal species (e.g. Schönswetter et al. 2002, 2003a, b, 2004a; Stehlik et al. 2002a; Tribsch et al. 2002; Margraf et al. 2007; Thiel-Egenter et al. 2009), is of four genetic groups localised in the south-western, western, central and eastern Alps (Fig. 3.2a). This pattern is assumed to have evolved in four glacial refugia in the lower and thus unglaciated parts of the south-western Alps, south of the western and central Alps as well as east of the eastern Alps, i.e. in southern peripheral refugia. Some species show a pattern of fewer refugia, with just an eastern and a western genetic group (Fig. 3.2b), and thus only two centres of survival (e.g. Pauls et al. 2006; Haubrich and Schmitt 2007; Schmitt and Haubrich 2008). However, some species even had peripheral refugia north of the Alps (e.g. *Erebia epiphron*; Schmitt et al. 2006) or survived (additionally or even exclusively) on nunataks (i.e. ice-free areas surrounded by the Alpine glaciers), as proven for several plant species (e.g. Stehlik et al. 2001, 2002b; Holderegger et al. 2002; Stehlik 2002).

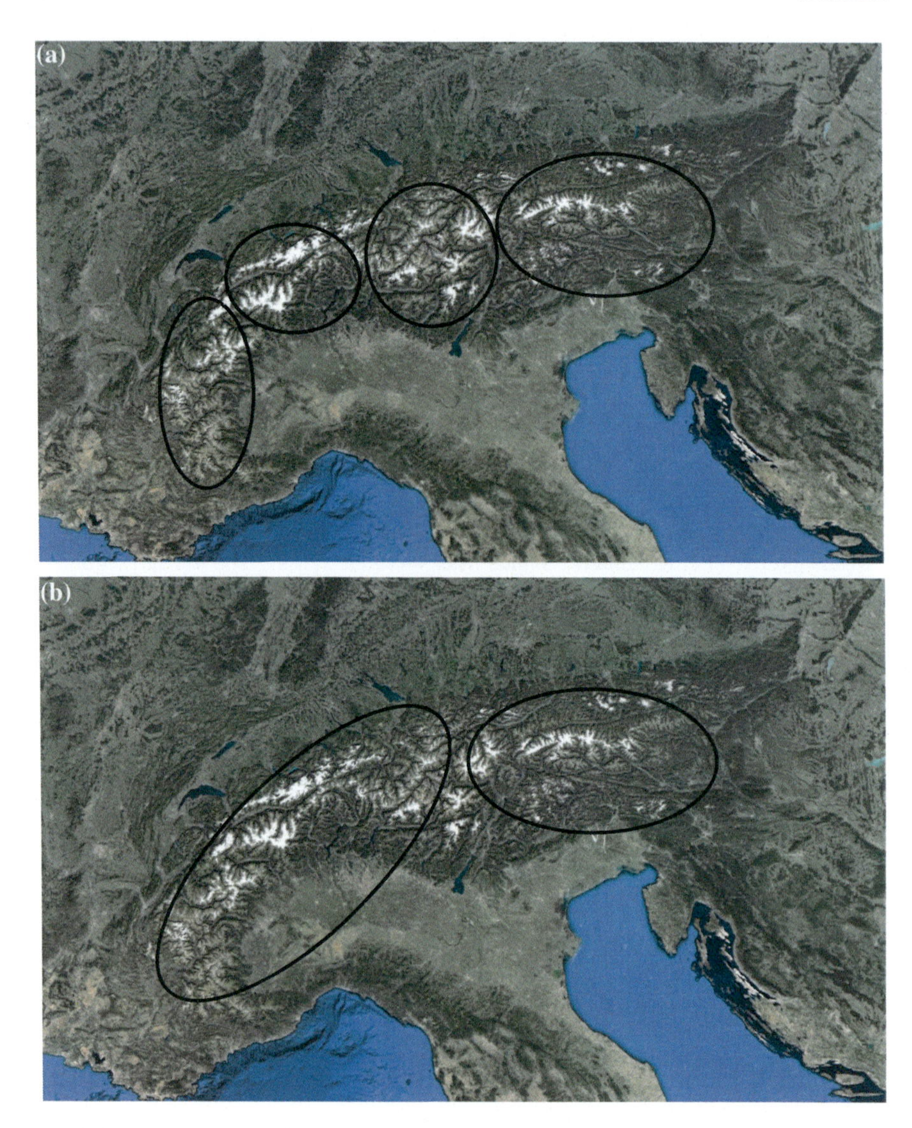

Fig. 3.2 Genetic structures in the Alps. **a** The most common pattern of genetic differentiation in the Alps is one of the four different lineages. **b** However, a reduction of this pattern, e.g. to just two lineages, is frequently observed. Maps based on Google Maps

In the Pyrenees, we also observe numerous species with more than one genetic lineage within these mountains. However, the most often observed number of different lineages in animals and plants found here is two (e.g. Kropf et al. 2002, 2003; Schmitt et al. 2006; Lauga et al. 2009), translating into only two glacial refugia (and differentiation centres) in close proximity to the Pyrenees and not four as in the much bigger Alps. Higher numbers of different lineages in the Pyrenees are rare, but have been shown, e.g. for *Rhododendron ferrugineum* with five (Charrier et al. 2014) and *Cardamine alpina* with a number of lineages that is difficult to quantify (Lihová et al. 2008).

Similarly to the Pyrenees, two genetic lineages were frequently observed in the Carpathians, with one of these in the mountains' northern part and the second in the eastern and southern regions (e.g. Pauls et al. 2006; Mráz et al. 2007; Ronikier et al. 2008a; Ujvárosi et al. 2010; Theissinger et al. 2012). An additional endemic genetic lineage in the Apuseni mountains (island mountains in the Carpathian Basin) was obtained for the stonefly *Arcynopteryx dichroa* (Theissinger et al. 2012). In *Soldanella* species, two major lineages were distinguished, but in contrast to the above pattern, the northern lineage also included the eastern Carpathians and thus was geographically more extended. Furthermore, some geographically restricted endemic species are known in this group (Zhang et al. 2001), hence underlining that the biogeographic history of the high mountain elements of the Carpathians can be more complex than the two-refugia theory.

The Balkan Peninsula presents a complex pattern of numerous small blocks of high mountain systems in relatively close geographic proximity to each other. Unfortunately, this area is still poorly studied phylogeographically. Morphological studies in butterflies, however, strongly support an east-west split for several high mountain species (Varga 1975), speaking for long lasting separation of the mountain areas east and west of the Central Balkan Depression. This split was supported by a genetic study of the stonefly *Arcynopteryx dichroa* (Theissinger et al. 2012), but the butterfly species *Erebia ottomana* and *Coenonympha rhodopensis* showed relatively uniform genetic constitutions throughout the area, thus supporting the hypothesis that in these cases gene flow connected populations throughout the Balkan mountains during glacial periods (Louy et al. 2013, 2014a).

3.3 Genetic Links Between High Mountain Systems

As Europe's most important high mountain system, the Alps have multiple biogeographic links to all neighbouring mountains (Fig. 3.3; Schmitt 2009). In a number of cases, in both animals and plants, identical genetic lineages exist in the south-western Alps and in the Pyrenees (e.g. Kropf et al. 2002; Martin et al. 2002; Schönswetter et al. 2002, 2004b; Gaudeul 2006; Schmitt et al. 2006; Reisch 2008). This often repeated pattern is in most cases thought to be the result of glacial distributions in the hilly areas of southern France and postglacial retreat into the adjoining mountain ranges, leading to a rather young vicariance event, which is still not mirrored in the genetic make-up of the now disconnected population groups. Quite similar phenomena, with similar biogeographic explanations, are known between the north-eastern Alps and the Tatra mountains (e.g. Kropf et al. 2003; Muster and Berendonk 2006; Schönswetter et al. 2006; Suda et al. 2007; Paun et al. 2008; Triponez et al. 2011; Schmitt et al. 2014) as well as the south-eastern Alps and the north-western Balkan mountains (Triponez et al. 2011; examples for

Fig. 3.3 Different high mountain systems share identical genetic lineages in many cases. *Bold arrows* indicate that such a sharing is very commonly observed. *Solid arrows* show that this pattern is frequent, while *dotted arrows* indicate relatively few known cases. The *arrows* indicate the dominating direction of exchange, with *two-sided arrows* assuming equilibrium of exchange between mountains. Map based on Google Maps

boreo-montane species: Ronikier et al. 2008b; Kramp et al. 2009). Although genetic data from the western Balkan mountains are generally scarce, it is worth noting that the south-eastern Alps—western Balkan mountains link is frequently supported by the existence of taxa that only occur in these two regions (e.g. Holdhaus 1954; Varga and Schmitt 2008; Schönswetter and Schneeweiss 2009; Tshikolovets 2011).

Focussing now on the smaller mountain ranges of Europe and their biogeographic links, we start with the Massif Central in France. This mountain range sometimes even possesses its own endemic genetic lineages, supporting the idea of independent glacial refugia in its geographic proximity without recent genetic exchange, neither with the western Alps nor with the Pyrenees (e.g. Pauls et al.

2006; Triponez et al. 2011; Kropf et al. 2012). Genetic links are known with the Pyrenees (e.g. Descimon 1995; Ronikier et al. 2008b (however, these two species are not typical alpine elements); Schmitt et al. 2014) and the western Alps (Kramp et al. 2009 (but referring to a boreo-montane species); Triponez et al. 2011). However, the latter link seems to be less frequent than the former, thus supporting the idea that the isolating power of the Rhone valley was stronger than that of the hilly regions between the Massif Central and the Pyrenees.

The Apennines, a long stretch of mountains with some interspersed insular high mountain areas, harbours some high mountain endemics, especially in its central and southern parts (e.g. the plants *Adonis distorta, Androsace mathildae, Aquilegia bertoloni, Soldanella calabrella*), hence supporting the evolutionary independence of this area. However, many taxa also show high similarity with counterparts in the south-western Alps and offer evidence of glacial gene flow between these mountain ranges during cold phases (Moore et al. 2013; Louy et al. 2014b). A particularly interesting case involves representatives of the beetle species complex *Oreina alpestris/speciosa*. In this example, the populations of the northern Apennines show much higher similarity with ones from the south-western Alps than with those of the central Apennines (Triponez et al. 2011), thus indicating several colonisation waves from the south-western Alps to the Apennines, with the older ones preserved in the more southern mountains, the younger ones more to the North.

The mountains north of the Alps, if they have high mountain species *sensu stricto* at all, in most cases share their genetic lineages with the Alps, i.e. they are derived from the same refugia as the nearby Alpine populations (e.g. Pauls et al. 2006; Schmitt et al. 2006; Mardulyn et al. 2009; Triponez et al. 2011; Alvarez et al. 2012; Charrier et al. 2014). Exceptions to this rule are the caddisfly *Drusus discolor* with a genetic lineage restricted to Jura, Vosges and Black Forest (Pauls et al. 2006) and the butterfly *Erebia manto* with the genetically strongly differentiated taxon *vogesiaca* endemic to the Vosges (Schmitt et al. 2014).

The Cantabrian mountains, being the westernmost range with alpine zonation, also harbour some endemic lineages, thus underlining their independent biogeographic status (e.g. Kropf et al. 2003; Pauls et al.2006). In the majority of cases, however, close genetic links with the Pyrenees exist, indicating the presence of glacial refugia between both mountain ranges and resultant genetic intermixing (e.g. Kropf et al. 2002; Vila et al. 2011).

The Carpathians and Balkan high mountain systems have in many respects distinct alpine floras and faunas. Nevertheless, they sometimes share identical typical species, such as the butterflies *Erebia melas* and *Coenonympha rhodopensis* (Tshikolovets 2011). However, even identical genetic lineages were recorded in some cases, thus presenting evidence for a recent (most likely Würm glacial) exchange between both regions. This was for example shown for the mountain forests butterfly *Erebia euryale* (Schmitt and Haubrich 2008) and the boreo-montane plant *Ranunculus platanifolius* (Stachurska-Swakon et al. 2013). For the stonefly *Arcynopteryx dichroa*, distinct genetic lineages exist in both areas, but rare occurrences of the Carpathian haplotype group in the Bulgarian high mountain systems call for a recent gene flow in North-South direction (Theissinger et al. 2012).

Although some similarities exist between the alpine faunas and floras of the Apennines and the Balkan high mountain systems (e.g. shared endemic plants such as *Aurinia rupestris, Leontopodium nivale, Saxifraga glabella*), the exchange between these two mountain areas might in most cases not be very recent, as in most of the examples referred to above (cf. Schönswetter and Schneeweiss 2009; Louy et al. 2013).

3.4 Arctic–Alpine Disjunction

In arctic–alpine species, i.e. species with occurrences in the high mountain ranges in the South and in the Arctic, wide zonal ice age distributions in the periglacial steppe region with postglacial retreats uphill and polewards have classically been assumed (Holdhaus 1954). Following this assumption, the disjunction between northern and southern populations dates only to the postglacial and should not have resulted in corresponding differentiations, so that mostly similar populations should be found in the North and the South. In support of this theory, highly similar genetic make-up of northern and southern populations has frequently been recorded (e.g. Schönswetter et al. 2003c, 2008; Albach et al. 2006; Muster and Berendonk 2006; Skrede et al. 2006; Ehrich et al. 2007; Reisch 2008; Schmitt et al. 2010).

In many arctic–alpine species, only part of the southern mountain populations is similar to the ones from the Arctic, while others are not. These patterns call for additional perialpine glacial refugia in close proximity to the mountain ranges (as exclusively in the alpine disjunct species) supplementing the extended zonal distributions. In most of these cases, the Alps and the Arctic share identical genetic lineages, which were thus apparently derived from this periglacial steppe region. Many species occurring in the Pyrenees and the Tatras also share these lineages (e.g. Muster and Berendonk 2006; Schmitt et al. 2010). However, lineages in the Alps do not always have a northern origin. For example, populations of *Gentiana nivalis* derived from the extended zonal ice age distribution were detected in the North, the Pyrenees and Carpathians, while three other lineages exist in the Alps, calling for three perialpine glacial refugia in addition to the zonal range (Alvarez et al. 2012). In the stonefly *Arcynopteryx dichroa*, the northern clade is limited in the South to the Black Forest, thus supporting the existence of an extended zonal range of the species, but its limited impact on the postglacial recolonisation of the southern mountains indicates that these were mostly colonised from glacial refugia in their foothills (Theissinger et al. 2012).

Acknowledgements I thank Andrew Liston (SDEI, Müncheberg) for linguistic improvements. Constructive comments of two anonymous referees are acknowledged.

References

Albach DC, Schönswetter P, Tribsch A (2006) Comparative phylogeography of closely related species of the *Veronica alpina* complex in Europe and North America. Mol Ecol 15:3269–3286

Alvarez N, Manel S, Schmitt T, IntraBioDiv Consortium (2012) Contrasting diffusion of Quaternary gene pools across Europe: the case of the arctic–alpine *Gentiana nivalis* L. (Gentianaceae). Flora 207:408–413

Charrier O, Dupont P, Pornon A, Escaravage N (2014) Microsatellite marker analysis reveals the complex phylogeographic history of *Rhododendron ferrugineum* (Ericaceae) in the Pyrenees. PLoS ONE 9:e92976

Descimon H (1995) La conservation des *Parnassius* en France: aspects zoogéographiques, écologiques, démographiques et génétiques. OPIE 1:1–54

Ehrich D, Gaudeul M, Assefa A, Koch M, Mummenhoff K, Nemomissa S, IntraBioDiv Consortium, Brochmann C (2007) Genetic consequences of Pleistocene range shifts: contrast between the Arctic, the Alps and the East African mountains. Mol Ecol 6:2542–2559

Gaudeul M (2006) Disjunct distribution of *Hypericum nummularium* L. (Hypericaceae): molecular data suggest bidirectional colonization from a single refugium rather than survival in distinct refugia. Biol J Linn Soc 87:437–447

Haubrich K, Schmitt T (2007) Cryptic differentiation in alpine-endemic, high-altitude butterflies reveals down-slope glacial refugia. Mol Ecol 16:3643–3658

Holderegger R, Stehlik I, Abbott RJ (2002) Molecular analysis of the Pleistocene history of *Saxifraga oppositifolia* in the Alps. Mol Ecol 11:1409–1418

Holdhaus K (1954) Die Spuren der Eiszeit in der Tierwelt Europas. Abh zool-bot Ges 18:1–493

Kramp K, Huck S, Niketić M, Tomović G, Schmitt T (2009) Multiple glacial refugia and complex postglacial range shifts of the obligatory woodland plant *Polygonatum verticillatum* (Convallariaceae). Plant Biol 11:392–404

Kropf M, Kadereit JW, Comes HP (2002) Late Quaternary distributional stasis in the submediterranean mountain plant *Anthyllis montana* L. (Fabaceae) inferred from ITS sequences and amplified fragment length polymorphism markers. Mol Ecol 11:447–463

Kropf M, Kadereit JW, Comes HP (2003) Differential cycles of range contraction and expansion in European high mountain plants during the Late Quaternary: insights from *Pritzelago alpina* (L.) O. Kuntze (Brassicaceae). Mol Ecol 12:931–949

Kropf M, Comes HP, Kadereit JW (2012) Past, present and future of mountain species of the French Massif Central—the case of *Soldanella alpina* L. subsp. *alpina* (Primulaceae) and a review of other plant and animal studies. J Biogeogr 39:99–112

Lauga B, Malaval S, Largier G, Regnault-Roger C (2009) Two lineages of *Trifolium alpinum* (Fabaceae) in the Pyrenees: evidence from random amplified polymorphic DNA (RAPD) markers. Acta Bot Gallica 156:317–330

Lihová J, Carlsen T, Brochmann C, Marhold K (2008) Contrasting phylogeographies inferred for the two alpine sister species *Cardamine resedifolia* and *C. alpina* (Brassicaceae). J Biogeogr 36:104–120

Louy D, Habel JC, Abadjiev S, Schmitt T (2013) Genetic legacy from past panmixia: High genetic variability and low differentiation in disjunct populations of the Eastern Large Heath butterfly. Biol J Linn Soc 110:281–290

Louy D, Habel JC, Abadjev S, Rakosy L, Varga Z, Rödder D, Schmitt T (2014a) Molecules and models indicate diverging evolutionary effects from parallel altitudinal range shift in two mountain Ringlet butterflies. Biol J Linn Soc 112:569–583

Louy D, Habel JC, Ulrich W, Schmitt T (2014b) Out of the Alps: The biogeography of a disjunctly distributed mountain butterfly, the Almond eyed ringlet *Erebia alberganus* (Lepidoptera, Satyrinae). J Hered 105:28–38

Mardulyn P, Mikhailov YE, Pasteels JM (2009) Testing phylogeographic hypotheses in a Euro-Siberian cold-adapted leaf beetle with coalescent simulations. Evolution 63:2717–2729

Margraf N, Verdon A, Rahier M, Naisbit RE (2007) Glacial survival and local adaptation in an alpine leaf beetle. Mol Ecol 16:2333–2343

Martin J-F, Gilles A, Lörtscher M, Descimon H (2002) Phylogenetics and differentiation among the western taxa of the *Erebia tyndarus* group (Lepidoptera: Nymphalidae). Biol J Linn Soc 75:319–332

Moore AJ, Merges D, Kadereit JW (2013) The origin of the serpentine endemic *Minuartia laricifolia* subsp. *ophiolitica* by vicariance and competitive exclusion. Mol Ecol 22:2218–2231

Mráz P, Gaudeul M, Rioux D, Gielly L, Choler P, Taberlet P (2007) Genetic structure of *Hypochaeris uniflora* (*Asteraceae*) suggests vicariance in the Carpathians and rapid post-glacial colonization of the Alps from an eastern Alpine refugium. J Biogeogr 34:2100–2114

Muster C, Berendonk TU (2006) Divergence and diversity: lessons from an arctic-alpine distribution (*Pardosa saltuaria* group, Lycosidae). Mol Ecol 15:2921–2933

Pauls SU, Lumbsch HT, Haase P (2006) Phylogeography of the montane caddisfly *Drusus discolor*: evidence for multiple refugia and periglacial survival. Mol Ecol 15:2153–2169

Paun O, Schönswetter P, Winkler M, Tribsch A (2008) A historical divergence versus contemporary gene flow: evolutionary history of the calcicole *Ranunculus alpestris* group (Ranunculaceae) in the European Alps and Carpathians. Mol Ecol 17:4263–4275

Reisch C (2008) Glacial history of *Saxifraga paniculata* (Saxifragaceae): molecular biogeography of a disjunct arctic-alpine species from Europe and North America. Biol J Linn Soc 93:385–398

Ronikier M, Cieslak E, Korbecka G (2008a) High genetic differentiation in the alpine plant *Campanula alpina* Jacq. (Campanulaceae): evidence for glacial survival in several Carpathian regions and long-term isolation between the Carpathians and the Eastern Alps. Mol Ecol 17:1763–1775

Ronikier M, Costa A, Fuertes Aguilar J, Feliner GN, Küpfer P, Mirek Z (2008b) Phylogeography of *Pulsatilla vernalis* (L.) Mill. (Ranunculaceae): chloroplast DNA reveals two evolutionary lineages across central Europe and Scandinavia. J Biogeogr 35:1650–1664

Scharff RF (1899) The history of the European fauna. Walter Scott, London

Schmitt T (2009) Biogeographical and evolutionary importance of the European high mountain systems. Front Zool 6:9

Schmitt T, Besold J (2010) Upslope movements and large scale expansions: the taxonomy and biogeography of the *Coenonympha arcania—C. darwiniana—C. gardetta* butterfly species complex. Zool J Linn Soc 159:890–904

Schmitt T, Haubrich K (2008) The genetic structure of the mountain forest butterfly *Erebia euryale* unravels the late Pleistocene and postglacial history of the mountain coniferous forest biome in Europe. Mol Ecol 17:2194–2207

Schmitt T, Hewitt GM, Müller P (2006) Disjunct distributions during glacial and interglacial periods in mountain butterflies: *Erebia epiphron* as an example. J Evol Biol 19:108–113

Schmitt T, Muster C, Schönswetter P (2010) Are disjunct alpine and arctic-alpine animal and plant species in the western Palearctic really "relics of a cold past"? In: Habel JC, Assmann T (eds) Relict species: phylogeography and conservation biology. Springer, Heidelberg, pp 239–252

Schmitt T, Habel JC, Rödder D, Louy D (2014) Effects of recent and past climatic shifts on the genetic structure of the high mountain Yellow-spotted ringlet butterfly *Erebia manto* (Lepidoptera, Satyrinae): a conservation problem. Glob Change Biol 20:2045–2061

Schönswetter P, Schneeweiss GM (2009) *Androsace komovensis* sp. nov., a long mistaken local endemic from the southern Balkan Peninsula with biogeographic links to the Eastern Alps. Taxon 58:544–549

Schönswetter P, Tribsch A, Barfuss M, Niklfeld H (2002) Several Pleistocene refugia detected in the high alpine plant *Phyteuma globulariifolium* in the European Alps. Mol Ecol 11:2637–2647

Schönswetter P, Tribsch A, Niklfeld H (2003a) Phylogeography of the high alpine cushion-plant *Androsace alpina* (Primulaceae) in the European Alps. Plant Biol 5:623–630

Schönswetter P, Tribsch A, Schneeweiss GM, Niklfeld H (2003b) Disjunctions in relict alpine plants: phylogeography of *Androsace brevis* and *A. wulfeniana* (Primulaceae). Bot J Linn Soc 141:437–446

Schönswetter P, Paun O, Tribsch A, Niklfeld H (2003c) Out of the alps: colonisation of the arctic by east alpine populations of *Ranunculus glacialis* (Ranunculaceae). Mol Ecol 12:3371–3381

Schönswetter P, Tribsch A, Stehlik I, Niklfeld H (2004a) Glacial history of high alpine *Ranunculus glacialis* (Ranunculaceae) in the European Alps in a comparative phylogeographical context. Biol J Linn Soc 81:183–195

Schönswetter P, Tribsch A, Niklfeld H (2004b) Amplified fragment length polymorphism (AFLP) reveals no genetic divergence of the eastern alpine endemic *Oxytropis campestris* subsp. *tiroliensis* (Fabaceae) from widespread subsp. *campestris*. Plant Syst Evol 244:245–255

Schönswetter P, Popp M, Brochmann C (2006) Rare arctic-alpine plants of the European Alps have different immigration histories: the snowbed species *Minuartia biflora* and *Ranunculus pygmaeus*. Mol Ecol 15:709–720

Schönswetter P, Elven R, Brochmann C (2008) Trans-Atlantic dispersal and large-scale lack of genetic structure in the circumpolar, arctic-alpine sedge *Carex bigelowii* s. lat. (Cyperaceae). Am J Bot 95:1006–1014

Skrede I, Eidesen PB, Portela RP, Brochmann C (2006) Refugia, differentiation and postglacial migration in arctic-alpine Eurasia, exemplified by the mountain avens (*Dryas octopetala* L.). Mol Ecol 15:827–1840

Stachurska-Swakon A, Cieslak E, Ronikier M (2013) Phylogeography of a subalpine tall-herb *Ranunculus platanifolius* (Ranunculaceae) reveals two main genetic lineages in the European mountains. Bot J Linn Soc 171:413–428

Stehlik I (2002) Glacial history of the alpine herb *Rumex nivalis* (Polygonaceae): a comparison of common phylogeographic methods with nested clade analysis. Am J Bot 89:2007–2016

Stehlik I, Schneller JJ, Bachmann K (2001) Resistance or emigration: response of the high-alpine plant *Eritrichium nanum* (L.) Gaudin to the ice age within the Central Alps. Mol Ecol 10:357–370

Stehlik I, Schneller JJ, Bachmann K (2002a) Immigration and *in situ* glacial survival of the low-alpine *Erinus alpinus* (Scrophulariaceae). Bot J Linn Soc 77:87–103

Stehlik I, Blattner FR, Holderegger R, Bachmann K (2002b) Nunatak survival of the high Alpine plant *Eritrichium nanum* (L.) Gaudin in the central Alps during the ice ages. Mol Ecol 11:2027–2036

Suda J, Weiss-Schneeweiss H, Tribsch A, Schneeweiss G, Trávníček P, Schönswetter P (2007) Complex distribution patterns of di-, tetra and hexaploid cytotypes in the European high mountain plant *Senecio carniolicus* Willd. (Asteraceae). Am J Bot 94:1391–1401

Theissinger K, Bálint M, Feldheim KA, Haase P, Johannesen J, Laube I, Pauls SU (2012) Glacial survival and post-glacial recolonization of an arctic–alpine freshwater insect (*Arcynopteryx dichroa*, Plecoptera, Perlodidae) in Europe. J Biogeogr 40:236–248

Thiel-Egenter C, Holderegger R, Brodbeck S, Intrabiodiv Consortium, Gugerli F (2009) Concordant genetic breaks, identified by combining clustering and tessellation methods, in two co-distributed alpine plant species. Mol Ecol 18:4495–4507

Tribsch A, Schönswetter P, Stuessy TF (2002) *Saponaria pumila* (Caryophyllaceae) and the ice-age in the Eastern Alps. Am J Bot 89:2024–2033

Triponez Y, Buerki S, Borer M, Naisbit RE, Rahier M, Alvarez N (2011) Discordances between phylogenetic and morphological patterns in alpine leaf beetles attest to an intricate biogeographic history of lineages in postglacial Europe. Mol Ecol 20:2442–2463

Tshikolovets V (2011) Butterflies of Europe and the Mediterranean area. Vadim Tshikolovets Publishing, Pardubice

Ujvárosi L, Bálint M, Schmitt T, Mészáros N, Ujvárosi T, Popescu O (2010) Divergence and speciation in the Carpathians area: patterns of morphological and genetic diversity of the crane fly *Pedicia occulta* (Diptera: Pediciidae). J North Am Benth Soc 29:1075–1088

Varga Z (1975) Geographische isolation und subspeziation bei den hochgebirgs-lepidopteren der Balkanhalbinsel. Acta Entomol Jugoslavica 11:5–40

Varga Z, Schmitt T (2008) Types of oreal and oreotundral disjunction in the western Palearctic. Biol J Linn Soc 93:415–430

Vila M, Marí-Mena N, Guerrero A, Schmitt T (2011) Some butterflies do not care much about topography: a single genetic lineage of *Erebia euryale* (Nymphalidae) along the northern Iberian mountains. J Zool Syst Evol Res 49:119–132

Zhang L-B, Comes P, Kadereit JW (2001) Phylogeny and Quaternary history of the European montane/alpine endemic *Soldanella* (Primulaceae) based on ITS and AFLP variation. Am J Bot 88:2331–2345

Chapter 4
The Beginning of High Mountain Occupations in the Pyrenees. Human Settlements and Mobility from 18,000 cal BC to 2000 cal BC

Ermengol Gassiot Ballbè, Niccolò Mazzucco, Ignacio Clemente Conte, David Rodríguez Antón, Laura Obea Gómez, Manuel Quesada Carrasco and Sara Díaz Bonilla

Abstract During the last two decades, the archaeological research carried out in the Pyrenees challenged the traditional images of the past in this mountain area. The archaeological sequence of the range goes back and sites like Balma Margineda, treated until recently as an exception, now are seen as part of more global process. Actual data suggest that main valleys of the Pyrenean frequented by humans at the end of the last glacial period, with sites slightly over 1000 o.s.l. After the Younger Dryas, the human presence ascended to alpine and subalpine areas, in accordance with current archaeological data. The Neolisitation process was early in some hillsides, with intense remains of farming and pastoralism in many sites from dated in the second half of the 6th millennia cal BC. Human settlements like Coro Tracito, Els Trocs and El Sardo confirm the full introduction of agrarian activity in the central part of the Pyrenees between 5300 and 4600 cal BC. After 3500/3300 cal BC the indices oh sheepherding rises to alpine areas, with an abrupt increase of known archaeological sites in alpine areas, above the current timberline. This phenomena, as well as the signs of anthropic disturbance of the alpine environment in sedimentary sequences, suggests a more stable and ubiquitous human presence, probably largely associated with the development of mobile herding practices.

E. Gassiot Ballbè (✉) · D. Rodríguez Antón · L. Obea Gómez · M. Quesada Carrasco · S. Díaz Bonilla
Department of Prehistory, Autonomous University of Barcelona, 08193 Cerdanyola del Vallès, Bellaterra, Spain
e-mail: ermengol.gassiot@uab.cat

N. Mazzucco
Préhistoire et Technologie UMR 7055, Université Paris Ouest Nanterre La Défense, Nanterre, France

I. Clemente Conte
Institució Milà I Fontanals, Spanish National Research Council (CSIC), C/Egipcíaques 15, 08001 Barcelona, Spain

J. Catalan et al. (eds.), *High Mountain Conservation in a Changing World*, Advances in Global Change Research 62, DOI 10.1007/978-3-319-55982-7_4

Keywords Mesolithic · Neolithic · High mountain human occupations · Farming · Pastoralism · Pyrenees

4.1 Introduction

The science of archaeology is undergoing constant change. In recent decades, our understanding of human populations in the late Pleistocene and the first half of the Holocene has increased considerably, owing both to new theoretical approaches and to an increase in the number of researchers. A greater volume of information now exists, for example about the changes of hunter-gatherer communities at the end of the last Ice Age and the development of the Neolithic in SW Europe (Bahn 1983; Galop 2006; Tzortzis and Delestre 2010). In connection with the latter period, in the last thirty years a considerable amount of data has accumulated about such diverse aspects as crop types, seasonality of the settlements, the mobility of flocks, people and objects, the impact of economic practices on the environment and the appearance of signs of social inequality.

Technical and methodological innovations have actively contributed to such a development; for example, the improvements in the methods of palaeocarpological remains recollection (and of palaeobotanical remains, in general) have altered the appearance and interpretations of many archaeological sites. Another is the introduction of new technologies and techniques for studying artefacts, such as the analysis of residues on pottery and lithic implements, the petrographic analysis of raw materials, traceological studies, etc. (Mazzucco et al. 2013). Also, the application of isotopic and DNA analysis to the study of animal and human osseous remains as well as of plants has improved our understanding of the processes of domestication of some species, livestock, and human mobility patterns, etc. (Edwards et al. 2007).

A further factor is related to the expansion of research, which has noticeably increased the number of empirical prehistoric records in southern Europe. New sites have been identified, some of which are now of great importance. One example is the early Neolithic site of La Draga, on the shore of Lake Banyoles (Antolín et al. 2014). Similarly, some geographic areas have received greater attention than in previous decades. This is the case of mountain areas in southern Europe, where numerous research projects are currently in progress. These programmes have largely embraced the growing sensibility of archaeological science towards understanding economic practices in the past, their relationship with the physical environment (geomorphology, climate, fauna and flora, etc.) and how this changed over time. In this context, it is important to identify the start of human presence in the different altitudinal zones in the mountain ranges after the last Ice Age and also the appearance of significant changes in the ways of life. One of these was undoubtedly the domestication of animals and plants in the first half of the Holocene. Recent publications reveal the current relevance of this archaeology of mountain areas such as the Pyrenees (Gassiot et al. 2016; Fortó and Vidal 2016; Clemente et al. 2014b; Tzortzis and Delestre 2010).

The archaeology of mountain areas in southern Europe is undergoing a kind of revolution that is very recent and yet cannot be adequately explained by general changes in the archaeological science. What is this phenomenon due to, therefore? Several factors have converged. On the one hand, some fortuitous finds have demonstrated that high mountain areas were occupied since very early times. The paradigmatic example is the find in summer 1991 in the Tissenjoch Glacier in South Tyrol, at 3,200 m altitude, of the remains of a man who was killed by an arrow wound between 3370 and 3100 cal BC (Kutschera and Miller 2003; Spindler 1995). Ötzi, as he is colloquially known, is currently one of the best known archaeological finds in Europe.

On the other hand, growing spatial sensibility of some social sciences (Santos 2000) has jelled in an archaeology that is increasingly aware that it needs to study how human societies live in a place, in an environment that influences them and which, at the same time, they transform (Redman et al. 2004; Folke 2006). This interest has been combined with a particular perception of the uniqueness of high mountain areas, which have slowly changed from being seen as practically pristine places to being viewed as areas full of life, including human life. In a certain way, the UNO's declaration of 2002 as "International Year of Mountains" attests this change in paradigm, which is also reflected in archaeology. Evidence of this is the proliferation of multidisciplinary programmes in the mountains of south-west Europe, studying the human occupations in long sequences, their impact on alpine and sub-alpine environments and the influence of climate variations (Galop 2006; Pèlachs et al. 2007, 2017; Bal et al. 2010; Palet et al. 2012; Cunill et al. 2012). In general, these different theoretical and methodological approaches share an interest in understanding past human activities in a particular physical space, high mountains, which they have shaped over time, together with climatic and ecological factors (Angelucci et al. 2013; Calastrenc et al. 2006; Fedele 1999; Le Couédic 2010; Leveau and Segard 2006; Orengo et al. 2014; Palet 2006; Palet et al. 2007, 2012; Rendu 2003; Rey 2006; Walsh et al. 2010).

The present study is a brief summary of archaeological data obtained in recent fieldwork in high altitude areas of the Pyrenees, especially in central and eastern parts of the mountain range. To delimit the long temporal sequence of the human population that is currently being documented, this paper will be focused on the first part of the sequence. This choice is connected with some of the questions posed by archaeological research in such regions in recent years. When did human occupation of high mountain areas begin? How was the livestock farming that currently characterises these areas introduced? In what way is this process related to the neolithisation process of nearby areas? From the beginnings of the twenty-first century, new approaches and techniques have been applied to answer these questions, obtaining results, as it will be shown below, that 10 or 12 years ago would have been unimaginable to most of the archaeologists.

4.2 The Start of Modern History: The Population of the Pyrenees in the Late Pleistocene and the First Half of the Holocene

The archaeology of hunter-gatherer groups in the late Pleistocene in southern France and the Iberian Peninsula has traditionally paid considerable attention to the Pyrenean–Cantabrian region. However, this interest has focused on the foothills of the two mountain chains or, at the most, their outer ranges, relatively distant from their higher areas. Archaeological deposits with Palaeolithic remains have been known in both regions since the late nineteenth or early twentieth centuries, and many of them have been studied archaeologically (Bahn 1983; Fullola 1995). The situation in the interior of the ranges is very different, as they have not attracted the interest of researchers, perhaps because it was assumed that type of environment, especially during the cold conditions of the Upper Palaeolithic, would have been unattractive to human groups. In recent years that outlook is beginning to change (Mangado et al. 2005, 2009).

4.2.1 *Human Presence in the Valleys in the Late Pleistocene*

Evidence of human settlement in the Pyrenees in the Pleistocene is very scanty. Numerous Upper Palaeolithic sites certify the presence of hunter-gatherer groups during the last 20,000 years in the outer ranges of the Pyrenees and the foothills (Fullola et al. 2004; Utrilla and Montes 2007; Martzluff et al. 2012). Most of these archaeological deposits are situated in limestone caves and rock-shelters on the southern and northern edges of the mountain range, located at mid and low altitude. However, evidence for the penetration of these hunter-gatherer groups into the inner mountain areas are virtually absent (not even some lithic scatter or some isolated find), despite surface surveying has been carried out extensively in the Axial Pyrenees and the pre-Pyrenees since at least 10–15 years ago (Gassiot et al. 2016). This lack could be explained in several ways. One possible explanation could be that mountain areas did not attract the communities of different human species living in south-west Europe during the Pleistocene; this would have been particularly likely during the coldest periods when glaciers were much more extensive than today and covered most of the mountain zones. Another explanation is that the erosion of hillsides and valleys caused by the advance of glaciers removed the evidence of previous occupations.

Two main sites indicate human presence in the central zone of the Pyrenees during the late Pleistocene (Fig. 4.1). Both are situated in valleys. The first of these is in the open air, on a ridge in the Segre valley, in the town of Montlleó (Cerdanya) at 1140 m a.s.l. Archaeological excavations have identified two hunter-gatherer occupation phases dated between 18,360 and 16,700 cal BC (Mangado et al. 2005, 2009). They both took place in a cold humid climate that gradually ameliorated.

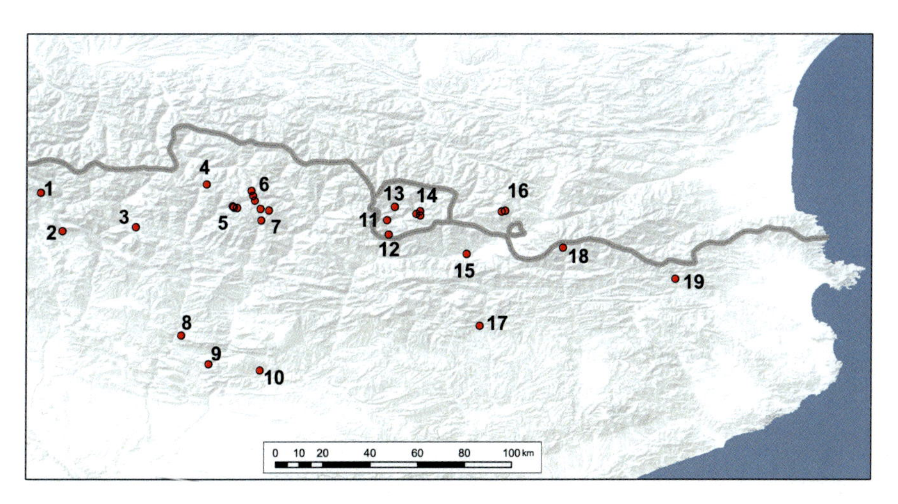

Fig. 4.1 Geographic localisation of the study area and the main archaeological sites mentioned in the text. *1* Coro Trasito, *2* La Puyascada, *3* Els Trocs, *4* Tuc deth Lac Redon, *5* Sites in the PNAESM (1): Covetes, Cova de Sarradé, Cova del Sardo, *6* Sites in the PNAESM (2): Abric del Lac Major de Saboredo II, Portarró, Obagues de Ratera, *7* Sites in the PNAESM (3): Girada Gran de Monestero, Abric de l'Estany de la Coveta I, Coma d'Espós, Dolmen de la Font dels Coms, *8* Cova Colomera, *9* Cova Gran de Santa Linya, *10* Cova del Parco, *11* Balma Margineda, *12* Juverri, *13* Cista de Segudet, *14* Sites in the VMPC, *15* Montlleó abd Sanavastre, *16* Sites in the Enveig mountains, *17* Font del Ros and Ca l'Oliaire, *18* Vall de Núria, *19* Balma del Serrat del Pont

The communities who lived there hunted mainly horses (*Equus caballus*) and to a lesser extent ibex (*Capra pyrenaica pyrenaica*) and red deer (*Cervus elaphus*). The implements found are knapped lithic fragments, with a significant representation of laminar debitage. The raw materials identified are flint from the head of the Llobregat river basin, in the southern Pre-Pyrenean range, jasper from the terraces of the River Tet in the north, and other materials from more distant sources, possibly in the Ebro valley or the Corberas and Narbonne–Sigean areas to the south and north of the range. Beads have been found made from marine mollusc shells obtained in the Mediterranean, except for one from an Atlantic source. The results are coherent with an occupation by nomadic communities who moved both north and south, possibly along the Segre valley. In this area, the pass across the mountains, Coll de la Perxa, is at 1570 m a.s.l. and may have been open at certain times of the year.

The second site is Balma Margineda (Fig. 4.1), a rock-shelter at 970 m altitude on the side of the River Valira, in Sant Julià de Lòria (Andorra). It was occupied from about 14,000 years ago to at least 7000 (Guilaine and Martzluff 1995, 2007) and thus its archaeological sequence covers the time from the Late Glacial period to the so-called Holocene Atlantic Optimum. The excavation identified a series of occupations by hunter-gatherer groups in the late Pleistocene (Levels 12–8) and early Holocene (Levels 7–4), covered by a thick Neolithic layer (Level 3).

The oldest levels in the sequence indicate mountain vegetation formed by pine and juniper, at a time when the terminus of the Valira glacier may have been at 1,300–1,400 m a.s.l. (Riera and Turu 2012). Both at that time and in later periods, the ibex was the main species hunted and consumed by the occupants of the rock-shelter. However, after 8000 cal BC, the diet was diversified, and subsistence activities included fishing trout, gathering snails and, gradually and more intensely in successive occupations, plants such as sloe, hazel and cherry. Hunting also diversified in the Holocene, and even though ibex still predominated, roe deer and wild boar were also hunted. In sum, Balma Margineda appears to have been a settlement used in different periods by hunter-gatherer groups in a context of increasingly mild climates. Changes in the groups' industry and diet follow the general patterns seen in hunter-gatherer communities in the late Pleistocene and early Holocene in Iberia (Gassiot 2001).

4.2.2 Initial Occupation of the High Zones in the Early Holocene

Unlike Montlleó, Balma Margineda displays a long occupation sequence lasting millennia. In the transition from the Pleistocene to the Holocene, the human occupations became more intense and in fact until recently this site was considered paradigmatic of the mountain occupation in the Pleistocene–Holocene transition. In the last few years, archaeological surveying in the Claror, Perafita and Madriu valleys, also in Andorra (Palet 2006) and Aigüestortes i Estany de Sant Maurici National Park (Gassiot et al. 2014) have altered this view significantly. This fieldwork has succeeded in documenting evidence of human occupations in the sub-alpine and alpine altitudinal zones, at above 1800 m a.s.l. (Fig. 4.2).

The oldest evidence is Level 5 in Dolmen de la Font dels Coms, corresponding to a possible phase prior to building the megalithic grave. This site is at the head of the Baiasca valley, 6 km to the south-east of the park boundary, at 1840 m a.s.l. and 150 m below the watershed marking the western side of the river basin. Excavated in 2002 and 2003 by the High Mountain Archaeological Group (GAAM, in Catalan), it revealed a more complex archaeological sequence than initially expected, with great re-use of the monument in later periods (Rappalino et al. 2007). The oldest occupation is related to a possible post-hole found when a trench was opened in the tumulus to document its building method. The post-hole was clearly underneath the base of the mound and was circular in cross-section, with a diameter of nearly 20 cm and about 20 cm deep (Fig. 4.3). The top of the hole was surrounded by pieces of schist, a local metamorphic rock, some of which were placed vertically, marking the section of the hole. The sediment inside it contained a large number of small pieces of charcoal, thus darkening it and making it easily differentiable from the utterly barren earth surrounding it. One piece was dated to 8746–8563 cal BC. The level containing the top of the post-hole yielded one of only two flint flakes found during the excavation of the site (Gassiot et al. 2014).

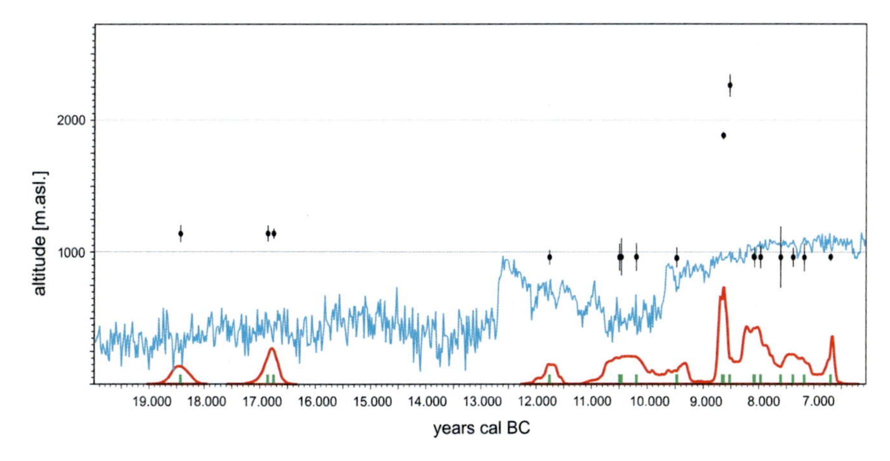

Fig. 4.2 Sum of probabilities of calibrated dates of Pleistocene and early Holocene human occupations in the axial Pyrenees. Individual dates are plotted in the x-axis (*green bars*). *Black circles* also indicate the individual dates with their standard deviation and the altitude of the site (in the *left* y-axis). Greenland Ice Core Chronology 2005—GICC05 (Andersen et al. 2006; Rasmussen et al. 2006) is indicated in *blue*

Fig. 4.3 Plant of the Dolmen of the Font dels Coms. The picture shows the open trench in the tomb. In *red* the hole that probably corresponds to a pole

A similar find was made in the medieval barn P009 at the Orrís de Perafita II site in Andorra. This site is in the Perafita valley, at 2,275 m altitude, in an area of pastures and bogs. The excavation of the architectonic structure showed that, below the walls visible on the surface, several occupation levels contained prehistoric remains. The lowest layer, Level 107, yielded flakes in local metamorphic rocks. This level was dated to 8764–8532 and 8517–8478 cal BC. The strata above it demonstrated the re-occupation of the site in later prehistoric periods. Lithic materials associated with this occupation are preliminarily classifiable as Notches and Denticulates Mesolithic (Orengo et al. 2014).

The third site with remains older than the Neolithisation process in the Pyrenees is Abric de l'Estany de la Coveta I (Fig. 4.4). This is a small rock-shelter formed by a mound of large granitc boulders left by glacial activity in the late Pleistocene, on a small hill between two lakes, at 2430 m a.s.l. The interior of the shelter, with a surface area of a little over 5.5 m^2, was occupied at several times in the past

Fig. 4.4 **a** General image of the Abric de l'Estany of Coveta I, located in an area of large blocks accumulated during the last glacial period. **b** Detail of the east section of the excavation. The *solid red line* marks the Mesolithic occupation of the shelter. *Dotted lines* indicate the Neolithic occupation and the historical period (late medieval or medieval), respectively

(Gassiot et al. 2014). The oldest phase of use was identified when a hearth was found in the place where the inner space connects with the small entrance passage. A pine charcoal fragment from this hearth, which was more or less circular in shape and a little over half a metre in diameter, was dated to 7001–6574 cal BC, over 3000 years before the Neolithic occupation.

A small quartz flake with thermal domes was found inside the hearth. Outside, in the layer that identified the occupation level, a further three chipped lithic fragments were found; together with the charcoal these form the only remains corresponding to this phase. All four lithic objects in the old level are small flakes less than 2 cm in size. Three of them were made in a dark brown flint that very probably comes from the Ebro Basin Oligocene–Miocene lacustrine formations. Microscopic study of one piece identified small notches and the characteristic polish caused by working dry hide. The action was clearly transversal and may indicate work connected with scraping the hide. As this small scraper is so small, it is likely that it would need to be hafted, otherwise it would have difficult to operate with it (Fig. 4.5).

Although the state of conservation of the charcoal in the hearth was quite poor, a sample was analysed. It practically all came from the wood of *Pinus*

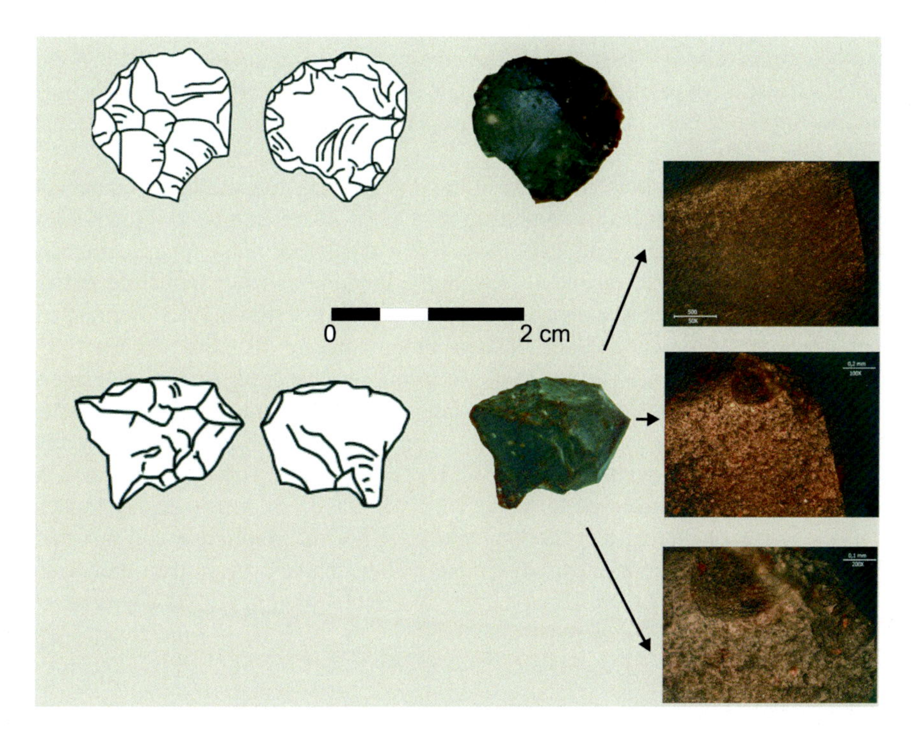

Fig. 4.5 Two of the lithic pieces recovered in the Mesolithic occupation of the Abric de l'Estany of Coveta I. At the *top* appears the photograph and drawing of the flint. At the *bottom*, the scraper shows the fingerprints of use from its distal edge, which was related to the processing of skin. From *top* to *bottom* 50×, 100× and 200×, respectively. In the sketches: *left*, the dorsal side; *right*, the ventral. Photographs and drawings by Virginia García

sylvestris/uncinata, thus suggesting the possible presence of arboreal vegetation in the proximity of the site. The growth rings displayed undulations caused by the pressure of the weight of snow. The rings were also very narrow, attesting the slow growth of trees in highland areas (Gassiot et al. 2010).

4.3 The Arrival of the Neolithic: The Pyrenees as Farming Land

Several centres of the initial domestication of animals and plants are now known on the different continents. Of these, the one that led to the adoption of domesticated species in the whole Mediterranean and Europe, in general, was located in the Near East. The gradual domestication of cereals, legumes, pigs, cattle and ovicaprids took place there about 10,000 or 9000 cal BC (Edwards et al. 2007; Haak et al. 2010; Zeder 2011). In the following 3000 years, these species rapidly spread beyond the "Fertile Crescent" to the Balkans, the Danube basin, the shores of the Mediterranean and North Africa, etc. The study of this process currently focus on: determining the time and mechanisms in which this expansion took place; the social and economic changes involved for the communities participating in the process; the relationship between groups of migrants or "colonists" and local communities; and the impact caused on the natural environment (Edwards et al. 2007; Haak et al. 2010; Zeder 2011).

Within these parameters, the study of the first agrarian communities in mountain areas, and in the Pyrenees in particular, is of great historical and anthropological importance. Until a few years ago, there was a somewhat non-explicit consensus that the range, like other mountain areas in the Iberian Peninsula, remained mostly on the sidelines of the first expansion of farming activities (Bahn 1983; Bertranpetit and Vives 1995; Jiménez 2006; Llovera 1986; Yáñez et al. 2002; Yáñez 2005; Orengo et al. 2014, Walsh et al. 2005). According to this concept, the adversity of mountain environments for the establishment of agriculture and their isolation would have marked the rate and form of the process, which to a great extent would have been much less intense and later than in the foothills. However, as it will be shown below, new data that is being collected thanks to recent archaeological projects is demonstrating that the Pyrenees were not marginal areas and that their Neolithisation has been a much more rapid and extensive process than previously thought.

4.3.1 The First Neolithic Occupations in the Pyrenees

The oldest evidence of prehistoric arable farming and animal husbandry in the Pyrenees and surrounding area is dated in the first half of the Holocene, at around

5500 cal BC. Practically all the known sites are situated outside the central part of the mountain range. The most representative are Cueva de Chaves in the Sierra de Guara (Aragon), the open-air settlement of Font de Ros (Berga, Catalonia) and Dourgne II rock-shelter, at the head of the Aude Valley, in France. All of these are in areas at a medium altitude, in the valleys, at between 200 and 700 m a.s.l. (Guilaine et al. 1993).

At this time, Cueva de Chaves was a stable settlement, characterised by its mixed subsistence based on crops and livestock, complemented with small contributions from hunting and gathering (Mazzucco et al. 2015). It is an ancient settlement, with a series of occupations dated from 5700 to 5000 cal BC, attesting an early Neolithisation of the outer ranges in the Pre-Pyrenees (Baldellou 2011; Mazzucco 2014). The cave probably mainly served as a place of human habitation as testified by the several hearths, pits and the other dwelling structures recovered. Palaeoenvironmental data suggest that human activity altered the surroundings of the cave by clearing land for pastures and crops in an area of mixed pine and oak woodland (López–García and López–Sáez 2000). The abundant archaeozoological evidence shows that domestic ovicaprids, and pigs and cattle in smaller proportions, were consumed inside the cave.

The small rock-shelter of Dourgne II is located at 710 m altitude in Roc de Dourgne. The excavations in the late 1970s and early 1980s revealed an occupation sequence dated between 6700 and 3500 cal BC (Guilaine et al. 1987). The start of the Neolithic occupations was dated to 5300 cal BC. Several hearths and areas for lithic debitage were documented, in addition to silos re-used as waste pits. The subsistence evidence indicates a predominance of the use of domestic and wild fauna: together with livestock of ovicaprids and pigs, hunting wild boar was important in the different occupations. Unlike Cueva de Chaves, human activity did not significantly alter the environment of mixed woodland (*Pinus sylvestris*, *Corylus* sp., *Juniperus* sp., *Taxus* sp., *Tilia* sp. and *Prunus* sp.). Evidence of farming activity in the surrounding area was not documented either.

The site of La Font de Ros is situated on the outskirts of Berga (Barcelona), in the contact zone between the outer Pre-Pyrenean ranges and the central Catalan depression, at about 650 m altitude. Remains of several dwellings and storage structures, belonging to both the Mesolithic and Neolithic, were documented in an area of over 1000 m^2. The latter were dated from 5500 to 4800 cal BC. The conservation of osseous remains was very poor although some domestic species were identified. In contrast, domestic cereals were recovered, mainly dressed barley (*Hordeum vulgare*) and emmer (*Triticum dicoccum*). Other sites dated in the early Neolithic are Balma del Serrat del Pont (in Girona) (Alcalde et al. 2005) and the crock shelters of Hueso Raso and Forcas I and II (Utrilla and Mazo 2007). At Forcas II was found what might be the earliest evidence of pottery in the Pre-Pyrenees, in a stratum dated between 5990 and 5670 cal BC. However, proof of a Neolithic economy, with signs of domesticated animals and plants, did not appear until about 5500 cal BC.

Until recently, Balma Margineda was the only site dated in the sixth-millennium cal BC in the central Pyrenees (Guilaine et al. 1985). Excavations in the rock-shelter documented several early Neolithic occupations dated from 5700 to 5500 cal BC, overlying the Mesolithic levels (Guilaine et al. 1995). These are surprisingly old, as they are also linked with clear evidence of domesticated goats and carpological remains of domestic cereals, as well as storage structures, hearths and pottery. The rock-shelter appears to have been used by communities that were "fully Neolithic", in the heart of the mountain range, at a very early date for the Neolithic in the whole of Iberia. This finding has provoked archaeological debates in which a possible route for the Neolithisation of the Iberian Peninsula through the Segre Valley has been proposed. However, oldest dates have often been questioned (Bernabéu 2006; Morales et al. 2010; Mazzucco 2014); a recent and detailed revision of the radiocarbon, stratigraphic and material culture data from the Balma Margineda, suggests that the first Neolithic occupation of the site should be dated not earlier than about 5500–5300 cal BC (Oms et al. 2016). These new dates appear more coherent with the chronological framework currently available for the Pyrenean area (Table 4.1).

Several recent finds have substantially confirmed this interpretation. On the one hand, archaeological surveying in high areas has located sites dated in the first half and middle of the sixth-millennium cal BC at over 1800 m a.s.l. On the other, the increase in research is revealing the existence of sites dated as well towards the last third of the sixth-millennium cal BC at between 1400 and 1600 m a.s.l. In the latter case, the most surprising aspect is that the existence of crops and livestock is widely found at these sites.

Two of the locations in the first group are SU 110 beneath Hut P009 at the site of Orris of the Perafita I peat bog, and Phase 9 at the Cova del Sardo. At the first of these, above a stratum with Mesolithic remains, the excavation under the medieval barn revealed a more recent prehistoric level that also contained some flakes. No structures were associated with these, and neither were potsherds or archaeozoological and archaeobotanical remains found. It may be a level representing visits to the area over which the buildings visible on the surface were built. A determination on a piece of charcoal dated the formation of this level between 5615 and 5475 cal BC (Orengo et al. 2014: 143). The second site, Cova del Sardo, is at 1,790 m a.s.l in the Sant Nicolau Valley, one of the central transversal valleys in the National Park. It is a small rock-shelter with a surface area of 19.3 m^2 produced by glacial erosion of the granite bedrock. It is 60 m above the modern river bed on the lower part of the hillside (Fig. 4.6). The deposit was excavated from 2006 to 2008 (Gassiot et al. 2014) when a first occupation defined by a small hearth in a pit dug in the sediment was documented above a large deposit of till probably resulting from the collapse of lateral moraines in the late Pleistocene or early Holocene. Chipped lithic fragments were found in association with the hearth, but no potsherds or any other type of archaeological object. A piece of charcoal from the hearth was dated to a time almost contemporary with the Andorran site; between 5609 and 5376 cal BC.

In neither of these two cases is there any evidence linking the occupations with farming activities, although this might be explained by the small size of the deposits and poor conservation of the osseous remains at both sites. However, this contrasts

Table 4.1 List of 14C dates from archaeological sites in the axial Pyrenees. All results have been calibrated using Oxcal 4.2 with the INTCAL13 curve

Lab. code	Site	Type	Intervention	Material	Result (bp)	±	Calibrated age 95.4% (cal BC)
Poz-28427	Els Estanys	Hut M176	Test pit	Charcoal	3685	30	2192–1970
Poz-32018	Els Estanys	Fenced M177	Test pit	Charcoal	3755	35	2285–2037
Poz-18812	Pleta de Bacives I	Fenced M151	Test pit	Charcoal	3755	35	2285–2037
Poz-32023	Els Estanys	Hut M217	Test pit	Charcoal	3760	35	2287–2039
Ly-8223	La Padrilla	Hut 49	Digging	Charcoal	3810	40	2458–2062
Poz-32017	Els Estanys	Fenced M218	Test pit	Charcoal	3885	35	2469–2212
Poz-28426	Els Estanys	Hut M175	Test pit	Charcoal	3885	30	2467–2284
KIA-32335	Cova de Sarradé	Cave	Test pit	Charcoal	3945	35	2567–2306
KIA-32341	Abric de les Covetes	Shelter	Test pit	Charcoal	3960	30	2569–2345
Beta-290113	Abric del Lac Major de Saboredo II	Shelter	Test pit	Charcoal	4010	40	2830–2459
KIA-32348	Cova del Sardo	Cave	Digging	Charcoal	4090	35	2862–2493
Poz-22584	Planells de Perafita I	Sòl P067	Test pit	Charcoal	4105	35	2864–2501
KIA-28280	Obagues de Ratera	Shelter	Test pit	Charcoal	4160	35	2878–2626
KIA-36936	Coma d'Espós	Hut 1	Test pit	Charcoal	4180	30	2885–2664
KIA-26251	Cova del Sardo	Cave	Test pit	Charcoal	4210	35	2899–2675
KIA-28276	Abric del Portarró	Shelter	Test pit	Charcoal	4255	40	3005–2692
Ly-6242	La Padrilla	Hut/Fenced 42	Digging	Charcoal	4370	30	3331–2885
Beta-377578	Tuc deth Lac Redon	Hut 2	Test pit	Charcoal	4400	30	3260–2915
Beta-316510	Els Trocs III	Cave	Digging	Seed	4410	40	3325–2913
Poz-22580	Orris de la Torbera de Perafita I	Hut P008	Test pit	Charcoal	4415	30	3312–2917
Poz-32012	Planells de Perafita I	Fenced P169	Test pit	Charcoal	4425	30	3321–2922
Poz-22561	Riu dels Orris I	Sòl M085	Test pit	Charcoal	4445	35	3333–2929
KIA-40850	Cova del Sardo 2	Cave	Test pit	Charcoal	4465	30	3335–3022
KIA-29816	Abric de l'Estany de la Coveta I	Shelter	Digging	Charcoal	4475	30	3337–3027
Mams-16167	Els Trocs III	Cave	Digging	Human bone	4512	25	3350–3101
Ly-7064	La Padrilla	Hut 75	Digging	Charcoal	4550	60	3497–3028
KIA-32351	Cova del Sardo	Cave	Digging	Charcoal	4555	30	3481–3103
KIA-37691	Cova del Sardo	Cave	Digging	Charcoal	4715	35	3630–3373
KIA-36934	Cova del Sardo	Cave	Digging	Charcoal	4765	40	3639–3379

(continued)

Table 4.1 (continued)

Lab. code	Site	Type	Intervention	Material	Result (bp)	±	Calibrated age 95.4% (cal BC)
Poz-22579	Orris de la Torbera de Perafita I	Hut P008	Test pit	Charcoal	4905	35	3763–3637
KIA-32342	Cova del Sardo	Cave	Digging	Charcoal	4945	35	3789–3650
KIA-40816	Cova del Sardo	Cave	Digging	Charcoal	5000	30	3937–3700
Mams-14856	Els Trocs III	Cave	Digging	Human bone	5005	27	3938–3706
Mams-16160	Els Trocs III	Cave	Digging	Human bone	5008	23	3933–3709
Mams-16165	Els Trocs III	Cave	Digging	Human bone	5035	23	3950–3760
KIA-26248	Cova del Sardo	Cave	Test pit	Charcoal	5060	40	3961–3764
KIA-32340	Cova del Sardo	Cave	Digging	Charcoal	5245	40	4227–3969
Beta-319513	Els Trocs II	Cave	Digging	Seed	5580	40	4490–4340
Beta-316511	Els Trocs II	Cave	Digging	Seed	5590	40	4500–4340
Beta-316515	Els Trocs II	Cave	Digging	Seed	5590	40	4500–4340
KIA-40815	Cova del Sardo	Cave	Digging	Charcoal	5635	35	4537–4367
KIA-41134	Cova del Sardo	Cave	Digging	Charcoal	5645	25	4541–4373
Poz-18807	Pleta de Bacives I	Hut M152	Test pit	Charcoal	5660	40	4590–4368
KIA-40817	Cova del Sardo	Cave	Digging	Charcoal	5685	35	4649–4447
KIA-36935	Cova del Sardo	Cave	Digging	Charcoal	5695	35	4651–4452
KIA-40878	Cova del Sardo	Cave	Digging	Charcoal	5715	35	4679–4461
CNA-2520.1.1.	Coro Trasito	Cave	Test pit	Seed	5830	35	4788–4590
KIA-37690	Cova del Sardo	Cave	Digging	Charcoal	5850	40	4824–4600
Beta-358571	Coro Trasito	Cave	Test pit	Charcoal	5990	40	4992–4786
Beta-316514	Els Trocs I	Cave	Digging	Seed	6050	40	5056–4836
Beta-295782	Els Trocs I	Cave	Digging	Bone	6060	40	5195–4842
Beta-284150	Els Trocs I	Cave	Digging	Seed	6070	40	5202–4844
Beta-316512	Els Trocs I	Cave	Digging	Seed	6080	40	5206–4847
Beta-366546	Coro Trasito	Cave	Test pit	Bone	6150	40	5216–4993
Mams-16161	Els Trocs I	Cave	Digging	Human bone	6217	25	5294–5066
Mams-16162	Els Trocs I	Cave	Digging	Human bone	6218	24	5294–5068
Mams-16166	Els Trocs I	Cave	Digging	Human bone	6234	28	5303–5075
Mams-16164	Els Trocs I	Cave	Digging	Human bone	6249	25	5310–5080
Mams-16168	Els Trocs I	Cave	Digging	Human bone	6249	28	5310–5078
CNA-2944,1,1	Coro Trasito	Cave	Test pit	Seed	6269	33	5323–5081
Mams-16159	Els Trocs I	Cave	Digging	Human bone	6280	25	5315–5215
Mams-16163	Els Trocs I	Cave	Digging	Human bone	6285	25	5315–5215

(continued)

Table 4.1 (continued)

Lab. code	Site	Type	Intervention	Material	Result (bp)	±	Calibrated age 95.4% (cal BC)
KIA-37689	Cova del Sardo	Cave	Digging	Charcoal	6525	45	5607–5374
Beta-285100	Orris de la Torbera de Perafita I	Hut P009	Test pit	Charcoal	6570	40	5613–5473
Ly-3288	Balma Margineda	Cave	Digging	Charcoal	6640	160	5885–5305
Ly-2839	Balma Margineda	Cave	Digging	Charcoal	6670	120	5806–5375
Ly-3290	Balma Margineda	Cave	Digging	Charcoal	6820	170	6040–5473
Ly-3289	Balma Margineda	Cave	Digging	Charcoal	6850	160	6023–5487
KIA-29818	Abric de l'Estany de la Coveta I	Shelter	Digging	Charcoal	7845	45	7001–6574
Ly-2840	Balma Margineda	Cave	Digging	Charcoal	8390	150	7724–7060
Ly-4402	Balma Margineda	Cave	Digging	Charcoal	8960	120	8447–7725
Ly-2842	Balma Margineda	Cave	Digging	Charcoal	9250	160	9236–8007
Poz-22583	Orris de la Torbera de Perafita I	Hut P009107	Test pit	Charcoal	9360	50	8764–8478
KIA-23142	Dolmen de la Font dels Coms	Open air	Digging	Charcoal	9375	35	8746–8563
Ly-3884	Balma Margineda	Cave	Digging	Charcoal	9900	110	9866–9174
Ly-4403	Balma Margineda	Cave	Digging	Charcoal	10340	130	10633–9692
Ly-3364	Balma Margineda	Cave	Digging	Charcoal	10630	190	11005–10031
Ly-2843	Balma Margineda	Cave	Digging	Charcoal	10640	260	11135–9767
Ly-5418	Balma Margineda	Cave	Digging	Charcoal	11230	170	11450–10800
Ly-4407	Balma Margineda	Cave	Digging	Charcoal	11320	120	11494–11006
Ly-4896	Balma Margineda	Cave	Digging	Charcoal	11690	90	11786–11393
Ly-4898	Balma Margineda	Cave	Digging	Charcoal	11870	110	12035–11517
OxA-9017	Montlleó	Open air	Digging	Charcoal	14440	80	15925–15415
OxA-14034	Montlleó	Open air	Digging	Molar	15550	140	17185–16576
OxA-X2234/52	Montlleó	Open air	Digging	Molar	16900	110	18717–18124

The source of the dating are mentioned in the main text for every site

Fig. 4.6 General image of the Cova del Sardo. Above, at the time of its location in 2004. *Bottom*, in the final phase of its excavation, with the bare levels of the 5th Millennium cal BC outside the cavity

with the evidence found at Coro Trasito and Els Trocs. These two sites, of which the second is paradigmatic, are in the Pyrenees in Huesca, in Sobrarbe and the Benasque Valley (Ribagorza), respectively. The first is a west-facing entrance of a cave filled with Holocene sediments at 1548 m a.s.l. It is now a large rock-shelter

Fig. 4.7 Location and detail of the entrance of the cave of Coro Trasito

with a surface area available for human occupation of over 300 m^2 (Figs. 4.7 and 4.8) A test excavation of 2.5 × 1.5 m performed in two stages in 2011 and 2103 revealed a thick sequence of *foumier* sediments (burned layers of animal wastes mixed with soil) dated from 5300 to 4600 cal BC (Clemente et al. 2014a, 2016). In this time, the cave entrance was used as a stable for livestock, for storage in small silos and as a dwelling. The accumulation of manure was so high that the rapid sedimentation blocked the passage to the inner part of the cave. The site was apparently abandoned in the mid fifth-millennium cal BC and not occupied again until nearly three millennia later, in about 1400 cal BC.

Both the test and the proper excavation, currently in progress, have recorded a large variety of archaeological remains. These include a considerable amount of

Fig. 4.8 Central section of the interior of Coro Trasito cornice, where excavations are currently being carried out. The survey of the year 2013 can be observed

pottery recipients in different sizes and shapes. While lithic objects are not very numerous, they reveal the use of raw materials from the Ebro basin, among other origins (Mazzucco et al. 2014). The type of sediment has enabled good conservation of osteological remains, resulting in the identification of a vast number of fragments of the animals consumed at the site. Their study is still in progress. The available radiocarbon dates show that in the oldest occupation phase currently known (5300–5100 cal BC), 90% of the remains that have been determined taxonomically come from domestic species: sheep, goat, pig and cattle. This proportion decreases in following Neolithic occupations to 70%. In the second millennium occupation cited above, the percentage of domestic fauna is still smaller; 55% of the identified remains. The wild animals consumed were mostly tortoise, red deer, roe deer and rabbit, among others.

During the test excavation a lot of carpological remains has been noted, despite sampling a relatively small amount of sediment; following the same trend as the fauna, the largest numbers of cultivated species and seeds came from the oldest occupation phase at the site. Naked barley (*Hordeum vulgare* var. *nudum*), free-threshing wheat (*Triticum aestivum/durum/turgidum*), and emmer (*Triticum* cf *dicoccum*) were documented as well as some hazelnuts and acorns. In later Neolithic levels, between 5000 and 4600 cal BC, the total number of seeds of domestic cereals decreases, but peas (*Pisum sativum*) have been found in some occupations and, above all, intense consumption of hazelnuts and, in some cases, of blackberries (*Rubus fructicosus*) has been documented in the most recent levels.

In short, the first occupants of Coro Trasito practised a subsistence economy in which animal husbandry and crops played a significant role in the supply of food. Their material culture is also comparable with that of other "Neolithic" deposits of the same period in less mountainous regions.

Another site excavated recently which has altered the view of the Neolithic in the Pyrenees is Cova dels Trocs (Sant Feliu de Veri, Bisauri) (Fig. 4.1). It is at 1530 m a.s.l in a small hill in the plateau between the Turbón Massif in the south and the peaks of Vallivierna to the north (Rojo et al. 2013 and 2014). It is an elongated cave, 15 m long and 6 m wide, with a narrow entrance. Inside, the excavations performed since 2009 have documented three Neolithic occupation phases dated in 5300–4850, 5050–4350 and 3940–2915 cal BC. In the oldest phase, the cave floor was paved with cobblestones and occasionally with large potsherds. This means that the number of sherds retrieved, especially in this occupation, is quite spectacular and comes to tens of thousands. Large fires were lit on these pavements, and these have left large stains of ashes. The identified bone remains include a large number of domestic species: pig, sheep, cattle and goats, which form a large part of the total. Domestic cereal grains (wheat and barley) were also recovered, although the excavators tend to believe that they were introduced from outside the area by the occupants of the cave, who were mainly transhumant shepherds. Whatever the economic model was, the deposit in Cova dels Trocs is another example of an entirely "Neolithic" site in the central Pyrenees above an elevation of 1500 m in the late sixth-millennium cal BC (Rojo et al. 2013). Supporting this impression, in a palaeoecological test sounding outside the cave, Uría (2013) documented the clearance of woodland possibly to create pastures between 5500 and 5400 cal BC.

Occupations with "Neolithic" evidence at higher altitudes are dated some centuries later. At the Cova del Sardo (Fig. 4.1), in the Aigüestortes i Estany de Sant Maurici National Park, the first potsherds and remains of domestic cereals appear in the second occupational phase, called Phase 8 (Gassiot et al. 2012 and 2014). This phase is formed by a series of repeated occupations of the rock-shelter, from about 4800 to 4400 cal BC; during this period a small terrace was built in front of the cave, a combustion area where pine and juniper woods were repeatedly burnt. Moreover, two hearths were lit under the roof of the shelter. A single grain of barley (*Hordeum vulgare*) was recovered from the outer terrace, together with seeds of spontaneous plants (*Galium aparine*) and a few fragments of hazelnut shells. The pollen record inside the rock-shelter is indicative of relatively open vegetation, with a low proportion of arboreal pollen and taxa of secondary woodland and shrubs, contrasting with the situation during the rest of the occupations in the rock-shelter until the Middle Ages. Unfortunately, the acidity of the sediment has deteriorated faunal remains and only the presence of ovicaprids, and medium-sized mammals have been documented among the animals consumed. Potsherds are scarce, like the lithic assemblage. However, this shows that the implements were nearly always made away from the site and a considerable part of the raw materials was not from the immediate surroundings of the rock-shelter. Some of the flints came from the Ebro valley, as was also the case at Els Trocs (Mazzucco 2014).

Another site that may have been occupied at a similar time is Covetes rock-shelter, near Cova del Sardo and also in the low part of the hillside in the Sant Nicolau Valley, at 1870 m a.s.l. A test excavation documented two phases of prehistoric occupations with pottery. The older one was covered by a thick barren layer, which was overlain by the more recent phase, dated to the early third-millennium cal BC (Gassiot et al. 2014).

However, for this first part of the Neolithic, evidence of occupation in the higher sub-alpine and alpine zones is practically absent. The only clear evidence is at the Cova del Sardo, in the bottom of a large valley and away from the cirques at the heads of the valleys and distant from the current timberline. In the Madriu valley in Andorra, Orengo (2014) cites a possible human occupation beneath a late Roman hut at the site of Pleta de Bacives I. However, the context is unclear, and the author provides no further information apart from the date. In any case, it does not alter the general view of the significant Neolithisation of the zone between 1500 and 1600 m a.s.l. in the late sixth-millennium cal BC, contrasting with lesser activity at higher altitudes in steeper valleys in the proximity of the modern alpine meadows.

4.3.2 Approaching the Heights in the Late Neolithic

In the fifth and fourth millennia, the presence of farming communities in medium-altitude mountains in the Pre-Pyrenees and Pyrenees becomes more intense and evident. Deposits are found in large caves, smaller sites and even in open-air structures. In the high Pyrenees, in addition to Coro Trasito (Clemente et al. 2014a, 2016) and Els Trocs (Rojo et al. 2013; Lancelotti et al. 2014), several sites are known at Juverri in Andorra, where numerous silos and evidence of farming activity have been documented (Fortó and Vidal 2016), all of them above 1200 m a.sl. At greater heights, the human use of Cova del Sardo continued in the fourth millennium, with an occasional construction outside the rock-shelter and the occupation of an adjacent rock-shelter at some point (Gassiot et al. 2014). In the Cerdanya valley, the remains of an open-air settlement dated in the second half of the fifth millennium have been documented at Sanavastre (1080 m a.s.l.). In Sierra Ferrera, in the Pre-Pyrenees of Aragon, the cave of Espluga de la Puyascada (1315 m a.s.l.) revealed occupations of a similar chronology (Baldellou 1987; Mazzucco et al. 2013). In the same way, numerous caves in the limestone areas in the outer mountain ranges of the southern Pre-Pyrenees were also used by human groups at this time, more intensely than in the early Neolithic.

A definite change took place in the human occupation of high mountain zones. This phenomenon has been documented very recently (Orengo et al. 2014; Rendu 2003) and has now begun to be addressed as a historical problem requiring an explanation (Gassiot et al. 2012, 2014). It can be seen in several aspects. First, the number of archaeological sites increases noticeably. In the area of Aigüestortes i Estany de Sant Maurici National Park, nine sites are known with dates in the centuries following 3400 cal BC (Gassiot et al. 2014, 2016), of which one site,

Cova del Sardo, contains two successive phases. This number contrasts with the single site dated in the previous phase of the Neolithic. The situation is similar in other high zones of the mountain range where systematic archaeological surveying has been carried out. In the Madriu-Perafita-Claror valley, six different sites are known for the period from 3300 to 2000 cal BC (Orengo et al. 2014). In the Enveig Mountains, in eastern Cerdanya, the three oldest occupations, all in the area of La Pradilla, are dated between 3500 and 2000 cal BC. Second, this increase in the number of sites is also seen in their altitude. They are all above 2000 m a.s.l., except for three in Aigüestortes i Estany de Sant Maurici National Park, which are at between 1780 and 1980 m. It, therefore, seems that this phenomenon can be extrapolated to at least the eastern half of the mountains.

In the Aigüestortes—Sant Maurici area, this increase in the number of occupations and their altitude takes place after 3400–3300 cal BC and becomes more noticeable after 3000 cal BC (Gassiot et al. 2014). The excavation at Abric de l'Estany de la Coveta I succeeded in documenting an occupation phase after the Mesolithic, in which a large hearth with abundant burnt pine wood was documented. Additionally, small soundings performed during surveying and surface archaeological documentation has identified the occupation of other rock-shelters between 3000 and 2400 cal BC. These are Abric del Portarró (2284 m) (Fig. 4.9), Abric del Lac Major de Saboredo II (2367 m) and Abric d'Obagues de Ratera (2312 m) (Fig. 4.10). The same period is documented in a hearth underneath Hut 2 in a settlement dated in Roman times, Despoblat del Tuc deth Lac Redon (2411 m) and in the construction with the base of a stone wall and about 30 m^2 of deposit at

Fig. 4.9 General view of the Portarró archaeological site that contains several shelters with fencing and walls of the historical period scattered between the scree. In the enlarged image, the Abric del Portarró with an occupation of the end of the Neolithic

Fig. 4.10 Image of the site (and detail, in the enlargement) of the Abric d'Obagues de Ratera

Coma de Espós (2230 m). In lower altitudes, apart from the Cova del Sardo, occupations have been documented in Covetes and Cova de Sarradé (1980 m).

At the Cova del Sardo, the occupations inside and outside the rock-shelter continued in the fourth millennium and first half of the third millennium (Gassiot et al. 2014). This situation is quite different from the pattern at higher sites which seem to have been occupied in a single phase. The sequence at Cova del Sardo appears to reflect periods when the occupation was more stable alternating with others when the site may have been occupied in passing. Potsherds increase in number, in particular between 3900 and 3500 cal BC. The lithic assemblage displays some continuity in the parameters of the previous phase, although an increase in the use of allochthones chert types is seen (Mazzucco 2014). These lithologies came from the outer river basins in the Pre-Pyrenees and the Ebro valley. Occasional findings also indicate the possible presence of raw materials from the Rhone valley, suggesting that the site was integrated within a broader context of people mobility and artefacts and raw materials exchange. In general, it can be observed that the circulation of these lithic objects is mainly organised on south-north axis, through the valleys of the Segre's affluents. Crossing mobility pattern with faunal and palaeoecological data, it seems that some form of mobile herding was carried out involving a movement of flocks and people from plain areas and valley bottoms to high-altitudes. It has often remarked that such a pattern of mobility (Geddes 1983; Rojo et al. 2014) is similar to the movements of transhumant flocks in historical times, although such issue should be further investigated to understand the differences between prehistoric and medieval/modern herding models, on a social, economic and demographic level. On the basis of current data, there is no evidence for herding specialisation during Pyrenean Neolithic, but it seems that pastoral activities were integrated within a mixed-farming system, in which both crop and animal husbandry were practised to some extent.

Further east in the Pyrenees, this increase in human settlement in the second half of the fourth-millennium cal BC is also seen in the Enveig mountains. In Level 2 in Hut 75, at the site of La Padrilla, Rendu (2003) documented an occupation dated between 3500 and 3030 cal BC. This stratum consisted of dark sediment with ash and charcoal which was followed over 11 m^2 under a construction visible on the surface and adjacent to some large boulders. The excavation was able to document four hearths delimited by stone circles or semicircles about 40 cm in diameter. This concentration of hearths might be the result of a repeated occupation of the site during a time that is still unknown. The use of the site in the late Neolithic was confirmed by another occupation. This is a stratum with abundant charcoal underneath Hut/Pen 42, 35 m away from the previous deposit. A piece of charcoal was dated to 3330–2885 cal BC.

In the Madriu-Perafita-Claror valley, the pattern is similar, with a clear increase in the number of archaeological sites after mid fourth-millennium cal BC. One of these is the level with charcoal identified underneath the early medieval hut P008 at Orris de la Torbera de Perafita I and which two radiocarbon determinations have situated between 3765–3640 and 3315–2920 cal BC (Orengo et al. 2014). In the same valley, at the nearby site of Planells de Perafita I, at 2223 m a.s.l., the same team of archaeologists identified the base of a stone wall around an area with an irregular shape, over 4 m in diameter, which they dated between 3325 and 2925 cal BC. Potsherds were found in the layer associated with this occupation.

In general terms, in the three areas of the Pyrenees studied here, the archaeological evidence tends to disappear in the second half of the third-millennium cal BC (Fig. 4.11). The exception is the site of Els Estanys, in the Madriu Valley, at 2525 m a.s.l. It consists of an architectonic ensemble composed of four rooms, a pen and a surrounding wall. The five test excavations performed indicate an occupation between 2470–2290 and 2195–1970 cal BC. A grain of wheat (*Triticum dicoccum*) was found in one of the rooms, next to a hearth which had pottery associated with it (Orengo et al. 2014). A little over 1 km away, at the site of Pleta de Bacives (2528 m a.s.l.) an animal pen has been dated to the same period as the Els Estanys pen; 2284–2041 cal BC.

Following the occupation of these two sites, evidence of human occupation practically disappears from the Madriu-Perafita-Claror valley during the second and first millennia cal BC (Orengo 2010). The situation is similar in Aigüestortes i Estany de Sant Maurici National Park. Although some mid-second millennium sites are known, the number of settlements that have been identified is considerably smaller (Gassiot et al. 2014, 2016). A similar pattern has been recognised in the Enveig area (Rendu 2003). It seems that this phenomenon can be extrapolated to other parts of the Pyrenees, although it is in contrast with some palaeoecological evidence that indicates enhanced human impact on the environment. Still, we are not able to fully explain such discrepancy; possibly it is due to a limit in the survey

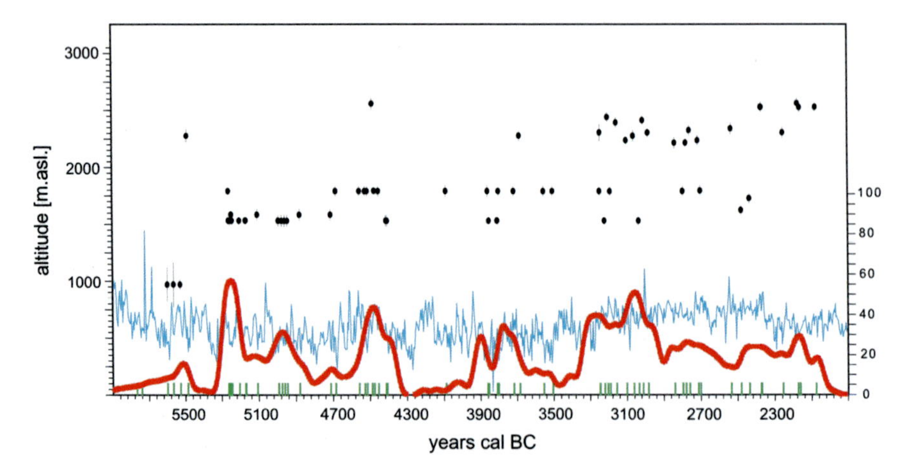

Fig. 4.11 Sum of probabilities of calibrated dates of Neolithic archaeological sites in the axial Pyrenees. Individual dates are plotted in the x-axis (*green bars*). *Black circles* also indicate the individual dates with their standard deviation and the altitude of the site (in the *left* y-axis). Greenland Ice Core Chronology 2005—GICC05 (Andersen et al. 2006; Rasmussen et al. 2006) is indicated in *blue*

techniques adopted, or it might be influenced by the reoccupation over time of the same site, making difficult to discern prehistoric phases from medieval and modern ones, especially in open-air sites. However, a detailed analysis of this aspect is beyond the scope of the present chapter.

4.4 Conclusions

This study can be summed up briefly because, in a certain way, in itself, it has presented a series of main findings of the research carried out in high areas of the Pyrenees during the last 15–20 years. The archaeological records that fieldworks have generated have been described in chronological terms; Upper Palaeolithic presence in the high mountain areas of the Pyrenees appears remarkably reduced. Most of the evidence suggests the occupation of large caves (*ca.* 19000–8000 year BC), rock-shelters and open-air sites in the peripheral areas of the mountain, at lower and mid-altitudes; higher altitudes were not occupied probably because of harsh climate conditions, glaciers presence and the subsequent lack of attractive food options. Starting from the Holocene, the situation would change with a gradually, even if discontinuously, increasing human pressure high on mountain environments. Three sites, a rock-shelter and two open-air occupations, testify the presence of Mesolithic groups (*ca.* 8000–6500 year BC) in the alpine areas of the Pyrenees. Despite being very partial and limited data, they might indicate the occasional exploitation of high-altitudes by groups of hunter-gatherers; evidence for hunting activities (at least for regular big-game hunting practices) are still scarce, so the option of overnight occupations associated to cross-Pyrenean routes should also be taken into account. With Neolithisation (*ca.* 5600–3000 year BC), human presence at both lower, mid- and high-altitudes gradually increase, with a diversification of sites of different duration and functionality associated with the extension of farming practices into mountainous zones. Around *ca.* 3500–2500 year BC, a sharp increasing in the number of sites and the signs of anthropic disturbance of the alpine environment is observed, suggesting a more stable and ubiquitous human presence, probably primarily associated with the development of mobile herding practices. Starting from the *ca.* 2000 year BC, a change in the settlement pattern probably occurs and, even if we are not still capable of fully explicate such process, the number of detected archaeological sites decrease quickly. Adopted survey techniques, medieval and modern reoccupation of prehistoric sites, changes in settlement strategy or the group's demography, lack of extensive research, all of them are factors that probably affect our site-detection capability. Future research might shed some light on this period.

Another consideration that can be drawn from the information obtained is that shortly after the end of the last important cold period, the Younger Dryas, human activity has been identified in sub-alpine and alpine zones. In these altitudes, the evidence of visits by hunter-gatherer groups is much weaker than in the valleys, and also more recent. However, these visits took place relatively quickly after the glaciers withdrew and these areas were re-colonised by vegetation. The information from the valleys is also of great interest and it suggests that they were occupied even in some of the colder periods in the late Ice Age. The results at Montlleó illustrate that the people who lived there were entirely familiar with the geography of the mountain range. Therefore, it was not an unplanned or occasional occupation.

A further aspect to be highlighted is the introduction of such essential practices in shaping the ecosystems as animal husbandry and arable farming. Until recently, information about sites with evidence of both activities in old chronologies, in the area of the western Mediterranean, was known in the outer ranges of the Pre-Pyrenees. Balma Margineda was, to some degree, an isolated site and it was not clear whether it was an isolated case. This perspective predominated, albeit not made explicit, and it was assumed that the expansion of the Neolithic, in social and economic terms, had largely been halted or slowed down in mountain areas. We now know that this picture was incorrect. At more recent times than the oldest sites in the foothills, but only a little more recent, sites like Coro Tracito, Els Trocs and El Sardo confirm the full introduction of agrarian activity in the central part of the Pyrenees. They indicate the intensity of Neolithisation on some hillsides, also used for arable farming, comparable with the situation in geographically proximate areas at much lower altitudes.

Finally, it is also interesting that the rates and characteristics of this process differ as a function of the altitude. In the sixth and fifth millennia, human presence in the alpine zone and upper part of the sub-alpine zone is rather tenuous, according to the current data. In contrast, from the mid-fourth millennium onwards, the number of sites and intensity of their occupation increases noticeably. This phenomenon, which has been fully described in the case of Aigüestortes i Estany de Sant Maurici National Park, appears to be comparable in other high mountain areas where fieldwork has also been intense and systematic.

In coming years, progress in our understanding of the human past of mountains in general and the Pyrenees, in particular, is sure to continue advancing rapidly. Ongoing research programmes at sites, the increase in archaeological surveying, greater cooperation between teams of archaeologists and also the increasingly interdisciplinary nature of the research are some of the factors that will contribute to this progress. Underpinning it is the growing awareness that high mountain areas have long been social spaces too. Empirical information is confirming it.

Acknowledgements This paper has been carried out in the frame of the GAAM/AGREST research activity. More specifically, this study is part of the projects "Análisi ecológico de la culturización del paisaje de alta montaña desde el Neolítico: los Parques Nacionales de montaña como modelo (CUL-PA)" funded by the Ministry of Agriculture, Food and Environment (Spain) and directed by J. Catalán and "Modelización de los espacios prehistóricos de montaña. Un SIG del patrimonio arqueológico y los territorios pastoriles" (HAR2015-66780-P, MINECO/FEDER) funded by the Ministry of Economy and Competitiveness (Spain) and directed by E. Gassiot.

References

Alcalde G, Molist M, Saña M (2005) Les ocupacions neolítiques de la Bauma del Serrat del Pont (La Garrotxa), Tribuna d'Arqueologia 2001–2002, 27–39

Andersen KA, Svensson A, Johnsen SJ, Rasmussen SO, Bigler M, Röthlisberger R, Ruth U, Siggaard-Andersen M-L, Steffensen JP, Dahl-Jensen D, Vinther BM, Clausen HB (2006) The greenland ice core chronology 2005, 15–42 ka. Part 1: constructing the time scale. Quat Sci Rev 25:3246–3257

Angelucci DE, Carrer F, Cavulli F, Delpero A, Foradori G (2013) Primi dati archeologici da una struttura pastorale d'alta quota in Val di Sole: il sito MZ005S (Mezzana, Trento) Mantova. In: Angelucci DE, Casagrande L, Colecchia A, Rottoli M (eds) APSAT 2. Paesaggi d'altura del Trentino. Evoluzione naturale e aspetti culturali. SAP - Società Archeologica, 2013, pp 141–162

Antolín F, Buxó R, Jacomet S, Navarrete V, Saña M (2014) An integrated perspective on farming in the early Neolithic lakeshore site of La Draga (Banyoles, Spain). Environ Archaeol 19(3): 241–255

Bahn P (1983) Pyrenean prehistory: a palaeoeconomic survey of the French sites. Aris & Phillips, London

Bal MC, Rendu C, Ruas MP, Campmajo P (2010) Paleosol charcoal: reconstructing vegetation history in relation to agro-pastoral activities since the Neolithic. A case of study in the Eastern French Pyrenees. J Archeol Sci 37:1785–1797

Baldellou V (1987) Avance al estudio de la Espluga de la Puyascada, Bolskan: Revista de arqueología del Instituto de Estudios Altoaragoneses 4:3–42

Baldellou V (2011) La Cueva de Chaves (Bastarás - Casbas, Huesca). Saguntum extra 12:141–144

Bernabéu J (2006) Una visión actual sobre el origen y difusión del Neolítico en la Península Ibérica c. 5600–5000 cal BC. In: O. García-Puchol y J. E. Aura (dir.), El Abric de la Falguera (Alcoi, Alacant). 8000 años de ocupación humana en la cabecera del río Alcoi, Ajuntament d'Alcoi, Dipt d'Alacant, Alicante, CAM:189–211

Bertranpetit J, Vives E (eds) (1995) Muntanyes i població. El passat dels Pirineus des d'una perspectiva multidisciplinària. I Simposi de Poblament dels Pirineus. Centre de Trobada de les Cultures Pirinenques, Andorra la Vella

Calastrenc C, Le Couédic M, Rendu C (2006) Archéologie pastorale en vallée d'Ossau. Problématiques, Méthodes et premiers resultats. Archéologie des Pyrénées Occidentales et des Landes 25:12–30

Clemente Conte I, Gassiot Ballbè E, Rey Lanaspa J, Mazzucco N, Obea Gómez L (2014a) "Cort o Transito"- Coro Trasito- o corral de tránsito: una cueva pastoril del Neolítico Antiguo en el corazón de Sobrarbe. In: Clemente Conte I, Gassiot Ballbè E, Rey Lanaspa J (eds) Sobrarbe antes de Sobrarbe: pinceladas de historia de los Pirineos (pp 11–32). Centro de Estudios de Sobrarbe (CES), Instituto de estudios Altoaragoneses (IEA) editores. Cometa S.A., Zaragoza

Clemente I, Gassiot E, Rey J (eds) (2014b) El Sobrarbe antes de Sobrarbe: Pinceladas de historia de los Pirineos. Centro de Estudios de Sobrarbe, Zaragoza

Clemente Conte I, Gassiot Ballbè E, Rey Lanaspa J, Obea Gómez L, Viñerta Crespo A, Saña Segui M (2016) Cueva de Coro Trasito (Tella-Sin, Huesca): Un asentamiento pastoril en el Pirineo central con ocupaciones del Neolítico Antiguo y del Bronce Medio. In: Lorenzo JI, Rodanés JM (eds) I Congreso Arqueologia y Patrimonio Aragonés. Colegio Oficial de Doctores y Licenciados en Filosofía y Letras y en Ciencias de Aragón, Zaragoza, pp 71–80

Cunill R, Soriano JM, Bal MC, Pèlachs A, Pérez-Obiol R (2012) Holocene high-altitude vegetation dynamics in the Pyrenees: a pedoanthracology contribution to an interdisciplinary approach. Quat Int 289(2013):60–70

Edwards CJ, Bollongino R, Scheu A, Chamberlain A, Tresset A, Vigne J-D, Baird JF, Larson G, Ho SYW, Heupink TH, Shapiro B, Freeman AR, Thomas MG, Arbogast RM, Arndt B, Bartosiewicz L, Benecke N, Budja M, Chaix L, Choyke AM, Coqueugniot E, Döhle HJ, Göldner H, Hartz S, Helmer D, Herzig B, Hongo H, Mashkour M, Özdogan M, Pucher E, Roth G, Schade-Lindig S, Schmölcke U, Schulting RJ, Stephan E, Uerpmann HP, Vörös I, Voytek B, Bradley DG, Burger J (2007) Mitochondrial DNA analysis shows a Near Eastern Neolithic origin for domestic cattle and no indication of domestication of European aurochs. Proc R Soc Lond B: Biol Sci 274:1377–1385

Fedele F (1999) Le ricerche del Pian dei Cavalli nel contesto del popolamento preistorico della Valchiavenna, Atti del II Convegno Archeologico Provinciale, vol 3, Grosio, pp 17–34

Folke C (2006) Resilience: the emergence of a perspective for social-ecological systems analyses. Glob Environ Change 16:253–267

Fortó A, Vidal À (eds) (2016) Comunitats agrícoles al Pirineu. L'ocupació humana a Juberri durant la segona meitat del V millenni cal AC (Feixa del Moro, Camp del Colomer i Carrer Llinàs 28, Andorra). Edicions del Govern d'Andorra, Andorra la Vella

Fullola JM, Garcia-Argüelles P, Serrat D, Bergadà MM (1995) El Paleolític i l'Epipaleolític al vessant meridional dels Pirineus catalans. Vint anys de recerca a la franja pirinenca sud; interrelacions amb les àrees circundants. In: Xè Colloqui Internacional d'Arqueologia de Puigcerdà. Cultures i Medi, de la Prehistòria a l'Edat Mitjana. Vint anys d'Arqueologia pirinenca. Institut d'Estudis Ceretans, Puigcerdà, pp 159–176

Fullola JM, Petit MA, Mangado X, Bartrolí R, Albert RM, Nadal J (2004) Occupation épipaléolithique microlamellaire de la grotte du Parco (Alòs de Balaguer, Catalogne, Espagne). Actes du XIV Congrès UISPP, section 7, Lieja 2001, Le Mésolithique, BAR International Series, 1302, Oxford, 2004, 121–128

Galop D (2006) La conquête de la montagne pyrénéenne au Néolithique. In: Guilaine J (ed) 2006, Populations Néolithiques et enivrements, Ed. Errance, Paris: 279–295

Gassiot E (2001) El canvi cap a l'explotació del litoral. PhD Dissertation, Departament de Prehistòria, UAB. http://www.tdx.cat/handle/10803/5498

Gassiot E, Pèlachs A, Bal MC, Garcia V, Julià R, Rodríguez-Antón D, Astrou ACh (2010) Dynamiques des activités anthropiques sur un milieu montagnard dans les píreneénne occidentales catalanes pendant la période de la préhistoire: une approche multidisciplinair. Archéologie de la Montagne Européenne, Bibliothèque d'Archéologie de la Méditerranéenne et Africaine 4:33–43

Gassiot E, Rodríguez Antón D, Burjachs F, Antolín F, Ballesteros A (2012) Poblamiento, explotación y entorno natural de los estadios alpinos y subalpinos del Pirineo central durante la primera mitad del Holoceno. Cuaternario y Geomorfología 26(3–4):26–42

Gassiot E, Rodríguez Antón D, Pèlachs A, Bal MC, Garcia V, Julià R, Pérez R, Mazzucco N (2014) La alta montaña durante la Prehistoria: 10 años de investigación en el Pirineo catalán occidental. Trabajos de Prehistoria 71(2):262–282

Gassiot E, Clemente I, Mazzucco N, Garcia D, Obea L, Rodríguez Antón D (2016) Surface surveying in high mountain areas, Is it possible? Some methodological considerations. Quat Int 402:35–45

Geddes DS (1983) Neolithic transhumance in the Mediterranean Pyrenees. World Archeol 15:51–66

Guilaine J, Martzluff M (1995) Les excavacions a la Balma de la Margineda (1979–1991) I–III. Ministeri d'Afers Socials i Cultura, Andorra

Guilaine J, Martzluff M (2007) Les excavacions a la Balma de la Margineda (1979–1991) IV. Ministeri d'Afers Socials i Cultura, Andorra

Guilaine J, Martzluff M, Geddes D, Coularou J, Le Gall O (1985) Postglacial Environments, Settlement and Subsistence in the Pyrenees: the Balma Margineda, Andorra. In: Bonsal C (ed) 1985: Mesolithic in Europe, Edinbourgh, pp 561–571

Guilaine J, Barbaza M, Gasco J, Geddes D, Jalut G, Vaquer J, Vernet J-L (1987) L'abri du Roc de Dourgne: Écologie des cultures du mésolithique et du néolithique ancien dans une vallée montagnarde des Pyrénées de l'est. In: Premières communautés paysannes en Méditerranée occidentale: Actes du Colloque International du CNRS (Montpellier, 26–29 avril 1983). Paris, CNRS Éditions

Guilaine J, Barbaza M, Gasco J, Geddès D, Coularou J, Vaquer J, André J, Jalut G, Vernet JL (1993) Dourgne: derniers chasseurs-collecteurs et premiers éleveurs de la Haute-vallée de l'Aude. Centre d'Anthropologie des Sociétés Rurales, Archeologie en Terre de l'Aude, Carcassone

Haak W, Balanovsky O, Sánchez JJ, Koshel S, Zaporozhchenko V, Adler CJ, Der Sarkissian CSI, Brandt G, Schwarz C, Nicklish N, Dresely V, Fritsch B, Balanovska E, Villems R, Meller H, Alt KW, Cooper A, The Genographic Consortium (2010) Ancient DNA from European early neolithic farmers reveals their near eastern affinities. PLoS Biol 8(11). doi:10.1371/journal.pbio.1000536

Jiménez J (2006) La imagen de los espacios de alta montaña en la prehistoria: El caso de los Pirineos Catalanes Occidentales, Treball d'investigació de doctorat, Dept. Prehistòria, UAB, Bellaterra. http://hdl.handle.net/2072/12393

Kutschera W, Miller W (2003) Isotope language of the Alpine Iceman investigated with AMS and MS. Nucl Instrum Methods Phys Res B 204:705–719

Lancelotti C, Balbo AL, Madella M, Iriarte E, Rojo-Guerra M, Royo JI, Tejedor C, Garrido R, García I, Arcusa H, Pérez-Jordà G, Peña-Chocarro L (2014) The missing crop: investigating the use of grasses at Els Trocs, a Neolithic cave site in the Pyrenees (1564 m asl). J Archaeol Sci 42(2014):456–466

Le Couédic M (2010) Les pratiques pastorales d'altitude dans une perspective ethnoarchéologique. Cabanes, troupeaux et territoires pastoraux pyrénéens dans la longue durée. PhD Dissertation, Université François-Rabelais de Tours. http://tel.archives-ouvertes.fr/tel-00543218/fr/

Leveau P, Segard M (2006) Le pastoralisme antique autourdu col du Petit-Saint-Bernard. In: Alpis Graia: archéologie sans frontières au col du Petit-Saint-Bernard. Seminario di chiusura, Aosta, 2–4 marzo 2006, Aosta, pp 153–161

Llovera X (1986) La Feixa del Moro (Juberri) i el Neolític Mig-Recent a Andorra. Tribuna d'Arqueologia, pp 14–24

Lopéz-García P, Lopéz-Sáez JA (2000) Le paysage et la phase Épipaléolithique-Mésolithique dans les Pré-Pyrénées Aragoinaises et le Bassin Moyen de l'Ebre à partir de l'analyse palynologique, Les derniers chaasseurs-cueilleurs d'Europe occidentale (13000-5500 av.j.c.). Actes du Colloque International de Desançon, 23–25 octobre 1998, Environment, societés et archéologie, 1, Besançon: Presses Universitaires-Fanc-Comtoises, Collection Annales Littéraires, pp 278–287

Mangado X, Mercadal O, Fullola JM, Esteve X, Langlais M, Nadal J, Estrada A, Bergadà MM (2005) Montlleó (La Cerdanya, Lleida), un yacimiento magdaleniense de alta montaña al aire libre en los Pirineos catalanes. In: Bicho N, Corchón, MS (eds) O Paleolítico: Actas do IV Congresso de Arqueologia Peninsular (Faro, 14 a 19 de Setembro de 2004). Faro: Centro de Estudos de Patrimonio, Departamento de Historia, Arqueologia e Patrimonio, Universidade do Algarve, pp 471–480

Mangado X, Fullola JM, Mercadal O, Bergadà MM, Langlais M, Esteve X, Estrada A, Nadal J, Tejero JM, Grimao J (2009) Montlleó. El primer poblament del Pirineu català. Cicle de Conferències. Patrimoni arqueològic i arquitectònic a les terres de Lleida 2009:48–61

Martzluff M, Martínez-Moreno J, Guilaine J, Mora R, Casanova J (2012) Transformaciones culturales y cambios climáticos en los Pirineos catalanes entre el Tardiglaciar y Holoceno antiguo: Aziliense y Sauveterriense en Balma de la Margineday Balma Guilanyà. Cuaternario y Geomorfología 26(3–4):61–78

Mazzucco N (2014) The Human Occupation of the Southern Central Pyrenees: a traceological approach to flaked stone assemblages. PhD Dissertation, Universitat Autònoma de Barcelona, 432 pp. http://www.tdx.cat/handle/10803/287893

Mazzucco N, Clemente-Conte I, Baldellou V, Gassiot E (2013) The management of lithic resources during the V millennium cal BC at Espluga de la Puyascada (La Fueva, Huesca). Preistoria Alpina 47:57–67

Mazzucco N, Ortega Cobos D, Clemente Conte I, Gassiot Ballbè E, Baldellou V, Rojo Guerra M (2014) Pautas de movilidad en el pirineo central durante el Neolítico antiguo: una aproximación a partir de los recursos líticos. In: Clemente Conte I, Gassiot Ballbè E, Rey Lanaspa J (eds) Sobrarbe antes de Sobrarbe, pinceladas de historia de los Pirineos, pp 107–126. Centro de Estudios de Sobrarbe (CES) e Instituto de Estudios Altoaragoneses (IEA) editores. Cometa S.A. Zaragoza

Mazzucco N, Clemente I, Gassiot E, Gibaja JF (2015) Insights into the economic organization of the first agro-pastoral communities of the NE of the Iberian Peninsula: a traceological analysis of the Cueva de Chaves flaked stone assemblage. J Archaeol Sci 2:353–366

Morales JI, Fontanals M, Oms FX, Vergès JM (2010) La chronologie du Néolithique ancien cardial du Nord-Est de la péninsule Ibérique. Datations, problématique et méthodologie. L'Anthropologie 114:427–444

Oms X, Gibaja JF, Mazzucco N, Guilaine J (2016) Revisión radiocarbónica y cronocultural del Neolítico Antiguo de la Balma Margineda (Aixovall, Andorra). Trabajos de Prehistoria 73 (1):29–46

Orengo H (2010) Arqueología de un paisaje cultural pirenaico de alta montaña. Dinámicas de ocupación del valle del Madriu-Perafita-Claror (Andorra). Tesi doctoral inèdita. Institut català d'Arqueologia Clàssica - Universitat Rovira i Virgili

Orengo HA, Palet JM, Ejarque A, Miras Y, Riera S (2014) Shifting occupation dynamics in the Madriu-Perafita-Claror valleys. Quat Int 353:140–152

Palet JM (2006) Stratégies de la recherche archéologique en haute montagne: les projets "Champsaur" (Alpes du sud) et "vallée du Madriu/ la Vansa - Serra del Cadi" (Pyrénées). In Alpis Graia: archéologie sans frontières au col du Petit-Saint-Bernard. Seminario di chiusura, Aosta, 2–4 marzo 2006, pp 381–385

Palet JM, Ejarque A, Miras Y, Riera S, Euba I, Orengo H (2007) Formes d'ocupació d'alta muntanya a la vall de la Vansa (Serra del Cadí - Alt Urgell) i a la vall del Madriu-Perafita-Claror (Andorra): estudi diacrònic de paisatges culturals pirinencs. Tribuna d'Arqueologia 2006–2007:229–253

Palet JM, Julià R, Riera S, Ejarque A, Orengo H, Miras Y, Garcia A, Allée Ph, Reed J, Marco J, Marqués MA, Furdada G, Montaner J (2012) Landscape systems and human land-use interactions in mediterranean highlands and littoral plains during the late holocene: integrated analysis from the InterAmbAr Project (North-Eastern Catalonia). J Ancient Stud 3:305–310

Pèlachs A, Soriano JM, Nadal J, Esteban A (2007) Holocene environmental history and human impact in the Pyrenees. Contrib Sci 3:423–431
Pèlachs A, Pérez-Obiol R, Soriano JM, Cunill R, Bal M-C, García-Codron JC (2017) The role of environmental geohistory in high mountain landscape conservation. In: Catalan J, Ninot JM, Aniz MM (eds) Challenges for high mountain conservation in a changing world. Springer, pp 107–129
Rappalino V, Marugan CM, Gassiot E, Font J, Cazanueve X, Ll Cases, Bringué JM, Adell JA (2007) Un passeig per la història de Llavorsí. Ajuntament de Llavorsí i Pagès, Lleida
Rasmussen SO, Andersen KK, Svensson AM, Steffensen SP, Vinther BM, Clausen HB, Siggaard-Andersen M-L, Johnsen SJ, Larsen LB, Dahl-Jensen D, Bigler M, Röthlisberger R, Fischer H, Goto-Azuma K, Hansson ME, Ruth U (2006) A new Greenland ice core chronology for the last glacial termination. J Geophys Res 111:D06102. doi:10.1029/2005JD006079
Redman ChL, Fish PR, James SR (eds) (2004) The archaeology of global change: the impact of humans on their environment. Smithsonian Institute Press
Rendu C (2003) La montagne d'Enveig, une estive pyrénéenne dans la longue durée. Canet-sur-mer, Trabucaire, p 333
Rey P-J (2006) Occupations et circulations pre-romaines autour du col du petit Saint-Bernard; methode et premiers resultats d'une etude archeologique et sedimentaire de la montagne alpine. In: Alpis Graia: archeologie sans frontieres au col du Petit-Saint-Bernard. Seminario di chiusura, Aosta, 2e4 marzo 2006, pp 77–117
Riera S, Turu V (2012) Cambios en el paisaje del Valle de Ordino al inicio del Holoceno: Evolución geomorfológica, paleovegetal e incendios de época mesolítica (NW del Principado de Andorra, Pirineos orientales). Cuaternario y Geomorfología 26:3–4
Rojo Guerra M, Peña Chocarro L, Royo Guillén JI, Tejedor Rodríguez C, Martínez García, de Lagrán I, Arcusa Magallón H, Garrido Pena R, Moreno M, Mazzuco N, Gibaja Bao JF, Ortega D, Kromer B, Alt KW (2013) Pastores trashumantes del Neolítico Antiguo en un entorno de alta montaña: secuencia crono-cultural de la Cova de Els Trocs (San Feliú de Veri, Huesca). BSAA, LXXIX, pp 9–54
Rojo M, Arcusa H, Peña L, Royo JI, Tejedor C, García I, Garrido R, Moreno-García M, Pimenta C, Mazzucco N, Gibaja JF, Pérez G, Jiménez I, Iriarte E. Art KW (2014) Los primeros pastores trashumantes de la Alta Ribagorza. In: Clemente I, Gassiot E, Rey J (eds) El Sobrarbe antes de Sobrarbe: Pinceladas de historia de los Pirineos, pp 127–151. Centro de Estudios de Sobrarbe (CES) e Instituto de Estudios Altoaragoneses (IEA) editores. Cometa S.A. Zaragoza
Santos M (2000) La naturaleza del espacio. Ariel, Barcelona
Spindler K (1995) The man in the ice: the discovery of a 5000-year-old body reveals the secrets of the stone age. Crown Publishers, Inc. (NYC)
Tzortzis S, Delestre X (eds) (2010) Archéologie de la Montagne Européenne. Bibliothèque d'Archéologie de la Méditerranéenne et Africaine 4. Aix-en-Provence, Errance
Uría Blanco N (2013) Registros sedimentarios como indicadores paleoambientales y de la actividad antrópica durante la Neolitización: La Cueva de Els Trocs y su entorno (Abella, Huesca). CKQ Estudios de Cuaternario/Leioa 3:123–134
Utrilla P, Montes L (2007) La période 19000-14000 BP dans le bassin de l'Èbre. Bulletin de la Société préhistorique française 104(4):797–807
Utrilla P, Mazo C (2007) La Peña de Las Forcas de Graus (Huesca). Un asentamiento reiterado desde el Magdaleniense Inferior hasta el neolítico antiguo. Saldvie 7:9–37
Walsh K, Mocci F, Tzortzis S, Bressy C, Talon B (2010) Les Écrins, un territoire d'altitude dans le contexte des Alpes occidentales de la Préhistoire récente à l'âge du Bronze (Hautes-Alpes, France). In: Tzortzis S & Delestre X (eds) Table ronde internationale "Archéologie de la montagne européenne", Sep 2008, Gap, France. Errance/Centre Camille Jullian, Paris/Aix-en-Provence, Bibliothèque d'Archéologie Méditerranéenne et africaine 4:211–225
Walsh K, Mocci F, Tzortis S, Bressy C, Talon B, Richer S, Court-Picon M, Dumass V, Palet Martinez JM (2005) Les Écrins, un territoire d'altitude dans le contexts des Alpes occidentales de la Préhistoire récente à l'âge du Bronze. In: Tzortzis S, Delestre X (eds) 2010. Archéologie

de la Montagne Européenne. Bibliothèque d'Archéologie de la Méditerranéenne et Africaine 4. Aix-en-Provence, Errance, pp 211–225

Yáñez C (2005) El Neolític. In: dins Belenguer E (eds) Història d'Andorra Barcelona, Edicions 62, pp 51–76

Yáñez C, Burjachs F, García C, Díaz N, Juan J, Malgosa A, Isidro A, Matamala JC (2002) El món funerari al final del V millenni a Andorra: la tomba de Segudet (Ordino), Cypsela, vol 14, pp 175–194

Zeder MA (2011) The origins of agriculture in the Near East. Curr Anthropol 52(4):221–235

Chapter 5
The Role of Environmental Geohistory in High-Mountain Landscape Conservation

Albert Pèlachs, Ramon Pérez-Obiol, Joan Manuel Soriano, Raquel Cunill, Marie-Claude Bal and Juan Carlos García-Codron

Abstract Proper management of the perceived value of any geographic space requires the capacity to interpret research results from spatial, temporal, and environmental points of view, applying the principles of environmental geohistory. Basic concepts such as baseline, threshold, or resilience are discussed from a long-term ecological perspective, with examples that explain the dynamics of fir forests as well as the changes in agricultural cover. Studying the changes in the altitudinal limit of the forest and surveying the wetlands dynamics on the southern slopes of the central Catalan Pyrenees have been shown to be effective tools to develop appropriate management tasks. The arguments presented are useful to enrich the public debate over management policies for natural protected spaces in high-mountain areas.

Keywords Environmental geohistory · Conservation · Pyrenees · *Abies alba* · Human perturbation · Pedoanthracology · Palynology · Sedimentary charcoals

---

A. Pèlachs (✉) · J.M. Soriano · R. Cunill
Facultat de Filosofia i Lletres, Department of Geography, Edifici B, Universitat Autònoma de Barcelona, 08193 Cerdanyola del Vallès, Bellaterra, Spain
e-mail: albert.pelachs@uab.cat

R. Pérez-Obiol
Botany Unit, Facultat de Biociències, Department of Animal Biology, Plant Biology, and Ecology, Edifici C, Universitat Autònoma de Barcelona, 08193 Cerdanyola del Vallès, Bellaterra, Spain

M.-C. Bal
Geolab UMR 6042 CNRS, Université de Limoges, 87000 Limoges, France

J.C. García-Codron
Department of Geography, Urban Studies and Land Planning, Avenida de Los Castros s/n, Universidad de Cantabria, 39005 Santander, Spain

J. Catalan et al. (eds.), *High Mountain Conservation in a Changing World*, Advances in Global Change Research 62, DOI 10.1007/978-3-319-55982-7_5

5.1 Palaeoenvironment, Biodiversity and Protected Areas

In recent decades, paleoenvironmental studies based on sedimentary records in mountain regions have contributed a large quantity of data that can help to explain the major environmental changes over time in different parts of the planet (Last and Smol 2001; Battarbe et al. 2005; Willis and Birks 2006; Catalan et al. 2013). The majority of these studies have focused on explaining the main vectors of global change, and have discussed the impact of human activities on planet Earth (Boada and Saurí 2002; Duarte 2006).

According to a recent international report, *Protected Planet 2016*, "there are 202,467 terrestrial and inland water protected areas recorded in the World Database on Protected Areas (WDPA), covering 14.7% (19.8 million km^2) of the world's extent of these ecosystems (excluding Antarctica)" (UNEP-WCMC and IUCN 2016: 30). The same source indicates that 19% of the world's mountain area has been declared Protected Areas, and in recent decades conservation policies have emphasized the role of protected areas in conserving biodiversity and cultural heritage. As Willis and Bhagwat (2010) point out, however, "since the first national park was established in 1872, [protected zones] are spatially fixed, meaning that migration beyond reserves in response to climate change may not be possible for many species" (Willis and Bhagwat 2010: 765). All of this occurs in a matrix that has been deeply affected by human impact and has experienced scenarios reflecting major climate changes in recent decades (IPCC 2014).

Taking all of this into account, Willis and Bhagwat ask, "how can we then create conditions that will protect native species beyond reserves and in novel ecosystems? This may require a whole new approach to conservation, restoring ecological processes and enhancing the quality of landscape matrix surrounding reserves" (Willis and Bhagwat 2010: 765). Is this reflection taking place? The answer is: at the theoretical level, yes—but only for the past decade—and at the practical level, no because of a lack of dialog between conservationists and the paleo-community (Sutherland et al. 2009). Some attribute this failure to communicate to a lack of awareness of scientific advances made by conservation managers, the absence of a long-term perspective that extends beyond 50 years, the priority given to the study of short time periods rather than the Holocene as a whole, and the self-imposed limitations within the scientific community when only the negative aspects of research attract attention (Froyd and Willis 2008). Other authors attribute the problem to differences between the palaeoecological descriptions used, without adjusting them to reflect the day-to-day management and conservation of biological diversity (Willis and Bhagwat 2010).

Richardson and Whittaker (2010) remind us that the conservation of biodiversity was formally defined during the past decade, when Whittaker et al. (2005) wrote about the biogeography of conservation and drew attention to four topics: (i) scale dependency; (ii) inadequacies in taxonomic and distributional data; (iii) developing improved understanding of the effects of model structure and parameterization, through increased sensitivity analyses; and (iv) areas in which applied theory derived from biogeographical science requires greater focused attention" (Whittaker et al. 2005: 318). That early 2000s, the expression 'ecosystem services' become formally stated as "the benefits people obtain from ecosystems" and highlighting "the consequences of the loss of biological diversity and degradation of ecosystem services for human well-being globally" (Sutherland et al. 2009: 560). Immediately thereafter, voices began to ask how to accomplish these things and what information was available to confront this challenge.

Willis and Birks (2006) pointed out the usefulness of studies based on sedimentary records in developing conservation policies. Their article, "*What Is Natural? The Need for a Long-Term Perspective in Biodiversity Conservation,*" shows the value to environmental managers of taking a "longer temporal perspective to address specific conservation issues relating to biological invasions, wildfires, climate change, and determination of natural variability" and considered a temporal perspective "essential for meaningful modeling, prediction, and development of conservation strategies in our rapidly changing Earth." They conclude that the answer to the question "what is natural" lies in the historical analysis of environmental changes, and thereby can "start to provide important guidance for long-term management and conservation at local, regional, and global scales" (Willis and Birks 2006: 1261).

After developing 2,291 questions from a review of literature, Sutherland et al. (2009) selected 100 key questions for conservation practices: "The questions are divided into 12 sections: ecosystem functions and services, climate change, technological change, protected areas, ecosystem management and restoration, terrestrial ecosystems, marine ecosystems, freshwater ecosystems, species management, organizational systems and processes, societal context and change, and impacts of conservation interventions" (Sutherland et al. 2009: 558). The broad range of topics shows the complexity of the system, which was refined in each study included in their review. Richardson and Whittaker (2010) use six keywords that exemplify this complexity: "Biological invasions, climate change, conservation planning, data requirements, invasion ecology, and species distribution modeling" (Richardson and Whittaker 2010: 313). Their research shows the need for multi-proxy studies to explain the changes in biodiversity and contribute to its long-term conservation. Willis et al. (2010) conclude their article masterfully with this idea:

> These archives indicate the complexity of responses to climate change over time, ranging from inherent variability through to rapid compositional turnover, broad-scale migrations, regime shifts, and the creation of novel ecosystems. They also indicate the dynamic interactions of biotic and abiotic processes that sometimes lead to thresholds and in other situations enable resilience and persistence. The record of these biotic responses obtained from paleoecological records provides a valuable resource for conservation strategies to conserve and manage ecological and evolutionary processes (Willis et al. 2010: 589).

In summary, biological conservation requires the analysis of a number of long-term states and processes in a world of extremely dynamic changes (Willis and Bhagwat 2010; Willis et al. 2010; Bradshaw et al. 2015) and a response to the following four questions: (1) What are the baseline or 'reference' conditions before recent times? (2) What is the range of natural variability? (3) Under what circumstances do negative impacts become apparent? (4) How can thresholds be determined beyond which specific management plans should be implemented? (Willis and Birks 2006: 1264). These questions lead us to various key concepts, as outlined below.

5.1.1 Baseline and Range of Natural Variability

A central aspect of any analysis is determining the 'baseline' or 'reference' conditions against which current changes can be assessed (Willis and Birks 2006; Froyd and Willis 2008; Willis and Bhagwat 2010; Willis et al. 2010; Bradshaw et al. 2015). Froyd and Willis (2008) agree with Lindbladh et al. (2007) that the "baseline is perceived as the ecosystem present before human influence became pronounced on the landscape" (Lindbladh et al. 2007). "Baselines are especially common in environmental regulation and ecosystem services" (Froyd and Willis 2008: 1724).

Closely related to this concept, Froyd and Willis (2008) also highlight the 'Range of Natural Variability' (RNV) of ecological systems; the central premise is that "resilient ecosystems will be maintained if land management activities operate within the range of conditions that would be expected under the natural disturbance regime" (Froyd and Willis 2008: 1725). This concept is promoted by managers with a vocation based on ecology, geography, etc. As these authors further explain, "application of the RNV incorporates the ideas of a historical baseline period or reference condition, but in addition the concept includes the implicit acknowledgement that ecological systems are not static and management applications should therefore vary, yet within a defined boundary range" (Froyd and Willis 2008: 1725). The question of how far back we should set the baseline is not exempt from dispute because it involves choosing a particular point on the space-time continuum, and this choice is not easy when we know that the landscape has never been static at any time in history. As Jackson and Hobbs (2009) point out, ecosystems are dynamic, fluid, and ever-changing, making it difficult to identify one moment in time to serve as a reference condition for the present situation.

5.1.2 Thresholds and Ecological Resilience and Persistence

In any case, it seems essential to set some thresholds, as Froyd and Willis (2008: 1726) indicated: "paleoecology holds great potential to inform conservation practice (…) through the identification of ecological thresholds, where the threshold is an abrupt change in an ecosystem and a switch from one stable state to another. The threshold concept is of particular interest to natural resource managers and policy makers because of the potential to define limits on the amount of 'acceptable' change an ecosystem can withstand."

This perspective need not distract our attention from moments of 'stability' in the natural system because, even if thresholds of change are abrupt, succession—whether progressive or regressive, will be much longer (Margalef 1991). According to Dietl and Flessa (2011) there are three main ways in which species respond to changes in their environment: "(1) they can move, tracking environmental changes; (2) they can stay and adapt to the changing environment; and (3) they can fail to track habitats or to adapt, thus becoming extinct" (Dietl and Flessa 2011: 32). This leads us to reflect on a typology of time concepts (change, evolution, transformation, process) (Mendizábal 2013), without neglecting another key concept: resilience, a term that is currently used in many fields. Froyd and Willis (2008: 1726) defined resilience as "the ability of systems to absorb disturbance and still maintain the same relationships between populations (…). Thus resilience is the magnitude of disturbance that can be tolerated before a system moves to another stable state." At present, the study of resilience is being applied to the effects of climate change and paleoecological archives that help us understand why some ecosystems are more resilient to climate change than other and identify different degrees of resilience in different systems. The number of available studies and the contributions of genetic diversity have been another important factor for the maintenance of resilience in studies on the conservation of genetic diversity (Willis et al. 2010).

The general debate has focused on the study of interactions between climate, ecological processes, and human activities in the past in order to better understand the behavior of ecosystems in the present and future (Dearing and Battarbe 2007). In this context, high-mountain zones have been considered among the most sensitive and vulnerable to the environmental changes predicted for the 21st century and one of the priority areas for attention to the value of natural attributes (Huber et al. 2005).

The reason for this is that human activity cannot be decoupled from landscape because they form part of it: "Over 75% of the Earth's terrestrial biomes now show evidence of alteration as a result of human residence and land use" (Willis and Bhagwat 2009: 807). Most western European forests today "have long and diverse histories of anthropogenic disturbance and current conservation values incorporate both natural and cultural features" (Bradshaw et al. 2015: 194). In this sense, pollen studies with high spatial resolution have shown that simple temporal concepts like 'natural baselines' and the continuity of forest cover underestimate the complexity of the past (Bradshaw et al. 2015). What is the main problem? The complex nature of the relations between climate change and human activities. Why? For the following reasons:

(a) The same evidence/variable may indicate different things, and these may change over time.
(b) The same effect may have multiple possible origins that are completely opposite and therefore can have multiple origins that are completely opposite and therefore antagonistic. For example, Willis and Birks (2006: 1263) wrote that it "is not unreasonable to assume that an increase in aridity would result in more fires; several studies indicate otherwise." They base their argument on the

> cases of Alaska and the Northern Great Plains grasslands of North America to explain how greater humidity brings with it increased the risk of forest fires, showing how the levels of complexity are increasing. In the study of forest fires, there is a complicated climate-fuel-fire relationship that determines the variability of wildfires (Willis and Birks 2006).

If ecological systems are dynamic at natural scale (i.e., long-term ecological records >50 years), why is management often carried out within a static framework, dealing with short-term changes (<50 years)? Various studies have shown that it is not possible to manage the natural environment without considering the dynamics beyond that 50-year timeline (Willis and Birks 2006; Willis and Bhagwat 2010). In a synthesizing review in the Applied Ecology journal, Froyd and Willis (2008) affirmed that the majority of publications in the conservation/applied ecological literature still focus on very short timescales (i.e., years or decades). Willis and Bhagwat commented: "Paleoecological records are replete with examples of biotic responses to past climate change and human impact, but how can we use these records in the conservation of current and future biodiversity?" (Willis and Bhagwat 2010: 759). The solution involves always considering the long-term perspective and providing a test of "predictions and assumptions of ecological processes that are directly relevant to management strategies necessary to retain biological diversity in a changing climate" (Willis and Bhagwat 2010: 759).

5.1.3 Why Environmental Geohistory and Not Only Environmental History?

The term 'environmental history' has been in use for years, despite other labels that have received a certain media notoriety (Fontana 1992; Pèlachs 2006), although some sectors of ecology have begun to take it into account more recently (Whittaker et al. 2005). Therefore, we share the view of the temporal dynamic described by Dietl and Flessa (2011), who understand historical ecology in a broader sense, based on two perspectives: one limited to relatively recent time intervals (i.e., the Pleistocene) and another that concentrates on ecological dynamics (e.g., changes in species distribution and abundance). They conclude the following: "(1) the temporal scope of conservation paleobiology also extends to the pre-Pleistocene record; and (2) conservation paleobiology, in addition to ecological dynamics, concentrates on evolutionary dynamics (e.g., adaptive responses of species to changing climates or ecological interactions)" (Dietl and Flessa 2011: 31). Our concept of environmental geohistory also considers the spatial variable as a key factor. The temporal dynamic, combined with the spatial variable, constitutes geohistory: time (dynamics) is as important a variable as is space (scale). Froyd and Willis (2008) remind us that time and space are concepts that do not always go hand in hand with paleoecology. Environmental geohistory only exists if both variables are treated at the same time, but not if they are considered separately. As Dietl and Flessa (2011:

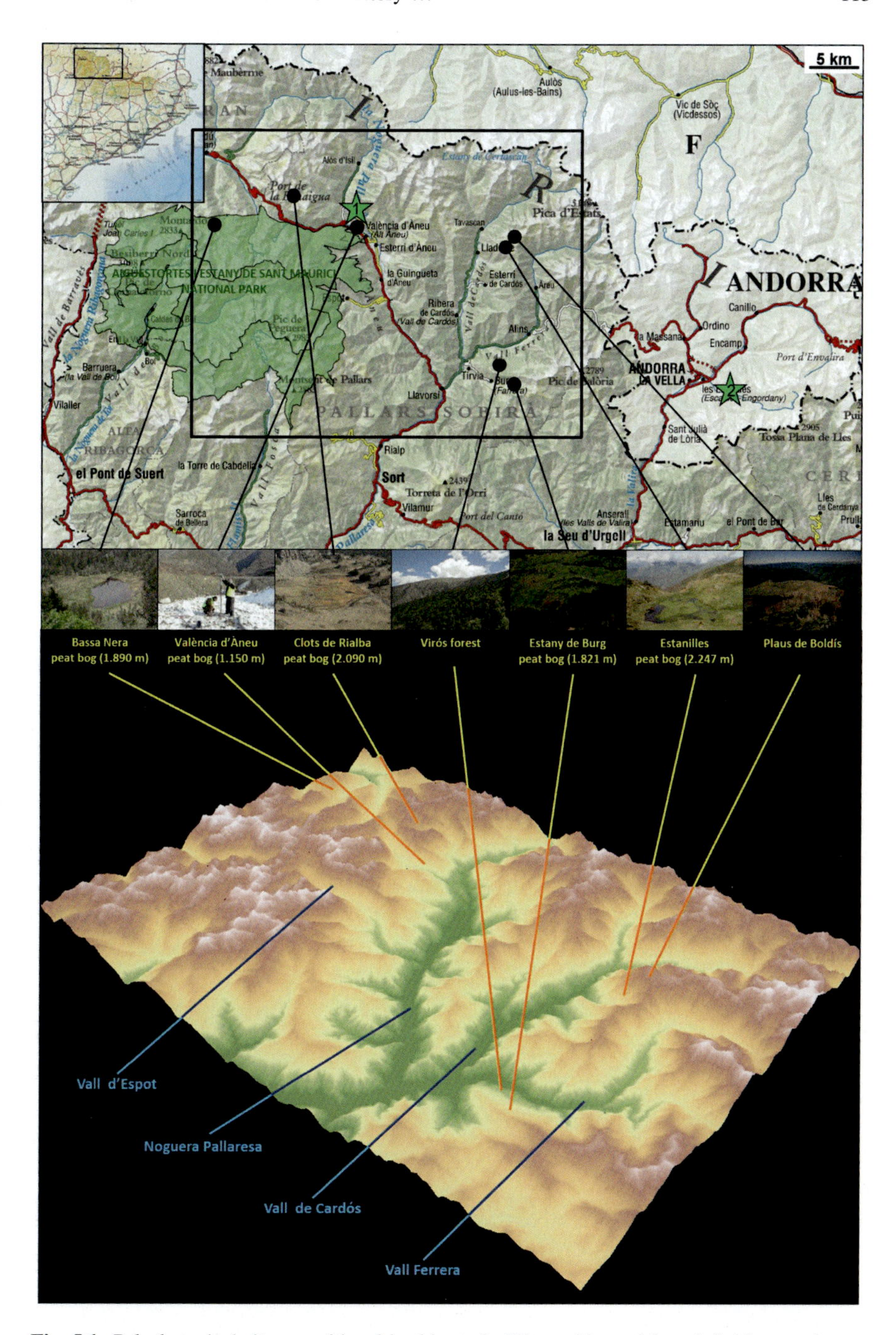

Fig. 5.1 Paleobotanical sites considered in this study [Sites with partial or definitive results are indicated with *black dots*. The two *green stars* (*1* Solana de Sorpe; *2* Vall de Madriu, Andorra) indicate works in progress (no results to date)]

30) put it, "our message is that the perspective provided by geohistorical data is essential for the development of successful conservation strategies in the midst of a constantly changing environment."

Environmental geohistory is an instrument with a focus on the future. Therefore, this chapter attempts to answer questions asked by managers by providing current examples that contribute information about diversity baselines, thresholds, resilience, and restoration of ecological processes (Willis and Bhagwat 2010). As Froyd and Willis (2008) point out, "the analysis of late Quaternary environmental change to key environmental issues of biodiversity and conservation management and examines areas which could be strengthened in the future including: (i) determination of baselines and natural ecosystem variability; (ii) understanding ecological thresholds and resilience; (iii) climate change conservation strategies; (iv) biological invasions; and (v) conservation and culture" (Froyd and Willis 2008: 1723). Willis et al. (2010) commented, "the message that emerges is that such paleo-records have much to offer not only regarding understanding the ecological and evolutionary processes responsible for biodiversity but also in guiding the management strategies necessary to ensure biodiversity conservation" (Willis et al. 2010: 584). Paleoecological records provide evidence of numerous climate changes and human impacts. The key to the question lies in how these records are used to benefit current and future conservation of biodiversity (Willis and Bhagwat 2010).

The main objective of this chapter is to contribute information and reflection on diversity baselines, thresholds, resilience, and restoration of ecological processes, based on three specific, structured examples in various settings with different levels of human intervention: (1) *Abies* and *Betula* as examples of forest changes permit a discussion of baseline and Range of Natural Variability at different stages of succession; (2) Alpine pastures at the tree line and agricultural mountain areas show the importance of thresholds and resilience in the management of more humanized spaces; (3) Peat bogs, the maximum expression of local environments, demonstrate that a lack of resilience requires maximum attention by managers. Both unpublished and previously published data will be presented to illustrate examples from the Pyrenees in areas surrounded by Natural Protected Spaces, both 'Aigüestortes i Estany de Sant Maurici National Park' and 'Alt Pirineu Natural Park' (Fig. 5.1).

5.2 Examples of Environmental Geohistory in the Pyrenees

5.2.1 The Baseline and Range of Natural Variability of Abies Alba Mill

In western Europe, *Abies alba* is distributed across the central and southern part of the continent; further north, *Picea abies* is the dominant genus. On the Iberian Peninsula, this species reaches its southwestern boundary in Europe, with its largest extension of forest area in the Pyrenees (approximately 50,000 ha); we find the southernmost populations in the Montseny massif (Costa et al. 1998). At present, the stands are highly fragmented and in some cases degraded by human activity (Alba–Sánchez et al. 2010). Their environmental needs include atmospheric humidity, dense cover for proper germination, and resistance to cold, placing this species below the altitude limits of *Pinus uncinata*, in other words at the subalpine and montane level. The species can also share the ecological range of *Fagus sylvatica*, with which it can form mixed stands of forest (Costa et al. 1998).

The available information about the paleogeographic history of *Abies alba* and its current distribution generates a long list of questions about its environmental history (genetic changes, a shift in distribution areas, etc.) and origins (colonization, refuge populations, etc.) in the Iberian peninsula. These questions must be answered to understand its current distribution, predict future developments within the current framework of global change, and support better management of the species.

Various paleogeographic studies at the European level, mainly in the Alps and Apennines, have shown the decrease in the distribution area of firs, compared to the mid-Holocene (ca. 5000–6000 years ago). At that time, the species would have achieved a dominant presence in the montane zones and low-altitude areas (Schneider and Tobolski 1985; Tinner et al. 1999; Wick and Möhl 2006). In the Pyrenees, environmental geohistory studies have confirmed this trend (Jalut 1988; Pèlachs et al. 2009b; Galop et al. 2013; Cunill et al. 2015). A study by Cunill et al. (2015) in fir populations at 300 m altitude on the north slope of the Pyrenees showed that, far from being anomalies or the fruits of repopulation efforts, these arboreal masses are witness to the greater importance of firs in the past in low pre-Pyrenees zones and a presence in the area for at least 5000 years.

It has not always been possible to apply the concepts of resilience, baseline, and RNV to managing these forests because we still lack knowledge of their paleoecological history and distribution. During the late glacial periods, *Abies alba* withdrew to refuge zones (still little-known to this day) and then at the beginning of the Holocene migrated progressively from east to west along the Pyrenees range. It appears earlier in the Mediterranean Pyrenees (in Estanilles about 8500 years cal BP, in the Burg Lake about 8000 years cal BP), and later in the Bassa Nera of the Atlantic-influenced Aran Valley (at about 6500 years cal BP) (Jalut et al. 1998; Esteban et al. 2003; Pèlachs et al. 2009b). In coming years, recently initiated pollen and genetic studies should begin to provide more information on this topic (Sancho–Knapik et al. 2014; Matías et al. 2016).

Palynological studies of five sedimentary records (Fig. 5.1) confirm the change over time in the presence of *Abies alba* and provide new data about the chronology of its dynamics (Pèlachs et al. 2009b; Bal et al. 2011; Pérez–Obiol et al. 2012; Cunill et al. 2013). In addition, other techniques of high spatial precision such as soil charcoal analysis or sedimentary macroremains analysis have confirmed this chronology and provided more information about the altitudinal and local distribution. Our group has found *Abies alba* on the shady and sunny sides of both slopes of the Pyrenees, at Montbrun–Bocage (Cunill et al. 2015) and Vall de Madriu et al. (work in progress).

At present, minimum *Abies* values have been found in all the diagrams studied. Human impact, as well as Holocene climatic changes leading to temporary moisture deficits, may have had a significant impact on the *Abies* forests. The species is considered to be less competitive on the sunny slope. Nevertheless, its rapid quantitative decline can only be understood if the silver fir was growing in habitats outside its current range. Millennia of land use could have eliminated *Abies alba* from the warmest portions of its potential range. Matías et al. (2016) recently suggested that the greater genetic diversity observed in the oldest populations of *Abies alba* provides greater resistance to drought and therefore the decline in silver fir could be a consequence of reduced genetic variation and the corresponding lack of adaptability.

The current presence of most firs is at the lowest level in history in the study area (Fig. 5.2). In an analysis of Holocene history between 4000 and 5000 years cal BP in Burg and Bassa Nera sequences and between 6000 and 7500 years cal BP in Estanilles peat bog, *Abies alba* would have had a much more important presence before the high human impacts observed. This time-point could provide the ecological baseline to be taken into account for conservation management studies. However, this abundance could be explained by their location on the sunny slope, as the values from Burg Lake, for example, seem to indicate. Beginning in 4200 years cal BP, the decline coincides with a period of rapid climate change (Bond Event 3) and with the beginning of the Bronze Age in the region (Pèlachs et al. 2011).

A period of recovery coincided with the influence of Roman period. After this phase, firs would never regain their previous coverage. In several pollen diagrams (Coma de Burg, Estanilles, and Bassa Nera), it seems that there was a selective management of firs or an unfavorable climate for these forests during this period. The *Abies* percentage curves behaved differently from the arboreal pollen curve (AP %) (Fig. 5.2). Some of our data are in good agreement with data from the Alps and descriptions of the Romanche Valley, which explain the selective exploitation of firs in the Roman period (Nakagawa et al. 2000).

Was the recovery of firs in the Pyrenees after the human perturbations a consequence of management? The data show marked declines after the Late Roman period that could be related to intensive, generalized exploitation across the last two millennia. Therefore, the RNV should, in any case, precede this decline, although it is hard to pinpoint because of the many human and climatic changes occurring throughout the entire time period.

From a forest management perspective, the question is whether or not fir should be harvested at present. There are two conflicting positions: one argues that selective cutting will encourage regeneration and the other contends that the trees should not be touched. The first is based on the notion that, if we have arrived at the current situation of a certain forest quality, it is thanks to the forest management efforts of the past century. The data, however, demonstrate that this is not the case. The interventionist management of recent centuries is precisely the reason for the *Abies alba* decline.

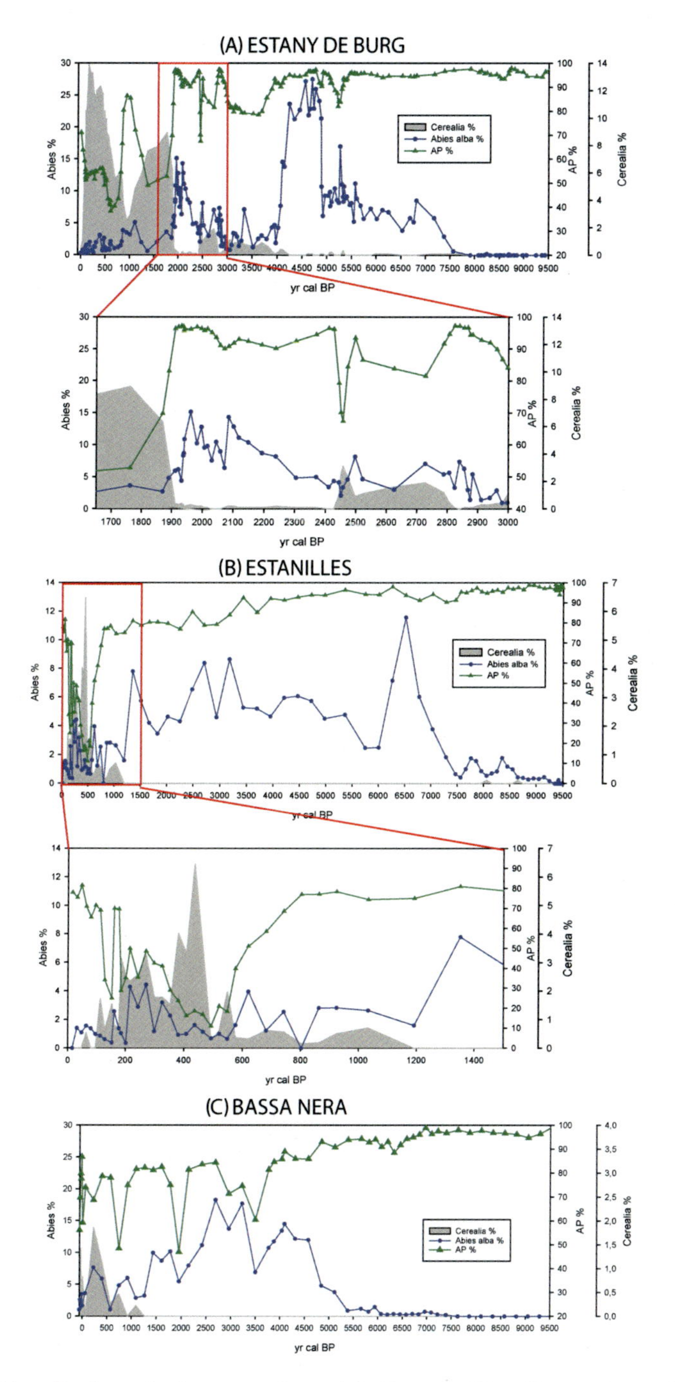

Fig. 5.2 *Abies alba* dynamics in some valleys of the Pyrenees [Estanilles results modified from Cunill et al. (2013) and Burg partially published in Pèlachs et al. (2011)]

What can we learn from the historical dynamics of *Abies alba?* Is there some type of human management that had previously benefited or harmed the firs? In any case, human impact, more than climate, appears to be the primary factor in local extinction and the loss of genetic diversity and, consequently, the loss of adaptability (Carcaillet and Muller 2005).

5.2.2 Are Baseline and Range of Natural Variability Appropriate Concepts for Secondary Communities?

Although Costa et al. (1998) reported that *Betula* is widely distributed across the northern and medial zones of the European subcontinent, in the south its presence becomes progressively weaker. In the Iberian peninsula, these trees often dot other forest formations, such as Eurosiberian or sub-Mediterranean, although it is also possible to find them in more or less pure stands. Therefore, from the viewpoint of current forest management, knowing the origin of the small birch woods that are scattered across the panorama of other forests or whether there are stable birch forests is of great interest to conservation efforts.

In the Pyrenees, we find two species: *Betula pendula* and *Betula pubescens.* Being heliophilous plants, they have serial behaviors, meaning that in the process of succession that allows the original forest to recover, they act as pioneers in the successional sequence after disturbances in other forest masses or other sites that are difficult to colonize. Birch is a eurioic genus, capable of rapid natural expansion under favorable conditions, i.e., heavy production of light, winged seeds, vegetative regeneration, and easy germination (Costa et al. 1998). Therefore, the discussion concerns whether birch should be considered a secondary formation or the optimum forest vegetation, which would not be unusual in the presence of rexistasia and geomorphological activity (landslides, avalanches, etc.).

From a paleoecological perspective (Fig. 5.3a), birch propagated rapidly at the beginning of the Holocene, when the last glaciation had relegated to far distant areas the various forest patterns that today occupy the territory, allowing a broader distribution than in present times. Later, its distribution was limited to small populations in more favorable local habitats, which contributed to its persistence during the rapid climate changes that followed. At present, when thinking about ecological succession after a system has been strongly modified by human action, we tend to consider changes in forest structure at human scale. That is why birch stands have not traditionally been considered a stable 'final' status but rather an intermediate phase of a succession. What happens if these secondary dynamics extend more than 100 years? During the beginning of the Holocene, birch woods have been stable formations in the Pallars Sobirà region, and it was not until recent millennia that they have adapted to specific, greatly disturbed areas. Therefore, the same species of this genus may have had two totally differentiated preferences over the course of time, and a species considered to be mainly secondary in the Pyrenees today might

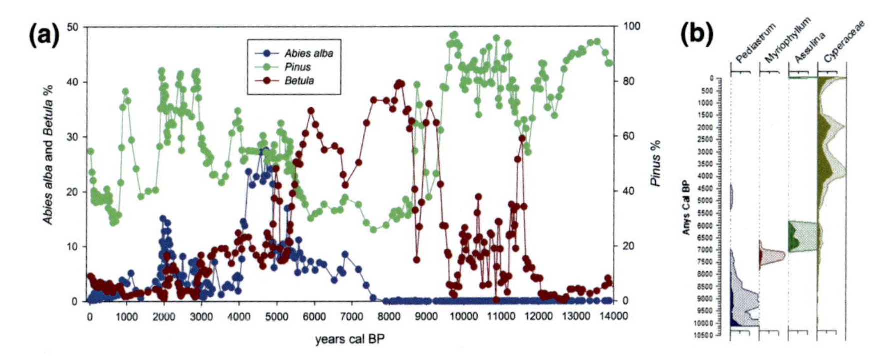

Fig. 5.3 **a** Holocene dynamic of the birch, the fir, and the pine at Burg Lake [partially published in Pèlachs et al. (2011)]. **b** Local environment in Bassa Nera during the Holocene [partially published in Pèlachs et al. 2016]

have had a different profile in the past (Fig. 5.3a). When these pioneer formations are being analyzed, the problem is greatest when attempting to define the ecological baseline and RNV. As indicated by Alagona et al. (2012: 65): "Restoration requires historical baseline targets, but all such targets are arbitrary for ecosystems that are constantly changing and have always been doing so (…) Recasting historical knowledge not as a narrow search for singular baseline conditions or specific population figures, but as a way to track multifaceted ecological changes over time, offers a middle ground where the past may inform but not determine the ecosystems of the future. The past may be imperfect as a model for the future, but it is an indispensable guide for understanding a world in flux."

At present, there is no special management of birch in the National Park. It is an opportunistic pioneer species that has adapted to colonizing unstable settings such as landslide channels and places with limited soil. Both species mentioned above, *Betula pendula* and *B. pubescens*, are found in this zone; the first is much more abundant now. We do not know whether this was also the case in the past, but it seems clear that *B. pubescens* must have a certain priority for conservation because it represents a relic of the past.

5.2.3 *How Should Open Spaces Cleared for Human Activities Be Managed When They Have a Semi-natural Function?*

The European Community's Directive 97/62/EC defines 198 habitats of community interest. In Catalonia, 94 habitats of community interest have been identified, of which 22 are prioritized. Among these habitats we find the acidic peat bogs of *Sphagnum* (see Sect. 2.5), natural and semi-natural grasslands (e.g., meadows of *Nardus stricta*, rich in flowers and sandy mountain soils of the mid-Atlantic or sub-Atlantic) and ecotonic bands of mountain with clusters of *Cytisus oromediterraneus* or *Juniperus communis*, among others, that colonize the meadows. In many cases, these areas are directly related to the open spaces and ecotonic areas that contain a great diversity of habitats.

At present, a topic of discussion is whether management must be based on a certain level of active human disturbance, or if open spaces can be maintained from a naturalist point of view. Fires and domestic cattle are indispensable elements for the first option, and there are wild plant-eating animals for the second. A second idea expressed by Willis and Bhagwat (2010: 765) is that "understanding the relationship between past herbivore densities and their role as 'ecosystem engineers' is an important future research challenge for long-term ecology." In both cases, the arboreal reforestation that could exist without either of these interventions is discarded. Just as we are told that not all meadows have an exclusively human origin, there are also natural origins, which quickly become the excuse to label the landscape as deteriorated and of little priority for conservation efforts, when it could mean exactly the opposite and have important endemic species of high conservation priority (Willis and Bhagwat 2010).

The altitudinal boundary of the forest is used to study climate change (Grau et al. 2013) and human impact (Catalan et al. 2013). As we have seen, however, the limits of the 'natural forested areas' have been substantially altered by the uses during recent centuries. That is why managing the forest for multiple uses, agricultural and livestock management and even improving animal habitat require knowledge of how the forest boundaries have varied to apply conservation policies.

A study by Cunill et al. (2013) carried out in the Vall de Cardós using pedoanthracology, palynology, and sedimentary charcoals deduced the transformation of the Pyrenean landscape, specifically the forest boundaries, during the Holocene (Fig. 5.4). Analysis of the data showed that human management has affected the forest boundaries. Fire, along with pastures and agriculture, has had a decisive role for thousands of years. Therefore, the current configuration of alpine and sub-alpine belts has been equally or more influenced by human action than by climate factors.

We could ask ourselves, why we value stability when studying vegetation succession but assign less value to systemic stability that results from human activity. Once again, the discussion must occur between two conflicting points of view: one side thinks that a certain dose of human disturbance contributes to the diversity of habitats and, therefore, to biodiversity; the others argue that the absence of human perturbations and the trend toward landscape homogenization that occurs with a high degree of maturity (e.g., in a mature forest) is the best way to encourage the natural system. The key question is, where is the limit? Does the management of protected natural spaces necessarily favor biodiversity by way of heterogeneity of habitats or will prolonged homogeneity lead to sufficient maturity and richness of species and biodiversity over time?

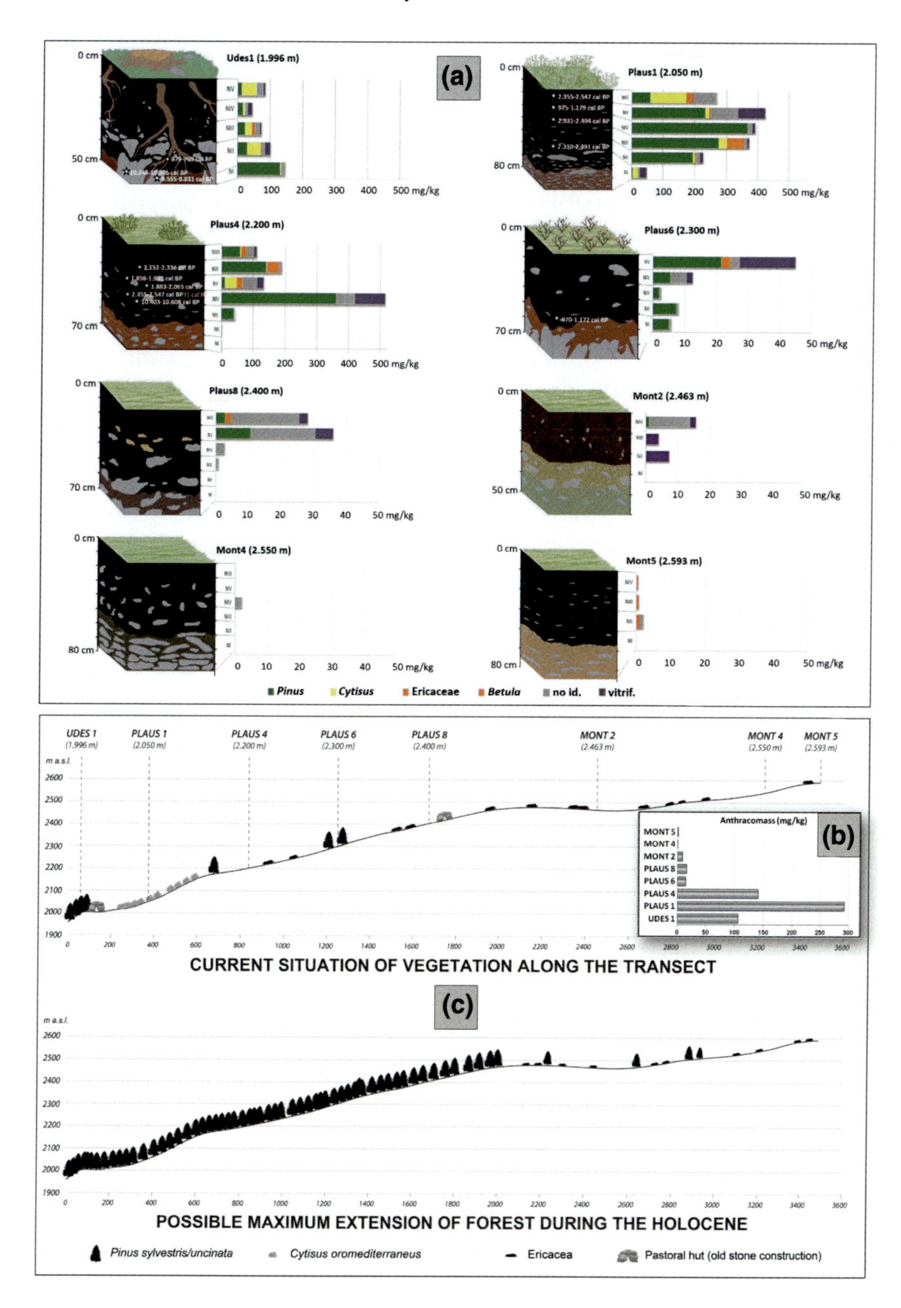

Fig. 5.4 **a** Anthracomass is defined as the quantity of charcoal per kilogram of dried soil and was calculated on the basis of the mass of charcoals larger than 0.8 mm (expressed in milligrams) and the total mass of the fraction of dry soil less than 5 mm (expressed in kilograms). The soil charcoal analysis showed that anthracomass was present at all sampling points. **b** Pedoanthracological diagrams (modified from Cunill et al. 2013). Taxon-specific anthracomass (TSA) by level in an elevational transect, expressed in mg/kg. **c** Theoretical profile of the upper forest boundary, based on the anthracomass at the time of maximum Holocene forestation, compared to the present

5.2.4 From the Late Roman to the Medieval Age Was Born the Open Landscape: Threshold Forestry Without Turning Back?

If we analyze the intensity of human disturbance in our mountains, agriculture would be among the most important. As Willis and Bhagwat (2010) point out, based on examples from around the world, "Many of the landscapes in so-called biodiversity hotspots have a long history of human habitation and have been under some form of cultivation in the past" (Willis et al. 2004). The socioeconomic changes of the second half of the 20th Century led to the abandonment of many traditional practices related to the primary sector, and the Pyrenees are no exception (García–Ruíz et al. 1996). The loss of biodiversity associated with farming and livestock mosaics is a subject of debate in the management of Natural Spaces, to the extent that in recent years some of them, such as the Alt Pirineu Natural Park, have put in place policies of recovering fields and planting cereal crops. According to Willis and Bhagwat (2009: 807), "It has long been assumed that in a fragmented landscape, the fragment size and its isolation are important factors in determining species persistence; the smaller and more isolated the fragment, the lower its occupancy." One idea that contradicts the forest recolonization that provoked the abandonment of the traditional farming and livestock system in the valleys and mid-slopes. Bradshaw et al. (2015) summed it up this way: "Long forest continuity may be of importance for the local survival of higher plants, but for the insects, fungi, lichens, and bryophytes that are so valued in contemporary European temperate and boreal forests, habitat diversity maintained by dynamic processes would appear to be of greater significance" (Bradshaw et al. 2015: 194). This issue is also under discussion. Should managers strengthen the farming and livestock mosaic? Should agriculture be incentivized? Managers ask for specific answers, but policy management must be more general (Dietl and Flessa 2011). The following is an example of a threshold from more than a thousand years ago.

To understand the agricultural system, we analyzed the curves of cereals and grasses on the pollen diagram for Burg Lake, observing that the starting point of the modern farming and livestock system began about 4250 years cal BP (Fig. 5.5). The trend has clearly been on the increase, especially over the last two millennia. In other words, in the past 2000 years, there has never been such a small farming and livestock system in that study area. During those two millennia, the landscape has almost always been more open than it is now and forest recovery has clearly been in deficit compared to the recovery of pastures and open spaces.

A paradigmatic case occurred in Estanilles (Fig. 5.2), where deforestation was at its maximum 500 years ago, with AP% below one-third, coinciding with a maximum curve of cereal pollen. The peculiarity of this case is that this process occurred at an altitude of 2247 m.a.s.l., which is usually associated only with pastures. Nonetheless, the data indicate an open farming and livestock landscape without trees.

Is the speed and intensity of these changes comparable throughout history? The current landscape often inhibits our ability to see beyond the human scale

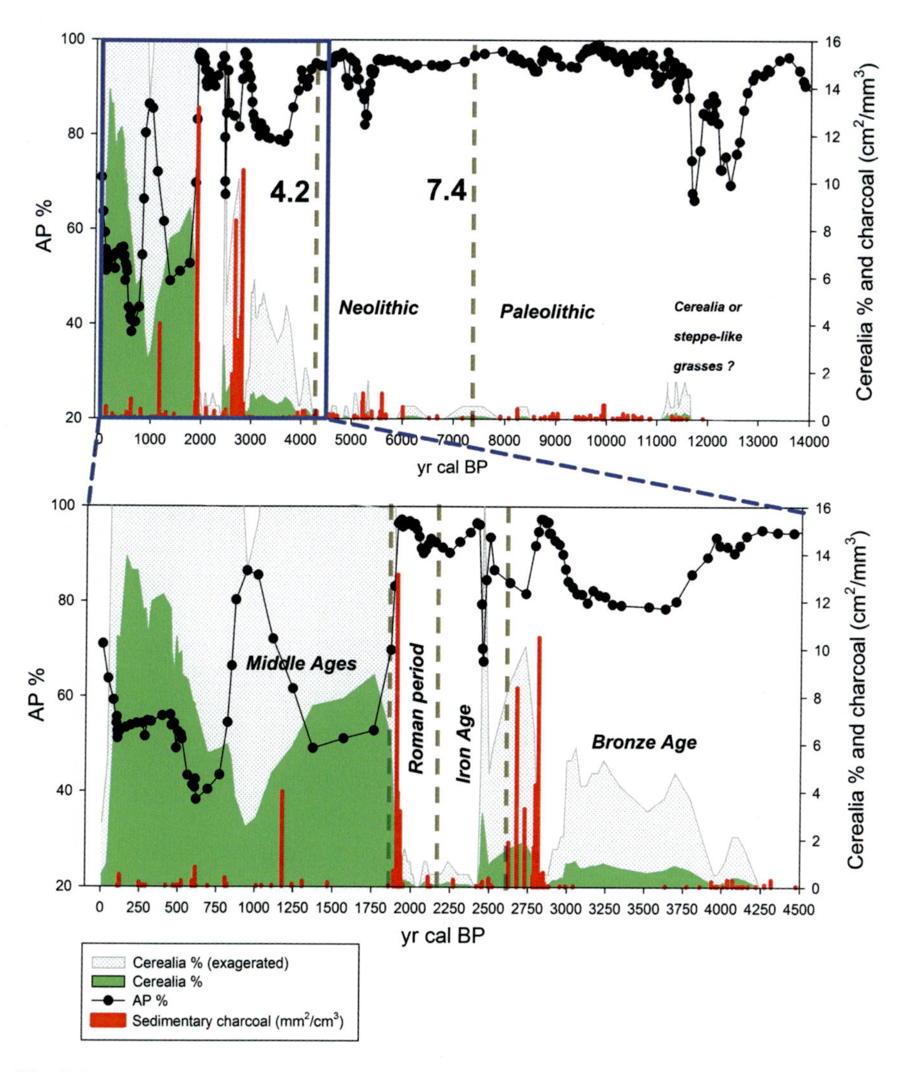

Fig. 5.5 The agricultural system at Coma de Burg [sedimentary charcoal partially published in Bal et al. (2011) and Gassiot et al. (2014)]

(decades or, in the best case, a few generations). There has been a highly accelerated and profound transformation as a result of human action during the 20th century, due to the exploitation of fossil fuels and technical and technological improvements. However, there also were notable changes in intensity level at the transition from the Roman period to the Late Roman or Late Middle Ages. From a forest point of view, the Pyrenees were more radically transformed during the first centuries of our era than what is now occurring—or at least the changes are comparable—and marked a threshold that represented the beginning of a deforestation of the space that will never again return to 'the way it was' (Fig. 5.5). In addition, we cannot uncouple the development of new agricultural spaces from the disappearance of forest species, such as the loss of beech from the Mata, València d'Àneu (Pèlachs et al. 2009b). Similarly, we cannot separate the use and management of fire as a means of creating clearings in the woods from other aspects of agricultural and livestock management over almost the past 3000 years (Bal et al. 2011).

It seems clear that management of natural spaces cannot dispense with an analysis of human influence at a very local scale. Also, the generalized fragmentation of habitats and the widespread creation of open spaces occurred 'only' during the past three millennia. The high temporal resolution seems to be one of the keys to studies of the dynamics of natural systems, but the high spatial resolution is the only way to test its effects. Among the examples from mountain areas in Spain, we would point out studies of the effects of agricultural abandonment around the Cadí (Soriano 1994; Molina 2000), the intensity of charcoal kilns and forest exploitation in the Viros forest of Vallferrera (Pèlachs et al. 2009a), or high-altitude vegetation dynamics in the Pyrenees (Cunill et al. 2013).

Nonetheless, this focus on details does not obscure the importance of each piece forming part of a larger picture; management also must be global because the key to good planning will always be found at medium scale and in the overall vision. At what scale, then, should we consider habitat diversity? As we have seen, geohistory does not encourage a generic response; on the contrary, accurate data are needed to be able to integrate different scenarios that combine various levels of intervention and fragmentation and guarantee biodiversity based on diverse formulas. Management of the natural surroundings in mountain areas can never be approached at the global level.

5.2.5 *Local Environments Appear to Have no Resilience!*

In analyzing local-level dynamics, we need to take into account the observation that "the search for the universal in the infinitesimally small" is a common theme in most cultures (Haskel 2014: 12). If we apply this concept to forest management and planning, the key question is the minimum surface area a species must occupy before it has 'value.' The maximum protection of nature in our study area occurs in the National Park. With the park, the most restrictive areas are the Integral reserves (Article 24, Natural Spaces Law 12/1985, 13 June), spaces preserved from any human intervention where only scientific research activities and the sharing of natural values. The National Parks at Aigüestortes and Estany de Sant Maurici are very localized spaces characterized by their exceptional values. Nevertheless, in a National Park, it is pertinent to ask if this is the proper management policy. Why are the integral reserves in a National Park small sites and not large protected areas?

For example, the integral reserves of peat bog, such as Bassa Nera or Trescuro in Aigüestortes i Estany de Sant Maurici National Park, contain a large part of the natural values recognized by the Natura 2000 network (Carreras and Farré 2014). Within the National Park they are small sites, highly localized, with very specific and unique characteristics. At Bassa Nera pond, as explained by Pérez-Haase (2016), 110 species of vegetation have been identified in 35 inventories, of which 11 are designated as rare species. Their singularity is based on their presence in ditches and borders, spaces where the humid ecosystem offers the most stressful conditions.

Environmental geohistory helps us to think about whether it has always been this way: How long ago was it a peat bog? What would be its dynamic? Does it make sense to preserve or protect it? The study of local vegetation demonstrates the extraordinary variability and mobility of the particular place. In this sense, the ombrotrophic raised bog at Bassa Nera has functioned for 60 years, alternating between emerging and submerging episodes for most of the Holocene with high environmental sensitivity (Fig. 5.3b). In other words, during the same period of time one of them could be an emerging peat bog but the other a submerging one. The point, in this sense, is to protect the Beta diversity. The substantial variability of these parts of the landscape, with a definite intra-annual seasonality (Catalan et al. 2013), convert the surrounding into places of extreme sensitivity, where resilience seems not to exist. Changes occur from one day to the next, and cannot be reversed. Therefore, the key is to adjudicate them in zones with maximum protection.

5.3 Final Considerations

Environmental geohistory is a useful tool for the understanding and management of natural systems in high-mountain areas. It contributes information at different scales but acquires its greatest usefulness at mid-scale and in explaining differences in various parts of the territory. High-mountain natural systems are sensitive to climate change, but also to the human management that clearly had begun by the Neolithic, became significant over the past three millennia, and were present everywhere starting in the Middle Ages.

The examples provided show the following:

(1) The relationship between environmental geohistory researchers and environmental managers is essential. Incorporation of a long-term perspective is necessary to making decisions for the future. The error is found in attempts to replicate the past because the choices made will always be arbitrarily based on a point in time and space that cannot be repeated. Therefore, the past may be an imperfect model for the future, but it is an essential consideration because it reveals the possibilities for biodiversity.

(2) Policies for the restoration of a natural environment that are based on concepts such as baseline, RNV, threshold, resilience, etc. are essential for theoretical reflection but should not be either the key factor nor the excuse for decision-making by managers. The modern key is found in the environmental values that are to be conserved, which requires that the natural dynamic of the

system be incorporated (i.e., the protected boundaries must be neither fixed nor stable) and that human impact be acknowledged, along with the degree of human disturbance that is necessary to conserve biodiversity in a climate scenario that is never the same. All options for the future must pass through the filters of environmental diversity and overall protection of the territory.

The specialization processes that resulted from exploiting the natural resources of each valley have influenced the configuration of the current landscape and require that managers make localized and global decisions. Therefore, the environmental geohistory of firs in the Pyrenees has shown that the current distribution is the least extensive since initial colonization. Human management of the forest, as well as climate factors, might have favored their presence during the Roman period.

The data indicate that this is the first time in the last 2000 years that the farming and livestock landscape opened by humans is being abandoned, giving way to rapid recovery of arboreal cover, on a human timescale: 60 years. Criteria should be established to determine whether historical evidence has been appropriately considered in making a decision to maintain a cultural landscape or return it to its wild state.

The fragmentation of the landscape that has persisted over the past two millennia is leading to forest homogeneity. The role and characterization of secondary communities and potential natural vegetation should be considered at different time scales as part of future studies analyzing the phases of stability and the periods of profound changes.

Acknowledgements This chapter was made possible by two coordinated project grants from Spain's Ministry of Economics and Competitiveness (MEC), "El uso del fuego y la conformació de los paisajes en la Montaña cantábrica y el Pirineo oriental: estudio comparado de su evolución historica y tendencias actuales" (CSO2012-39680-C02-01), awarded to the Department of Geography, Urban Studies and Land Planning, Universidad de Cantabria and "Geohistoria ambiental del fuego en el Holoceno. Patrones culturales y gestión territorial desde el inicio de la ganadería y la agricultura en la montañana Cantábrica y Pirineo," awarded to the Department of Geography, Universitat Autònoma de Barcelona (CSO2012-39680-C02-02). In addition, the project was funded by the Catalan government's applied geography program, 'Grup de Geografia Aplicada' (AGAUR, Generalitat de Catalunya, 2014 SGR 1090). Raquel Cunill gratefully acknowledges the Government of the Principality of Andorra, for the research grant on Andorran Issues, APTA007-AND/2014.

The authors appreciate the English language review by Elaine M. Lilly, Ph.D.

References

Alagona PS, Sandlos J, Wiersma YF (2012) Past imperfect: using historical ecology and baseline data for conservation and restoration projects in North America. Environ Philos 9(1):49–70

Alba-Sánchez F, López-Sáez JA, Benito-de Pando B, Linares JC, Nieto-Lugilde D, López-Merino L (2010) Past and present potential distribution of the Iberian Abies species: a phytogeographic approach using fossil pollen data and species distribution models. Divers Distrib 16:214–228

Bal MC, Pèlachs A, Pérez-Obiol R, Cunill R (2011) Fire history and human activities during the last 3300 cal yr BP in Spain's Central Pyrenees: the case of the Estany de Burg. Palaeogeogr Palaeocl 300:179–190

Battarbe RW, Gasse F, Stickley C (eds) (2005) Past climate variability through Europe and Africa, vol 6. Springer, Dordrecht

Boada M, Saurí D (2002) El canvi global. Rubes, Barcelona

Bradshaw RHW, Jones CS, Edwards SJ, Hannon GE (2015) Forest continuity and conservation value in Western Europe. Holocene 25:194–202

Carcaillet C, Muller S (2005) Holocene tree-limit and distribution of *Abies alba* in the inner French Alps: anthropogenic or climatic changes? Boreas 34:468–476

Carreras C, Ferré A (eds) (2014) Cartografia dels hábitats de Catalunya versió 2. Manual d'interpretació, Barcelona, El Tinter

Catalan J, Pla-Rabés S, Wolfe AP, Smol JP, Rühland KM, Anderson NJ, Kopácek J, Stuchlík E, Schmidt R, Koinig KA, Ll Camarero, Flower RJ, Heiri O, Kamenik C, Korhola A, Leavitt PR, Psenner R, Renberg I (2013) Global change revealed by palaeolimnological records from remote lakes: a review. J Paleolimnol 49:513–535

Costa M, Morla C, Sainz H (eds) (1998) Los bosques ibéricos. Una interpretación geobotánica, Barcelona, Planeta

Cunill R, Soriano JM, Bal MC, Pèlachs A, Rodríguez JM, Pérez-Obiol R (2013) Holocene high-altitude vegetation dynamics in the Pyrenees: a pedoanthracology contribution to an interdisciplinary approach. Quat Int 289:60–70

Cunill R, Métailié JP, Galop D, Poublanc S, Munnik N (2015) Palaeoecological study of Pyrenean lowland fir forests: exploring mid-late Holocene history of *Abies alba* in Montbrun (Ariège, France). Quat Int 366:37–50

Dearing J, Battarbee R (2007) Past human-climate-ecosystem interactions (PHAROS). PAGES News 15(1):8–10

Dietl GP, Flessa KW (2011) Conservation paleobiology: putting the dead to work. Trends Ecol Evol 26:30–37

Duarte CM (2006) Cambio Global: Impacto de la actividad humana sobre el sistema Tierra. Consejo Superio de Investigaciones Científicas, Madrid

Esteban A (Coord), Oliver J, Còts P, Pèlachs A, Mendizàbal E, Soriano JM, Nasarre E, Matamala N (2003) La humanización de las altas cuencas de la Garona y las Nogueras (4500 aC–1955 dC). Servicio Nacional de Parques Nacionales, Madrid

Fontana J (1992) La història després de la fi de la història. Reflexions i elements per a una guia dels corrents actuals. Vic, Institut Universitari d'Història Jaume Vicens Vives i Eumo editorial

Froyd CA, Willis KJ (2008) Emerging issues in biodiversity and conservation management: the need for a palaeoecological perspective. Quat Sci Rev 27:1723–1732

Galop D, Rius D, Cugny C, Mazier F (2013) A history of long-term human-environment interactions in the French Pyrenees inferred from the pollen data. Continuity and change in cultural adaptation to mountain environments. Springer, New York, pp 19–30

García-Ruiz JM, Lasanta T, Ruiz-Flano P, Ortigosa L, White S, González C, Martí C (1996) Land-use changes and sustainable development in mountain areas: a case study in the Spanish Pyrenees. Landscape Ecol 11:267–277

Gassiot E, Rodríguez D, Pèlachs A, Pérez-Obiol R, Julià R, Bal MC, Mazzucco N (2014) La alta montaña durante la Prehistoria: 10 años de investigación en el Pirineo catalán occidental. Trabajos de Prehistoria 71(2):261–281

Grau O, Ninot JM, Cornelissen JC, Callaghan TV (2013) Similar tree seedling responses to shrubs and to simulated environmental changes at Pyrenean and subarctic treelines. Plant Ecol Divers 6(3–4):329–342

Haskel DG (2014) En un metro de bosque: Un año observando la naturaleza. Turner, Madrid

Huber UM, Bugmann HKM, Reasoner MA (2005) Global change and mountain regions. An overview of current knowledge. Springer, Dordrecht

IPCC (2014) Climate change 2014: mitigation of climate change. In: Edenhofer O, Pichs-Madruga R, Sokona Y, Farahani E, Kadner S, Seyboth K, Adler A, Baum I, Brunner S, Eickemeier P,

Kriemann B, Savolainen J, Schlömer S, von Stechow C, Zwickel T, Minx JC (eds) Contribution of working group III to the fifth assessment report of the intergovernmental panel on climate change. Cambridge University Press, Cambridge, United Kingdom and New York, NY, USA
Jackson ST, Richard JH (2009) Restoration in the light of ecological history. Science 325:567–568
Jalut G (1988) Les principales etapes de l'histoire de la forêt pyrénénne française depuis 15,000 ans. Monografías del Instituto Pirenaico de Ecología 4:609–615
Jalut G, Galop D, Belet JM, Aubert S, Esteban A, Bouchette A, Dedoubat JJ, Fontugne M (1998) Histoire des forêts du versant nord des Pyrénées au cours des 30000 dernières années. J Bot Soc Bot Fr 5:73–84
Last WM, Smol P (eds) (2001) Tracking environmental change using lake sediments, Volume 1: Basin analysis, coring, and chronological techniques, vol 1. Kluwer Academic Publishers, New York, Boston, Dordrecht, London, Moscow
Lindbladh M, Brunet J, Hannon G, Niklasson M, Eliasson P, Eriksson G, Ekstrand A (2007) Forest history as a basis for ecosystem restorationda multidisciplinary case study in a south Swedish temperate landscape. Restor Ecol 15:284–295
Margalef R (1991) Teoría de los sistemas ecológicos. Barcelona, Univ, de Barcelona
Matías L, Gonzalez-Díaz P, Quero JL, Camarero JJ, Lloret F, Jump AS (2016) Role of geographical provenance in the response of silver fir seedlings to experimental warming and drought. Tree Physiol 00:1–11
Mendizábal E (2013) ¿Hay alguna geografía humana que no sea geografía histórica? Revista de Geografía Norte Grande 54:31–49
Molina D (2000) Conservació i degradació de sòls a les àrees de muntanya en procés d'abandonament. La fertilitat del sòl al Parc Natural del Cadí-Moixeró. Bellaterra, Departement de Geografia, Universitat Autònoma de Barcelona
Nakagawa T, Edouard JL, de Beaulieu JL (2000) A scanning electron microscopy (SEM) study of sediments from Lake Cristol, southern French Alps, with special reference to the identification of Pinus cembra and other Alpine Pinus species based on SEM pollen morphology. Rev Palaeobot Palyno 108(1–2):1–15
Pèlachs A (2006) Algunes reflexions sobre geografia, paisatge i geohistòria ambiental. Documents d'Anàlisi Geogràfica 48:179–192
Pèlachs A, Nadal J, Soriano JM, Molina D, Cunill R (2009a) Changes in Pyrenean woodlands as a result of the intensity of human exploitation: 2,000 years of metallurgy in Vallferrera, northeast Iberian Peninsula. Veg Hist Archaeobot 18(5):403–416
Pèlachs A, Pérez-Obiol R, Ninyerola M, Nadal J (2009b) Landscape dynamics of Abies and Fagus in the southern Pyrenees during the last 2200 years as a result of anthropogenic impacts. Rev Palaeobot Palyno 156(3–4):337–349
Pèlachs A, Julià R, Pérez-Obiol R, Soriano JM, Bal MC, Cunill R, Catalan J (2011) Potential influence of bond events on mid-holocene climate and vegetation in southern Pyrenees as assessed from Burg lake LOI and pollen records. The Holocene 21(1):95–104
Pèlachs A, Pérez-Obiol R, Soriano JM, Pérez-Haase A (2016) Dinàmica de la vegetació, contaminació ambiental i incendis durant els últims 10.000 anys a la Bassa Nera (Val d'Aran). In: La investigació al Parc Nacional d'Aigüestortes i Estany de Sant Maurici. X Jornades sobre Recerca al Parc Nacional d'Aigüestortes i Estany de Sant Maurici. [Boí]: Parc Nacional d'Aigüestortes i Estany de Sant Maurici; Barcelona: Generalitat de Catalunya. Departament de Territori i Sostenibilitat, 2016, pp 75–87
Pérez-Haase A (2016) Patrons estructurals, ecològics i biogeogràfics en vegetació de molleres i de torberes d'esfagnes. Barcelona, Facultat Biologia de la Universitat de Barcelona. Unpublished doctoral dissertation
Pérez-Obiol R, Bal MC, Pèlachs A, Cunill R, Soriano JM (2012) Vegetation dynamics and anthropogenically forced changes in the Estanilles peat bog (southern Pyrenees) during the last seven millennia. Veg Hist Archaeobot 21(4–5):385–396
Richardson DM, Whittaker RJ (2010) Conservation biogeography—foundations, concepts and challenges. Divers Distrib 16:313–320

Sancho-Knapik D, Peguero-Pinal JJ, Cremer E, Camarero JJ, Fernández-Cancio Á, Ibarra N, Konnert M, Gil-Pelegrín E (2014) Genetic and environmental characterization of *Abies alba* Mill. Populations at its western rear edge. Pirineos 169 Enero-Diciembre, e007
Schneider R, Tobolski K (1985) Lago di Ganna-Late-glacial and Holocene environments of a lake in the Southern Alps. Diss Bot 87:229–271
Soriano JM (1994) Efectes del despoblament sobre el medi físic d'un territori de muntanya (Tuixén, Parc Natural Cadí-Moixeró). Estudi de la variació de la fertilitat del sòl en camps de conreu abandonats. Bellaterra, Departament de Geografia, Universitat Autònoma de Barcelona
Sutherland WJ, Adams WM, Aronson RB, Aveling R, Blackburn TM, Broad S, Ceballos G, Cote IM, Cowling RM, Da Fonseca GA, Dinerstein E, Ferraro PJ, Fleishman E, Gascon C, Hunter M Jr, Hutton J, Kareiva P, Kuria A, Macdonald DW, Mackinnon K, Madgwick FJ, Mascia MB, McNeely J, Milner-Gulland EJ, Moon S, Morley CG, Nelson S, Osborn D, Pai M, Parsons EC, Peck LS, Possingham H, Prior SV, Pullin AS, Rands MR, Ranganathan J, Redford KH, Rodriguez JP, Seymour F, Sobel J, Sodhi NS, Stott A, Vance-Borland K, Watkinson AR (2009) One hundred questions of importance to the conservation of global biological diversity. Conserv Biol 23:557–567
Tinner W, Hubschmid P, Wehrli M, Ammann B, Conedera M (1999) Long-term forest fire ecology and dynamics in Southern Switzerland. J Ecol 87(2):273–289
UNEP-WCMC and IUCN (2016) Protected planet report 2016 UNEP-WCMC and IUCN, Cambridge UK and Gland, Switzerland
Whittaker RJ, Araújo MB, Jepson P, Ladle RJ, Watson JEM, Willis KJ (2005) Conservation biogeography: assessment and prospect. Divers Distrib 11:3–23
Wick L, Möhl A (2006) The mid-Holocene extinction of silver fir (*Abies alba*) in the Southern Alps: a consequence of forest fires? Palaeobotanical records and forest simulations. Veg Hist Archaeobot 15:435–444
Willis KJ, Birks HJB (2006) What is natural? The need for a long-term perspective in biodiversity conservation. Science 314:1261–1265
Willis KJ, Bhagwat SA (2009) Biodiversity and climate change. Science 326:806–807
Willis KJ, Bhagwat SA (2010) Questions of importance to the conservation of biological diversity: answers from the past. Clim Past 6:759–769
Willis KJ, Gillson L, Brncic TM (2004) How "virgin" is virgin rainforest? Science 304:402–403
Willis KJ, Bailey RM, Bhagwat SA, Birks HJB (2010) Biodiversity baselines, thresholds and resilience: testing predictions and assumptions using palaeoecological data. Trends Ecol Evol 25:583–591

Chapter 6
The Multiple Factors Explaining Decline in Mountain Forests: Historical Logging and Warming-Related Drought Stress is Causing Silver-Fir Dieback in the Aragón Pyrenees

J. Julio Camarero

Abstract The drivers and patterns of drought-related forest dieback are not as well understood in mountain conifer forests. Most studies have obviated the role of historical use as a predisposing factor of forest dieback. Here I focus on the recent silver-fir (*Abies alba*) dieback observed since the 1980s in the Aragón Pyrenees (NE Spain) as study case. I argue that such dieback was predisposed by past historical logging and incited by warming-induced drought. I analyzed environmental, structural and tree-ring data from 32 sites with contrasting degrees of dieback at the tree and stand levels. I found that a peak in late-summer water deficit observed in 1985 caused a severe growth reduction in 1986, resulting in subsequent crown defoliation, dieback and increased mortality. Dieback was more severe and widespread in western low-elevation mixed forests dominated by smaller trees with low growth rates. These marginal sites receive less late-summer rainfall, which is a key climatic variable controlling silver-fir growth, than eastern sites. Declining sites showed more frequent growth releases induced by historical logging than non-declining sites. Silver-fir growth is becoming more dependent on climatic conditions of previous September, which may be connected with changing modes of atmospheric variability affecting Iberian climate. Historical logging and warming-induced drought stress during late summer are the most likely predisposing and inciting factors driving silver-fir dieback, respectively. A sustainable management of mountain forests shaped by past historical use requires changing their current structure and composition to make them more resilient to climate warming.

Keywords *Abies alba* · Climate warming · Dendroecology · Radial growth · Releases

---

J.J. Camarero (✉)
Instituto Pirenaico de Ecología (IPE-CSIC), Avda. Montañana 1005,
50192 Zaragoza, Spain
e-mail: jjcamarero@ipe.csic.es

J. Catalan et al. (eds.), *High Mountain Conservation in a Changing World*,
Advances in Global Change Research 62, DOI 10.1007/978-3-319-55982-7_6

6.1 Introduction

Forests store almost half of terrestrial carbon, but the long-term net carbon uptake by forests is a slow process mainly controlled by the growth rate of woody tissues (Bonan 2008). Contrastingly, huge carbon emissions may occur rapidly from sudden or widespread mortality episodes often preceded by forest dieback and growth decline (Breshears and Allen 2002). In water-limited regions, climate warming may intensify drought stress and lead to growth decline and forest dieback (Allen et al. 2010, 2015). This dieback phenomenon is linked to rapid defoliation and selective mortality of overstory trees (McDowell et al. 2008). However, in mountain and temperate forests, the factors causing forest dieback under more mesic conditions are not as well understood (van Mantgem and Stephenson 2007).

Mountains are characterised by: (i) a high heterogeneity in local climate conditions which change over short distances as a function of altitude and topography and create steep ecological gradients (Barry 2008), and (ii) by preserving large forested areas subjected to an extended historical management, particularly in Europe (Kirby and Watkins 1998). It is predicted for the twenty-first century a greater warming in the mountains than in the lowlands of the Northern Hemisphere (Kohler et al. 2014). Mountain forests face climatic gradients which limit their productivity, cold stress towards the uppermost treeline but also drought stress downwards, and make them sensitive to climate warming but also to changes in management. The management and use of these forests in Europe have changed rapidly during the past century. Mountain forests provide many ecosystem services (carbon uptake, regulation of water cycles, protection from snow avalanches, biodiversity conservation, etc.) but they also represent a source of timber, biomass for energy production and non-woody goods (e.g. mushrooms, hunting). During the twentieth century, timber extraction was still profitable in some European forests as those formerly exploited at the Spanish Pyrenees (Cabrera 2001). However, currently, many of these mountain forests are not managed for timber production because commercial forestry is no longer profitable. In this chapter, I explore how changes in historical management interact with climate warming and intensified drought stress to trigger dieback in mountain forests.

Here I argue that the historical use of forests strongly interacts with current climatic trends as the rise in temperatures to determine the current fate of forests, which in some case can jeopardise its future. I will illustrate how past historical logging and recent warming-related drought stress contribute to silver-fir dieback in Pyrenean mountain forests subjected to mesic conditions. The processes leading to forest dieback are still poorly understood because of the interaction of several stress factors acting at different spatiotemporal scales, which complicates the disentangling of lagged cause–effect relationships (Pedersen 1999). Many dieback episodes have been studied following Manion's (1981) conceptual model, which includes predisposing, inciting and contributing stress factors causing a loss in tree vigour. Predisposing factors such as site conditions reduce a tree's vigour over the long term (Suarez et al. 2004), whereas inciting factors such as drought lead to a sharp and short-term reduction in tree growth and vitality (Bigler et al. 2006). Other factors (mainly pathogens and insects) may contribute to dieback acting as secondary stress factors. According to this model, the most sensitive trees to short-term inciting stress factors will be those that were previously most strongly weakened by long-term predisposing factors. Here I use this conceptual model to assess the roles played by long-term predisposing (historical management) and short-term inciting (drought stress) factors as drivers of dieback in mountain forests.

Most studies on forest dieback have obviated the potential role of historical land use (e.g. past logging) as a predisposing factor (but see Linares et al. 2009). Furthermore, the different nature of interacting factors such as land-use legacies (e.g. past logging) and climatic extremes (e.g. severe droughts) have precluded considering the interactions between them. In Europe, historical effects have persisted for decades and centuries shaping the current structure of most mountain conifer forests (Kirby and Watkins 1998). Therefore, past forest use should be considered as an additional driver of dieback and its role should be assessed.

Here I focus on dieback episodes of silver-fir (*Abies alba* Mill.) forests reported since the 1980s in the Western and Central Spanish Pyrenees (Navarra and Aragón Pyrenees; see Fig. 6.1), near the south-western limit of the species' distribution area (Camarero et al. 2002). Silver-fir dieback has been systematically reported in the 1970s and 1980s across central Europe (Skelly and Innes 1994). In the Aragón Pyrenees silver-fir dieback was more severe in western stands located at medium elevation than in eastern high-elevation sites. In the most affected stands, up to 30–50% of trees showed severe defoliation, which was related to the occurrence of severe summer droughts in the 1980s (Camarero et al. 2002). Therefore, it may be hypothesised that drought stress has recently increased by climate warming and precipitation regime shifts causing silver-fir dieback. In addition, most of these forests were logged to extract timber up to the 1970s when their management ceased due to rural migration to cities (Cabrera 2001). So, it remains to be answered if those forests that were more intensively logged in the past were also more predisposed to drought-triggered dieback during the 1980s and more recently. In this study, I address the following questions: (1) How did silver-fir growth change in the Aragón Pyrenees during the twentieth century and how was it affected by the recent patterns of defoliation? (2) Did historical logging and warming-induced drought cause the recent silver-fir dieback? To answer these questions, I will focus on the retrospective analyses of tree-ring data.

6.1.1 Geographical and Climatic Backgrounds

The Pyrenees constitute a transitional mountainous area between more humid conditions in their northern margin, where Eurosiberian vegetation is dominant, and drier conditions southwards (Vigo and Ninot 1987). This gradient overlaps with a similarly relevant longitudinal gradient caused by the location of the range between the Atlantic Ocean and the Mediterranean Sea. According to meteorological data

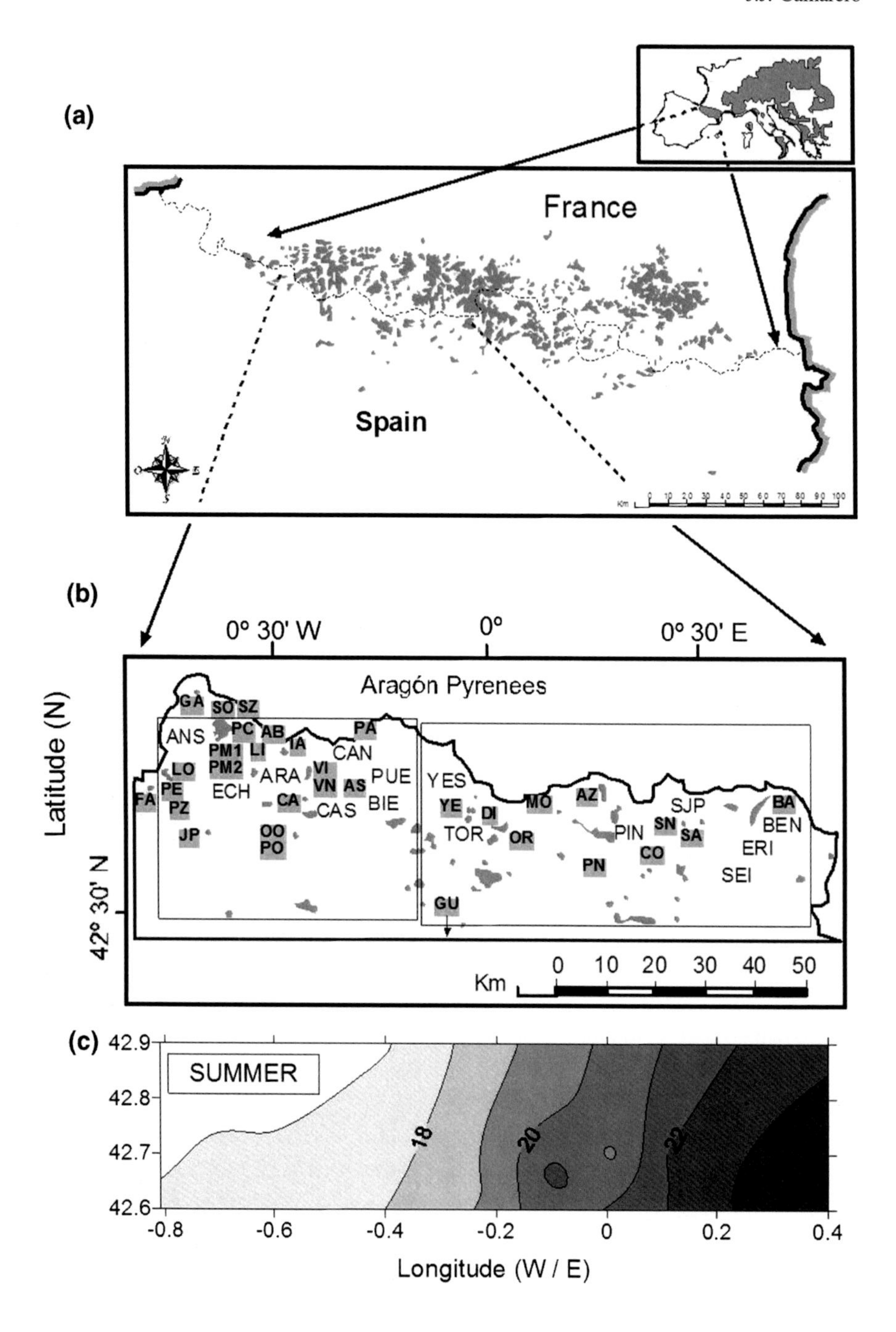
(a)
France
Spain
(b)
0° 30' W
0°
0° 30' E
Aragón Pyrenees
Latitude (N)
42° 30' N
GA
SO
SZ
ANS
PC
AB
IA
PA
PM1
LI
CAN
LO
PM2
ARA
VI
PUE
YES
FA
PE
ECH
VN
AS
MO
AZ
SJP
BA
PZ
CA
CAS
BIE
YE
DI
SN
BEN
JP
OO
PO
TOR
OR
PIN
SA
ERI
CO
PN
SEI
GU
Km
0 10 20 30 40 50
(c)
SUMMER
18
20
22
42.9
42.8
42.7
42.6
-0.8
-0.6
-0.4
-0.2
0
0.2
0.4
Longitude (W / E)

◂**Fig. 6.1** Geographical and climatic context of the study silver-fir forests. Distribution of silver fir in Europe (**a**) and study sites in the Aragón Pyrenees, northeastern Spain (sites' codes are indicated with *bold letters*) (**b**). The *two rectangles* in the *lower figure* delineate the two homogeneous climatic sub-regions based on local data from the displayed climatic stations (indicated by *three-letter codes*): Western (WAP) and Eastern (EAP) Aragón Pyrenees. **c** The longitudinal gradient of percent summer precipitation in the study area interpolated through kriging

from nearby stations, the climate in the study area can be described as continental with oceanic (western sites) or Mediterranean (eastern and southern sites) influences. The westward oceanic influence leads to greater precipitation in winter and a smaller temperature range than eastwards. This oceanic influence decreases eastwards where the Mediterranean influence prevails being characterised by higher precipitation in summer than westwards (Fig. 6.1c).

The studied silver-fir populations are located in the Aragón Pyrenees, NE Spain (Fig. 6.1; ca. 5600 km^2). The main geographic and topographic characteristics of the 32 sampled stands are described in Table 6.1. In the Aragón Pyrenees, silver-fir stands are usually found at humid sites on north-facing slopes where they form pure or mixed stands with European beech (*Fagus sylvatica* L.) or Scots pine (*Pinus sylvestris* L.). Silver-fir forests in the study area may experience summer drought stress in August, despite they receive a total annual precipitation between 900 and 2000 mm, which usually increases with elevation. Most studied stands are located on marls and limestones, which generate basic soils, or on moraine deposits with rocky but deep soils.

Because data from local meteorological stations are often inadequate to study the spatiotemporal variation of mountain climates, regional records averaging the longest and most complete local data available from the study area were created (see Camarero et al. 2011). I used monthly climatic data (mean temperature, total precipitation) to delineate two homogeneous and distinct climatic areas within the Aragón Pyrenees, hereafter abbreviated as WAP (western Aragón Pyrenees) and EAP (eastern Aragón Pyrenees) sub-regions. Mean temperature and precipitation data for each sub-region were obtained for the period 1940–1999. Finally, I calculated annual and cumulative monthly water deficits for both climatic sub-regions using a modified Thornthwaite water-budget procedure (see Camarero et al. 2011).

6.1.2 Field Sampling

Field sampling was conducted between 1999 and 2015. At least one site was sampled in each 10-km^2 grids covering the Aragón Pyrenees where silver-fir formed forests (Fig. 6.1). Because I was primarily interested in discerning the causes of forest dieback, more stands in the most affected areas were sampled (e.g. Camarero et al. 2015). In total, 21 and 11 sites were sampled within the WAP and EAP sub-regions, respectively, (Table 6.1). At each site, 10–15 dominant trees were selected for sampling within a 500-m long and 20-m wide transect randomly

Table 6.1 Main characteristics of the sampled Pyrenean silver-fir forests (means ± SE). Sites' codes are as in Fig. 6.1

Code	Elevation (m)	Dbh (cm)	Basal area (m^2 ha^{-1})	Age at 1.3 m (years)	Dead trees (%)	No. trees (No. radii)	Tree-ring width (mm)	Climate-R^2 (%)[a]
FA	918 ± 3	61.3 ± 1.6	45.9	96 ± 3	0.0	12 (24)	2.90 ± 0.12	49.14
PE	1232 ± 1	35.0 ± 2.3	10.1	88 ± 5	14.3	15 (39)	1.70 ± 0.09	50.83
PZ	1073 ± 3	43.0 ± 1.2	24.7	114 ± 7	8.3	11 (25)	1.67 ± 0.10	47.84
LO	1009 ± 3	38.1 ± 2.7	24.8	104 ± 6	12.0	10 (23)	1.48 ± 0.11	51.82
GA	1400 ± 10	64.2 ± 2.2	55.8	129 ± 8	0.0	15 (26)	2.19 ± 0.20	41.96
SZ	1272 ± 5	58.2 ± 3.1	38.3	115 ± 9	4.7	13 (26)	2.57 ± 0.25	41.13
SO	1195 ± 6	66.6 ± 3.0	51.7	152 ± 14	5.0	12 (25)	2.31 ± 0.20	43.92
JP	1393 ± 33	46.0 ± 2.3	17.9	95 ± 9	0.0	13 (28)	2.37 ± 0.19	55.52
PM1	1353 ± 2	45.8 ± 2.3	31.7	97 ± 3	10.0	10 (22)	2.01 ± 0.13	49.29
PM2	1313 ± 13	54.8 ± 4.1	34.2	104 ± 8	9.3	10 (21)	2.50 ± 0.22	58.64
PC	1248 ± 3	47.5 ± 3.0	43.3	64 ± 5	5.0	13 (27)	3.67 ± 0.24	47.34
LI	1222 ± 3	74.1 ± 4.1	87.1	96 ± 6	0.0	11 (22)	3.08 ± 0.23	43.63
OO	1604 ± 15	59.1 ± 2.8	43.8	95 ± 4	0.0	10 (24)	2.88 ± 0.18	53.73
PO	1587 ± 17	46.5 ± 3.9	34.6	77 ± 6	0.0	11 (23)	2.62 ± 0.11	54.39
AB	1403 ± 9	75.0 ± 4.4	63.8	65 ± 6	0.0	12 (24)	4.70 ± 0.20	41.76
CA	1175 ± 15	41.2 ± 2.0	30.5	131 ± 9	0.0	10 (20)	1.31 ± 0.11	52.72
IA	1478 ± 5	69.0 ± 5.6	56.0	103 ± 15	4.7	13 (25)	3.23 ± 0.25	40.28
VN	1270 ± 2	37.4 ± 3.0	20.2	100 ± 7	2.0	10 (12)	1.88 ± 0.16	34.78
VI	1234 ± 4	42.7 ± 2.5	41.6	96 ± 4	0.0	14 (30)	2.08 ± 0.13	47.04
AS	1327 ± 3	60.6 ± 3.6	37.2	87 ± 7	5.0	10 (20)	3.26 ± 0.22	41.42
PA	1280 ± 4	71.0 ± 2.7	56.3	117 ± 9	0.0	12 (23)	2.50 ± 0.13	43.15
YE	1399 ± 4	48.1 ± 3.1	31.7	64 ± 4	0.0	12 (24)	3.52 ± 0.15	49.18
GU	1428 ± 9	52.5 ± 2.6	13.5	80 ± 9	0.0	10 (23)	3.11 ± 0.16	65.08

(continued)

Table 6.1 (continued)

Code	Elevation (m)	Dbh (cm)	Basal area (m^2 ha^{-1})	Age at 1.3 m (years)	Dead trees (%)	No. trees (No. radii)	Tree-ring width (mm)	Climate-R^2 (%)[a]
DI	1528 ± 4	56.5 ± 2.2	45.4	98 ± 6	0.0	12 (24)	2.75 ± 0.13	47.14
OR	1370 ± 5	42.3 ± 1.5	39.5	108 ± 4	0.0	11 (22)	1.73 ± 0.03	46.80
MO	1400 ± 30	46.1 ± 1.6	29.6	117 ± 9	0.0	15 (30)	1.61 ± 0.09	42.04
AZ	1613 ± 17	68.0 ± 3.6	33.3	90 ± 5	0.0	11 (22)	3.41 ± 0.17	42.51
PN	1519 ± 22	45.1 ± 2.2	28.5	78 ± 5	2.0	11 (22)	2.77 ± 0.18	45.01
CO	1474 ± 10	56.0 ± 4.4	29.9	83 ± 10	0.0	12 (27)	3.01 ± 0.21	47.50
SN	1431 ± 7	49.5 ± 3.5	29.1	74 ± 3	5.0	14 (29)	2.90 ± 0.14	48.27
SA	1789 ± 5	60.1 ± 2.8	36.5	117 ± 24	0.0	12 (29)	2.38 ± 0.25	33.16
BA	1600 ± 4	49.6 ± 2.3	48.6	107 ± 5	0.0	11 (29)	2.12 ± 0.09	48.60

Declining sites are indicated with underlined codes

[a]R^2: percentage of growth variance of the 1950–1999 residual ring-width chronologies explained by climate. No vigour data were taken in site MO where growth was studied using stem sections from recently felled trees

located within the stand. Several topographical variables were obtained for each site and tree. Elevation, aspect and slope steepness were measured at the tree level.

The diameter at 1.3 m (diameter at breast height, dbh) of each tree located within the transect was also measured, and I assessed their vigour using a semi-quantitative scale based on the percentage of crown defoliation (Müller and Stierlin 1990): class 0, 0–10% defoliation (healthy tree); (1) 11–25% (slight damage); (2) 26–50% (moderate damage); (3) 51–75% (severe damage); (4) 76–90% (dying tree); (5) dead trees with >91% defoliation or only retaining red needles. Since estimates of percent crown defoliation may vary among observers and places, I used as a reference a tree with the maximum amount of foliage at each site. Declining trees were considered as those with crown defoliation greater than 50%, and declining sites were regarded as those with more than 25% trees with such degree of defoliation. Dead trees were regarded as those whose crowns showed complete defoliation or only retained red needles and whose most recently formed rings corresponded to years prior to the sampling year. Lastly, the number and dbh of all neighbouring trees found within a circular plot of 7.62 m in radius placed around each subject tree was measured to estimate the basal area (m^2 ha^{-1}) of the silver-fir neighbourhood. Values are given as means ± standard errors throughout the text.

6.1.3 Tree-Ring Data

I followed established dendrochronological methods to analyse tree-ring data (Fritts 1976). Two or three cores were taken from each tree at breast height (1.3 m) using an increment borer. In the field, sapwood length was estimated visually whenever possible (n = 92 trees from 22 sites). The wood samples were air-dried and polished with a series of successively finer sandpaper grits. Then, wood samples were visually cross-dated. Tree rings were measured to the nearest 0.01 mm using a binocular scope and an LINTAB measuring device (Rinntech, Heidelberg, Germany). Cross-dating of the tree rings was checked using the program COFECHA (Holmes 1983). To calculate tree age at 1.3 m, in the case of cores without pith, a geometric method based on the curvature of the innermost tree ring was used to estimate the missing distance to the pith. Stem sections and cores with pith (n = 120) were used to calculate regressions between the distance to the pith and the number of tree rings ($r > 0.98$ and $P < 0.05$ in all cases).

The percentage growth change (GC) filter of Nowacki and Abrams (1997) was applied to identify abrupt and sustained increases or decreases in radial growth (i.e. releases or suppressions, respectively). First, I calculated the ring-width medians of subsequent 10-year periods along all the growth series because medians are more robust estimators of central tendency than means. The M1 and M2 values are defined as the preceding and subsequent 10-year ring-width medians of a given dated ring, respectively. For instance, the periods M1 = 1946–1955 and

M2 = 1956–1965 are used to calculated the percentage growth change at the year 1955. The percentage of positive (PGC) and negative (NGC) growth changes were calculated in yearly increments as:

$$\mathrm{PGC} = [(\mathrm{M2} - \mathrm{M1})/\mathrm{M1}] \cdot 100 \tag{6.1}$$

$$\mathrm{NGC} = [(\mathrm{M1} - \mathrm{M2})/\mathrm{M2}] \cdot 100 \tag{6.2}$$

Growth releases were then defined as those periods with at least five consecutive years showing PGC values greater than 75%.

Basal area increment (BAI, cm^2 $year^{-1}$) is assumed to be a more meaningful indicator of tree growth than tree-ring width because it removes variation in growth attributable to increasing circumference. Therefore, ring widths were converted to BAI assuming a circular outline of stem cross sections and using the formula:

$$\mathrm{BAI} = \pi\left(R_t^2 - R_{t-1}^2\right) \tag{6.3}$$

where R is the radius of the tree and t is the year of tree-ring formation. In dominant trees, BAI series usually show an early phase of low growth followed by a rapid increase and a final stable phase. Mean annual values of tree-ring width, growth change and BAI were obtained for declining and non-declining sites throughout the twentieth century.

6.1.4 Climate-Growth Analyses

To assess the growth-climate relationships, a tree-ring width chronology was established for each site (Table 6.1). For each tree, its ring-width series was double detrended using a negative linear or exponential function and a cubic smoothing spline with a 50% frequency response cutoff of 30 years to preserve high- and medium-frequency variability. A spline flexible enough to maximise the tree-to-tree shared growth variance and its response to climatic variability was selected following Macias et al. (2006). Autoregressive modelling was performed on each detrended ring-width series, which were finally averaged using a biweight robust mean to obtain residual site chronologies. All chronologies were built using the program ARSTAN (Cook and Krusic 2005). All further climate-growth analyses were performed using residual chronologies. The spatial and temporal relationships among these site chronologies for the period 1900–1999 were summarised using Principal Component Analysis (PCA).

I calculated 32 correlation functions relating each site chronology to the corresponding sub-regional climate dataset for the period 1950–1999. Climate-growth relationships were calculated using monthly mean temperature and total precipitation from the previous January up to September of the growth year.

Finally, to summarise these results and to define the main climatic response of declining versus non-declining sites, I performed a PCA on the matrix of the correlation function coefficients.

6.2 Warmer Climate Conditions and the 1980s Peak in Water Deficit

Overall, the climate is becoming warmer in the study area with a peak in drought stress during the 1980s. A strong warming trend was observed in the WAP during the late 1940s and in the 1980s (Fig. 6.2a). Precipitation decreased in the 1980s and increased during the 1960s. Furthermore, the relative contribution of summer rainfall to the total annual precipitation has significantly declined in the WAP area during the last half of the twentieth century (trend = −0.09; $P < 0.01$). The mean annual water deficit during the 1940–1999 period was significantly higher in the WAP (82.3 ± 8.3 mm) than in the EAP (49.6 ± 7.0 mm) sub-region. The estimated annual water deficit peaked in 1985 and 1967 in both sub-regions but always reaching a greater annual deficit in the WAP sub-region than eastwards (Fig. 6.2b).

6.3 Structural Features of Silver-Fir Stands Presenting Dieback: Low Growth Rates

Most sites showing severe defoliation and dieback were located in the WAP sub-region at mid elevation (Table 6.1; Fig. 6.3b). These stands were dominated by silver-fir trees with low dbh and ages ranging 90–155 years old and corresponded to mixed forests with a low basal area. On average, we found a significantly higher frequency (Mann–Whitney test, $U = 7.5$, $P < 0.001$) of dead trees in declining (10.8 ± 1.1%) than in non-declining sites (1.9 ± 0.4%; Table 6.1).

The low productivity of declining sites is confirmed by the fact that radial growth as a function of cambial age was consistently lower there than in non-declining sites (Fig. 6.4). The most defoliated trees from declining sites also tended to show less sapwood area ($r_s = -0.26$, $P < 0.05$). Declining sites were also those where more releases were detected since we found a significant positive relationship ($r_s = 0.88$, $P < 0.001$) between the stand mean defoliation and the average number of releases per tree (Fig. 6.5a). For instance, 58% of defoliated (crown defoliation > 50%) trees from declining sites showed at least two releases before the 1980s, whereas only 35% of trees without defoliation from non-declining sites did (Fig. 6.5b).

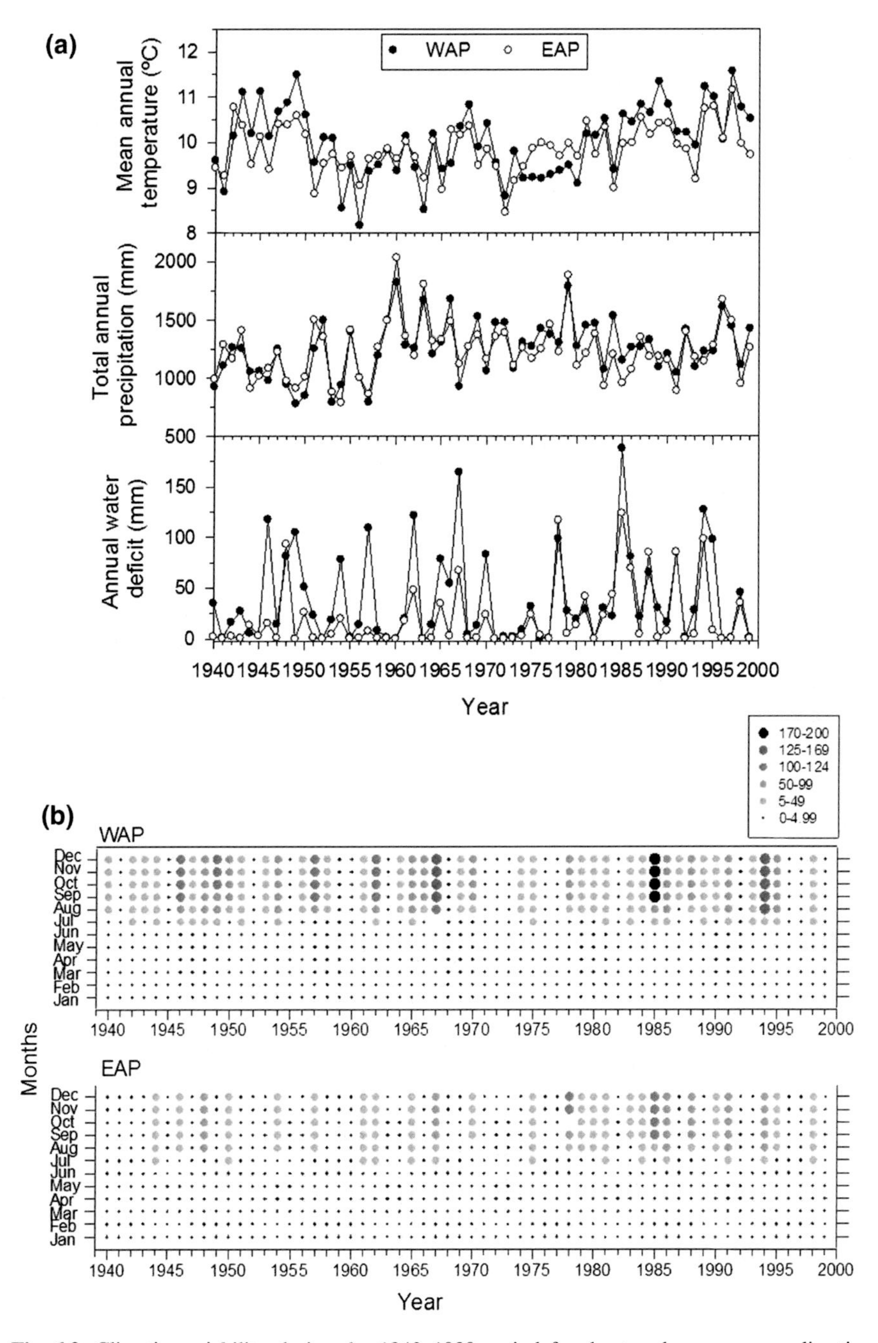

Fig. 6.2 Climatic variability during the 1940–1999 period for the two homogeneous climatic sub-regions: western (WAP) and eastern (EAP) Aragón Pyrenees. **a** Mean annual temperature, total annual precipitation and annual water deficit. **b** Cumulative monthly water deficits (the scale is in mm)

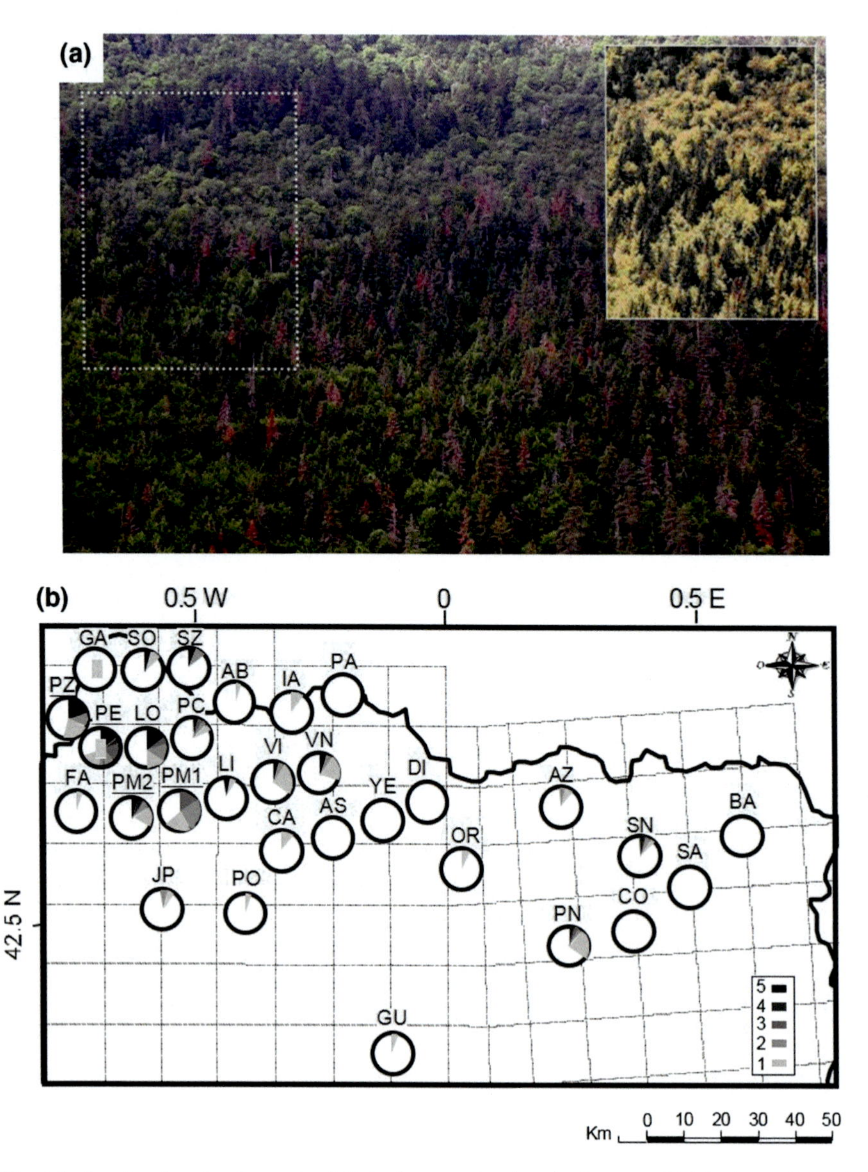

Fig. 6.3 Local and regional aspects of dieback across the Aragón Pyrenees, NE Spain. **a** View of a declining silver-fir stand situated in Paco Ezpela (abbreviated as PE). *Red* and defoliated trees are declining silver firs. The *dark-* and *clear-green* trees are healthy firs or pines and beeches, respectively. Note also the logged areas in the *upper area* of the image. The photograph was taken in June 2006, and the *upper inset* photograph corresponding to the area outlined by *dashed line* was taken in August 2001. Forest dieback has been widespread on this site since 1986. **b** Geographical patterns of crown defoliation in Aragón silver-fir forests. The *graph* shows the percentage of trees in each stand with different defoliation levels (class (1) 11–25% crown defoliation; (2) 26–50%; (3) 51–75%; (4) 76–90%; (5) >91%, dead tree). Sites' codes are as in Fig. 6.1. Declining sites correspond to *underlined codes* (sites with more than 25% trees with crown defoliation > 50%). The grid corresponds to 10 $km^2 \times 10$ km^2

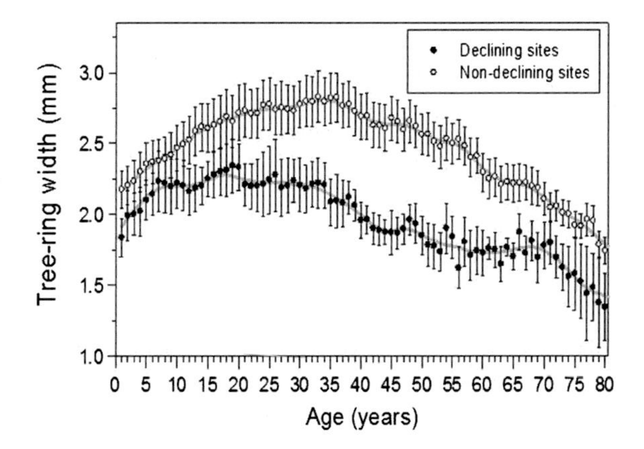

Fig. 6.4 The mean radial growth of silver-fir trees from declining and non-declining sites according to the cambial age of tree-ring formation. The smoothed growth curves (*grey lines*) were obtained using a loess function with a 0.1 smoothing parameter (span)

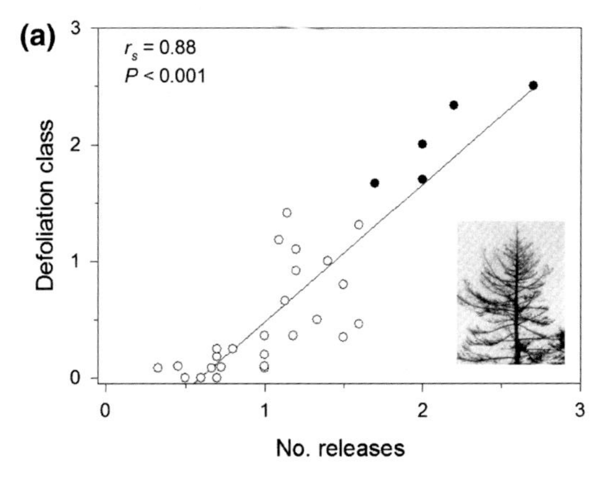

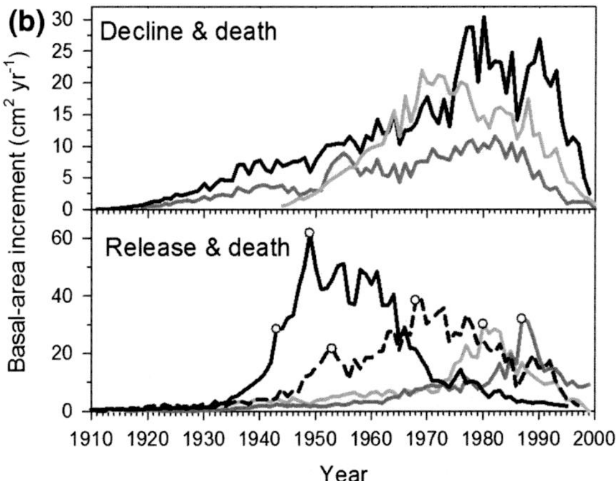

Fig. 6.5 Declining sites were characterised by defoliated trees with growth releases. **a** The positive relationship between the stand mean defoliation and the mean number of releases detected per tree for the 1900–1980 period. Defoliation classes are as in Fig. 6.3. **b** Characteristic radial-growth (basal area increment) patterns of declining silver-fir trees, which died recently, without (*upper graph*) and with (*lower graph*) releases previous to the 1980s dieback. The *white symbols* indicate releases

6.4 Growth Trends of Silver Fir Indicate that Dieback is Predisposed by Past Logging

Tree-ring width was significantly ($F = 6.17$, $P = 0.02$) lower (1.9 ± 0.2 mm) in declining than in non-declining (2.7 ± 0.1 mm) sites. In addition, the percentage of growth variability explained by climate, calculated using multiple linear regressions, was significantly ($F = 4.55$, $P = 0.04$) higher in declining (52.3 ± 1.6%) than in non-declining sites (45.9 ± 1.3%; Table 6.1). Such percentage was highest (57%) in the southernmost sites (e.g. GU) which experience a greater drought stress than more northerly sites (Table 6.2). Thus, declining and southern sites showed a greater responsiveness to climatic stress than the rest of sites. However, no southern site showed signs of recent dieback such as severe defoliation or reduced radial growth. Crown defoliation was negatively associated with site longitude and elevation, and with tree dbh (Table 6.2).

The basal area increment of declining sites diverged from that of non-declining sites since the 1940s (Fig. 6.6). Nevertheless, both types of sites showed similar growth trends and short-term responses to climatic stress such as a very narrow ring in 1986 when the NGC reached minimum values everywhere. Such severe growth reduction was unprecedented during the twentieth century.

Declining sites showed a greater frequency of trees with releases than non-declining sites during several decades (e.g. the 1950s). It is inferred that such growth releases were the result of intense and widespread logging during that decade with many silver-fir forests affected across the Aragón Pyrenees (Cabrera 2001). In the Pyrenees, historical logging has mostly affected fast-growing and big trees thus promoting the persistence of small-diameter slow-growing trees, which might be more susceptible to drought stress. In agreement with this, we detected

Table 6.2 Relationships among the variables measured at the study silver-fir forests (values are Pearson correlation coefficients, excepting those related to defoliation that are Spearman coefficients)

	Latitude	Longitude[a]	Elevation	Defoliation	Dbh	Age	Tree-ring width
Latitude							
Longitude	−0.57**						
Elevation	−0.53**	0.69**					
Defoliation	0.42*	−0.43*	−0.52**				
Dbh	0.20	0.04	0.20	−0.48*			
Age	0.43*	−0.17	-0.20	−0.02	0.10		
Tree-ring width	−0.14	0.13	0.29	−0.33	0.63**	-0.65**	
Climate R^2	−0.42*	−0.20	−0.14	0.13	−0.33	-0.24	-0.08

Defoliation refers to the percentage of trees in each stand with more than 50% crown defoliation.
Significance levels: *0.01 < $P \leq 0.05$, **$P \leq 0.01$
[a]Negative and positive longitude values correspond to western and eastern sites, respectively

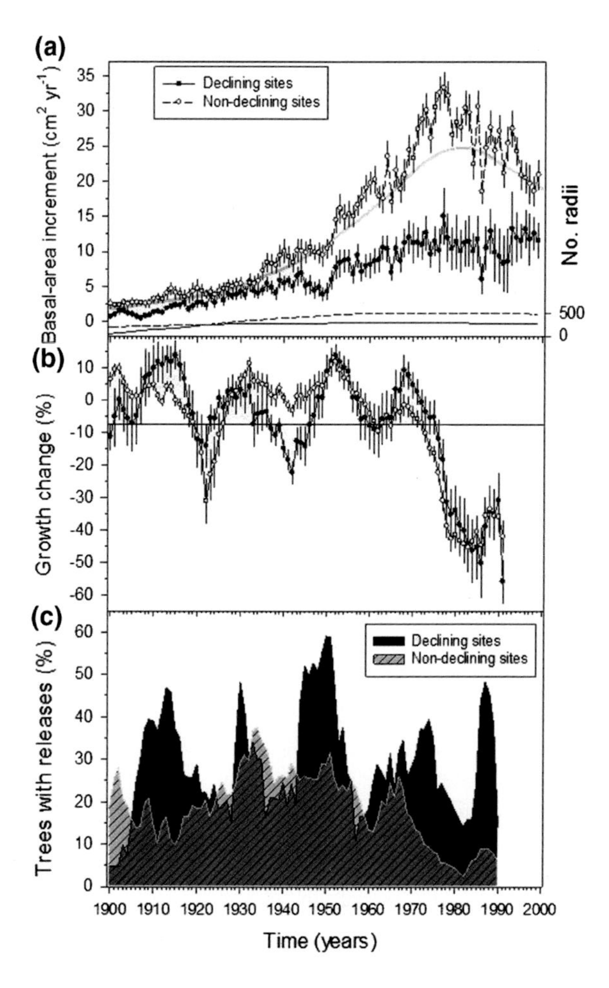

Fig. 6.6 Trends in basal area increment of declining and non-declining silver-fir stands (**a**), mean negative growth changes (**b**), and frequency of trees showing releases (**c**). Releases were defined as those years with growth changes greater than 75%, respectively. The sample size is displayed in the *upper graph*. The smoothed curve in the *upper graph* (*grey line*) shows the long-term trend of basal area increment for all trees and was obtained using a loess function with a 0.1 span. The *dark-grey area* in the *lower graph* shows the common percentage of trees showing releases in both types of stands

that tree dbh and growth along the cambial age of a tree were consistently lower in declining than in non-declining sites (Fig. 6.4). Other factors may explain why logging increases the sensitivity of silver fir to drought stress. For instance, there may be a boost in resources after logging, usually expressed as a growth release, which favours the acquisition of hydraulic traits (e.g. tracheids with wider lumen area) that increase hydraulic conductivity but also make trees more prone to drought damage. Logging is also related to the presence of root-rot fungi, but their abundance was not greater in declining than in non-declining Pyrenean sites (Oliva and Colinas 2007; Sangüesa–Barreda et al. 2015). Lastly, mistletoe infestation and insects whose outbreaks cause punctual silver-fir defoliation (e.g. *Epinotia subsequana*) have been identified in some affected forests, but they are not considered the primary drivers of silver-fir dieback but secondary factors contributing to tree death (Camarero et al. 2002, 2003, 2011).

Reams and Huso (1990) also noted that declining red spruce stands in Maine were released one to three decades before dieback started. Historical logging may thus have caused sudden changes in the growth dynamics of surviving trees leading to drought-induced hydraulic failure. Also, microclimatic conditions (e.g. air and soil humidity) change drastically in logged open stands as compared with closed forests which may also affect silver-fir performance (Aussenac 2002). For instance, silver-fir defoliation increases in habitats with low soil-water holding capacity (Thomas et al. 2002). The 1980s releases found only in the declining sites were caused by the felling of dying trees, but many of the surviving trees did not improve their growth in the long term and, in many cases, died. These facts suggest that their performance was permanently affected by drought stress leading to an irreversible reduction in growth and suggesting a loss in stomatal regulation of declining trees despite the recent rise in atmospheric CO_2 concentrations (Linares and Camarero 2012). Additional research is required to establish the links between soil-water availability, growth trends and dieback if these processes are to be used as monitors of the effects of climate warming on mountain forests.

6.5 Climate-Growth Associations: The Critical Role Played by Late-Summer Water Deficit

Silver-fir growth was enhanced by wet spring conditions and by cool and wet conditions during the summer prior to tree-ring formation (Fig. 6.7a). The first two axes of the PCA based on the correlation coefficients between climatic variables and site chronologies accounted for 26.7% and 17.0% of the total variance, respectively, (Fig. 6.7b). These analyses detected a positive and stronger response of tree growth in declining than in non-declining sites to current June precipitation, and a lower response to previous September precipitation and February temperature (negative relationship). The positive effect of current June (previous September) precipitation on growth in declining sites was significantly stronger (lower) than in non-declining sites (June, $F = 6.57$; September, $F = 6.30$, $P < 0.02$ in both cases). The cumulative water deficits of the previous spring and the current growing season (January–May) were also negatively related to growth in declining sites, and these effects were more marked westwards. Silver-fir growth was also negatively (positively) associated with higher September (April) temperatures of the previous (current) year, and this association increased eastwards (westwards).

Climate-growth relationships were unstable through time since most growth-climate relationships changed in the 1980s according to moving correlations functions (Fig. 6.8). The negative influence of previous February and September temperatures and the positive influence of previous September precipitation on silver-fir growth strengthened since 1985 in the non-declining sites, whereas the positive influence of current July precipitation decreased. Declining sites showed a similar temporal instability of their growth-climate relationships,

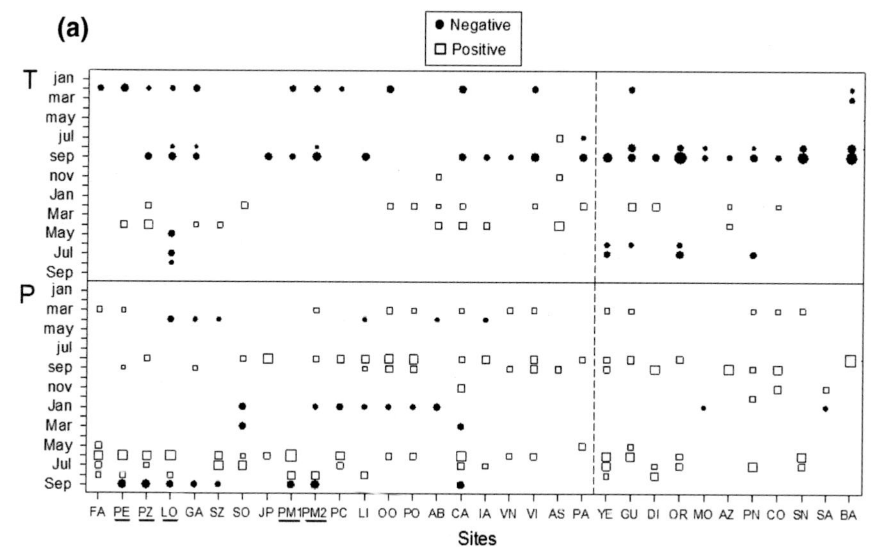
(a)
Negative
Positive
T
P
jan
mar
may
jul
sep
nov
Jan
Mar
May
Jul
Sep
FA PE PZ LO GA SZ SO JP PM1 PM2 PC LI OO PO AB CA IA VN VI AS PA YE GU DI OR MO AZ PN CO SN SA BA
Sites

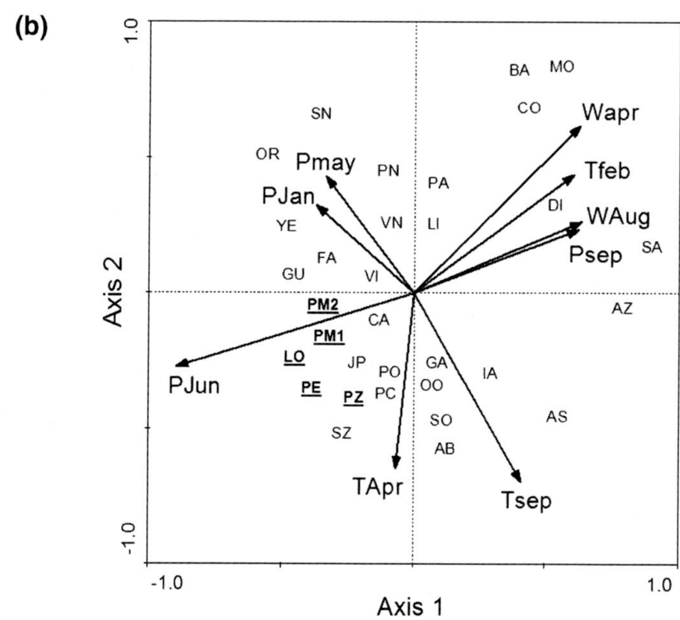
(b)
1.0
-1.0
-1.0
1.0
Axis 1
Axis 2
Wapr
Tfeb
WAug
Psep
Pmay
PJan
PJun
TApr
Tsep
BA
MO
CO
SN
OR
PN
PA
DI
VN
LI
YE
SA
FA
GU
VI
PM2
AZ
CA
PM1
LO
JP
GA
IA
PO
OO
PE
PZ
PC
SO
AS
SZ
AB

◀**Fig. 6.7** Climate-growth relationships in Pyrenean silver-fir forests. **a** The *upper graphs* show the significant ($P < 0.05$) bootstrap correlation coefficients calculated between monthly climatic data (*T* mean temperature; *P* total precipitation) and the residual chronologies for the period 1950–1999. The strength of the correlation is indicated by the size of the symbol. *Open squares* indicate a positive correlation, and *solid circles* indicate a negative correlation. The months studied go from previous January to current September (months abbreviated by *capital letters* correspond to the year of tree-ring formation). The climatic data were calculated for the two climatic sub-regions (Western and Eastern Aragón Pyrenees), which are divided by the *vertical dashed line*. The sites are arranged from the west (*left*) to the east (*right*). The codes of declining sites are *underlined*. **b** Relative positions of silver-fir correlation functions based on the first two components of a Principal Component Analysis (axes 1 and 2 correspond to the first and second Principal Components, respectively) calculated on the matrix of climate-growth correlations shown in (**a**). Only the most significant climatic variables (*arrows*) are represented, and they are abbreviated using a *three-letter code*. Climatic variables starting with "W" refer to cumulative monthly water deficit (e.g. WAug, cumulative water deficit from January up to August of the year of growth). The months studied go from previous January to current September (months abbreviated by *lower-/uppercase letters* correspond to the previous/current year of tree-ring formation; e.g. TApr stands for April temperature of the year of tree-ring formation). The climatic data were calculated for the two climatic sub-regions (Western and Eastern Aragón Pyrenees). Declining sites are shown as *underlined bold codes*

but with some characteristic features. First, the negative response to previous February temperatures became stronger during the late twentieth century in the declining sites, while a similar response to current July temperatures almost disappeared. Second, the positive response to previous March precipitation and current June precipitation also turned out to be stronger during the past decades. These changes coincided with warmer February conditions and the drier September conditions detected during the last decades.

The maximum growth reduction of silver fir occurred in 1986, which was preceded by the highest water deficit of the available climatic record. Defoliation and mortality were widespread in western low-elevation sites related to the warming trend since the 1980s and the outstanding water deficit in 1985, which were higher there than in the eastern study area. Macias et al. (2006) also suggested that silver fir is experiencing a greater late-summer drought stress in the Spanish Pyrenees and that the effects of water shortage on growth were more intense in low-elevation stands subjected to higher water deficit than elsewhere. The detected increase in drought stress was not only due to a decrease in precipitation, since similar dry periods occurred in the 1940s, and suggest a link with the 1980s warming (Vicente–Serrano et al. 2015) when the water-use efficiency of declining trees also increased sharply (Linares and Camarero 2012). The most pronounced warming in the western than in the eastern Aragón Pyrenees indicates that warming-induced drought stress triggered forest dieback westwards (Fig. 6.2). In addition, climate-growth relationships support a key role of late-summer water deficit controlling the silver-fir growth and dieback patterns in the Aragón Pyrenees (Figs. 6.3 and 6.7).

These results agree with dendroecological studies which found that silver-fir growth is very sensitive to water deficit during August and September before ring formation, responding negatively to high temperatures in those months (Bert 1993;

Rolland et al. 1999; Tardif et al. 2003). This fact is consistent with the low stomatal regulation of gas exchange in silver fir and its dieback in sites with high leaf-to-air vapour pressure difference in response to warmer air temperatures (Peguero–Pina et al. 2007; Vicente–Serrano et al. 2015). Other factors such as nitrogen deposition might also be involved as occurred in the Vosges, where silver-fir dieback was also related to acidification (Pinto et al. 2007). However, nitrogen deposition is much higher there than in the Aragón Pyrenees (de Vries et al. 2003), and most of the study sites were located on basic soils and showed N-NW exposure.

The negative response of silver-fir growth to the growing-season cumulative water deficit increased westwards and was more marked in declining than in non-declining sites. Radial growth was also enhanced by current June precipitation and this effect also increased westwards as summer precipitation decreased (Figs. 6.3 and 6.7). Accordingly, growth in the western declining sites was more constrained by climatic conditions during the year of tree-ring formation than in the eastern non-declining sites. However, growth in declining sites was also affected by distinctive factors such as the negative relationship with previous February temperature.

Contrastingly, an increasing response of growth to previous September climatic conditions eastwards was detected, whereas the negative (positive) relationship with temperature (precipitation) increased during the second half of the past century (Fig. 6.8). These results seem to be counter intuitive since the 1986 dieback onset in the western study area was preceded by an extremely dry and warm September. Indeed, 1985 was the year with lowest September precipitation in the western study area during the twentieth century, and such dryness was caused by the presence of two high pressures over the northern Atlantic.

In the second half of the past century, silver-fir growth was positively and significantly related to the Scandinavian ($r = 0.41$) and the North Atlantic Oscillation ($r = 22$) indices of the previous September and November, respectively. The associations with these two dominant modes of atmospheric variability in SW Europe (Sáenz et al. 2001) were unstable since the correlations with the Scandinavian index peaked in the 1967–1986 period, whereas the NAO effects have changed from positive to negative. This transition was clearer for the western declining sites than elsewhere which suggests some connection between atmospheric variability and silver-fir dieback (Camarero 2011). The lower responsiveness to previous September climate in the westernmost area was linked to an earlier worsening of late-summer conditions there, probably through an intensification of warming-induced drought stress.

The presented findings have implications for silver-fir growth and persistence near the rear xeric edge of the species' distribution area. First, the year-to-year variation in tree growth of silver fir has increased in the second half of the twentieth century in response to a greater warming-induced drought stress and the occurrence of more frequent dry spells in the last decades as compared with earlier more favourable conditions in the first half of the past century (Tardif et al. 2003). It is also evidenced that this instability also affected the associations between growth and atmospheric variability since 1950. Second, it is expected that, in some

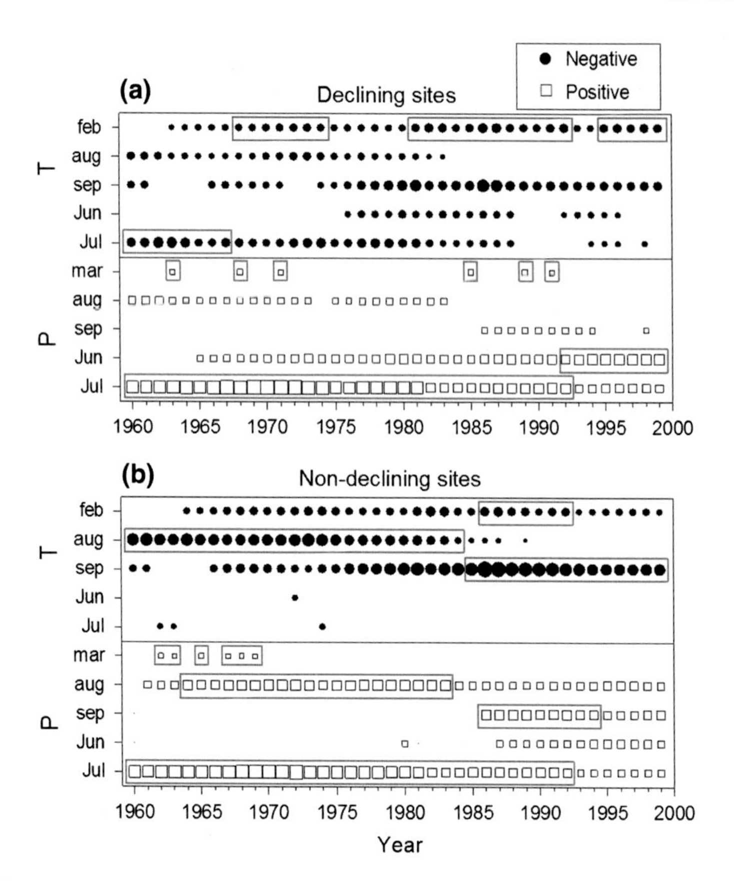

Fig. 6.8 Temporal instability of the climate-growth relationships for selected climatic variables. Moving-interval correlations functions show the significant ($P < 0.05$) bootstrap correlation coefficients based on the relationships between monthly climatic data (*T* mean temperature; *P* total precipitation) and the mean chronologies for declining (**a**) and non-declining (**b**) sites. Months abbreviated by *upper case letters* correspond to the year of tree-ring formation, and months abbreviated by *lower case letters* correspond to the previous year. The years shown in the *x*-axis correspond to the last year of 50-year moving intervals lagged by 1 year (1911–1960, …, 1950–1999). The strength of the correlation is indicated by the size of the symbol. *Open squares* and *solid circles* indicate positive and negative correlations, respectively. *Boxes* enclose periods whose coefficients were significant ($P < 0.05$)

low-elevation forests showing intense dieback, there will be a replacement of silver fir by beech and Scots pine in mesic and xeric sites, respectively. In fact, these findings reveal that declining sites were characterised by a higher basal area of beech and Scots pine than non-declining sites. We predict that the replacement of silver fir by co-occurring species will proceed faster in the western declining sites than eastwards. Finally, paleoecological evidence supports the contention that similar past abrupt climatic changes may have caused analogous drought-induced diebacks of tree species thus leading to rapid (ca. 500 years) changes in forest composition (Foster et al. 2006).

6.6 Lessons for Forest Use and Conservation

The reconstructed history of growth releases in Pyrenean silver-fir forests suggests that dieback is the product of both predisposing and inciting factors (sensu Manion 1981) such as historical logging and warming-induced drought stress, respectively. Historical land-use changes have persistent effects on forest dynamics on decadal and even millennial time scales (Dupouey et al. 2002). However, historical legacies have not always been considered when explaining the causes of forest dieback; although several authors have illustrated how changes in management cause over-stocking and increase tree-to-tree competition for water making tree species, including firs, more susceptible to drought-induced damage (Becker et al. 1989; see some cases in the reviews by Allen et al. 2010, 2015).

The Pyrenean silver-fir dieback was triggered by pronounced late-summer water deficit (inciting factor) due to the rapid temperature rise observed in the Aragón Pyrenees since the 1980s. Historical logging is the most likely predisposing factor of this dieback process. The geographic pattern of forest dieback was a response to a regional climatic gradient with decreasing summer precipitation westwards, whereas dieback and crown defoliation and growth decline were highest in marginal low-elevation sites. Declining stands were characterised by being mixed forests with silver-fir trees of low size and growth rate. Silver-fir growth is becoming more dependent on previous September climatic conditions which may be connected with changing modes of atmospheric variability affecting the Iberian climate. Finally, I partly concur with Auclair (2005) that some dieback cases might be regarded as an additional disturbance factor of forest dynamics. However, both historical management and warming-induced drought stress are altering these dynamics. First, management might set the stage for forest dieback through the selection of particular trees or cohorts highly vulnerable to climatic stress. Second, inciting climatic stressors as warming-related droughts may become more frequent in a warmer world. Thus, the recurrence and severity of forest dieback episodes may be exacerbated leading to unprecedented growth drops outside the historical range of growth variability (sensu Veblen 2003).

To manage mountain forests in a more sustainable way so as to preserve biodiversity and provide ecosystem services including timber production it must be considered that their past use and history constrain their current structure and how they will respond to climate warming. Mountain forests are erroneously perceived as intact and wild ecosystems, but they have been shaped by centuries of exploitation, at least in Europe. If this past use has lead to uniform stands of trees vulnerable to climate warming, more dieback processes are expected across southern European mountains holding the southernmost limits of several tree species as is the case of silver fir in the Pyrenees (Gazol et al. 2015). A more sustainable use and conservation of mountain forests will require managing their structure to make them more resilient and less vulnerable to climate warming. Such management should consider the enhancement of functional and structural diversities which can contribute to increasing forest resilience and promote the post-drought recovery of the most affected forests (Gazol and Camarero 2016). Achieving such effective management measures to prevent or buffer some of the adverse effects of drought-induced dieback or to enhance forest resilience is a challenge for managers and researchers.

References

Allen CD, Breshears DD, McDowell NG (2015) On underestimation of global vulnerability to tree mortality and forest die-off from hotter drought in the Anthropocene. Ecosphere 6:129

Allen CD, Macalady AK, Chenchouni H, Bachelet D, McDowell N, Vennetier M, Kitzberger T, Rigling A, Breshears DD, Hogg EH, Gonzalez P, Fensham R, Zhang Z, Castro J, Demidova N, Lim J-H, Allard G, Running SW, Semerci A, Cobb N (2010) A global overview of drought and heat-induced tree mortality reveals emerging climate change risks for forests. For Ecol Manag 259:660–684

Auclair AND (2005) Patterns and general characteristics of severe forest dieback from 1950 to 1995 in the northeastern United States. Can J For Res 35:1342–1355

Aussenac G (2002) Ecology and ecophysiology of circum-mediterranean firs in the context of climate change. Ann For Sci 59:823–832

Barry RG (2008) Mountain weather and climate. Cambridge University Press, Cambridge, UK

Becker M, Landmann G, Lévy G (1989) Silver fir decline in the Vosges mountains (France): role of climate and silviculture. Water Air Soil Pollut 48:77–86

Bert GD (1993) Impact of ecological factors, climatic stresses, and pollution on growth of the silver fir (*Abies alba*) in the Jura mountains: an ecological and dendrochronological study. Acta Oecol 14:229–246

Bigler C, Braker OU, Bugmann H, Dobbertin M, Rigling A (2006) Drought as an inciting mortality factor in Scots pine stands of the Valais, Switzerland. Ecosystems 9:330–343

Bonan GB (2008) Forests and climate change: forcings, feedbacks, and the climate benefits of forests. Science 320:1444–1449

Breshears DD, Allen CD (2002) The importance of rapid, disturbance-induced losses in carbon management and sequestration. Glob Ecol Biogeogr 11:1–5

Cabrera M (2001) Evolución de abetares del Pirineo aragonés. Cuadernos de la Sociedad Española de Ciencias Forestales 11:43–52

Camarero JJ (2011) Direct and indirect effects of the North Atlantic Oscillation on tree growth and forest decline in northeastern Spain. In: Vicente-Serrano SM, Trigo RM (eds) Hydrological, socioeconomic and ecological impacts of the north atlantic oscillation in the mediterranean region. Advances in global change research, vol 46. Springer, pp 129–152

Camarero JJ, Padró A, Martín-Bernal E, Gil-Pelegrín E (2002) Aproximación dendroecológica al decaimiento del abeto (*Abies alba* Mill.) en el Pirineo aragonés. Montes 70:26–33

Camarero JJ, Martín-Bernal E, Gil-Pelegrín E (2003) The impact of a needleminer (*Epinotia subsequana*) outbreak on radial growth of silver fir (*Abies alba*) in the Aragón Pyrenees: a dendrochronological assessment. Dendrochronologia 21:1–10

Camarero JJ, Bigler C, Linares JC, Gil-Pelegrín E (2011) Synergistic effects of past historical logging and drought on the decline of Pyrenean silver fir forests. For Ecol Manag 262:759–769

Camarero JJ, Gazol A, Sangüesa-Barreda G, Oliva J, Vicente-Serrano SM (2015) To die or not to die: early-warning signals of dieback in response to a severe drought. J Ecol 103:44–57

Cook ER, Krusic PJ (2005) Program Arstan, a tree-ring standardization program based on detrending and autoregressive time series modeling, with interactive graphics. Tree-Ring Laboratory, Lamont Doherty Earth Observatory, Columbia University, Palisades, NY

de Vries W, Reinds GJ, Vel E (2003) Intensive monitoring of forest ecosystems in Europe-2: atmospheric deposition and its impacts on soil solution chemistry. For Ecol Manag 174:97–115
Dupouey JL, Dambrine E, Laffite JD, Moares C (2002) Irreversible impact of past land use on forest soils and biodiversity. Ecology 83:2978–2984
Foster DR, Oswald WW, Faison EK, Doughty ED, Hansen BCS (2006) A climatic driver for abrupt mid-Holocene vegetation dynamics and the hemlock decline in New England. Ecology 87:2959–2966
Fritts HC (1976) Tree rings and climate. Academic Press, London, UK
Gazol A, Camarero JJ (2016) Functional diversity enhances silver fir growth resilience to an extreme drought. J Ecol 104:1063–1075
Gazol A, Camarero JJ, Gutiérrez E, Popa I, Andreu-Hayles L, Motta R, Nola P, Ribas M, Sangüesa-Barreda G, Urbinati C, Carrer M (2015) Distinct effects of climate warming on populations of silver fir (*Abies alba*) across Europe. J Biogeogr 42:1150–1162
Holmes RL (1983) Computer-assisted quality control in tree-ring dating and measurement. Tree-Ring Bull 43:69–78
Kirby KJ, Watkins C (eds) (1998) The ecological history of European forests. CABI, Wallingford, UK
Kohler T, Wehrli A, Jurek M (eds) (2014) Mountains and climate change: a global concern. Sustainable mountain development series. CDE-SDC, Bern, Switzerland
Linares JC, Camarero JJ (2012) From pattern to process: linking intrinsic water-use efficiency to drought-induced forest decline. Glob Change Biol 18:1000–1015
Linares JC, Camarero JJ, Carreira JA (2009) Interacting effects of climate and forest-cover changes on mortality and growth of the southernmost European fir forests. Glob Ecol Biogeogr 18:485–497
Macias M, Andreu L, Bosch O, Camarero JJ, Gutiérrez E (2006) Increasing aridity is enhancing silver fir (*Abies alba* Mill.) water stress in its south-western distribution limit. Clim Change 79:289–313
Manion PD (1981) Tree disease concepts. Englewood Cliffs, Prentice Hall, NJ, USA
McDowell N, Pockman WT, Allen CD, Breshears DD, Cobb N, Kolb T, Plaut J, Sperry J, West A, Williams DG, Yepez EA (2008) Mechanisms of plant survival and mortality during drought: why do some plants survive while others succumb to drought? New Phytol 178:719–739
Müller EHR, Stierlin HR (1990) Sanasilva tree crown photos with percentages of foliage loss. Swiss Federal Institute for Forest, Snow and Landscape Research, Birmensdorf, Switzerland
Nowacki GJ, Abrams MD (1997) Radial-growth averaging criteria for reconstructing disturbance histories from presettlement-origin oaks. Ecol Monogr 67:225–249
Oliva J, Colinas C (2007) Decline of silver fir (*Abies alba* Mill.) stands in the Spanish Pyrenees: role of management, historic dynamics and pathogens. For Ecol Manag 252:84–97
Pedersen BS (1999) The mortality of Midwestern overstory oaks as a bioindicator of environmental stress. Ecol Appl 9:1017–1027
Peguero-Pina JJ, Camarero JJ, Abadía A, Martín E, González-Cascón R, Morales F, Gil-Pelegrín E (2007) Physiological performance of silver-fir (*Abies alba* Mill.) populations under contrasting climates near the south-western distribution limit of the species. Flora 202:226–236
Pinto PE, Gégout J-C, Hervé J-C, Dhôte J-F (2007) Changes in environmental controls on the growth of *Abies alba* Mill. in the Vosges Mountains, north-eastern France, during the 20th century. Glob Ecol Biogeogr 16:472–484
Reams GA, Huso MMP (1990) Stand history: an alternative explanation of red spruce radial growth reduction. Can J For Res 20:250–253
Rolland C, Michalet R, Desplanque C, Petetin A, Aime S (1999) Ecological requirements of *Abies alba* in the French Alps derived from dendro-ecological analysis. J Veg Sci 10:297–306
Sáenz J, Rodríguez-Puebla C, Fernández J, Zubillaga J (2001) Interpretation of interannual winter temperature variations over Southwestern Europe. J Geophys Res 106:20641–20652
Sangüesa-Barreda G, Camarero JJ, Oliva J, Montes F, Gazol A (2015) Past logging, drought and pathogens interact and contribute to forest dieback. Agric For Meteorol 208:85–94

Skelly JM, Innes JL (1994) Waldsterben in the forests of Central Europe and Eastern North America: Fantasy or reality? Plant Dis 78:1021–1032
Suarez ML, Ghermandi L, Kitzberger T (2004) Factors predisposing episodic drought-induced tree mortality in *Nothofagus*—site, climatic sensitivity and growth trends. J Ecol 92:954–966
Tardif J, Camarero JJ, Ribas M, Gutiérrez E (2003) Spatiotemporal variability in tree ring growth in the Central Pyrenees: climatic and site influences. Ecol Monogr 73:241–257
Thomas AL, Gegout JC, Landmann G, Dambrine E, King D (2002) Relation between ecological conditions and fir decline in a sandstone region of the Vosges mountains (northeastern France). Ann For Sci 59:265–273
van Mantgem PJ, Stephenson NL (2007) Apparent climatically induced increase of tree mortality rates in a temperate forest. Ecol Lett 10:909–916
Veblen TT (2003) Historic range of variability of mountain forest ecosystems: concepts and applications. Forest Chron 79:223–226
Vicente-Serrano SM, Camarero JJ, Zabalza J, Sangüesa-Barreda G, López-Moreno JI, Tague CL (2015) Evapotranspiration deficit controls net primary production and growth of silver fir: implications for circum-mediterranean forests under forecasted warmer and drier conditions. Agric For Meteorol 206:45–54
Vigo J, Ninot JM (1987) Los Pirineos. In: Peinado Lorca M, Rivas-Martínez S (eds) La vegetación de España. Publicaciones de la Universidad de Alcalá de Henares, Alcalá de Henares, Spain, pp 351–384

Part III
Emerging Values in Mountain Conservation

Chapter 7
Towards a Microbial Conservation Perspective in High Mountain Lakes

Emilio O. Casamayor

Abstract Microorganisms are fundamental components to maintain the ecological integrity of any ecosystem. Microscopic organisms have been, however, mostly excluded in conservation studies and microbiology has been developed as a scientific discipline lacking a natural history background. The detailed genetic studies carried out in the Aigüestortes i Estany de Sant Maurici National Park and recent works in the mostly scarce literature, show that the mostly oligotrophic and highly diluted waters in high mountain lakes hold a larger microbial phylogenetic uniqueness than expected and are reservoirs of large evolutionary potential, providing an overall natural history perspective for alpine archaea, bacteria, fungi and protists. Microbes arise as an important part of the biological richness of these environments that should be considered as a fundamental component of the natural heritage. Microbial ecologists are now closer than ever to deal with conservation biology concepts such as biological richness, extinction, biotic interactions, and ecosystems management. First insights emerge for establishing the microbial tolerance to different environmental conditions, for estimating which is the potentiality of survival and dispersal abilities in the different species, and for highlighting how the underappreciated microbiota will respond to stresses and disturbances brought by the global change. Warming and eutrophication may jeopardise the most idiosyncratic microbial populations that have found in these (ultra)oligotrophic and diluted systems the most appropriate conditions to thrive. Environmental managers and lawyers, citizen, and stakeholders, in general, have now access to scientifically informed advice for the unseen microbial life in the unexpectedly rich high mountain microbial ecosystems.

Keyword Bacteria · Archaea · Protists · Fungi · Biodiversity · Conservation biology · Lakes

---

E.O. Casamayor (✉)
Integrative Freshwater Ecology Group, Center for Advanced Studies of Blanes (CEAB), Spanish Council for Research (CSIC), Girona, Spain
e-mail: casamayor@ceab.csic.es

J. Catalan et al. (eds.), *High Mountain Conservation in a Changing World*,
Advances in Global Change Research 62, DOI 10.1007/978-3-319-55982-7_7

7.1 Introduction

Conservation biology is the scientific study of biodiversity oriented to protect species, habitats and ecosystems from unsustainable exploitation, uncontrolled extinctions and the increasing weakening of biological interactions. Conservation biology involves the interaction among apparently unrelated disciplines such as social and natural sciences, economics and computational and political sciences, among others, promoting integrative and transdisciplinary views on ecosystem health and aiming to formulate the scientific basis for the best practices in natural resource management. Unlike with plants and animals, microscopic organisms have been mostly excluded in conservation studies and microbiology has been developed as a scientific discipline lacking a natural history background (Margulis et al. 1986). Apparently, the study of microorganisms lacks naturalistic attractive and conservation-orientated perspectives because of microbial inconspicuousness, low probabilities of extinction and potential widespread distribution. Therefore, environmental managers and lawyers, citizens and stakeholders, in general, have obviated the fate of natural microbial communities among their daily worries and strategic planning beyond pathogens. Threats on microorganisms have been not considered as part of the natural resources management policies or in the estimation of the influence of human activities in nature. Ecosystem functioning carried out by microorganisms appears to rely on a high functional redundancy and its maintenance could be largely uncoupled to microbial biodiversity erosion (Wertz et al. 2006). Microorganisms, however, are the most abundant and widespread forms of life on Earth and encompass the highest taxonomic, metabolic, genetic, and functional diversity. They rule the whole biodiversity on Earth and their activities have a major ecosystem effect (Madigan et al. 2015). Conservation biology perspectives where the fundamental unit in the conservation of biodiversity is not the species but the habitat would probably apply very well for microorganisms.

In the early 90s of the past century, the International Programme of Biodiversity Science DIVERSITAS, a programme promoted among others by the International Council of Scientific Unions (ICSU) and now migrated to both Future Earth (http://www.futureearth.org/) and the Intergovernmental Science-Policy Platform on Biodiversity and Ecosystem Services (IPBES), tried to reverse this lack of knowledge highlighting the crucial role that microbial biodiversity plays in the maintenance of many ecosystem services. The term "ecosystem services" describes ecosystem resources and processes that benefit human society, and its identification and value quantification can provide additional arguments for the protection of species and ecosystems that could easily reach public opinion and policy decisions (Daily et al. 2009). DIVERSITAS emphasised the immense genetic diversity of microorganisms and their crucial and unique roles as essential components of food chains and biogeochemical cycles and included "microbial biodiversity" within the nine fundamental cross-cutting research themes of critical importance for biodiversity science. The Microbial Biodiversity programme called to develop innovative methods and techniques to accelerate the discovery and characterization of microbial diversity, and to establish reliable databases to collect and exchange information on the biological characteristics of microorganisms. It was particularly encouraged to improve the knowledge in freshwater microbial diversity to increase the available understanding of the effects of microbial diversity on aquatic ecosystem functioning, suppression of diseases organisms, provision of clean water through the respiration of organic carbon, denitrification and other metabolic processes, and to gain insights into the functioning and microbial regulation of biogeochemical cycles. The programme also urged to explore major issues concerning the conservation, origin and maintenance of microbial biodiversity.

In the past 20 years, and mainly over the past decade, microbial ecologists have developed, optimised and standardised powerful methods to capture microbial taxonomic and functional diversity. They have successfully combined multidisciplinary approaches from different scientific disciplines such as microbiology, ecology, molecular biology, bioinformatics and computational science and have efficiently linked available information on microbial diversity within a worldwide network. This sustained effort circumvented some of the methodological and conceptual concerns that had strongly limited the general perception of how important microbes are for Earth biodiversity and initiated the effective transplantation of concepts and basic knowledge of the general ecology grounded on plants and animals to microbial ecology. Efforts are now addressed to establish a predictive framework for the distribution and diversity of microorganisms, blurring the disciplinary boundaries that traditionally separated ecologists of tall and tiny. Currently, global initiatives such as the International Census of Marine Microbes (ICoMM), the Earth Microbiome (Gilbert et al. 2014) and the Human Microbiome projects (Methé et al. 2012), as well as international marine initiatives to profusely explore the microbial component in the surface (Tara Oceans Cruise, JCVI Global Ocean Survey, Ocean Sampling Day) and in the deep ocean (Malaspina Expedition) have been successfully promoted. The creation of an interdisciplinary Unified Microbiome Initiative to understand and harness the capabilities of the set of Earth's microbial ecosystems has been recently proposed (Alivisatos et al. 2015; Dubilier et al. 2015). Although, nothing equivalent is found for the freshwater realm despite the fact that the wide repertory of inland waters on Earth (Downing et al. 2006) contains a large, novel and unexplored microbiota (e.g. Hahn 2006; Esteban and Finlay 2010; Barberán and Casamayor 2011; Newton et al. 2011).

In the case of aquatic ecosystems from remote high-altitude mountain areas, the gap of knowledge is still big. Both the oligotrophic and highly diluted nature of the lake waters and the difficulties to collect samples in such painful to reach places have limited the interest among microbiologists to explore these isolated environments. Comparatively, such remote inland waters have remained less explored than lowland freshwaters. Although, aquatic microbes are abundant in the plankton of these lakes (Felip et al. 1999), and the environmental heterogeneity at the local scale is larger than expected (Catalan et al. 1992, 2006). In addition, mountain lakes have been traditionally studied by limnologists with a background of ecology and phyto- and zooplankton biology (e.g. Catalan et al. 2006 and references therein, Tolotti et al. 2009) rather than by microbial ecologists with a background in microbiology and genetics. Consequently, the genetic diversity, taxonomic identity and ecological distribution of the unseen majority in genuinely high mountain systems have remained mostly unknown. A few investigations in the past years are however helping to fill this gap, primarily in the Alps (Pernthaler et al. 1998; Pérez and Sommaruga 2011), the Himalayas (Liu et al. 2006; Sommaruga and Casamayor 2009; Kammerlander et al. 2015), the high mountains of west USA (Nelson 2009; Hayden and Beman 2016) and, specially, in the Central Pyrenees within and around the Aiguestortes i Estany de Sant Maurici National Park (Catalan et al. 2006). Meteorologic variability, catchment inputs, and warming are key factors structuring microbial communities (Nelson 2009). These lakes are very sensitive to detect an excess of reactive N of human origin circulating through the atmosphere (Camarero and Catalan 2012). Some of these lakes are glacier-fed ecosystems and hold specific physical conditions, biodiversity and ecological functioning (Edwards et al. 2013; Peter and Sommaruga 2016). High mountain lakes formed by glaciers erosion are very comparable worldwide ecosystems, and the new lakes that are currently appearing after the glacial retreat in mountain areas offer great opportunities for ecological and limnological studies (Catalan and Rondon 2016). Microbial diversity in high-altitude aquatic ecosystems from highlands such as Tibetan lakes (Zhang et al. 2013) and Andean lakes (Ordoñez et al. 2009) are not considered here because of the consistent limnological, physicochemical and environmental differences mostly characteristic from the relatively flat terrain of plateau areas.

Probably, one of the most intensive and extensive limnological studies of alpine lakes to date have been carried out in the Central Pyrenees (Catalan et al. 2006). Extensive studies on microscopically conspicuous organisms have been already done (Catalan et al. 2015; Felip et al. 1999; Pla et al. 2003) and will be obviated here. Using environmental ribosomal RNA genes sequencing, a sample set representative of the lacustrine Pyrenean landscape heterogeneity has been studied in detail for bacteria, archaea and mostly inconspicuous microbial eukaryotes biodiversity, as will be presented further in this chapter. This alpine area contains the main freshwater lake district of south-west Europe, and constitutes a mosaic of highly diverse water bodies mimicking the geological diversity of the catchments (Catalan et al. 1992) that usually remain ice-covered for 4–7 months every year (typically from December to April). Very low concentrations of nutrients and salts (i.e. ultraoligotrophic and hypotonic waters), persistent extreme conditions (high UV exposure and water transparency, and low temperatures), and isolation might promote an adapted microbiota with a large number of specialist populations (Catalan et al. 2006). Although, lakes located at high altitudes efficiently integrate information about changes in the catchment and land use and in the atmosphere and are very sensitive to environmental and global changes (Adrian et al. 2009), mainly remote depositions and global warming. These changes may induce rapid changes in the microbial community composition and potentially erode the most idiosyncratic populations leading to species decline by ecosystem degradation. Experimental evidence is however needed to test these hypotheses. New approaches to predict the impacts of global change at the microbial community level are also required for scientifically informed conservation and management of biodiversity and ecosystem services (Bellard et al. 2012).

7.2 A Biodiversity Unit for the Microbial World

Systematics and taxonomy are fundamental tools to track biological history understanding life forms origins and relationships and to help to organise biological complexity knowledge to support biological conservation, respectively (Cotterill 1995). Estimating the number of microbial species even at the order of magnitude is, however, a great challenge in biology and matter of intense debate in microbiology. Two of the main problems microbiologists have to successfully face this challenge are the own definition of the species concept and the little success bringing into a culture most of the wild microbes, respectively (Fig. 7.1). The species definition for bacteria requires individuals previously isolated and grown in culture and needs highly standardised laboratory protocols, comparative genomic information and a dataset of physiological and other phenotypical features. Thus, a pragmatic phylophenetic species concept for microbial taxonomists is only useful if pure cultures are available in the laboratory (Rosselló-Mora and Amann 2001). Currently, c. 13,000 bacterial and archaeal species are available in culture (Amann and Rosselló-Móra 2016). The last estimation using >20,000 microbial molecular surveys and mathematical modeling and scaling laws predicts the existence of between 10^{11}–10^{12} microbial species on Earth (Locey and Lennon 2016). If true, and according to the current number of catalogued microorganisms, that would mean that 99.999% of total microbial species are missing still. However, the lack of consensus ranges of several orders of magnitude. Estimations based on the empirical analysis of the bacterial and archaeal 16S rRNA gene sequences available in curated databases (c. 1.5 million full-length sequences), reduce the molecular census to a few millions Operational Taxonomic Units (OTU, 97% sequence identity "species-level" cutoff) (Schloss et al. 2016). In addition, it has also been predicted that environmental 16S rRNA gene sequences of the highest novelty are reaching a plateau and that most of the high microbial taxa will be discovered within a few years (Yarza et al. 2014). Still, to carry out the inventory and interpretation of the most abundant organisms on earth is a vast enterprise that will keep microbiologists busy for many decades. Telling apart microbial species is still highly controversial and difficult, and a consensus definition of what should be the most operative and predictive unit to measure biological diversity in the microbial world is still under discussion (range 97–99% identity in 16S rRNA gene sequence, Kim et al. 2014). Luckily, diversity can be studied with any coherent and well-defined standardised unit as far as it is defined in a simple, clear, and unambiguous way. The ribosomal RNA genes are still the most successful proxy for a combined view of systematics and taxonomy in microorganisms circumventing the limitation that only a very small number of microbes can grow in culture media. Although, two intrinsic limitations should be considered. First, the use of a proxy too conservative for the "specics level" that underestimates species diversity. Second, the fact that it fails to detect fast speciation processes in natural communities by horizontal gene flow. These genetic processes can maintain adaptability and provide ecological success to colonise particular environments (e.g. Llorens-Marés et al. 2017), leading to the emergence of a new "ecological species" nearly identical in the 16S rRNA gene sequence to a former population.

Altogether, microbial ecology is still in its infancy predicting species distributions across landscapes and identifying areas of high and low species richness, or highlighting whether or not vulnerable groups of microbial species or microbial processes exits, missing useful information for land management. There is considerable diversity to be explored yet, but the rate of new full-length sequences deposited in databases has consistently declined in the last years since next-generation sequencing (NGS) and high throughput has been expanded (Schloss et al. 2016). Although microbes can currently be reasonably well identified and classified in relation to each other allowing fast and proper universal

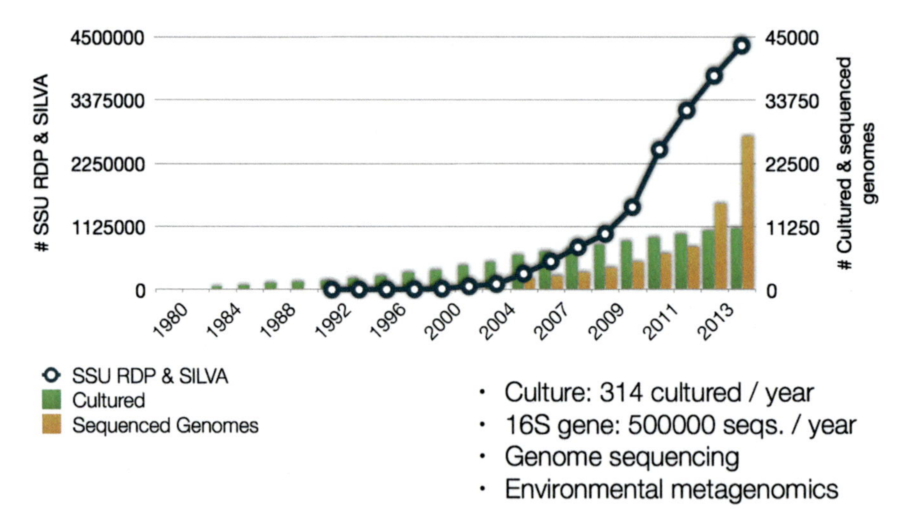

Fig. 7.1 Temporal trend and annual rates for the available number of pure cultures of bacteria and Archaea (*green label*), 16S rRNA gene sequences in curated databases (RDP and SILVA) and genomes (obtained from cultured strains and metagenomic surveys) (*orange label*). Data from Llorens-Marés 2015, Ph.D. thesis, University of Barcelona

communication among microbial ecologists, a high-quality census needs full-length 16S rRNA gene sequences and the short sequence lengths provided by current NGS methodologies need to be substantially improved (Amann and Rosselló-Mora 2016). Genomes and metagenomes currently available are growing exponentially (Fig. 7.1), and single cell genomics permits to reach the genetic potential of a microorganism without culturing. For the microbial world, the inadequacy of methods and conceptual separation of microbiology from natural sciences with strong ecological and evolutionary background such as zoology and botany should not be an unaffordable challenge anymore. In fact, microbial systems may push classical natural sciences disciplines and theoretical ecology forward to new unexplored frontiers. An integrative approach prevails today in the environmental sciences with an inclusive view on biological interactions networks (Faust and Raes 2012; Fuhrman et al. 2015), and on the integration of molecular biology at the community and ecosystem levels (Raes and Bork 2008). Merging community ecology and phylogenetics among co-occurring species can provide a new view for the study of microbial assemblages in situ (Barberán and Casamayor 2014). The use of phylogenetic approaches as a measure of biodiversity based on the phylogenetic difference between species (i.e. phylogenetic diversity, PD), offers new perspectives without previously fixing an operational taxonomic unit definition, and reduces to a single value the whole community complexity (see below). This approach may help to find patterns and to develop hypotheses based on the coexistence and adaptation of closely related species and to try to unveil the processes that shape community structure and composition.

7.3 A Natural History Perspective for Microorganisms in High Mountain Lakes

High-altitude mountain lakes hold a larger microbial biodiversity than could be initially expected in such very diluted waters. Typically, several hundred million prokaryotic cells and around a million of microscopic eukaryotes are present per litre of alpine lake water (Felip et al. 1999). In general, freshwater archaea in high mountain lakes show one to two orders of magnitude lower abundances than bacterial cells, and the richness within the Bacteria domain is substantially higher than within the Archaea. For instance, the analysis of the plankton in three connected shallow Pyrenean lakes within the Aiguestortes i Estany de Sant Maurici National Park estimated a bacterial richness of c. 2500 OTUs and an archaeal richness of c. 900 OTUs (Fig. 7.2, upper panel). These estimations required a sampling effort of c. 250,000 bacterial and c. 20,000 archaeal 16S rRNA gene sequences (Fig. 7.2, lower panel), only accessible through recent NGS technologies.

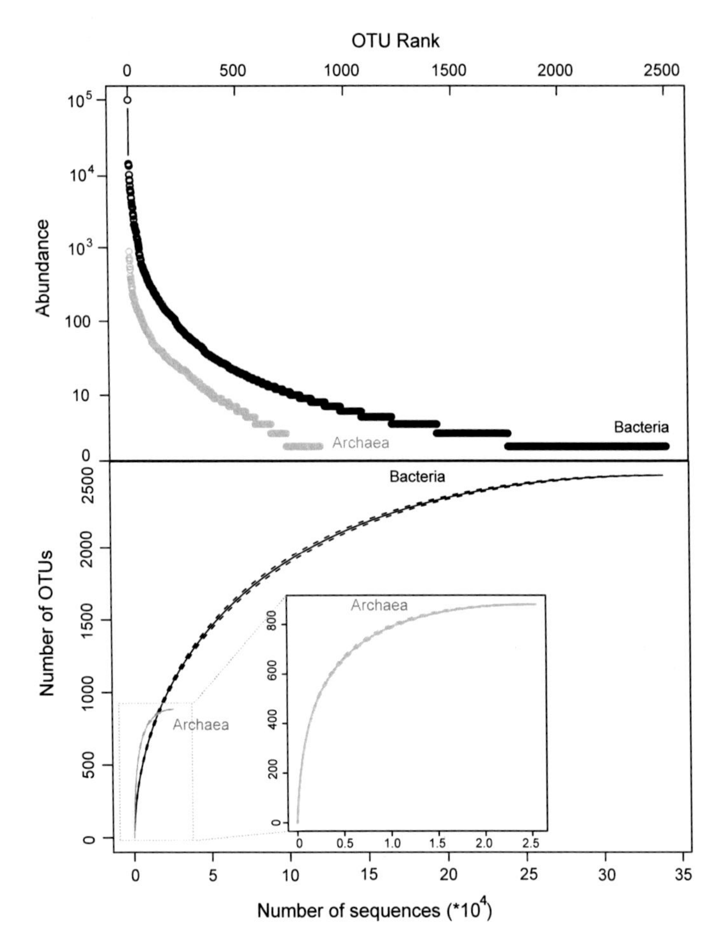

Fig. 7.2 Rank abundance (*top*) and rarefaction curves (*bottom*) for bacterial and archaeal OTUs inhabiting three connected shallow Pyrenean lakes communities within the Aigüestortes i Estany de Sant Maurici National Park. From Vila-Costa et al. (2013) with kind permission from Oxford University Press

7.3.1 Bacteria

The major bacterial taxa present in high mountain Pyrenean lakes are characteristic of the plankton present in worldwide freshwater environments and different from the oceans (Fig. 7.3, data from Barberán and Casamayor 2010). Bacteroidetes, Betaproteobacteria and Actinobacteria are the predominant groups in high-altitude lakes, with a significantly higher proportion of Betaproteobacteria and a lower proportion of Alphaproteobacteria than in lowland freshwaters (Barberán and Casamayor 2010). Also, the lakes' ecosystems can be synoptically sorted according to the evolutionary history contained in the whole bacterial assemblages without the

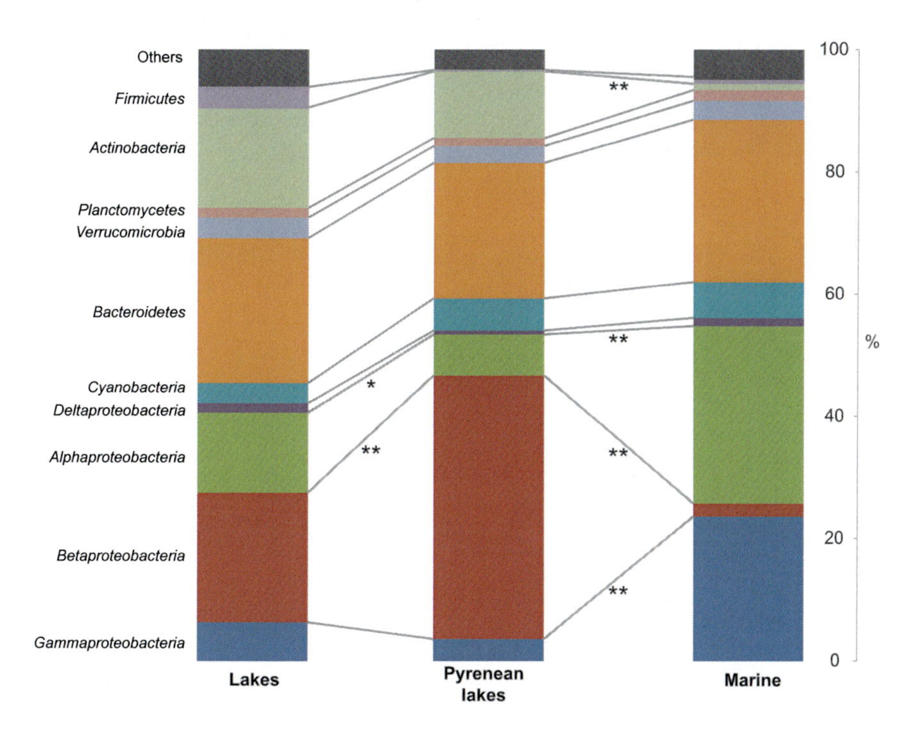

Fig. 7.3 Bacterial taxonomic composition of Pyrenean lakes, and comparison with a global meta-analysis of lakes and seas (Barberán and Casamayor 2010). Significant differences are shown (*$P < 0.05$, **$P < 0.01$, t-test). From Barberán et al. (2012), Ph.D. thesis, University of Barcelona

need of any particular taxonomic level using the phylogenetic dispersion of communities, through the phylogenetic diversity index (PD) (Barberán and Casamayor 2014). High or low levels of PD did not necessarily match those Pyrenean lakes with the highest or lowest OTUs richness, respectively, adding relevant information for stakeholders and managers and showing where most of the biological diversity accumulates (see Supporting information in Barberán and Casamayor 2014). Higher PD might result in higher functional diversity and versatility of the bacterial assemblages, but this assumption remains to be tested. This integrative tool also provides proper interbiome comparison. Interestingly, when the Pyrenean dataset is compared with a similar study of the plankton from surface waters in different seas and oceans (first 5 m sampled), and after correcting for unequal number of sequences between the studies, it arises that freshwater bacterial communities accumulated higher genetic dispersion than the very surface marine assemblages (Fig. 7.4). This result nicely contextualises any bacterial community circumventing culturing and species definitions limitations and agrees with the highest environmental heterogeneity present in the Pyrenean dataset analysed (see below). It also indicates that microbial diversity is a direct reflection of habitat diversity.

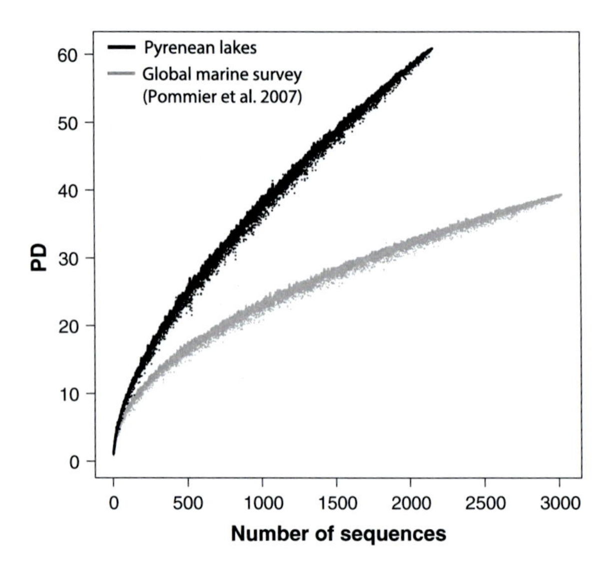

Fig. 7.4 The potential of phylogenetic diversity (PD) measures to direct intrabiome and interbiome comparisons. From Barberán and Casamayor (2014) with kind permission from John Wiley & Sons, Inc.

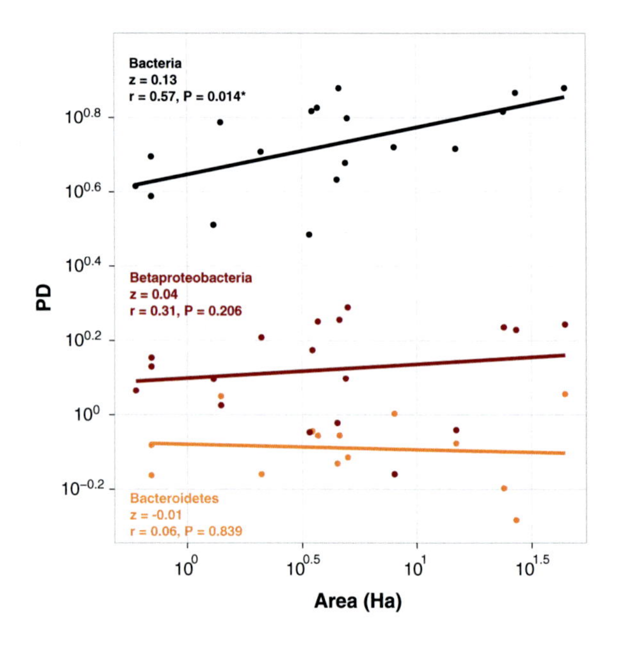

Fig. 7.5 Relationship between phylogenetic diversity (PD) and lake area. Significant linear regression line for the log-transformed relationship. From Barberán and Casamayor (2014) with kind permission from John Wiley & Sons, Inc.

A significant relationship between Pyrenean lake area and PD of bacterial assemblages also exists (Fig. 7.5). Interestingly, the slope (z) of the log-transformed relationship (0.130) was similar to the value found in a study carried out in mountain lakes in Sierra Nevada (0.161) using OTUs obtained from a bacterial genetic fingerprinting profile (Fig. 7.6 panel A). The slope z is usually <1 meaning that, although the number of species increases with area, larger areas have in proportion fewer species per unit area. In the case of bacteria, the z value (the slope of the species area relationship) is closer to the lower end, similar to the values reported for other planktonic organisms (Fig. 7.6 panel B). The species–area relationship (SAR) has been successfully used in macroecology and conservation biology to predict extinction according to the habitat reduction. In the case of microorganisms, how reliable are the extrapolation of the ends of the slope can be now better explored using massive sequencing technologies. Most microorganisms likely show long-distance dispersal abilities and large population sizes modulating the relative importance of niche, stochastic and historical processes that shape the structure of microbial communities (Barberán et al. 2014a), and first insights show that bacterial ubiquity may be a common pattern in high-altitude lakes worldwide (Sommaruga and Casamayor 2009). Cell dormancy, high persistence, and the fact that a single cell can generate a new population in a short time are microbial attributes that all together should be considered to accurately address the mechanisms that generate this pattern. Most probably, the presence of more available niches in larger lakes, a higher number of interactions, and more complex food webs may play a major role in determining bacterial richness and phylogenetic

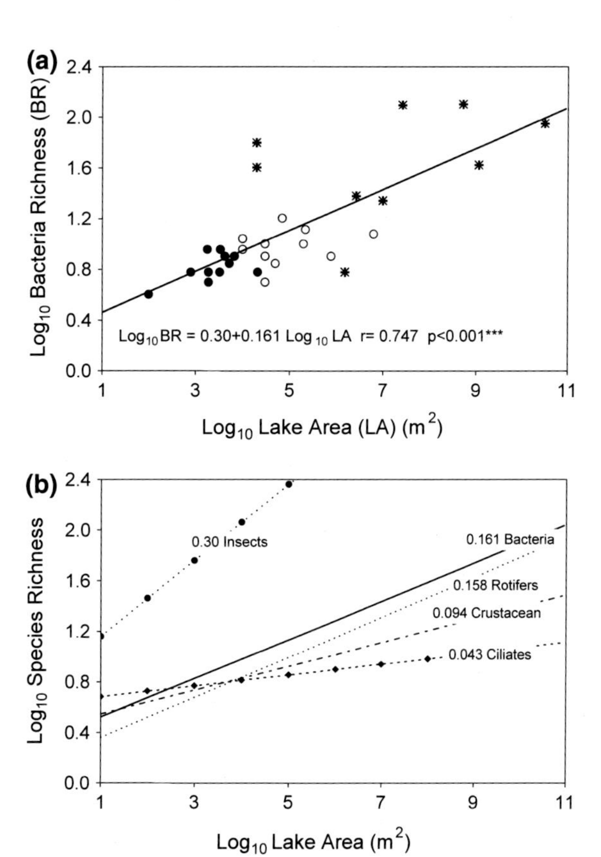

Fig. 7.6 **a** Significant relationship (z value, the slope of the species area relationship) found between bacterial OTU richness and alpine lakes area in Sierra Nevada, SE Spain. **b** Comparison of z values among taxa. From Reche et al. (2005) with kind permission from John Wiley & Sons, Inc.

dispersion. Additional studies exploring the mechanisms that generate this pattern, the interbiome and empirical variability of the z values mentioned above, the variations with taxonomic resolution (Horner-Devine et al. 2004) and the limits and spatial scaling, are topics that microbial ecologists may be interested in developing further and high mountain lakes offer a very convenient natural scenario to explore it. If microorganisms are globally dispersed and cosmopolitan, the OTUs locally present in Pyrenean lakes will represent a large fraction of the cumulative species pool present in high mountain lakes around the world. Because of the high similarity of this type of environment across continents, this is a hypothesis that can be reasonably well tested.

The quantification of community similarity based on phylogenetic relatedness also provides new perspectives to bridge the gap between evolutionary and ecological analyses (Barberán et al. 2014a). Patterns capturing how phylogenetic community similarity is distributed along environmental gradients emerge without relying on any particular operational taxonomic unit definition (Barberán and Casamayor 2014). After testing for biogeographical patterns in Pyrenean lakes using distance matrix based on both environmental variables and geographical distance, environmental filtering and not spatial distance is most probably shaping the phylogenetic structure of the freshwater bacterial assemblages (Fig. 7.7). The high environmental richness within small distances in the Pyrenean lacustrine area, covered a gradient of environmental and trophic conditions of more than four units of pH, one order of magnitude of conductivity (from highly diluted to typical freshwater values), 10-fold phosphorous concentration, and 20-fold chlorophyll-a

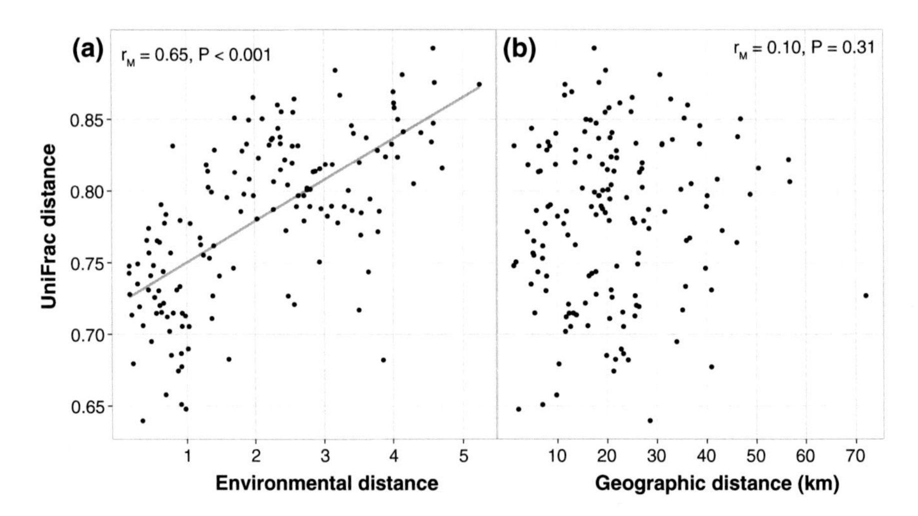

Fig. 7.7 Relationship between the UniFrac distance matrix and **a** the environmental Euclidean matrix (E) or **b** the spatial distance matrix (S). UniFrac is a β-diversity metric that quantifies community similarity based on phylogenetic relatedness. From Barberán and Casamayor (2014) with kind permission from John Wiley & Sons, Inc.

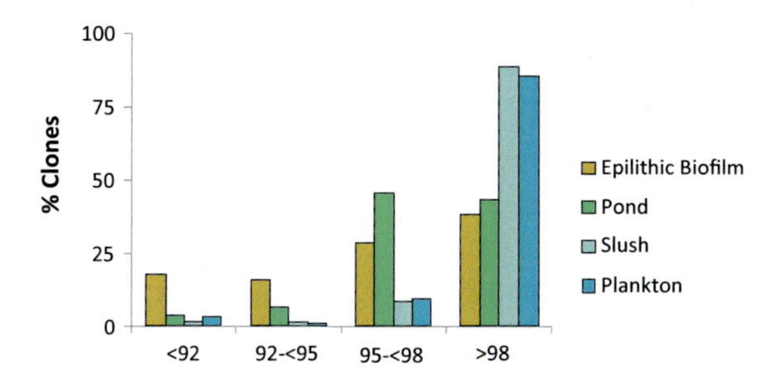

Fig. 7.8 Taxonomic novelty (percentage of identity with DNA sequences available in GenBank) for the bacterial 16S rRNA gene sequences found in the epilithon biofilm, plankton, slush (mixture of water and snow) and snow melting ponds in lakes of the Pyrenees. From Bartrons et al. (2012) with kind permission from Springer Publisher

content, being most of the lakes ultraoligotrophic or oligotrophic. Again, the higher than expected habitat richness provides a rich background for diverse planktonic bacterial communities.

Bacteria are also highly abundant and diverse in the epilithic biofilms of streams and lakes, although no consistent global elevational patterns in biodiversity for stream bacteria exist (Wang et al. 2017). These biofilms play, however, a relevant and specific biogeochemical role in mountain lakes (Vila-Costa et al. 2014). Bacteroidetes and Cyanobacteria are the most common groups found in the epilithon, whereas Actinobacteria were not detected and Betaproteobacteria were present in low abundances (Bartrons et al. 2012). Interestingly, most of the epilithic Bacteroidetes form distinct phylogenetic clusters and may represent particular poorly known ecotypes with a potentially major role in the organic matter cycling. In fact, the taxonomic novelty analysis for the bacterial 16S rRNA gene sequences showed that only 40% of the epilithon bacteria had been previously reported at the "species" level. Such value reached >80% in the slush (Llorens-Marès et al. 2012) or the plankton (Fig. 7.8). Interestingly, >25% of the epilithon species may potentially represent new bacterial families or even orders (identity in 16S rRNA gene < 95%). This idiosyncratic assemblage may be related to the large heterogeneity operating at the microscale, closer microbial interactions and coexistence of different physiologies and aerobic, anaerobic, phototrophic and chemotrophic metabolisms (Bartrons et al. 2012; Vila-Costa et al. 2014) a highly remarkable unexpected feature under such (ultra)oligotrophic prevailing conditions that deserves further studies. Potential for nitrogen fixation (i.e. presence of *nifH* genes) was also detected in the biofilms (Vila-Costa et al. 2014). Altogether, epilithic biofilms from mountain lakes and streams could hold a hotspot of microbial diversity, very rich in poorly known microbial species. The different taxa are substantially different from the bacterioplankton species and from previously reported gene sequences in databases, adding relevant spatial heterogeneity for the microorganisms in these environments.

7.3.2 Archaea

Archaea are commonly present in freshwater plankton, but most have remained unseen to aquatic ecologists and limnologists. For many years, archaea strains available in the laboratory were restricted to methanogens and microorganisms adapted to extreme temperature, pH and salinity. Apparently, the archaeal metabolic diversity and ecological distribution appeared more limited than their bacterial counterparts. In the most recent years, the environmental ribosomal RNA surveys unveiled Archaca as ubiquitous in freshwaters. Most of them but methanogens are distantly related to any laboratory strain, and there is ample room for new discoveries related to archaea and cold water habitats in lakes (Auguet et al. 2010). Curiously, freshwater archaeal richness and diversity appeared higher than in other biomes such as the oceans and soils after a meta-analysis of globally distributed clone libraries of the 16S rRNA gene (Auguet et al. 2010).

In the Pyrenean high mountain lakes, archaea are widespread and diverse. Archaea could be detected in 90% of surface waters in a large dataset of lakes examined (n = 313), with relative abundances generally up to 10% of the bacterioplankton sequences (Ortiz-Alvarez and Casamayor 2016). Alpine archaea belong to 13 different lineages (Fig. 7.9), with Pacearchaeota and Woesearchaeota as the most common groups, followed by Micrarchaeota–Diapherotrites (Euryarchaeota MEG cluster), Methanogens, Thermoplasmata and planktonic AOA (ammonia-oxidising Thaumarchaeota). Minor groups are related to the SM1K20 cluster, Aenigmarchaeota (Euryarchaeota DSEG cluster), MCG (Miscellaneous Crenarchaeotic Group, currently Bathyarcheota) and soil AOA. In subsurface and bottom waters of deeper lakes, accumulations of AOA and Aenigmarchaeota are however detected (Auguet et al. 2012; Restrepo-Ortiz and Casamayor 2013). This extensive study in the Pyrenean lacustrine district unveiled the environmental preferences and habitat breadth for the different lineages. The species with wide niche breadth, i.e. generalists, were related to methanogens and Aenigmarchaeota, whereas the most specialists were Thermoplasmata, Micrarchaeota and AOA. Pacearchaeota and Woesearchaeota, the most abundant and widespread taxa, showed intermediate values (Ortiz-Alvarez and Casamayor 2016).

The metabolic potential of most lacustrine archaea and the impact in freshwater biogeochemical cycles are largely unknown. In some cases, the biogeochemical activities of Archaea can be environmentally traced by the study of functional genes coding for reactive enzymes such as the ammonia monooxygenase (Amo) present in AOA (Fernàndez-Guerra and Casamayor 2012). The Amo plays a fundamental role in the interconnection between N fixation and N losses, catalysing the oxidisation of NH_4^+ to NO_2^-. Nitrification helps to remove excessive ammonium nitrogen and prevent lakes from eutrophication. Thus, increasing evidence suggests that Archaea may play a significant role in ammonia oxidation in freshwaters in general, and specifically in alpine lakes with submerged vegetation (Vila-Costa et al. 2016). In fact, the interaction between microbial ecology and macrophytes ecology determines the ecosystem-level denitrification and the submerged vegetation landscape has a

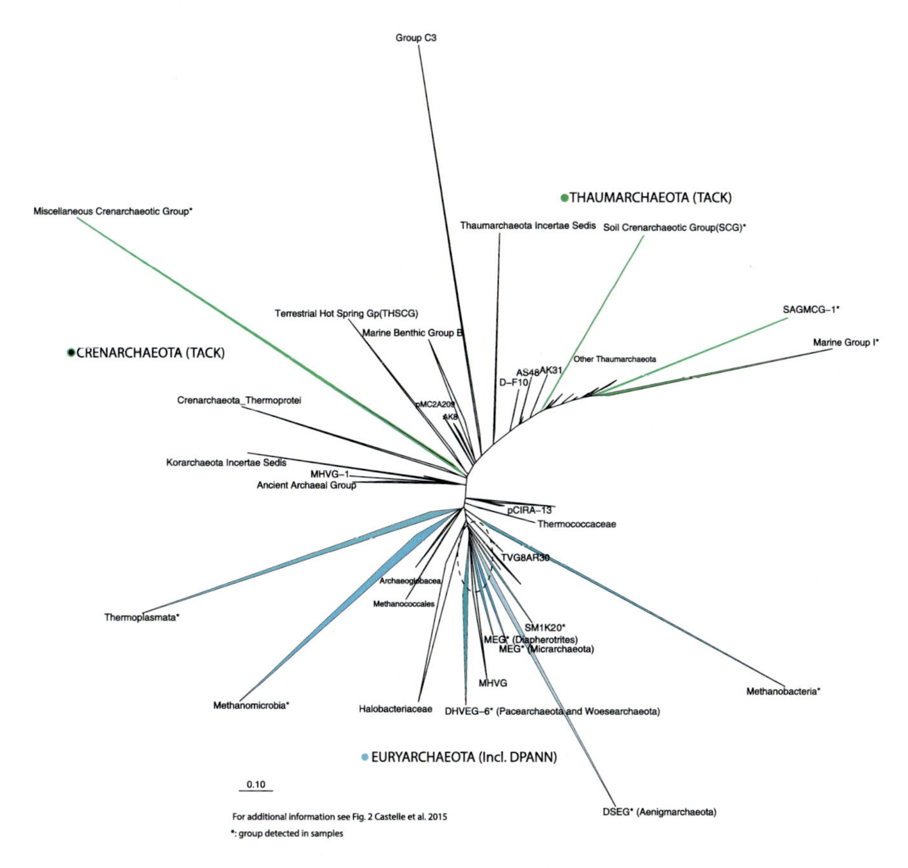

Fig. 7.9 Archaeal lineages found in the Pyrenean lakes data set within the TACK, DPANN and Euryarchaeota superphyla. From Ortiz-Alvarez and Casamayor (2016) with kind permission from John Wiley & Sons, Inc.

major role in creating heterogeneity within the lakes that promotes microbial diversity (Vila-Costa et al. 2016). A symbiotic or parasitic lifestyle even with Bacteria has been suggested for Pacearchaeota and Woesearchaeota, probably related to small genomes sizes and limited metabolic capabilities (Spang et al. 2015; Ortiz-Alvarez and Casamayor 2016) but the gap of knowledge is certainly large.

The comparison of PDs for the whole archaeal assemblage and the AOA of Pyrenean lakes with globally distributed seas and soils unveiled again a phylogenetically rich freshwater assemblage in the Pyrenees, approximately twofold higher than in other biomes (Fig. 7.10). The regional diversity of AOA in the Pyrenees is of identical magnitude that the diversity globally observed in marine and soil habitats (amoA gene phylogenetic diversity by clone libraries). Apparently, Pyrenean lakes promote the growth of very diverse and distantly related archaeal communities probably because of the specific combination of persistently cold (ultra)oligotrophic waters, highly diluted and isolated water bodies, and

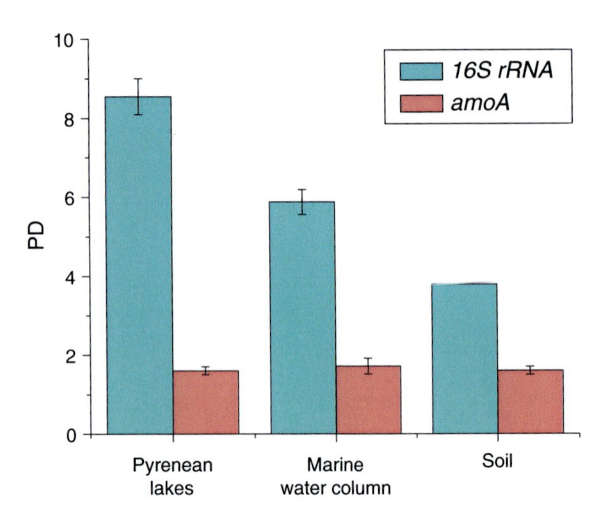

Fig. 7.10 Averaged PD for the archaeal 16S rRNA and amoA genes obtained in the Pyrenean study lakes compared with globally distributed marine and soil environmental studies. From Auguet and Casamayor (2013) with kind permission from Oxford University Press

heterogeneous landscape. High mountain lakes are therefore natural laboratories of great interest to improve the current knowledge of archaeal biology and ecology, but perturbations in the trophic status and dissolved organic matter content may induce substantial changes in the archaeal community composition (Auguet and Casamayor 2013). Field experimentation should confirm whether or not warming and eutrophication are major threats for such idiosyncratic assemblages.

7.3.3 *Protists and Fungi*

Microeukaryotes are essential components of microbial food webs, and morphological criteria have been traditionally used to tell apart the different species and to successfully study the biology and ecology of the different taxa (e.g. diatoms, cryptophytes, chrysophytes, among others). Inconspicuous forms usually of the smallest sizes are however abundant and difficult to study because the cells lack morphological features for identification. Microscopic eukaryotes constitute much of the genetic diversity within the domain Eukarya and the development of genetic approaches mostly based on the 18S rRNA gene sequencing have revealed new previously undescribed clades and large diversity than expected present in nearly all the Eukarya lineages. As it occurs with their prokaryotic counterparts, most of them are uncultivable, but new molecular approaches combined with high-throughput sequencing technologies and bioinformatics analyses provide for the first time the possibility for detailed species inventories, although not free of limitations and biases (e.g. Stoeck et al. 2014; Triadó-Margarit and Casamayor 2012). More accurate preliminary estimations on how many different eukaryotic species exist, on their functions, and on the environments with high and low microbial eukaryotic diversity can be obtained by comparison of both traditional and molecular technologies.

In a preliminary study of 11 selected high mountain Pyrenean lakes, the microeukaryotes major taxa found belonged to 9 high-rank taxonomic groups and 26 eukaryal classes. Predominant groups, both in abundance and in occurrence, were Chrysophyceae, Cryptophyta, uncultured Alveolata, pennate diatoms of the class Fragilariophyceae, Chlorophyceae, Dinophyceae and Fungi of the Chytridiomycota clade among others (Triadó-Margarit and Casamayor 2012) with most of the OTUs found exclusively affiliated to clusters formed by uncultured microorganisms. The genetic diversity within the Cryptophyta and Chlorophyceae was low and these groups also have the highest relatedness to cultured species. Cryptophyta contained the OTUs with the highest ubiquity in the dataset, but in general, most of the microeukaryote OTUs (>75% of them) were found at only one lake, highlighting the high potential of the whole Pyrenean lacustrine district to contain a high number of new microeukaryotes taxa. Overall, this study and a recent study in the Alps and Himalayan mountains (Kammerlander et al. 2015) unveiled the high mountain lakes habitat as an important biodiversity reservoir of genetically rich Stramenopiles (mostly Chrysophyceae), Alveolata (Ciliophora) and Opisthokonta (Fungi). A comparison with the community composition of marine and freshwater molecular samples (Fig. 7.11) shows the consistent dominance of Chrysophyceae in high-altitude lakes and Artic lakes (Charvet et al. 2012). Overall, Chrysophyceae were more widely distributed in lakes with high oligotrophic conditions. Trophic status modulates the changes in freshwater eukaryote community composition, and eutrophic lakes are less species-rich. Perturbation such as higher availability of reactive nitrogen introduced by atmospheric deposition may also change the community structure of the most sensible species (Kammerlander et al. 2015). Thus, preserving the cold and (ultra)oligotrophic characteristics of the high mountain lakes environment may be of great interest for the study of the ecology and evolution of such idiosyncratic protists but, again, field experimentation should be carried out to confirm these findings.

In fact, the genetic novelty level after a GenBank database search (Triadó-Margarit and Casamayor 2012) showed that many of the 18S rRNA gene sequences recovered in the Pyrenean survey were below the species-level cut-off most widely accepted for microeukaryotes (i.e. 98% identity, Caron et al. 2009). Mostly in the case of Rhizaria-Cercomonads, a set of small (about 10 μm) free-living heterotrophic flagellates difficult to identify under the microscope for species identity, for which c. 90% of the sequences would be new species, and Stramenopiles-Chrysophyceae with 30–40% of the sequences potentially as new species. Conversely, Pyrenean Cryptophyta and Chlorophyceae showed very low genetic novelty (<3% and <5% of new species, respectively). Other taxa with substantial novelty were found within the Opisthokonta (Fungi). Fungi have been invoked as a target group to develop a microbial perspective on conservation biology because both it is important by itself and the fact that fungi biodiversity and their ecosystems roles can benefit conservation in general (Griffith 2012; Heilmann-Clausen et al. 2015).

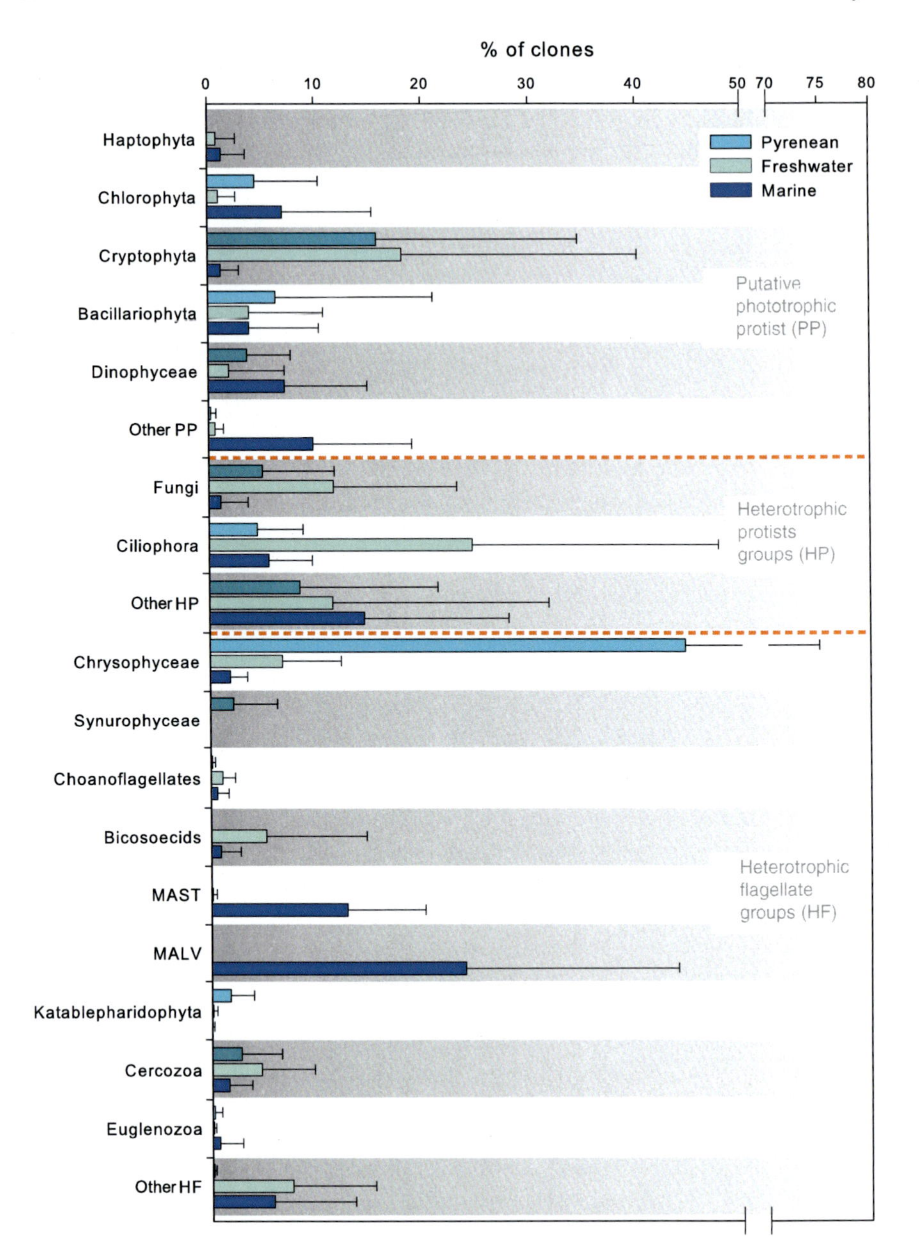

Fig. 7.11 Relative abundances of different microeukaryotes (i.e. <40 μm size) taxa identified by 18S rRNA gene sequencing in Pyrenean lakes and comparative analysis with marine and freshwater systems. *HF* (putative) heterotrophic flagellate groups; *HP* heterotrophic protists groups; *MAST* marine stramenopiles; *MALV* marine alveolates; *PP* (putative) phototrophic protists groups. From Triadó-Margarit and Casamayor (2012) with kind permission from John Wiley & Sons, Inc.

7.4 Towards a Microbial Conservation Perspective in High Mountain Lakes

Microorganisms are fundamental components to maintain the ecological integrity of any ecosystem but the exercise to define challenges and strategies (if needed) for microbial diversity conservation have only been seldom considered. Through the detailed studies carried out in the past years in the Aigüestortes i Estany de Sant Maurici National Park and recent few works available in the literature from other high mountain lakes districts such as the Alps, Himalayas and the western mountains of North America mentioned above, it arises that high mountain lakes hold a high microbial phylogenetic uniqueness and are reservoirs of large evolutionary potential. Microbes are therefore an important part of the biological richness of these environments that should be considered as a fundamental component of the natural heritage. However, we are in the very beginning for understanding microbial tolerance to different environmental conditions and how such underappreciated microbiota will respond to stresses and disturbances brought by the global change. How is the potentiality of survival and dispersal abilities for the different species, and whether or not the distribution of a single microbial population is restricted by the interactions established with other biological entities are also mostly unknown. Despite the advent of laboratory-friendly molecular methods and approaches, this incomplete knowledge and the lack in databases of enough microbial biogeographical studies of the high mountain realm to compare with, strongly limits the scientifically informed advice microbial ecologists can provide to managers dealing with the challenges for conservation of the whole high mountain landscape. Our basic knowledge of the biology and ecology of microorganisms will only improve after intensive inventories and worldwide-distributed field studies (Cotterill et al. 2007).

Whether or not there is any worry for microbes conservation in the high mountain ecosystem (and, in fact, in any other ecosystem) is yet a controversial and difficult to answer question, but global-scale warming and eutrophication may jeopardise the most idiosyncratic microbial populations that found in these areas the most appropriate conditions to grow. Conservation biology perspectives that rely mostly on habitat conservation apply very well for microorganisms. The microbial world may also nicely merge the compositionalist (i.e. entity-oriented through the lens of evolutionary ecology) and the functionalist (i.e. process-oriented through the lens of ecosystem ecology) views on conservation biology (Callicott et al. 1999). Conservation biology main concepts such as species, extinction, biotic interactions and management are however difficult to apply or poorly explored for microbes in situ. As mentioned above, microbial ecologists have managed somehow to sort out the problem of the species definition using either an operational and pragmatic biodiversity unit or phylogenetic approaches. Now they can sort out species rich and poor environments and explore how is the diversity of microbial life distributed around the planet and the relationships between microbial diversity, ecosystem health and biotic interactions (Faust and Raes 2012). The general concept that biological diversity is threatened with extinction when an element is rare or when it is in decline do not fit very well for the microbial world for several reasons. For instance, the presence of resting or dormant life forms in the rare microbial biosphere with the potential to rapid response to environmental changes and to become an important member of the community. And also the low probability of local extinction because bacteria do not need to find a partner to reproduce and can remain viable in the rare biosphere for very long periods. Finally, the apparently high connectivity among ecosystems, in general, may provide continuous species transfers although there is a lack of experimental information on dispersal and extinction rates in the context of conservation. Airborne supply of species from adjacent or remote ecosystems to alpine areas (Hervàs and Casamayor 2009; Hervàs et al. 2009; Barberán et al. 2014b) may circumvent such extinction through global-scale dynamics, high dispersibility and high speciation in the local communities (Barberán et al. 2014a).

Fortunately, microbial ecologists and theoretical ecologists are now analysing in parallel large temporal datasets of microbial species using stochastic community assembly models and establishing distinctive extinction–colonisation signatures at the microbial taxa level. The concept of "colonisation" in this context is understood as the ability of a particular microbial taxon to appear within the most abundant members of the community, whereas "extinction" would be the process to fall into the rare biosphere and become undetectable in the best case. The method identifies for the first time those microbial taxa with an ephemeral dynamic (higher colonisation and higher extinction rates), and those that are more stable (lower colonisation, lower extinction) and it is a promising approach to applying a closer conservation view on both microbial populations and microbial habitats with a more dynamic perspective (Jiménez-Ontiveros et al. 2017). The inadequacy of knowledge on the microbial world needs the development of major educational programs (Barberán et al. 2016) with emphasis on the engagement of ecosystems managers and citizens to better understand the whole picture of the biosphere and the key role of microorganisms within it. Criteria such as microbial species richness, phylogenetic diversity, exclusive microbial species occurrences and the microbial conservation value of a habitat, are now available to support environmental managers and lawyer decisions, and to citizens and stakeholders in general. Examples of threatened areas of high microbial interest in which efforts should be more intense to preserve their biodiversity are now available (e.g. Casamayor et al. 2013). Disturbance and loss of habitats or massive habitat change by anthropogenic activities may indeed threaten microbes and led to local extinction as frequently as in macroorganisms. Ecosystem management in alpine areas should promote to maintain spots of habitat heterogeneity within the landscape matrix between lakes and within the lake to retain such phylogenetically rich and diverse natural heritage.

Acknowledgements I am thankful to the Authorities of the Aigüestortes and Estany de St Maurici National Park for sampling facilities in the protected areas and continuous support, and Centre de Recerca d'Alta Muntanya, Universitat de Barcelona, Vielha for laboratory and logistics facilities. Technical support, fieldwork assistance, and fruitful discussions with many students, friends and colleagues along the works reviewed in this chapter are highly appreciated. I thank X. Triadó for help with figures, the support of the projects AERBAC (Ref. 079/2007), AERBAC-2 (Ref. 178/2010) and DISPERSAL (Ref. 829/2013) from the Spanish National Parks research program (OAPN-MAGRAMA), and PIRENA (Ref. CGL2009-13318) and BRIDGES (Ref. CGL2015-69043-P) from the Spanish Office for Science (MINECO), and the positive feedback and helpful comments from anonymous reviewers.

References

Adrian R, O'Reilly CM, Zagaresc H, Baines SB, Hessen DO, Keller W et al (2009) Lakes as sentinels of climate change. Limnol Oceanogr 54:2283–2297

Alivisatos AP, Blaser MJ, Brodie EL, Chun M et al (2015) MICROBIOME. A unified initiative to harness Earth's microbiomes. Science 350:507–508

Amann R, Rosselló-Móra R (2016) After all, only millions? mBio 7(4):e00999-16. doi:10.1128/mBio.00999-16

Auguet JC, Casamayor EO (2013) Partitioning of Thaumarchaeota populations along environmental gradients in alpine lakes. FEMS Microbiol Ecol 84:154–164

Auguet JC, Barberán A, Casamayor EO (2010) Global ecological patterns in uncultured Archaea. ISME J 4:182–190

Auguet JC, Triadó-Margarit X, Nomokonova N, Camarero L, Casamayor EO (2012) Vertical segregation and phylogenetic characterization of ammonia-oxidizing Archaea in a deep oligotrophic lake. ISME J 6:1786–1797

Barberán A, Casamayor EO (2010) Global phylogenetic community structure and beta-diversity patterns of surface bacterioplankton metacommunities. Aquat Microb Ecolo 59:1–10

Barberán A, Casamayor EO (2011) Euxinic freshwater hypolimnia promote bacterial endemicity in continental areas. Microb Ecol 61:465–472

Barberán A, Casamayor EO (2014) A phylogenetic perspective on species diversity, β-diversity, and biogeography for the microbial world. Mol Ecol 23:5868–5876

Barberán A, Bates ST, Casamayor EO, Fierer N (2012) Using network analysis to explore co-occurrence patterns in soil microbial communities. ISME J 6:343–351

Barberán A, Casamayor EO, Fierer N (2014a) The microbial contribution to macroecology. Front Microbiol 5:203. doi:10.3389/fmicb.2014.00203

Barberán A, Henley J, Fierer N, Casamayor EO (2014b) Structure, inter-annual recurrence, and global-scale connectivity of airborne microbial communities. Sci Tot Environ 487:187–195

Barberán A, Hammer TJ, Madden AA, Fierer N (2016) Microbes should be central to ecological education and outreach. J Microbiol Biol Educ 17:23–28

Bartrons M, Catalan J, Casamayor EO (2012) High bacterial diversity in epilithic biofilms of mountain lakes. Microb Ecol 64:860–869

Bellard C, Bertelsmeier C, Leadley P, Thuiller W, Courchamp F (2012) Impacts of climate change on the future of biodiversity. Ecol Lett 15:365–77

Callicott JB, Crowder LB, Mumford K (1999) Current normative concepts in conservation. Conserv Biol 13:22–35

Camarero L, Catalan J (2012) Atmospheric phosphorus deposition may cause lakes to revert from phosphorus limitation back to nitrogen limitation. Nat Comm 3:1118

Caron DA, Countway PD, Savai P, Gast RJ, Schnetzer A, Moorthi SD, Dennett MR, Moran DM, Jones AC (2009) Defining DNA-based operational taxonomic units for microbial-eukaryote ecology. Appl Environ Microbiol 75:5797–5808

Casamayor EO, Triadó-Margarit X, Castañeda C (2013) Microbial biodiversity in saline shallow lakes of the Monegros Desert, Spain. FEMS Microbiol Ecol 85:503–518

Catalan J, Rondon JCD (2016) Perspectives for an integrated understanding of tropical and temperate high-mountain lakes. J Limnol 75:215–234

Catalan J, Ballesteros E, Camarero L, Felip M, Gacia E (1992) Limnology in the Pyrenean Lakes. Limnetica 8:27–38
Catalan J, Camarero L, Felip M, Pla S, Ventura M, Buchaca T et al (2006) High mountain lakes: extreme habitats and witnesses of environmental changes. Limnetica 25:551–584
Catalan J, Barbieri MG, Bartumeus F et al (2015) Ecological thresholds in European alpine lakes. Fresh Biol 54:2494–2517
Charvet S, Vincent WF, Lovejoy C (2012) Chrysophytes and other protists in High Arctic lakes: molecular gene surveys, pigment signatures and microscopy. Polar Biol 2012:733–748
Cotterill FPD (1995) Systematics, biological knowledge and environmental conservation. Biodivers Conserv 4:183–205
Cotterill FPD, Al-Rasheid K, Foissner W (2007) Conservation of protists: is it needed at all? Biodivers Conserv 17:427–443
Daily GC, Polasky S, Goldstein J, Kareiva PM, Mooney HA, Pejchar L, Ricketts TH, Salzman J, Shallenberger R (2009) Ecosystem services in decision making: time to deliver. Front Ecol Environ 7:21–28
Downing JA, Prairie YT, Cole JJ, Duarte CM et al (2006) The global abundance and size distribution of lakes, ponds, and impoundments. Limnol Oceanogr 51:2388–2397
Dubilier N, McFall-Ngai M, Zhao L (2015) Microbiology: create a global microbiome effort. Nature 526:631–634
Edwards A, Douglas B, Anesio AM et al (2013) A distinctive fungal community inhabiting cryoconite holes on glaciers in Svalbard. Fungal Ecol 6:168–176
Esteban GF, Finlay BJ (2010) Conservation work is incomplete without cryptic biodiversity. Nature 463:293–293
Faust K, Raes J (2012) Microbial interactions: from networks to models. Nat Rev Microbiol 10 (8):538–550
Felip M, Bartumeus F, Halac S et al (1999) Microbial plankton assemblages, composition and biomass, during two ice-free periods in a deep high mountain lake (Estany Redo, Pyrenees). J Limnol 58:193–202
Fernàndez-Guerra A, Casamayor EO (2012) Habitat-associated phylogenetic community ecology of microbial ammonia oxidizers. PLoS One 7:e47330. doi:10.1371/journal.pone.0047330
Fuhrman JA, Cram JA, Needham DM (2015) Marine microbial community dynamics and their ecological interpretation. Nat Rev Microbiol 13:133–146
Gilbert JA, Jansson JK, Knight R (2014) The earth microbiome project: successes and aspirations. BMC Biol 12:69
Griffith GW (2012) Do we need a global strategy for microbial conservation? Trends Ecol Evol 27:1–2
Hahn MW (2006) The microbial diversity of inland waters. Curr Opin Biotechnol 17:256–261
Hayden CJ, Beman JM (2016) Microbial diversity and community structure along a lake elevation gradient in Yosemite National Park, California. Environ Microbiol 18:1782–1791
Heilmann-Clausen J, Barron ES, Boddy L, Dahlberg A, Griffith GW, Nordén J, Ovaskainen O, Perini C, Senn-Irlet B, Halme P (2015) A fungal perspective on conservation biology. Conserv Biol 29:61–68
Hervàs A, Casamayor EO (2009) High similarity between bacterioneuston and airborne bacterial community compositions in a high mountain lake area. FEMS Microbiol Ecol 67:219–228
Hervàs A, Camarero L, Reche I, Casamayor EO (2009) Viability and potential for immigration of airborne bacteria from Africa that reach high mountain lakes in Europe. Environ Microbiol 11:1612–1623
Horner-Devine MC, Lage M, Hughes JB, Bohannan BJM (2004) A taxa–area relationship for bacteria. Nature 432:750–753
Jiménez-Ontiveros V, Capitan JA, Casamayor EO, Alonso D (2017) Effective parameters for complex temporal dynamics of ecological communities: the Island R package. Meth Ecol Evol (submitted)

Kammerlander B, Breiner HW, Filker S, Sommaruga R, Sonntag B, Stoeck T (2015) High diversity of protistan plankton communities in remote high mountain lakes in the European Alps and the Himalayan mountains. FEMS Microbiol Ecol 9: fiv010

Kim M, Oh HS, Park SC, Chun J (2014) Towards a taxonomic coherence between average nucleotide identity and 16S rRNA gene sequence similarity for species demarcation of prokaryotes. Int J Syst Evol Microbiol 64:346–351

Liu Y, Yao T, Jiao N, Kang S, Zeng Y, Huang S (2006) Microbial community structure in moraine lakes and glacial meltwaters, Mount Everest. FEMS Microbiol Lett 265:98–105

Llorens-Marès T, Auguet JC, Casamayor EO (2012) Winter to spring changes in the slush bacterial community composition of a high mountain lake (Lake Redon, Pyrenees) Environ Microbiol Repo 4:50–56

Llorens-Marès T, Liu Z, Allen LZ, Rusch DB, Craig MT, Dupont CL, Bryant DA, Casamayor EO (2017) Speciation and ecological success in dimly lit waters: horizontal gene transfer in a green sulfur bacteria bloom unveiled by metagenomic assembly. ISME J 10:201–211. doi:10.1038/ismej.2016.93

Locey KJ, Lennon JT (2016) Scaling laws predict global microbial diversity. Proc Nat Acad Sci USA 113:5970–5975

Madigan MT, Martinko JM, Bender KS, Buckley DH, Stahl DA (2015) Brock biology of microorganisms, 14th edn. Pearson, Boston

Margulis L, Chase D, Guerrero R (1986) Microbial communities: invisible to the scrutiny of naturalists, most microbial communities have escaped description. Bioscience 36:160–170

Methé BA, Nelson KE, Pop M, Creasy HH et al (2012) A framework for human microbiome research. Nature 486:215–221

Nelson CE (2009) Phenology of high-elevation pelagic bacteria: the roles of meteorologic variability, catchment inputs and thermal stratification in structuring communities. ISME J 3:13–30

Newton RJ, Jones SE, Eiler A, McMahon KD, Bertilsson S (2011) A guide to the natural history of freshwater lake bacteria. Microbiol Mol Biol Rev 75:14–49

Ordoñez OF, Flores MR, Dib JR, Paz A, Farias ME (2009) Extremophile culture collection from Andean Lakes: extreme pristine environments that host a wide diversity of microorganisms with tolerance to UV radiation. Microb Ecol 58:461–473

Ortiz-Alvarez R, Casamayor EO (2016) High occurrence of Pacearchaeota and Woesearchaeota (superphylum DPANN) in surface waters of oligotrophic high-altitude lakes. Environ Microbiol Rep 8:210–217

Pérez MT, Sommaruga R (2011) Temporal changes in the dominance of major planktonic bacterial groups in an Alpine Lake: discrepancy with their contribution to bacterial production. Aquat Microb Ecol 63:161–170

Pernthaler J, Glockner FO, Unterholzner S, Alfreider A, Psenner R, Amann R (1998) Seasonal community and population dynamics of pelagic bacteria and archaea in a high mountain lake. Appl Environ Microbiol 64:4299–4306

Peter H, Sommaruga R (2016) Shifts in diversity and function of lake bacterial communities upon glacier retreat. ISME J 10:1545–1554

Pla S, Camarero L, Catalan J (2003) Chrysophyte cyst relationships to water chemistry in Pyrenean lakes (NE Spain) and their potential for environmental reconstruction. J Paleolimnol 30:21–34

Raes J, Bork P (2008) Molecular eco-systems biology: towards an understanding of community function. Nat Rev Microbiol 6:693–699

Reche I, Pulido-Villena E, Morales-Baquero R, Casamayor EO (2005) Does ecosystem size determine aquatic bacterial richness? Ecology 86:1715–1722

Restrepo-Ortiz C, Casamayor EO (2013) Environmental distribution of two widespread uncultured freshwater Euryarchaeota clades unveiled by specific primers and quantitative PCR. Environ Microbiol Rep 5:861–867

Rosselló-Mora R, Amann R (2001) The species concept for prokaryotes. FEMS Microbiol Rev 25:39–67

Schloss PD, Girard R, Martin T, Edwards J, Thrash JC (2016) Status of the archaeal and bacterial census: an update. mBio 7(3):e00201–16. doi:10.1128/mBio.00201-16

Sommaruga R, Casamayor EO (2009) Bacterial 'cosmopolitanism' and importance of local environmental factors for community composition in remote high-altitude lakes. Fresh Biol 54:994–1005

Spang A, Saw JH, Jorgensen SL, Zaremba-Niedzwiedzka K, Martijn J, Lind AE et al (2015) Complex archaea that bridge the gap between prokaryotes and eukaryotes. Nature 521:173–179

Stoeck T, Breiner HW, Filker S et al (2014) A morpho-genetic diversity survey on ciliate plankton from a mountain lake pinpoints the necessity of protist barcoding in microbial ecology. Environ Microbiol 16:430–444

Tolotti M, Forsström L, Morabito G et al (2009) Biogeographical characterization of phytoplankton assemblages in high altitude, and high latitude European lakes. Adv Limnol 62:55–75

Triadó-Margarit X, Casamayor EO (2012) Genetic diversity of planktonic eukaryotes in high mountain lakes (Central Pyrenees, Spain). Environ Microbiol 14:2445–2456

Vila-Costa M, Barberan A, Auguet JC, Shalabh S, Moran MA, Casamayor EO (2013) Bacterial and archaeal community structure in the surface microlayer of high mountain lakes examined under two atmospheric aerosol loading scenarios. FEMS Microbiol Ecol 84:387–397

Vila-Costa M, Bartrons M, Catalan J, Casamayor EO (2014) Nitrogen cycling genes in epilithic biofilms of oligotrophic high altitude lakes (Central Pyrenees, Spain). Microbial Ecol 68:60–69

Vila-Costa M, Pulido C, Chappuis E, Casamayor EO, Gacia E (2016) Macrophytes landscape modulates ecosystem-level nitrogen losses through tightly coupled plant-microbe interactions. Limnol Oceanogr 61:78–88

Wang J, Meier S, Soininen J, Casamayor EO, Pan F, Yang X, Zhang Y, Wu Q, Zhou J, An Z, Shen J (2017) Regional and global elevational patterns of microbial species richness and evennes. Ecography 40:393–402. doi:10.1111/ecog.02216

Wertz S, Degrange V, Prosser JI, Poly F, Commeaux C, Freitag T, Guillaumaud N, Roux XL (2006) Maintenance of soil functioning following erosion of microbial diversity. Environ Microbiol 8:2162–2169

Yarza P, Yilmaz P, Pruesse E, Glöckner FO et al (2014) Uniting the classification of cultured and uncultured bacteria and archaea using 16S rRNA gene sequences. Nat Rev Microbiol 12:635–645

Zhang R, Wu Q, Piceno YM et al (2013) Diversity of bacterioplankton in contrasting Tibetan lakes revealed by high-density microarray and clone library analysis. FEMS Microbiol Ecol 86: 277–287

Chapter 8
Why Should We Preserve Fishless High Mountain Lakes?

Marc Ventura, Rocco Tiberti, Teresa Buchaca, Danilo Buñay, Ibor Sabás and Alexandre Miró

Abstract High mountain lakes are originally fishless, although many have had introductions of non-native fish species, predominantly trout, and recently also minnows introduced by fishermen that use them as live bait. The extent of these introductions is general and substantial often involving many lakes over mountain ranges. Predation on native fauna by introduced fish involves profound ecological changes since fish occupy a higher trophic level that was previously inexistent. Fish predation produces a drastic reduction or elimination of autochthonous animal groups, such as amphibians and large macroinvertebrates in the littoral, and crustaceans in the plankton. These strong effects raise concerns for the conservation of high mountain lakes. In terms of individual species, those adapted to live in larger lakes have suffered a higher decrease in the size of their metapopulation. This ecological problem is discussed from a European perspective providing examples from two study areas: the Pyrenees and the Western Italian Alps. Species-specific studies are urgently needed to evaluate the conservation status of the more impacted species, together with conservation measures at continental and regional scales, through regulation, and at local scale, through restoration actions, aimed to stop further invasive species expansions and to restore the present situation. At different high mountain areas of the world, there have been restoration projects aiming to return lakes to their native fish-free status. In these areas autochthonous species that disappeared with the introduction of fish are progressively recovering their initial distribution when nearby fish-free lakes and ponds are available.

M. Ventura (✉) · T. Buchaca · D. Buñay · I. Sabás · A. Miró
Integrative Freshwater Ecology Group, Center For Advanced Studies of Blanes (CEAB-CSIC), Accés a la Cala St. Francesc, 14, Blanes, Girona, Catalonia, Spain
e-mail: ventura@ceab.csic.es

R. Tiberti
Dipartimento di Scienze della Terra e dell'Ambiente (DSTA), University of Pavia, Via Ferrata 9, 27100 Pavia, Italy

R. Tiberti
Alpine Wildlife Research Centre, Gran Paradiso National Park, Degioz 11, 11010 Valsavarenche, Aosta, Italy

J. Catalan et al. (eds.), *High Mountain Conservation in a Changing World*, Advances in Global Change Research 62, DOI 10.1007/978-3-319-55982-7_8

Keywords Global change · Introduced species · Trout · Minnows · Amphibians · Restoration · Conservation · High mountain lakes

8.1 Introduced Species, a Global Threat to Freshwater Ecosystems

Among the biomes of the planet, freshwaters have suffered the strongest decline in the number of vertebrate species during 1970–2010 (76% in freshwaters compared to 39% in the other two biomes; McRae et al. 2014). This great decline is a result, on a global scale, of factors related to habitat destruction, pollution, water-level reductions and invasive species among others (Collen et al. 2014). Although these factors are affecting all species within the habitats, amphibians are amongst the most threatened freshwater groups (Collen et al. 2014).

In Europe, freshwater fish are one of the animal groups with a higher number of invasive species (Hulme et al. 2009). The introduction of freshwater fish is closely related to human activities (Gido et al. 2004; Marchetti et al. 2004) and especially to fishing in the case of salmonids (Cambray 2003; Cowx and Gerdeaux 2004; Granek et al. 2008). The circumstances and timing of recent arrival of some species have been documented worldwide (FAO 2003) and in some particularly well-studied lakes (Villwock 1994; Pringle 2005; Volta and Jepsen 2008). In other cases, it has been possible to describe the introductions that have suffered different lakes using historical reconstructions (Emery 1985; Garcia-Berthou and Moreno-Amich 2000; Brancelj et al. 2000; Buchaca et al. 2011). Previous studies show that the introduction of freshwater fish can often have significant negative ecological consequences (Vitule et al. 2009).

High mountain lakes are usually isolated from lower streams by physical barriers that have prevented natural colonisation of fish (Pechlaner 1984; Knapp et al. 2001a; Miró and Ventura 2013). In these ecosystems, the introduction of fish species is mainly related to recreational fishing and promoted by different administrations (Pister 2001; Schindler and Parker 2002). There is detailed information on the historical evidences of the spread of salmonids in some high mountain areas of the western United States (Christenson 1977; Bahls 1992; Knapp 1996; Wiley 2003) and the Canadian Rockies (Schindler 2000). In these areas, introductions are described chronologically from the end of nineteenth and beginning of twentieth centuries and they were initially made by individual fishermen, and a few decades later by the Governmental agencies (Christenson 1977; Schindler 2000). In contrast to these areas, the information about the stocking history in other mountain ranges is difficult to access. In particular the information from the high mountain ranges of Asia (Petr 1999) and South America (Vigliano and Alonso 2007; Martín-Torrijos et al. 2016) are insufficient, because of uncontrolled stocking (Vigliano and Alonso 2007; Ortega et al. 2007).

In contrast to the American continent, in European high mountain lakes the colonisation process has not been studied in detail (Gliwicz and Rowan 1984;

Pechlaner 1984; Sostoa and Lobón-Cerviá 1989). The first introductions in the Alps were carried out at the end of the sixteenth century (Pechlaner 1984) and in the Tatra mountains at the end of the nineteenth century (Brancelj 2000; Gliwicz and Rowan 1984). In the Cantabric mountains (Iberian Peninsula) introductions also occurred at the end of the nineteenth century (Terrero 1951) and even more recently in the Sistema Central and Sistema Ibérico, also in the Iberian Peninsula (Martinez-Solano et al. 2003; Toro et al. 2006). In the Pyrenees, the first fish introductions took place at least during the middle ages, being the first written evidence of fish presence from the fourteenth century (Miró 2011).

In conclusion, we know that introduced fish are present in high-altitude aquatic habitats in many mountain ranges all around the world, but we are not aware of the exact spatial extent of their distribution (e.g. the percentage and features of stocked lakes at a regional scale). We also know that except a very few medieval and some pre-industrial introductions occurred in Europe, the majority of the introductions started during the nineteenth century and in particular in the second half of the twentieth. Indeed in the past decades, a critical acceleration of the invasion has been promoted by different governmental agencies, supporting the growing popularity of recreational angling (Pister 2001; Schindler and Parker 2002).

8.2 The Process of Species Introductions in the Alps

The first documented fish introductions in the Alps were carried out during the middle Age, at the end of the sixteenth century, probably during the reign of Kaiser Maximilian I (Pechlaner 1984). However, similarly to most mountain ranges across Europe, most of the fish introductions took place in the recent decades to support recreational angling. In Italy, the phenomenon became particularly evident from the 1960s onwards when, recreational angling became a popular activity (Cantonati et al. 2006). The presence of introduced species such as *Salmo trutta* L 1758 or species from across the Atlantic Ocean, such as *Oncorhynchus mykiss* (Walbaum 1792) and *Salvelinus fontinalis* (Mitchill 1814) is common all around the Alps. Moreover *Salvelinus umbla* (L 1758) have been introduced across the Italian Alps (Zerunian 2003). Restocking with *S. umbla* is considered a conservation measure in some Alpine regions, even if its status as an indigenous species in Italy has given rise to serious doubts (Pechlaner 1984; Machino 1999; Piccinini et al. 2004) and the fact that this species does not require ex situ conservation actions. Also the marble trout (*Salmo marmoratus* Cuvier, 1829), an alpine endemism, was sometimes introduced in alpine lakes, as well as *Salvelinus namaycush* (Walbaum 1792) from North America. Finally cyprinids, used as live baits by anglers, are accidentally or voluntary released in many high-altitude lakes, and minnows (*Phoxinus* sp.) are able to establish reproductive populations.

For the Alps, a detailed and coherent history of the fish introductions is not available: information is often insufficient and still scattered in many archives and grey literature. Also, the distribution of introduced fish has rarely been assessed over large regions. In a study conducted in the Eastern Alps, Jersabek et al. (2001) report that the 41% of the high-altitude lakes (>0.5 ha, above 1300 m a.s.l.) present fish fauna; in Valle d'Aosta (Western Alps), where the first documented fish introductions date back to 1926 (Mammoliti Mochet 1995), an accurate catalogue of all the ponds and lakes (Frezet 2003) indicate that the 43% of the lakes (>0.5 ha, above 1000 m a.s.l.) present fish fauna and this percentage increase to the 50% excluding those lakes above 2900 m a.s.l., where fish survival is challenged by the extreme temperatures and glacial influence; in the Gran Paradiso National Park (Western Alps) 35% of all the lakes (>0.5 ha) is occupied by fish (RT personal observation). These percentages are similar to those observed in other mountain ranges (Miró and Ventura 2013; Bahls 1992).

8.3 The Process of Species Introductions in the Pyrenees

The different fish species described to be introduced in the Pyrenees include the salmonids *S. trutta*, *O. mykiss* and *S. fontinalis*, and the cyprinid *Phoxinus* sp. in the southern (Spanish) Pyrenees (Miró and Ventura 2013, 2015) and these species together with *S. umbla* and *S. namaycush* in the northern (French) Pyrenees (Delacoste et al. 1997). At a global scale, *S. trutta* and *O. mykiss* are considered among the 100 most invasive alien species in the world (Lowe et al. 2000). Within Europe, *O. mykiss*, *S. fontinalis* and *S. namaycush* have been introduced from North America, while *S. trutta* and *S. umbla* that are native European species have been widely introduced beyond their native range, mostly in high mountain areas. Therefore, they are all classified as European alien species (Hulme et al. 2009).

8.3.1 Trout Introductions

The oldest written documents describing fishing rights in high mountain lakes of the Pyrenees date back to the fourteenth and fifteenth centuries (Miró 2011). These initial introductions for traditional exploitation resulted in 26.5% of the lakes of the southern slope of the Pyrenees having introduced trout by 1900 (Fig. 8.1; Miró and Ventura 2013). Similarly, on the northern side of the Pyrenees, it has been described that ca. 25% of the lakes had fish before the onset of widespread introductions after 1936 (Delacoste et al. 1997), which might also be attributed to traditional fishing activities by local fishermen. Miró and Ventura (2013) findings suggest that human exploitation of some of these lakes might have originated further back in time, possibly back to prehistoric times, when primitive residents already used the high-altitude pastures (Miró 2011). However, written evidence suggests that the first major historical introductions occurred within the medieval warm period (1000–1300 AD) when the human population in the Pyrenees was highest (Miró 2011).

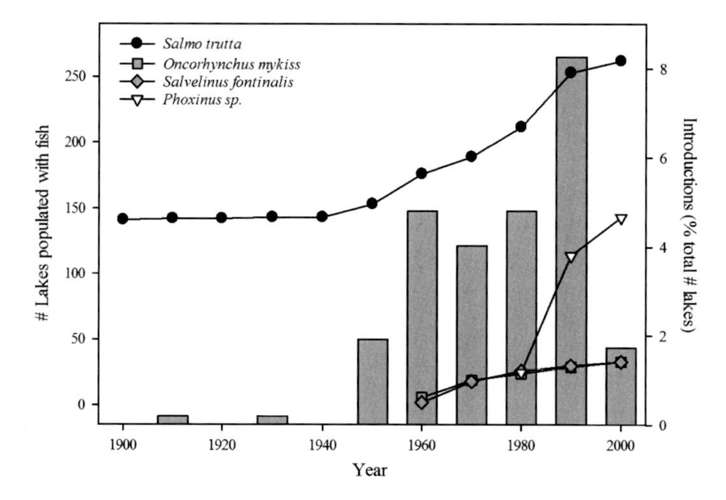

Fig. 8.1 Introduction process of non-native fish species in the southern Pyrenees during the twentieth century. *Vertical grey bars* are the decadal total number of lakes that have suffered fish introductions as a percentage of the total number of lakes >0.5 ha of the study area (n = 520). Data from Miró and Ventura (2013, 2015)

With the onset of modern fish management that took place during the period 1960–2000, the most significant factors explaining trout distribution were related to management practices (Miró and Ventura 2013). However, different factors were important for the different species. For *S. trutta*, the probability of occurrence was highest in areas where fish were stocked by helicopter, and in fish-managed areas. The repeated use of helicopter stocking in recent decades has resulted in an extensive occurrence of trout in these areas and, as a result, in more than half of the lakes with fish. This result is similar to that found in other parts of the world where helicopters and aeroplanes have been used for stocking trout, such as several mountain areas of West USA, where trout is present in ca. 60% of lakes (Bahls 1992).

In contrast to *S. trutta*, *S. fontinalis* was preferentially found in lakes with forestry road access, while *O. mykiss* was also found in lakes with road access but where active fish management has been carried out. These two last species are present in a much lower number of lakes, and also were introduced mainly between 1960 and 1980 (Fig. 8.1). Their presence is closely linked to the construction of the majority of hydroelectric power station infrastructure at high altitudes in the Pyrenees (Catalan et al. 1997). Hydroelectric companies compensated local citizens by developing local fisheries that were used to introduce these two species to the lakes around the area, where hydroelectric power station infrastructure was constructed. The same hydroelectric infrastructure construction was responsible for the forestry roads that were, in turn, the main routes facilitating the introduction of these two species. Amongst the species, *S. fontinalis* was mainly introduced in lakes with vehicle access. In the 1980s, administrative changes shifted stocking responsibilities to local fishermen's societies together with governmental agencies. This fact favoured the closure of *S. fontinalis* and *O. mykiss* hatcheries and the introduction of the north and central European lineages of *S. trutta* in official fish hatcheries. As a result, there has been a shift to using lineages from different parts of central and northern Europe (Araguas et al. 2009). Modern management practices have therefore resulted in increased fish introductions during the past few decades with a maximum during the 1990s (Fig. 8.1). As a result, trout have been introduced in most lakes with higher fishing interest (e.g. lower altitude or bigger surface area). The low number of lakes remaining without fish after the 1990s is likely the reason why fish introductions dropped considerably during 2000.

8.3.2 Collateral Introductions

The salmonid introductions described above have resulted in collateral introductions of other fish species mainly used as live bait (Miró and Ventura 2015). This fact may eventually lead to stronger ecological consequences than the introduction of trout alone. In the Pyrenees, Miró and Ventura (2015) found that the introduction of minnows in high mountain lakes was mediated by a preceding invasive species and facilitated by human activity. They found that the introduction of the minnow is a more recent and faster process than that of salmonids (Fig. 8.1). Since 1970, when the first minnow introduction took place, it has now spread to 27% of the lakes of the southern Pyrenees with an introduction rate of 4.7 lakes per year, compared to those of trout, at 2.2 lakes per year for the period between 1940 and 2000 (Miró and Ventura 2015).

Differences in minnow life history characteristics compared to those of trout give them a higher acclimation success and therefore high invasive potential. The minnow, like other small widely distributed freshwater fish, displays a remarkable variability in its life history depending on the local temperature. For example, minnows have maximum age ranges between 3 and 13 years reaching maturity between 1–2 and 5–7 years in optimal and sub-optimal conditions respectively, and show significantly lower growth increments in cold summers (Mills 1988). This plasticity in their life history is what has allowed the species to easily acclimate to high mountain lakes, showing higher resistance to harsh conditions than trout. Miró and Ventura (2015) have not found any lake where minnows have disappeared once established. In contrast, in lakes with trout, between 10 and 44% of the trout populations have become extinct after 20–30 years due to a lack of favourable conditions (Knapp 2005; Armstrong and Knapp 2004; Miró and Ventura 2013). Moreover, in the Pyrenees, Miró and Ventura (2015) have found that in twenty lakes and ponds, pre-existing trout populations have disappeared after minnow introductions. Some of these lakes and ponds had brown trout introduced centuries ago while others were stocked recently.

Miró and Ventura (2015) also found that minnow presence was more likely in lakes with lower elevation, greater surface areas and higher temperatures. This distribution could be a result of the pattern of introductions (e.g. fishermen using live bait do not fish so often at the upper altitude lakes) or to fair habitat conditions for released minnows to become established. Taking into account the high plasticity of minnows, it seems more likely that their presence was a result of the pattern of introduction. As well as minnows, other species might be used as live bait, increasing the number of invasive species to lakes. This is the case for gudgeon (*Gobio* spp.), which is now found in some lakes of the northern slope of the Pyrenees (Miró 2011). This fish has similar size and flexible life history features to those of minnows (Tang et al. 2011). Thus, the ecological effects of fish introductions in high mountain lakes can result in stronger unpredicted consequences.

Live bait-related introductions of minnows have occurred in lower, boreal and arctic lakes where trout is present. The distribution of minnow expanded considerably throughout the twentieth century in the north European lakes of Scotland and Norway, especially in mountain areas, due mainly to the use of minnows as live bait for angling (Maitland and Campbell 1992; Museth et al. 2007). When minnow is introduced in lakes with autochthonous trout, it reduces recruitment and annual growth rates of trout, causing a decrease of the trout abundance by 35% on average; however, the effect on other native fauna takes place primarily in the shallow littoral areas (Museth et al. 2007).

Angling practices with live bait represents a worrying pathway for alien species introductions (Kerr et al. 2005; Webb 2007; DiStefano et al. 2009; Ward et al. 2012). The largest organisms used as live bait are several species of small fishes, but other animals such as amphibians, earthworms, crayfishes, grubs and insects are also used (Lindgren 2006; Keller and Lodge 2007). Improper disposal of live bait has been attributed as the source of introduction of at least 14 species of fishes in Ontario (Kerr et al. 2005). In the English Lake District, individuals of at least 12 native and non-native fish species have been brought to Windermere for the purpose of live baiting (Winfield et al. 2011). Live bait use was also responsible for the introduction of 47 known freshwater species in United States Mid-Atlantic slope drainage systems, among which are at least 5 non-native fishes, 4 non-native crayfishes and 9 non-native earthworm species (Kilian et al. 2012). Live baits may be released to the medium by accidental escape or, more often, by purpose at the end of the fishing trip (Winfield et al. 2011; Kilian et al. 2012). In some cases, it has been shown that bait-related introductions have resulted in established populations of invasive species (e.g. Callaham et al. 2006; Migge-Kleian et al. 2006).

In high mountain lakes, trout are the only species group authorised for introduction by governmental agencies worldwide (e.g. Sostoa and Lobón-Cerviá 1989; Wiley 2003), mainly associated with recreational fishing (Cambray 2003). Unlike trout, minnow introductions are in general not authorised by governmental authorities, and their introduction is often an illegal angling practice. As mentioned, the final result in many cases is that fish unused as live bait are released at the end of the fishing expedition (Maitland and Campbell 1992; Kerr et al. 2005; Winfield et al. 2011; Kilian et al. 2012). This practice has been quantified to be done by 36% of the fishermen in Michigan and Wisconsin, 41% in Ontario and 65% in Maryland (Litvak and Mandrak 1993; Kerr et al. 2005; Kilian et al. 2012). To prevent the widespread release of non-native species used as live bait, many US states and Canadian territories have restricted the use, sale or transport of bait (Kerr et al.

2005; Peters and Lodge 2009). A similar situation exists on the southern slope of the Pyrenees, where the release of any organism to the environment without government authorization is also strictly prohibited (Miró 2011). Nevertheless, similar to the findings from the southern valleys of the Pyrenees, in some regions of North America a large proportion of anglers appear to be unaware of, or choose to ignore, the current regulation forbidding the release of live organisms because they believe their actions are compassionate and that the released unused bait is a suitable food for angling fishes (Kerr et al. 2005; Kilian et al. 2012). The results described above strongly suggest the need to intensify preventive actions by giving accurate information of the potentially adverse effects on the local environment of the release of non-native organisms to fisheries boards and local communities. These actions are one of the best guiding principles to prevent the spread of invasive species together with regulation and legislation (Simberloff et al. 2013).

8.4 Ecological Consequences

The implications of stocking into high mountain lakes derive primarily from the fact that fish occupy a higher trophic level that was previously inexistent leading to profound ecological changes. Their introduction in fishless lakes is commonly associated with extirpation or reduction of native aquatic species (e.g. invertebrates and amphibians) and can have indirect effects on the whole ecosystem, and on its linkage with the surrounding terrestrial habitats (Eby et al. 2006). From high altitude lakes, fish can colonise otherwise inaccessible downstream aquatic habitats often propagating the ecological impact at a whole basin scale (Adams et al. 2001).

The studies concerning the ecological impact of introduced fish in high-altitude lakes are strongly biased towards the impact of salmonids, even if minnows and other small species are likely to become—or already are—a serious conservation problem in the near future. Since trout are visual predators, most of the direct impacts are attributable to their size selective predation strategy, affecting only larger non-fossorial taxa (Knapp et al. 2001b; Tiberti et al. 2014a, b). Published studies from different mountain regions of the world have focused mainly on the effects of trout introductions that produce a drastic reduction or elimination of those autochthonous animal groups of bigger size. Threatened groups live mostly in the littoral and benthic zones and include amphibians (Knapp 2005; Orizaola and Braa 2006; Pope 2008; Pope et al. 2008; Pilliod et al. 2010; Tiberti and von Hardenberg 2012; Bosch et al. 2006; in particular those with aquatic larval stages; Bradford et al. 1993; Martinez-Solano et al. 2003; Tiberti et al. 2014b) and conspicuous macroinvertebrates (Knapp et al. 2001b; Pope et al. 2009; de Mendoza et al. 2012; Pope and Hannelly 2013; Tiberti et al. 2014b; Fig. 8.2). The effects on amphibians are of special concern, since they are one of the most threatened animal groups worldwide (Beebee and Griffiths 2005).

The effects of trout on the plankton are less predictable, basically due to the smaller size of the planktonic animals, and the biology of the introduced

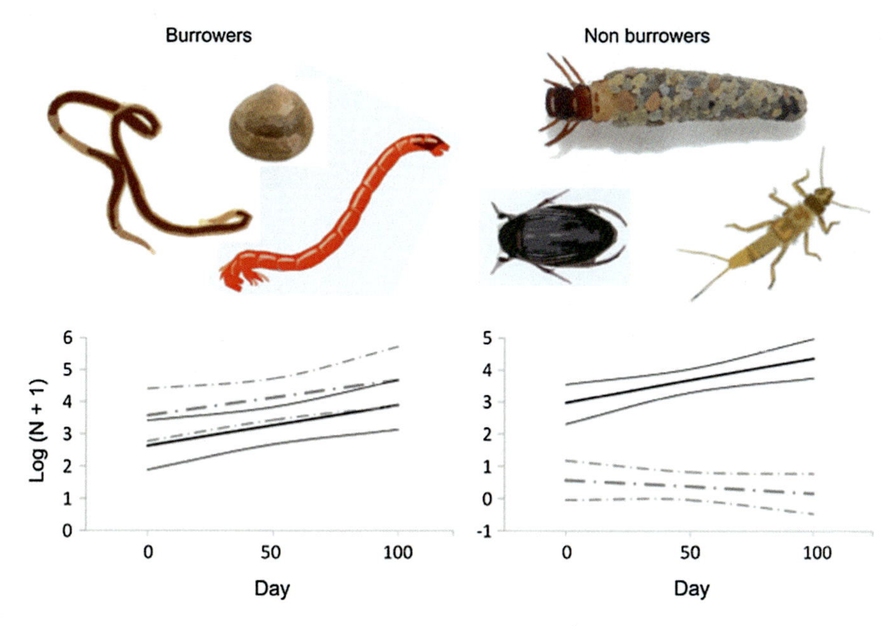

Fig. 8.2 Seasonal trend of abundance from day 0 (15 June) to day 100 (23 September) of littoral macroinvertebrates in naturally fishless (*solid line*) and stocked (*dashed lines*) lakes. While borrowing taxa are unaffected or even indirectly favoured by fish presence, nectonic and benthonic taxa (non-borrowers) are very sensitive to fish predation. Modified from Tiberti et al. (2014b)

species/populations, which can be more or less planktivorous. However, larger crustaceans have also decreased their abundance and body size, or disappeared from some high mountain lakes with stocked fish (Fig. 8.3; Brancelj 2000; Knapp et al. 2001b; Tiberti et al. 2014b). Among salmonids, this impact has been described in particular for *S. fontinalis*, which is a generalist predator and which can rely on large zooplankton species as a temporary or seasonal resource, especially during the winter (Knapp et al. 2001b; Schabetsberger et al. 2009; Tiberti et al. 2014b; Dawidowicz and Gliwicz 1983).

Direct predation of more visible taxa can produce a series of indirect ecological effects (top-down and cascading effects) which affect the entire ecosystem (Hall et al. 1976; Schindler et al. 2001; Sarnelle and Knapp 2005; Eby et al. 2006). The decrease of large zooplankton species can produce a competitive release of small zooplankton species (e.g. rotifers), which are less efficient grazers (Brooks and Dodson 1965). Therefore, a decrease in crustaceans together with enhanced nutrient cycling have led to changes in primary producers (Sarnelle and Knapp 2005; Magnea et al. 2013).

The effect on benthic macroinvertebrates is also important during their emergency phase, when passing through the water column, which results in a substantial reduction or alteration of their emergency rates (Pope et al. 2009; Epanchin et al. 2010; Tiberti et al. 2016b). This is also linked with indirect effects on surrounding habitats through resource depletion for terrestrial insectivores (Epanchin et al. 2010;

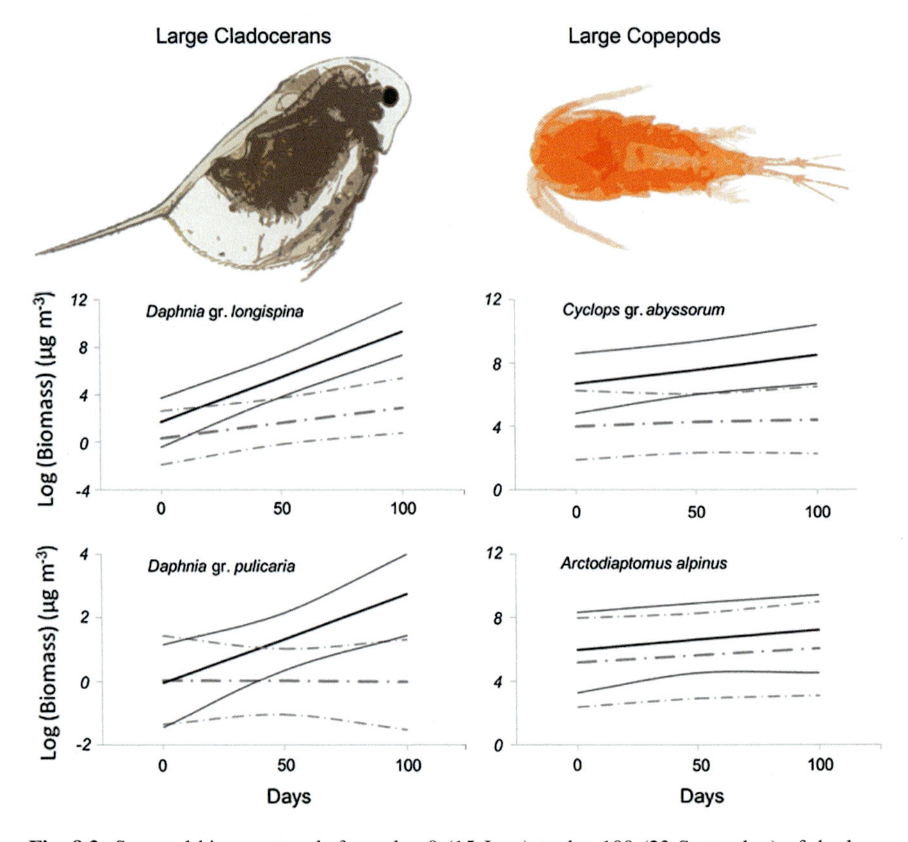

Fig. 8.3 Seasonal biomass trends from day 0 (15 June) to day 100 (23 September) of the large (>1 mm) zooplankton species in fishless (*solid line*) and stocked (*dashed lines*) lakes. With the exception of *Arctodiaptomus alpinus*, the biomass of large zooplankton species is significantly larger and its increment is significantly faster in fishless lakes. Modified from Tiberti et al. (2014b)

Eby et al. 2006; Finlay and Vredenburg 2007). Similarly, a reduction of the amphibians' emergence can affect terrestrial habitats and predators (Matthews et al. 2002). In the complex, aquatic subsidies depletion by introduced fish can have deep and detectable, indirect cascading effects in terrestrial habitats (Knight et al. 2005; Gratton et al. 2008).

On the other hand, the effects of minnow introductions on high mountain lakes have scarcely been studied in comparison with those of trout. Regardless of their smaller size, minnows are also occupying the top of the food chain (Hesthagen et al. 1992; Borgstrom et al. 1985; Museth et al. 2010), and when co-occurring with salmonids remain mainly in the littoral to escape trout predation (Museth et al. 2003). However, minnows are also the main cause for a reduction in trout reproduction (Borgstrøm et al. 1996) that leads to trout disappearances in some high mountain lakes (Miró and Ventura 2015). In this case, minnow seem to have a strong impact on amphibians and macroinvertebrates, but also on crustacean biomass (Schabetsberger et al. 1995), that might result in a strong trophic cascade, as has been described for other lakes such as those from the north-temperate area (Carpenter et al. 2011).

Even if the effects listed above (see Fig. 8.4 for a schematic summary) are general and common findings in invaded ecosystems, the magnitude of the ecological impact of introduced fish can change a lot among invaded lakes. While scientists devoted a sufficient amount of energy to understand the general ecological consequences of fish presence, there are still many overlooked aspects of the

Fig. 8.4 Stocking fishes into originally fishless high mountain lakes results in a series of effects that cascade through the food web: (*a*) direct predation can affect large planktonic crustaceans and produce indirect top-down and cascading effects altering the communities and biomasses of small zooplankton species and phytoplankton (Knapp et al. 2001b; Sarnelle and Knapp 2005; Tiberti et al. 2014b); (*b*) nektonic and (*c*) benthonic macroinvertebrates often undergo local extinction after fish introduction, while (*d*) fossorial macroinvertebrates are usually unaffected or indirectly favoured by introduced fish (Knapp et al. 2001b; Tiberti et al. 2014b); (*e*) fish introduction is often a factor of ecological exclusion for amphibians (Bradford et al. 1993; Vredenburg 2004; Tiberti and von Hardenberg 2012; Knapp et al. 2016); (*f*) introduced fish can alter the aquatic nutrient subsidy (in the form of emerging insects and amphibians) entering the terrestrial environment (Pope et al. 2009; Tiberti et al. 2016b) and indirectly affect (*g*) terrestrial predators, such as birds, reptiles, amphibians, spiders and bats (Matthews et al. 2002; Finlay and Vredenburg 2007; Epanchin et al. 2010; Benjamin et al. 2011; Joseph et al. 2011; Gruenstein 2014); (*h*) salmonid species are usually introduced in high mountain lakes to sustain recreational angling (Bahls 1992; Miró and Ventura 2013); (*k*) small fish species, such as *Phoxinus* sp., are used as live baits by anglers and often released in high mountain lakes with overlooked, but probably important additive and interactive ecological impacts (Miró and Ventura 2015)

invasion ecology of high-altitude lakes. The reasons why some lakes show stronger or weaker resistance to fish invaders are still largely speculative or need to be studied in more detail. Indeed understanding the characteristics of the lakes which determine a higher resistance to fish predation and resulting indirect ecological impacts can help high-altitude lakes conservation and management and buffer such a widespread conservation problem. It is generally accepted that ecological complexity and the existence of antipredatory refugia (e.g. aquatic vegetation and the dark refugia in the deeper parts of deep lakes) enhance the resistance of native aquatic communities (Knapp et al. 2001b; Pope et al. 2009). However little is known about many aspects related to:

- The biology of introduced populations: the impact of different trout species has not been compared in detail as well as the dependence between the fish density and the magnitude of the impact.
- The ecology of invaded populations: the existence of metapopulations and the vicinity to fishless lakes could probably contrast the effects of fish predation on many prey species through immigration of new individuals, but, except for amphibians (Vredenburg 2004), this issue has rarely been studied. Also, the egg banks and propagules could subsidise native populations with new individuals and buffer fish predation, but also, in this case, their roles have not been well studied (but see Parker et al. 1996 and Latta et al. 2010). There are also many overlooked aspects concerning the cascading effects of fish introduction (e.g. trophic cascades) and their influence on the ecological connection between invaded lakes and terrestrial habitats (reciprocal terrestrial and aquatic subsidies).
- The role of different fish management practices: the consequences of different fish management practices (fishing bans vs. fishing enabled, periodic fish stocking vs. fish stocking halt) on the magnitude of the ecological impact has not been assessed.
- The evolutionary consequences of fish invasion: introduced species can produce evolutive changes in natives, and the evolutionary component of native/non-native species interactions are likely to be a cutting edge field of research in invasion ecology, with important conservation consequences (Lambrinos 2004; Schoener 2011).
- The interacting threats: the effects of fish can be exacerbated by interaction with other stressors. Airborne pesticides (Davidson and Knapp 2007), infections by moulds (chytridiomycosis; Walker et al. 2010; Rosa et al. 2013; Martel et al. 2013; Vredenburg et al. 2010) and viruses (Price et al. 2014; Teacher 2010), climate changes (Bosch et al. 2007) or increased ultraviolet radiation by ozone layer thinning (Adams et al. 2005), water exploitation and water-level fluctuations, point source of organic pollutants, can interact with introduced fish possibly exacerbating the poor conservation status and the resilience potential of many high altitude lakes at a local and regional scale. For example maintaining fishless lakes or eradicating fish can be considered as a measure to contrast the biodiversity loss due to climate warming, restoring a safe stepping stone habitat for many aquatic organisms forced to find cooler condition with an altitudinal shift.

8.5 A Serious Problem for Conservation

The effects of fish introductions on high mountain lake ecosystems summarised above and at Fig. 8.4 demonstrate unambiguously that there are substantial effects at the lake scale, but it is also a problem at a regional and biogeographic scale, either affecting some particular lakes or individual species.

In the Pyrenees, the number of lakes with introduced fish increases with lake size, with more than half of the lakes >0.5 ha and nearly all lakes >2.5 ha with introduced fish, representing a 72% of the lake's total surface area (Fig. 8.5a). Also, half of these lakes have both trout and minnows, and a few dozen with minnow alone (Fig. 8.5b). Fish are usually more present in larger lakes in different high mountain areas (e.g. Matthews and Knapp 1999), since smaller lakes and ponds may freeze completely during winter, and therefore fish disappear naturally. Previous studies at Pyrenean or whole European high mountain region scale show that lake size is one of the main variables explaining organism distributions (Kernan et al. 2009; Catalan et al. 2009), being lakes of larger size (surface area >10 ha) those with the most distinctive species groups. Therefore, fish introductions in the Pyrenees have affected more seriously the conservation of larger lakes. Although they are relatively scarce, they occupy more than 75% of the lake surface area of the Pyrenean lentic waters (Fig. 8.5a). Furthermore, these lakes are also those with a higher proportion of water-level oscillations due to hydroelectrical power stations activity (Catalan et al. 1997), a second stressor that also affects negatively lake ecosystems (Miró 2016).

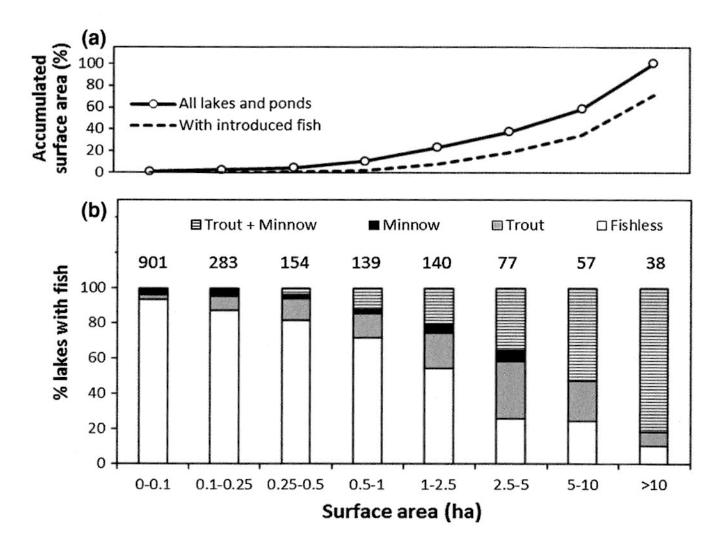

Fig. 8.5 **a** Percentage of the surface area occupied by lakes and ponds of different size categories in the Pyrenees (*straight line*) and those with introduced fish (*dotted line*). **b** Percentage of lakes without fish (*white bars*), with introduced fish, distinguishing those with only minnow, trout or both. The proportions include all lakes >0.5 ha from (Miró and Ventura 2013, 2015) and ca. 70% of the ponds (with surface <0.5 ha) within the southern slope of the Pyrenees

The introduction of fish might also affect the conservation status of some species at the metapopulation level depending on their preference for lakes over ponds (Denoël et al. 2016). Most animal groups inhabiting alpine lakes have species better adapted to colonise lakes rather than ponds or vice versa, for example, because lacking desiccation resistance. Figure 8.6 shows the ecological preferences (lakes vs. ponds) of some species of planktonic crustaceans and amphibians. Those preferentially inhabiting lakes are those more threatened by the introduction of fish. Amongst them, the conservation status of the Pyrenean newt [*Calotriton asper* (Dugès 1852)] is of particular concern since it is an endemic species of the Pyrenees and surrounding areas. For this species, the reduction of its habitat has likely been substantial. Despite it can also be found in fishless upland streams (Oromi et al. 2014), these upland streams have also been affected by modern fish introductions and therefore the reduction in habitat size at its metapopulation scale has been drastically reduced during the 20th century. To what extent this might threaten the future survival of the species is unknown, but its classification within the UICN red list as "nearly threatened" due to habitat loss (Bosch et al. 2009) shows that

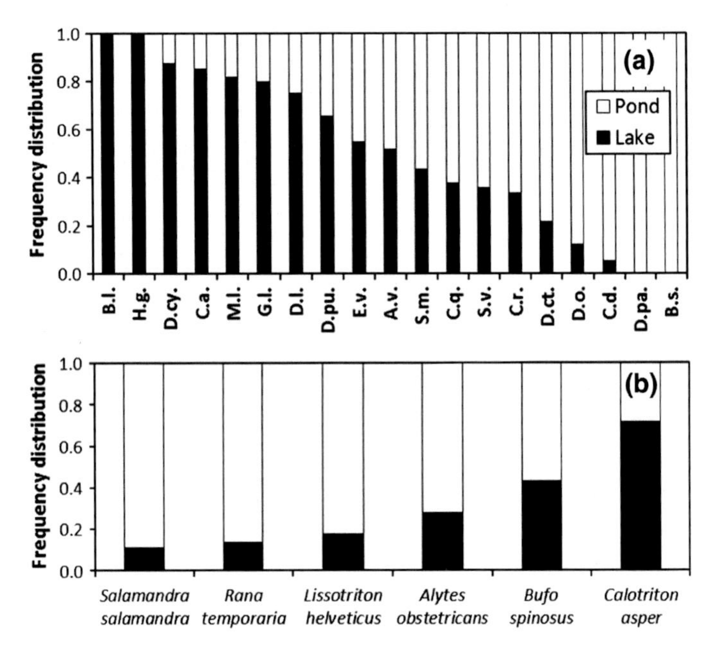

Fig. 8.6 **a** Probability of finding the different species of planktonic crustaceans and **b** amphibians in lakes and ponds (water bodies > or <0.5 ha respectively). B.l. *Bosmina longirostris*, H.g. *Holopedium gibberum*, D.cy. *Diaptomus cyaneus*, C.a. *Cyclops abyssorum*, M.l. *Mixodiaptomus laciniatus*, G. l. *Gammarus*, *cf. alpinus*, D.l. *Daphnia longispina*, D.pu. *Daphnia pulicaria*, E.v. *Eudiaptomus vulgaris*, A.v. *Acanthocyclops vernalis*, S.m. *Scapholeberis mucronata,* C.q. *Ceriodaphnia quadrangula*, S.v. *Simocephalus vetulus*, C.r. *Ceriodaphnia reticulata*, D.ct. *Diaptomus castaneti*, D.o. *Daphnia obtusa*, C.d. *Chirocephalus diaphanus*, D.pa. *Daphnia parvula*, B.s. *Branchipus schaefferi*

invasive fish can seriously threaten the conservation of some of these species. This can also be the case of other species with wider distribution, either because their distribution is restricted to high mountain areas (fish introductions in mountain areas are a global threat; Hulme et al. 2009) or because a synergy with other stressors also affecting the remaining distribution area (e.g. Davis et al. 2010; Ban et al. 2014; Matthaei et al. 2010). Recent genetic studies have also shown that for other species with wide geographic distribution there are genetically differentiated lineages in high mountain areas (Ventura et al. 2014; e.g. Bellati et al. 2014), that might also be at risk of disappearance.

To what extent this substantial habitat loss is indeed affecting the survival of some of the native species is not well known for most of them. Theoretical studies point out that species-specific empirical studies are required with great urgency. If the decrease of habitats is fast, as occurred with the habitat destruction due to fish introductions during the twentieth century in the Pyrenees (see above), there may be a time lag before colonisation and extinction dynamics of the native species reach equilibrium with current habitat distributions (defined as "non-equilibrium dynamics" by Harrison 1991). As a result, species may persist for some time in habitat networks, although they might go extinct some years later even without further landscape change due to a time lag delay (defined as "extinction debt" by Tilman et al. 1994). The apparent survival of species in networks that are insufficient for their long-term persistence may cause underestimates of the area and quality of habitat needed to preserve threatened taxa (Hanski et al. 1996), and overestimates of the species richness that landscapes can support in the long term (Tilman et al. 1994). Therefore, the prevalence of nonequilibrium systems among rare species or lineages, and the implications for their conservation need to be determined as soon as possible (Bulman et al. 2007). However, urgent conservation measures are needed, meanwhile these species-specific studies are not available.

8.6 Conservation and Restoration, What Has Been Done so Far?

8.6.1 Protection Measures

Different initiatives have been implemented to stop species invasions at various administrative scales. At European scale, the EU regulation 1141/2016 provides a list of invasive alien species considered to be of Union concern ("the Union list"), which does not include any fish species commonly used for stocking mountain lakes. However the regulation contains an invitation for the Member States to counteract the negative effects of some invasive alien species of national or regional concern also cooperating at a regional scale (e.g. mountain ranges). The alien invasive species of national or regional concern can be native in some European Member States and, therefore, Member States could correctly undertake actions to contrast all the species commonly used to stock mountain lakes. At present, there are different legislations at the different EU Member States that regulate some introduced species (mainly those introduced from America). However, stocking several salmonid species is currently allowed and, surprisingly, it is usually sustained or approved by governmental agencies (Miró and Ventura 2013), being in clear contradiction with conservation legislation.[1] The most ambitious initiative existent at European scale is the Natura 2000 network, aiming to protect all species and habitats of special conservation concern (included in the habitats directive). Member States are committed to guarantee the conservation of Natura 2000 habitats and species which, for the case of high mountain lakes and some of their species (e.g. some amphibians, aquatic plants and invertebrates), is incompatible with allowing fish introductions.

Data from the western USA showed that areas with different angling management practices have a different probability of finding introduced trout in high mountain lakes, which is lower in National Parks where fishing is prohibited (Knapp 1996; Wiley 2003). Within the Pyrenees, Miró and Ventura (2013, 2015) compared the spread of new fish introductions among different areas of the Aigüestortes i Estany de Sant Maurici National Park and found that in the area where fishing has been forbidden since 1988, the number of lakes with fish stopped increasing just at the time of the prohibition, while at the other areas, it continued to increase (Fig. 8.7). These common findings in several mountain areas over the world illustrate that the only management practice that has had a positive effect up to present in the protection of lakes is the prohibition of fishing.

Although the most obvious route to avoid introductions would be for governmental agencies to stop stocking trout, this is complicated by the fact that there have been several other agents involved in these introductions in addition to governmental agencies through the history of trout introductions (e.g. local citizens in the past or local fishermen's societies in the past decades). As a result, Miró and Ventura (2013, 2015) found some lakes that have been stocked recently without the collaboration of the administration.

In this context, protected areas sometimes provide the favourable local context (i.e. the presence of a surveillance personnel) which is essential to enforce the protection measures (fishing ban, stocking halt) and restoration projects (fish eradications). Therefore, a new and general attitude of mountain protected areas towards less permissive policies concerning fish management is the first and most urgent measure to stem the ecological consequences of the fish invasion.

---

[1]Council Decision 93/626/EEC of 25 October 1993 concerning the conclusion of the Convention on Biological Diversity (OJ L 309, 13.12.1993, p. 1); Regulation 1143/2014 of the European Parliament and of the Council of 22 October 2014 on the prevention and management of the introduction and spread of invasive alien species (OJ L 317, 4.11.2014, p. 1); Directive 2000/60/EC of the European Parliament and of the Council of 23 October 2000 establishing a framework for Community action in the field of water policy (OJ L 327, 22.12.2000, p. 1); Commission Implementing Regulation (EU) 2016/1141 of 13 July 2016 adopting a list of invasive alien species of Union concern pursuant to Regulation (EU) No 1143/2014 of the European Parliament and of the Council C/2016/4295 (OJ L 189, 14.7.2016, pp. 4–8).

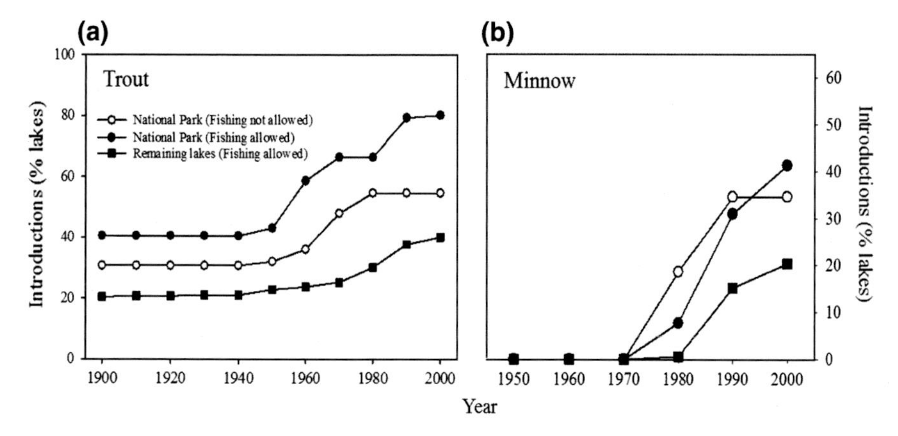

Fig. 8.7 **a** Effect of not allowing fishing in Aigüestortes i Estany de Sant Maurici National Park on the spread of trout and **b** minnows in high mountain lakes of the southern Pyrenees. *Circles* are the lakes within the National Park, and *squares* are lakes outside the National Park. *White circles* are the area of the National Park where fishing is not allowed, and *black circles* and squares are the lakes in fishing allowed areas. Data from Miró and Ventura (2013, 2015)

The social implications of prohibiting fishing would not be relevant if the prohibition is implemented in those lakes that are at present fishless (at the moment there is the paradox that some lakes have fishing rights while never being stocked with fish). It could also be implemented in some particular lakes or groups of lakes that are especially important for the conservation of some animal groups (e.g. Pyrenean newt *C. asper*). Finally, prohibition on fishing a proportion of the lakes has already been performed in the Aigüestortes i Estany de Sant Maurici National Park, with no documented adverse social impact: basically local citizens realised that the social benefits of nature preservation (through higher tourism) have been greater than those that were obtained via fishing.

8.6.2 Restoration Projects: Techniques Used and Successful Cases

Due to the substantial reduction in habitat size directly associated with fish introductions, different protected areas of the world have started restoration projects with the aim to eradicate trout in high altitude lakes. The method most widely used in lentic systems is gill netting (e.g. Knapp and Matthews 1998; Parker et al. 2001; Vredenburg 2004; Knapp et al. 2007; Pacas and Taylor 2015). The most successful ones at removing non-native fishes have focused in relatively small lakes (e.g. Knapp and Matthews 1998; Parker et al. 2001; Vredenburg 2004; Knapp et al. 2007), but this technique demonstrated to be effective also in relatively large high altitude lakes (Pacas and Taylor 2015). The use of gill netting is often complemented with electrofishing (Pacas and Taylor 2015), and in one case with local fishermen (Tiberti et al. 2016a). Piscicides (e.g. rotenone and antimycin) have also been used to eradicate fish (e.g. Sanni and Wærvågen 1998; Gresswell 1991). However, their use has ecological effects on native fauna such as invertebrates (e.g. Dalu et al. 2015; Kjærstad and Arnekleiv 2011), and legal and social complications, especially in protected areas because the very notion of a protected area implies that it remains chemical free. It is important to highlight that eradication actions are a "second step", which should come when protection measures (fishing ban and the prohibition of fish stocking) are already in place and effectively enforced; if these fundamental guarantees are absent or weak, the local context (i.e. concrete risk of sabotage of the conservation actions) might suggest deferring the project to future times.

The most ambitious projects have been carried out in North America both USA and Canada (e.g. Knapp and Matthews 1998; Parker et al. 2001; Vredenburg 2004; Knapp et al. 2007; Pacas and Taylor 2015). In Europe, removal of brook trout has also been achieved in a small alpine lake in the central Iberian Peninsula (Toro et al. 2006). Furthermore, two recent European Commission funded projects, the LIFE+ Bioaquae (www.bioaquae.eu) and LIFE+ LimnoPirineus (www.lifelimnopirineus.eu) are removing fish from different high mountain lakes. The former started removing brook trout from four lakes and their surrounding streams of the Gran Paradiso National Park (Italian Alps) in 2013 (Tiberti et al. 2013), having already removed fish from all lakes (R. Tiberti, Pers. Comm.). The latter started at 2015 in the Pyrenees (Aigüestortes i Estany de Sant Maurici National Park and Alt Pirineu Natural Park) and is aiming at removing brook trout, rainbow trout, brown trout and European minnow from eight lakes and their surrounding streams. Three of them have minnow only, two both minnow and brown trout, two brook trout and one rainbow trout.

Studies of the recovery of restored lakes have shown the relatively fast recovery of lake populations when surrounding refugees are available (Knapp et al. 2001b, 2005, 2007; Sarnelle and Knapp 2004), being amphibians and macroinvertebrates those with faster recovery (Knapp et al. 2001b, 2005, 2007), and crustacean zooplankton the later (Knapp and Sarnelle 2008). These latter authors have also shown that some species producing resistance eggs are not able to survive for many years in the egg bank (e.g. calanoid copepods). Fish residence times (years that fish have been in the lake) above *ca.* 50 years greatly reduce the fast recovery of these species from the egg bank.

Acknowledgements The authors would like to acknowledge those people that provided information about fish introductions, two anonymous reviewers and the editors for giving us the opportunity to participate in the workshop and writing this paper. This book chapter is a joint contribution of the LIFE+ projects BIOAQUAE (Biodiversity Improvement of Aquatic Alpine Ecosystems, LIFE11 BIOIT000020) and LIMNOPIRINEUS (Restoration of lentic habitats and aquatic species of Community interest in high mountains of the Pyrenees, LIFE13 NAT/ES/001210). RT thanks the University of Pavia and Bruno Bassano and Giuseppe Bogliani for their support. DB benefitted from a scholarship from Government of Ecuador (SENESCYT 20090187-20130946) and IS from a Catalan Government grant (2015 FI_B 01147).

References

Adams MJ, Hossack BR, Knapp RA, Corn PS, Diamond SA, Trenham PC, Fagre DB (2005) Distribution patterns of lentic-breeding amphibians in relation to ultraviolet radiation exposure in western North America. Ecosystems 8:488–500

Adams SB, Frissell CA, Rieman BE (2001) Geography of invasion in mountain streams: consequences of headwater lake fish introductions. Ecosystems 4:296–307

Araguas RM, Sanz N, Fernández F, Utter FM, Pla C, García-Marín J-L (2009) Role of genetic refuges in the restoration of native gene pools of brown trout. Conserv Biol 23:871–878

Armstrong TW, Knapp RA (2004) Response by trout populations in alpine lakes to an experimental halt to stocking. Can J Fish Aquat Sci 61:2025–2037

Bahls P (1992) The status of fish populations and management of high mountain lakes in the Western United States. Northwest Sci 66:183–193

Ban SS, Graham NAJ, Connolly SR (2014) Evidence for multiple stressor interactions and effects on coral reefs. Glob Change Biol 20:681–697

Beebee TJC, Griffiths RA (2005) The amphibian decline crisis: a watershed for conservation biology? Biol Conserv 125:271–285

Bellati A, Tiberti R, Cocca W, Galimberti A, Casiraghi M, Bogliani G, Galeotti P (2014) A dark shell hiding great variability: a molecular insight into the evolution and conservation of melanic *Daphnia* populations in the Alps. Zool J Linn Soc 171:697–715

Benjamin JR, Fausch KD, Baxter CV (2011) Species replacement by a nonnative salmonid alters ecosystem function by reducing prey subsidies that support riparian spiders. Oecologia 167:503–512

Borgstrom R, Garnas E, Salveit SJ (1985) Interactions between brown trout, *Salmo trutta* (L.), and minnow, *Phoxinus phoxinus* (L.), for their common prey, *Lepidurus arcticus* (Pallas). Ver Internat Verein Limnol 22:2548–2552

Borgstrøm R, Brittain JE, Hasle K, Skjølås S, Dokk JG (1996) Reduced recruitment in brown trout *Salmo trutta*, the role of interactions with minnow *Phoxinus phoxinus*. Nord J Freshw Res 72:30–38

Bosch J, Rincon PA, Boyero L, Martinez-Solano I (2006) Effects of introduced salmonids on a montane population of Iberian frogs. Conserv Biol 20:180–189

Bosch J, Carrascal LM, Duran L, Walker S, Fisher MC (2007) Climate change and outbreaks of amphibian chytridiomycosis in a montane area of Central Spain; is there a link? Proc Roy Soc Lond B Bio 274:253–260

Bosch J, Tejedo M, Lecis R, Miaud C, Lizana M, Edgar P, Martínez-Solano I, Salvador A, García-París M, Recuero Gil E, Marquez R, Geniez P (2009) *Calotriton asper* (Publication no. http://dx.doi.org/10.2305/IUCN.UK.2009.RLTS.T59448A11943040.en)

Bradford DF, Tabatabai F, Graber DM (1993) Isolation of remaining populations of the native frog, *Rana muscosa*, by introduced fishes in Sequoia and Kings Canyon National-Parks, California. Conserv Biol 7:882–888

Brancelj A (2000) The extinction of *Arctodiaptomus alpinus* (Copepoda) following the introduction of charr into a small alpine lake Dvojno Jezero (NW Slovenia). Aquat Ecol 33:355–361

Brancelj A, Sisko M, Brancelj IR, Jeran Z, Jacimovic R (2000) Effects of land use and fish stocking on a mountain lake-evidence from the sediment. Period Biol 102:259–268

Brooks JL, Dodson SI (1965) Predation, body size, and composition of plankton. Science 150:28–35

Buchaca T, Skov T, Amsinck SL, Gonçalves V, Neto Azevedo JM, Andersen TJ, Jeppesen E (2011) Rapid ecological shift following piscivorous fish introduction to increasingly eutrophic and warmer Lake Furnas (Azores Archipelago, Portugal): a paleoecological approach. Ecosystems 14:458–477

Bulman CR, Wilson RJ, Holt AR, Bravo LG, Early RI, Warren MS, Thomas CD (2007) Minimum viable metapopulation size, extinction debt, and the conservation of a declining species. Ecol Appl 17:1460–1473

Callaham MA Jr, Gonzalez G, Hale CM, Heneghan L, Lachnicht SL, Zou X (2006) Policy and management responses to earthworm invasions in North America. Biol Invasions 8:1317–1329

Cambray JA (2003) Impact on indigenous species biodiversity caused by the globalisation of alien recreational freshwater fisheries. Hydrobiologia 500:217–230

Cantonati M, Lapini L, Paradisi S, Stoch F (2006) Conservation and management. In: Stoch F (ed) High-altitude lakes. Quaderni habitat vol 14. Ministero dell'Ambiente e della Tutela del Territorio. Museo Friulano di Storia Naturale. English Edition: Italian Habitats, pp 123–135

Carpenter SR, Cole JJ, Pace ML, Batt R, Brock WA, Cline T, Coloso J, Hodgson JR, Kitchell JF, Seekell DA, Smith L, Weidel B (2011) Early warnings of regime shifts: a whole-ecosystem experiment. Science 332:1079–1082

Catalan J, Vilalta R, Weitzman B, Ventura M, Comas E, Pigem C, Aranda R, Ballesteros E, Camarero L, García Serrano J, Pla S, Sáez A, Aiguabella P (1997) The hydraulic industry in the Pyrenees: evaluation, correction and prevention of the environmental impact at the Aigüestortes i Estany de Sant Maurici National Park. La Caixa, Barcelona

Catalan J, Barbieri MG, Bartumeus F, Bitusik P, Botev I, Brancelj A, Cogalniceanu D, Garcia J, Manca M, Marchetto A, Ognjanova-Rumenova N, Pla S, Rieradevall M, Sorvari S, Stefkova E, Stuchlik E, Ventura M (2009) Ecological thresholds in European alpine lakes. Freshw Biol 54:2494–2517

Collen B, Whitton F, Dyer EE, Baillie JEM, Cumberlidge N, Darwall WRT, Pollock C, Richman NI, Soulsby A-M, Böhm M (2014) Global patterns of freshwater species diversity, threat and endemism. Glob Ecol Biogeogr 23:40–51

Cowx IG, Gerdeaux D (2004) The effects of fisheries management practises on freshwater ecosystems. Fish Manag Ecol 11:145–151

Christenson DP (1977) History of trout introductions in California high mountain lakes. In: Hall A (ed) A symposium on the management of high mountain lakes in California's National Parks, San Francisco, California, 9–16 May 1977, vol 1. California Trout Inc., pp 9–15

Dalu T, Wasserman RJ, Jordaan M, Froneman WP, Weyl OLF (2015) An assessment of the effect of rotenone on selected non-target aquatic fauna. PLoS ONE 10:e0142140

Davidson C, Knapp RA (2007) Multiple stressors and amphibian declines: dual impacts of pesticides and fish on yellow-legged frogs. Ecol Appl 17:587–597

Davis J, Sim L, Chambers J (2010) Multiple stressors and regime shifts in shallow aquatic ecosystems in antipodean landscapes. Freshw Biol 55:5–18

Dawidowicz P, Gliwicz ZM (1983) Food of brook charr in extreme oligotrophic conditions of an alpine lake. Environ Biol Fishes 8:55–60

de Mendoza G, Rico E, Catalan J (2012) Predation by introduced fish constrains the thermal distribution of aquatic Coleoptera in mountain lakes. Freshw Biol 57:803–814

Delacoste M, Baran P, Lascaux JM, Abad N, Besson JP (1997) Evaluation of salmonid introductions in high-elevation lakes and streams of the Hautes-Pyrenees region. Bull Fr Peche Piscic 205–219

Denoël M, Scimè P, Zambelli N (2016) Newt life after fish introduction: extirpation of paedomorphosis in a mountain fish lake and newt use of satellite pools. Curr Zol 62:61–69

DiStefano RJ, Litvan ME, Horner PT (2009) The bait industry as a potential vector for alien crayfish introductions: problem recognition by fisheries agencies and a Missouri evaluation. Fisheries 34:586–597

Eby LA, Roach WJ, Crowder LB, Stanford JA (2006) Effects of stocking-up freshwater food webs. Trends Ecol Evol 21:576–584

Emery L (1985) Review of fish species introduced into the Great Lakes, 1819–1974, vol 45. Great Lakes Fishery Commission Technical Report. Great Lakes Fisheri Commission, Ann Arbor, MI

Epanchin PN, Knapp RA, Lawler SP (2010) Nonnative trout impact an alpine-nesting bird by altering aquatic insect subsidies. Ecology 91:2406–2415

FAO (2003) Fishery records collections. FIGIS data collection, Inland Water Resources and Aquaculture Service (FIRI). http://www.fao.org/figis/servlet/static?dom=collection&xml=dias.xml

Finlay JC, Vredenburg VT (2007) Introduced trout sever trophic connections in watersheds: Consequences for a declining amphibian. Ecology 88:2187–2198
Frezet C (2003) Catasto dei laghi valdostani. ARPA Valle d'Aosta
Garcia-Berthou E, Moreno-Amich R (2000) Introduction of exotic fish into a Mediterranean lake over a 90-year period. Arch Hydrobiol 149:271–284
Gido KB, Schaefer JF, Pigg J (2004) Patterns of fish invasions in the Great Plains of North America. Biol Conserv 118:121–131
Gliwicz ZM, Rowan MG (1984) Survival of *Cyclops abyssorum tatricus* (Copepoda, Crustacea) in alpine lakes stocked with planktivorous fish. Limnol Oceanogr 29:1290–1299
Granek EF, Madin EMP, Brown MA, Figueira W, Cameron DS, Hogan Z, Kristianson G, De Villiers P, Williaims JE, Post J, Zahn S, Arlinghaus R (2008) Engaging recreational fishers in management and conservation: global case studies. Conserv Biol 22:1125–1134
Gratton C, Donaldson J, Vander Zanden MJ (2008) Ecosystem linkages between lakes and the surrounding terrestrial landscape in northeast Iceland. Ecosystems 11:764–774
Gresswell RE (1991) Use of antimycin for removal of brook trout from a tributary of yellowstone lake. N Am J Fish Manag 11:83–90
Gruenstein E (2014) The response of bats to introduced trout in naturally fishless lakes Sierra Nevada. San José State University, California Master's Theses
Hall D, Threlkeld ST, Burns CW, Crowley PH (1976) The size-efficiency hypothesis and the size structure of zoplankton communities. Annu Rev Ecol Syst 7:177–208
Hanski I, Moilanen A, Gyllenberg M (1996) Minimum viable metapopulation size. Am Nat 147:527–541
Harrison S (1991) Local extinction in a metapopulation context: an empirical evaluation. Biol J Linn Soc 42:73–88
Hesthagen T, Hegge O, Skurdal J (1992) Food choice and vertical distribution of European minnow, *Phoxinus phoxinus*, and young native and stocked brown trout, *Salmo trutta*, in the littoral zone of a subalpine lake. Norw J Freshwat Res 67:72–76
Hulme PE, Nentwig W, Pyšek P, Vilà M (eds) (2009) DAISIE Handbook of alien species in Europe, vol 3. Invading nature. Springer series in invasion ecology. Springer, Dordrecht
Jersabek CD, Brancelj A, Stoch F, Schabetsberger R (2001) Distribution and ecology of copepods in mountainous regions of the Eastern Alps. Hydrobiologia 453–454:309–324
Joseph MB, Piovia-Scott J, Lawler SP, Pope KL (2011) Indirect effects of introduced trout on Cascades frogs (*Rana cascadae*) via shared aquatic prey. Freshw Biol 56:828–838
Keller RP, Lodge DM (2007) Species invasions from commerce in live aquatic organisms: problems and possible solutions. Bioscience 57:428–436
Kernan M, Ventura M, Brancelj A, Clarke G, Raddum G, Stuchlík E, Catalan J (2009) Regionalisation of remote European mountain lake ecosystems according to their biota: environmental versus geographical patterns. Freshw Biol 54:2470–2493
Kerr SJ, Brousseau CS, Muschett M (2005) Invasive aquatic species in Ontario: a review and analysis of potential pathways for introduction. Fisheries 30:21–30
Kilian JV, Klauda RJ, Widman S, Kashiwagi M, Bourquin R, Weglein S, Schuster J (2012) An assessment of a bait industry and angler behavior as a vector of invasive species. Biol Invasions 14:1469–1481
Kjærstad G, Arnekleiv JV (2011) Effects of rotenone treatment on lotic invertebrates. Int Rev Hydrobiol 96:58–71
Knapp RA (1996) Non-native trout in natural lakes of the Sierra Nevada: an analysis of their distribution and impacts on native aquatic biota. In: Sierra Nevada Ecosystem Project: Final Report to Congress, 1996. UC Davis
Knapp RA (2005) Effects of nonnative fish and habitat characteristics on lentic herpetofauna in Yosemite National Park, USA. Biol Conserv 121:265–279
Knapp RA, Matthews KR (1998) Eradication of nonnative fish by gill netting from a small mountain lake in California. Restor Ecol 6:207–213
Knapp RA, Sarnelle O (2008) Recovery after local extinction: factors affecting re-establishment of alpine lake zooplankton. Ecol Appl 18:1850–1859

Knapp RA, Corn PS, Schindler DE (2001a) The introduction of nonnative fish into wilderness lakes: good intentions, conflicting mandates, and unintended consequences. Ecosystems 4:275–278

Knapp RA, Matthews KR, Sarnelle O (2001b) Resistance and resilience of alpine lake fauna to fish introductions. Ecol Monogr 71:401–421

Knapp RA, Hawkins CP, Ladau J, McClory JG (2005) Fauna of Yosemite National Park lakes has low resistance but high resilience to fish introductions. Ecol Appl 15:835–847

Knapp RA, Boiano DM, Vredenburg VT (2007) Removal of nonnative fish results in population expansion of a declining amphibian (mountain yellow-legged frog, *Rana muscosa*). Biol Conserv 135:11–20

Knapp RA, Fellers GM, Kleeman PM, Miller DAW, Vredenburg VT, Rosenblum EB, Briggs CJ (2016) Large-scale recovery of an endangered amphibian despite ongoing exposure to multiple stressors. Proc Natl Acad Sci USA. doi:10.1073/pnas.1600983113

Knight TM, McCoy MW, Chase JM, McCoy KA, Holt RD (2005) Trophic cascades across ecosystems. Nature 437:880–883

Lambrinos JG (2004) How interactions between ecology and evolution influence contemporary invasion dynamics. Ecology 85:2061–2070

Latta LC, Fisk DL, Knapp RA, Pfrender ME (2010) Genetic resilience of *Daphnia* populations following experimental removal of introduced fish. Conser Genet 11:1737–1745

Lindgren CJ (2006) Angler awareness of aquatic invasive species in Manitoba. J Aquat Plant Manag 44:103–108

Litvak MK, Mandrak NE (1993) Ecology of fresh-water baitfish use in Canada and the United-States. Fisheries 18:6–13

Lowe S, Browne M, Boudjelas S, De Poorter M (2000) 100 of the worlds worst invasive alien species A selection from the global invasive species database. World Conservation Union (IUCN)

Machino Y (1999) History and status of Arctic charr introductions in southern Europe. Int Soc Artic Charr Fanatics, Drottningholm, Sweden Inf Ser 7:33–39

Magnea U, Sciascia R, Paparella F, Tiberti R, Provenzale A (2013) A model for high-altitude alpine lake ecosystems and the effect of introduced fish. Ecol Model 251:211–220

Maitland PS, Campbell RN (1992) Freshwater fishes of the British Isles, vol Book. Whole, Harper Collins, London

Mammoliti Mochet A (1995) Analisi dell'ittiofauna in Valle d'Aosta allo stato attuale e allo stato pregresso. Università degli Studi di Torino

Marchetti MP, Light T, Moyle PB, Viers JH (2004) Fish invasions in California watersheds: testing hypotheses using landscape patterns. Ecol Appl 14:1507–1525

Martel A, Spitzen-van der Sluijs A, Blooi M, Bert W, Ducatelle R, Fisher MC, Woeltjes A, Bosman W, Chiers K, Bossuyt F, Pasmans F (2013) *Batrachochytrium salamandrivorans* sp. nov. causes lethal chytridiomycosis in amphibians. Proc Natl Acad Sci USA 110:15325–15329

Martín-Torrijos L, Sandoval-Sierra JV, Muñoz J, Diéguez-Uribeondo J, Bosch J, Guayasamin JM (2016) Rainbow trout (*Oncorhynchus mykiss*) threaten Andean amphibians. Neotrop Biodiver 2:26–36

Martinez-Solano I, Barbadillo LJ, Lapena M (2003) Effect of introduced fish on amphibian species richness and densities at a montane assemblage in the Sierra de Neila, Spain. Herpetol J 13:167–173

Matthaei CD, Piggott JJ, Townsend CR (2010) Multiple stressors in agricultural streams: interactions among sediment addition, nutrient enrichment and water abstraction. J Appl Ecol 47:639–649

Matthews KR, Knapp RA (1999) A study of high mountain lake fish stocking effects on the U.S. Sierra Nevada wilderness. Int J Wilderness 5:24–26

Matthews KR, Knapp RA, Pope KL (2002) Garter snake distributions in high-elevation aquatic ecosystems: is there a link with declining amphibian populations and nonnative trout introductions? J Herpetol 36:16–22

McRae L, Freeman R, Deinet S (2014) The Living Planet Index. In: McLellan R, Iyengar L, Jeffries B, Oerlemans N (eds) Living planet report 2014. Species and spaces, people and places. World Wild Fund International, Gland, Switzerland

Migge-Kleian S, McLean MA, Maerz JC, Heneghan L (2006) The influence of invasive earthworms on indigenous fauna in ecosystems previously uninhabited by earthworms. Biol Invasions 8:1275–1285

Mills CA (1988) The effect of extreme northerly climatic conditions on the life history of the minnow, *Phoxinus phoxinus* (L.). J Fish Biol 33:545–561

Miró A (2011) Trout in Pyrenean Lakes: tradition, history and conservation implications (In Catalan). Pagès Editors, Lleida

Miró A (2016) Fish as local stressors of Pyrenean high mountain lakes: arrival process and impact on amphibians and other organisms. PhD Thesis, University of Barcelona, Barcelona

Miró A, Ventura M (2013) Historical use, fishing management and lake characteristics explain the presence of non-native trout in Pyrenean lakes: Implications for conservation. Biol Conserv 167:17–24

Miró A, Ventura M (2015) Evidence of exotic trout mediated minnow invasion in Pyrenean high mountain lakes. Biol Invasions 17:791–803

Museth J, Borgstrom R, Hame T, Holen LA (2003) Predation by brown trout: a major mortality factor for sexually mature European minnows. J Fish Biol 62:692–705

Museth J, Hesthagen T, Sandlund OT, Thorstad EB, Ugedal O (2007) The history of the minnow *Phoxinus phoxinus* (L.) in Norway: from harmless species to pest. J Fish Biol 71:184–195

Museth J, Borgstrom R, Brittain JE (2010) Diet overlap between introduced European minnow (*Phoxinus phoxinus*) and young brown trout (*Salmo trutta*) in the lake, Øvre Heimdalsvatn: a result of abundant resources or forced niche overlap? Hydrobiologia 642:93–100

Orizaola G, Braa F (2006) Effect of salmonid introduction and other environmental characteristics on amphibian distribution and abundance in mountain lakes of northern Spain. Anim Conserv 9:171–178

Oromi N, Amat F, Sanuy D, Carranza S (2014) Life history trait differences between a lake and a stream-dwelling population of the Pyrenean brook newt (*Calotriton asper*). Amphib-Reptilia 35:53–62

Ortega H, Guerra H, Ramírez R (2007) The introduction of nonnative fishes into freshwater systems of Peru. In: Ecological and genetic implications of aquaculture activities. Springer, pp 247–278

Pacas C, Taylor M (2015) Nonchemical eradication of an introduced trout from a headwater complex in Banff National Park, Canada. N Am J Fish Manag 35:748–754

Parker BR, Wilhelm FM, Schindler DW (1996) Recovery of *Hesperodiaptomus arcticus* populations from diapausing eggs following elimination by stocked salmonids. Can J Zool 74:1292–1297

Parker BR, Schindler DE, Donald DB, Anderson RS (2001) The effects of stocking and removal of a non-native brook trout on the plankton populations of an alpine lake. Ecosystems 4:334–345

Pechlaner R (1984) Historical evidence for the introduction of *Arctic charr* into high-mountain lakes of the Alps by man. In: Burns BL (ed) Johnson J. University of Manitoba Press, Winnipeg, pp 449–557

Peters JA, Lodge DM (2009) Invasive species policy at the regional level: a multiple weak links problem. Fisheries 34:373–381

Petr T (1999) Fish and fisheries at higher altitudes: Asia, vol 385. Food & Agriculture Org

Piccinini A, Nonnis Marzano F, Gandolfi G (2004) Il Salmerino alpino (*Salvelinus alpinus*): prove storiche alla sua introduzione sul territorio italiano. Nonnis Marzano F, Maldini M & Gandolfi G (eds.). Atti 9:259–264

Pilliod DS, Ricciardi A, Hossack BR, Bahls PF, Bull EL, Corn PS, Hokit G, Maxell BA, Munger JC, Wyrick A (2010) Non-native salmonids affect amphibian occupancy at multiple spatial scales. Divers Distrib 16:959–974

Pister EP (2001) Wilderness fish stocking: history and perspective. Ecosystems 4:279–286

Pope KL (2008) Assessing changes in amphibian population dynamics following experimental manipulations of introduced fish. Conserv Biol 22:1572–1581

Pope KL, Hannelly EC (2013) Response of benthic macroinvertebrates to whole-lake, non-native fish treatments in mid-elevation lakes of the Trinity Alps, California. Hydrobiologia 714:201–215

Pope KL, Garwood JM, Welsh HH, Lawler SP (2008) Evidence of indirect impacts of introduced trout on native amphibians via facilitation of a shared predator. Biol Conserv 141:1321–1331
Pope KL, Piovia-Scott J, Lawler SP (2009) Changes in aquatic insect emergence in response to whole-lake experimental manipulations of introduced trout. Freshw Biol 54:982–993
Price SJ, Garner TWJ, Nichols RA, Balloux F, Ayres C, Mora-Cabello de Alba A, Bosch J (2014) Collapse of amphibian communities due to an introduced ranavirus. Curr Biol 24:2586–2591
Pringle RM (2005) The origins of the Nile perch in Lake Victoria. Bioscience 55:780–787
Rosa GM, Anza I, Moreira PL, Conde J, Martins F, Fisher MC, Bosch J (2013) Evidence of chytrid-mediated population declines in common midwife toad in Serra da Estrela, Portugal. Anim Conserv 16:306–315
Sanni S, Wærvågen SB (1998) Oligotrophication as a result of planktivorous fish removal with rotenone in the small, eutrophic, Lake Mosvatn, Norway. Hydrobiologia 200:263–274
Sarnelle O, Knapp RA (2004) Zooplankton recovery after fish removal: limitations of the egg bank. Limnol Oceanogr 49:1382–1392
Sarnelle O, Knapp RA (2005) Nutrient recycling by fish versus zooplankton grazing as drivers of the trophic cascade in alpine lakes. Limnol Oceanogr 50:2032–2042
Schabetsberger R, Jersabek CD, Brozek S (1995) The impact of alpine newts (*Triturus alpestris*) and minnows (*Phoxinus phoxinus*) on the microcrustacean communities of two high altitude karst lakes. Alytes (Paris) 12:183–189
Schabetsberger R, Luger MS, Drozdowski G, Jagsch A (2009) Only the small survive: monitoring long-term changes in the zooplankton community of an Alpine lake after fish introduction. Biol Invasions 11:1335–1345
Schindler DE, Parker BR (2002) Biological pollutants: alien fishes in mountain lakes. Water Air Soil Pollut Focus 2:379–397
Schindler DE, Knapp RA, Leavitt PR (2001) Alteration of nutrient cycles and algal production resulting from fish introductions into mountain lakes. Ecosystems 4:308–321
Schindler DW (2000) Aquatic problems caused by human activities in Banff National Park. Ambio 29:401–407
Schoener TW (2011) The newest synthesis: understanding the interplay of evolutionary and ecological dynamics. Science 331:426–429
Simberloff D, Martin J-L, Genovesi P, Maris V, Wardle DA, Aronson J, Courchamp F, Galil B, García-Berthou E, Pascal M, Pysek P, Sousa R, Tabacchi E, Vila M (2013) Impacts of biological invasions: what's what and the way forward. Trends Ecol Evol 28:58–66
Sostoa A, Lobón-Cerviá J (1989) Fish and fisheries of the River Ebro: actual state and recent history. In: Petts GE, Mller H, Roux AL (eds) Historical change of large alluvial rivers: Western Europe. Wiley, Chinchester, pp 233–247
Tang KL, Agnew MK, Chen W-J, Hirt MV, Raley ME, Sado T, Schneider LM, Yang L, Bart HL, He S, Liu H, Miya M, Saitoh K, Simons AM, Wood RM, Mayden RL (2011) Phylogeny of the gudgeons (Teleostei: Cyprinidae: Gobioninae). Mol Phylogenet Evol 61:103–124
Teacher AGF (2010) Assessing the long-term impact of Ranavirus infection in wild common frog populations. Anim Conserv 13:514–522
Terrero D (1951) Historia local de piscicultura. Montes 39:165–171
Tiberti R, von Hardenberg A (2012) Impact of introduced fish on Common frog (*Rana temporaria*) close to its altitudinal limit in alpine lakes. Amphib-Reptilia 33:303–307
Tiberti R, Acerbi E, Iacobuzio R (2013) Preliminary studies on fish capture techniques in Gran Paradiso alpine lakes: towards an eradication plan. J Mountain Ecol 9:61–74
Tiberti R, Brighenti S, Iacobuzio R, Pasquini G, Rolla M (2014a) Behind the impact of introduced trout in high altitude lakes: adult, not juvenile fish are responsible of the selective predation on crustacean zooplankton. J Limnol 73:593–597
Tiberti R, von Hardenberg A, Bogliani G (2014b) Ecological impact of introduced fish in high altitude lakes: a case of study from the European Alps. Hydrobiologia 724:1–19
Tiberti R, Ottino M, Brighenti S, Iacobuzio R, Rolla M, von Hardenberg A, Bassano B (2016a) Involvement of recreational anglers in the eradication of alien brook trout from alpine lakes. J Mountain Ecol 10:13–26

Tiberti R, Rolla M, Brighenti S, Iacobuzio R (2016b) Changes in the insect emergence at the water–air interface in response to fish density manipulation in high altitude lakes. Hydrobiologia 779:93–104
Tilman D, May RM, Lehman CL, Nowak MA (1994) Habitat destruction and the extinction debt. Nature 371:65–66
Toro M, Granados I, Robles S, Montes C (2006) High mountain lakes of the Central Range Iberian Pennsula): regional limnology & environmental changes. Limnetica 25:217–252
Ventura M, Petrusek A, Miró A, Hamrová E, Buñay D, De meester L L, Mergeay J (2014) Local and regional founder effects in lake zooplankton persist after thousands of years despite high dispersal potential. Mol Ecol 23:1014–1027
Vigliano PH, Alonso MF (2007) Salmonid introductions in Patagonia: a mixed blessing. In: Ecological and genetic implications of aquaculture activities. Springer, pp 315–331
Villwock W (1994) Consequences of exotic fish introductions on the naitive Lake Titicaca species (in Spanish). Ecología en Bolivia 23:49–56
Vitule JRS, Freire CA, Simberloff D (2009) Introduction of non-native freshwater fish can certainly be bad. Fish Fish 10:98–108
Volta P, Jepsen N (2008) The recent invasion of *Rutilus rutilus* (L.) (Pisces: Cyprinidae) in a large South-Alpine lake: Lago Maggiore. J Limnol 67:163–170
Vredenburg VT (2004) Reversing introduced species effects: Experimental removal of introduced fish leads to rapid recovery of a declining frog. P Natl Acad Sci USA 101:7646–7650
Vredenburg VT, Knapp RA, Tunstall TS, Briggs CJ (2010) Dynamics of an emerging disease drive large-scale amphibian population extinctions. Proc Natl Acad Sci USA 107:9689–9694
Walker SF, Bosch J, Gomez V, Garner TWJ, Cunningham AA, Schmeller DS, Ninyerola M, Henk DA, Ginestet C, Arthur C-P, Fisher MC (2010) Factors driving pathogenicity vs. prevalence of amphibian panzootic chytridiomycosis in Iberia. Ecol Lett 13:372–382
Ward JM, Cudmore B, Drake DAR, Mandrak NE (2012) Summary of a survey of baitfish users in Canada. Can Manuscr Rep Fish Aquat Sci 2972:i–v, 1–23
Webb AC (2007) Status of non-native freshwater fishes in tropical northern Queensland, including establishment success, rates of spread, range and introduction pathways. J Proc R Soc N S W 140:63–78
Wiley RW (2003) Planting trout in Wyoming high-elevation wilderness waters. Fisheries 28:22–27
Winfield IJ, Fletcher JM, James JB (2011) Invasive fish species in the largest lakes of Scotland, Northern Ireland, Wales and England: the collective UK experience. Hydrobiologia 660:93–103
Zerunian S (2003) Piano d'azione generale per la conservazione dei pesci d'acqua dolce italiani, vol 17. Ministero dell'ambiente e della tutela del territorio, Direzione per la protezione della natura

Chapter 9
Are Soil Carbon Stocks in Mountain Grasslands Compromised by Land-Use Changes?

Jordi Garcia-Pausas, Joan Romanyà, Francesc Montané, Ana I. Rios, Marc Taull, Pere Rovira and Pere Casals

Abstract Mountain grasslands are generally rich in soil organic C, but the typical high spatial variability of mountain environments, together with the different management systems, makes their soil C content particularly variable. Socio-economic changes of the past decades have caused a progressive abandonment of the traditional use for grazing of some areas, while grazing pressure at easily accessible grasslands have increased. Here, we analyse the effect of these land-use changes on the factors regulating the soil C accumulation and stocks. Overgrazing generally leads to a reduction above- and below-ground litter inputs and a decrease in soil C stocks, affecting some soil physicochemical and biological properties. Additionally, the labile C inputs coming from animal faeces may accelerate the mineralisation of organic matter. Grazing abandonment causes a reduction of aboveground productivity, but the lack of consumption causes a short-term accumulation of organic matter. Its effect on belowground biomass and productivity is less clear. At longer term, grazing abandonment causes a change in the plant community composition, having the shrub encroachment the strongest effect on C storage. The low biochemical quality of shrub litter delays its decomposition and allows higher organic matter accumulation in the topsoil. But the effect of shrub proliferation at the deeper soil is less clear. The low root turnover of shrubs compared to grasses may reduce the C inputs to the soil. But, at the same time, the reduction of the root exudates may also reduce the microbial activity and the organic matter mineralisation.

J. Garcia-Pausas (✉) · F. Montané · A.I. Rios · M. Taull · P. Rovira · P. Casals
Forest Sciences Centre of Catalonia, CEMFOR-CTFC, Ctra de St. Llorenç de Morunys, km 2, 25280 Solsona, Catalonia, Spain
e-mail: jordi.gpausas@gmail.com

J. Romanyà
Department of Biology, Health and Environment, Universitat de Barcelona, Joan XXIII s/n, 08028 Barcelona, Catalonia, Spain

F. Montané
School of Natural Resources and the Environment, University of Arizona, Tucson, AZ 85721, USA

J. Catalan et al. (eds.), *High Mountain Conservation in a Changing World*, Advances in Global Change Research 62, DOI 10.1007/978-3-319-55982-7_9

Keywords Grassland abandonment · Land-use changes · Mountain grasslands · Grazing intensification · Shrub encroachment · Soil organic carbon dynamics · Soil organic carbon stocks

9.1 Introduction

Soil organic matter plays essential roles in terrestrial ecosystems. It maintains the soil structure, favours water infiltration and reduces the risk of soil erosion. It also increases the water holding capacity of soils and, through its decomposition by soil biota, provides nutrients to the plants.

Carbon (C) comprises about 45% of the mass of soil organic matter. Plant photosynthetic activity produces organic matter using atmospheric CO_2, which is then accumulated in soil mainly by incorporating plant residues into the soil organic matter. Although this is the primary pathway by which atmospheric CO_2–C is incorporated into the soil, some additional atmospheric CO_2–C can also be sequestered in soil in inorganic forms by rock weathering and precipitation of Ca- and Mg-carbonates. Then, the oxidation of organic matter by soil microorganisms is the main process causing a release of carbon as CO_2 to the atmosphere, leaving less decomposable organic compounds, which are accumulated in the soil. Together with this biotic process, a significant amount of soil C can also be exported from the soil by leaching. Overall soil C sequestration results from the balance between the C flux from the atmosphere into the soil and the C release back to the atmosphere through microbial decomposition. This balance determines if soil behaves as a net sink for removing CO_2 from the atmosphere or a net source that contributes to rising atmospheric CO_2.

Soils represent the main compartment of organic C in most terrestrial ecosystems, containing globally about 1550 Pg C (1 Pg = 10^{15} g), which roughly is twice the amount of C in the atmosphere (760 Pg C) and three times the amount in the biomass (550 Pg C) (Lal 2008). Given the large magnitude of these soil C stocks, potential reductions as little as 10% of the soil C content would equal to the anthropogenic CO_2 emitted over 30 years (Kirschbaum 2000), meaning significant changes in the atmospheric CO_2 concentrations and the reinforcement of the current global warming trend. So, there is a strong interest in avoiding C losses from soils and, if possible, to promote the C sequestration to mitigate the current greenhouse gases (GHG) emissions.

In this chapter, we summarise the special features of mountain alpine soils that contribute to explaining the organic C content and explore the challenges for soil C conservation due to changes in land management and use.

9.2 Mountain Soils and Their C Stocks

Mountain soils are generally steep, shallow, with relatively high erosion rates and influenced by harsh climatic conditions. Despite mountain ecosystems have much in common with those in high latitude, mountain soils are markedly different. These differences originate from both climate and soil formation processes. High-altitude mountain ranges generally receive much higher rainfall, both in quantity and intensity (high torrentiality) than lowlands. Moreover, sunshine incidence in mountain slopes is usually higher than in high latitudes and largely depends on the aspect. In south-facing slopes solar radiation is high, even in winter time, thus reducing the snow cover and the chance of frost layers. In mountain ranges of temperate areas, even on north-facing slopes, winter temperatures are warmer than in high latitude areas and because of the higher precipitation the snow cover is thicker. Consequently, mountain soils are better insulated, with high solar radiation and thus their frost layer is less thick and not permanent in most cases. The reduction or lack of permafrost of the mountain soils contributes to their general good drainage and thus wet soils (i.e. peatlands) in mountain landscapes are mainly confined to bottom areas and depressions, and they are not widespread. Conversely, in mountain slopes high rainfall and good drainage speeds up soil formation processes. However, natural disturbances also linked to the slopes such as soil erosion, rock fall, landslides, avalanches and snow ablation play an important role in rejuvenating mountain soils. As a result of these complex interactions and because of its diverse geomorphology mountain landscapes hold a large spatial variability that is depicted in both soils and vegetation.

Soils of mountain areas tend to be young and highly influenced by their bedrock and physiographic properties. The wide range of soil types occurring in mountain regions is driven by microtopography, slope and aspect which, as stated above, define the snowpack and melting patterns that influence soil temperature, nutrient leaching and soil moisture (Stöhr 2007). Young mountain soils occur in well-drained areas and are classified as Leptosols or Regosols. Leptosols are thin soils, extremely gravelly and/or stony and with strong limitations to rooting. Regosols are weakly developed mineral soils in unconsolidated materials that occur in less stony areas and are typically highly erodible. On calcareous areas Rendzic or Chromic Leptosols dominate. Rendzic leptosols have a surface layer with high accumulation of organic matter and calcium carbonate. Chromic leptosols have a red surface layer and low or no calcium carbonate content. On siliceous bedrock, Regosols and the extremely thin Lithic leptosols are commonly found. But in stable and well-drained surfaces soils are often more developed, being common Dystric Cambisols and different types of Podzols. These latter two soil types show a thick and well-developed acidic horizon, but Podzols contain a subsurface horizon with illuvial amorphous organic matter and/or Al and Fe oxides. Finally, Histosols occur in poorly drained areas. These last soils evolve from incompletely decomposed plant remains and thus their features are quite independent of the bedrock type (IUSS Working Group WRB 2015).

Mountain soils in temperate areas usually have a high organic matter content, as shown in some regional soil C maps (Baritz et al. 2010; Doblas-Miranda et al. 2013). Although plant biomass in alpine grasslands is much lower than in forests, their soil C stocks are also generally high (Table 9.1) and comparable to forested areas (Berninger et al. 2015). The large amount of soil C in alpine environments is related to the high residence time of organic matter in the soil compared to the

Table 9.1 Some examples of soil organic C stocks in grasslands of the European mountains

Mountains	C stocks (Mg ha^{-1})	n	Altitude (m a.s.l.)	Annual mean air temperature (°C)	Plant formation	References
Iberian Central System	72–324	5	1653–2051	6.5–7.0	Mesic grasslands	(1)
Pyrenees	45–365	16	1704–2092	3.2–6.1	Mesic grasslands	(1)
Pyrenees	65–300	35	1845–2900	−0.7 to 5.0	Alpine and subalpine grasslands	(2)
Austrian Alps	260 and 130	2	1700 and 1900	2.1[a]	Alpine grassland, pine bushes and shrubs	(3)
Swiss Alps	53–116	8	810–2200	0.9–8.9	Grasslands	(4)
Tatra Mountains	20–250	25	1725–2368	−2.0 to 1.6	Alpine meadows	(5)
Eastern Swiss Alps	100	6	2616–2674	−2.6	Alpine tundra with permafrost	(6)
Eastern Swiss Alps	150	6	2577–2695	−2.6	Alpine tundra without permafrost	(6)

(1) Montané et al. (2007); (2) Garcia-Pausas et al. (2007); (3) Djukic et al. (2010); (4) Leifeld et al. (2009); (5) Kopáček et al. (2006); (6) Zollinger et al. (2013)
[a]Annual mean air temperature at 2277 m a.s.l.

living biomass (Körner 2003). This fact is caused by the harsh climatic conditions of the alpine environments that slow down the degradation of organic matter. Also, while forest soils receive large amounts of organic matter coming from aboveground biomass, in grasslands the primary organic matter inputs to the soil mainly come from root turnover and deposition. Consequently, the vertical distribution of organic C along the soil profile is typically shallower in forest soils than in grasslands (Jobbágy and Jackson 2000).

Mountain soils show a great variability in their characteristics. Thus, far from being evenly distributed, soil organic C content in mountain areas is particularly variable. Not only temperature reduces and precipitation increases with the elevation, but also the significant differences in solar radiation between north- and south-facing slopes, create environments that can be highly variable over relatively short distances. This feature, together with the high variability of soil depth and the natural diversity of substrates in mountain areas, makes the alpine landscapes a mosaic of different local conditions to the development of plant growth. This variability results in a considerable heterogeneity in plant community composition and structure, which in turn shape the distribution patterns of other organisms such as arthropods, fungi and soil bacterial communities. Different land management practices (e.g. various grazing pressures in pasturelands, timber removal in subalpine forests, etc.) is another source of variation that affects the C cycling and storage. All these factors result in soils with a highly heterogeneous amount of stored C, making difficult the prediction of current stocks and its response to the expected climate and land-use changes.

9.3 Factors Controlling Soil Organic C Stocks in Mountain Grasslands

The development of soil is a complex and continuous process, driven by parent material, climate and soil biota. In the mountains, the topography also plays an important role modifying the climate and creating different landforms for soil development. All these factors determine the physical, chemical and biological properties of soils and control their capacity to accumulate organic C.

9.3.1 Bedrock Type

Mountain areas are often geologically complex, as a result of past volcanism, compression and tension faults, plate subduction and uplift. The parental material from which a soil develops determines many hydrological, ecological and pedogenic processes, having implications for the capacity of soils to store C, the C accumulation rates and its persistence in the soil. Differences in the lithology determine the differences in the mineral composition of soils and influence their texture, chemistry and weathering processes. Changes in plant composition and structure are also frequently associated with changes in the bedrock type through its effect on nutrient status and physical characteristics of the soil.

Soil texture is the most relevant characteristic that is determined by the bedrock type. For instance, soils developed on sandstones or granites usually have coarser textures than those developed on limestones or slates. Soil texture is particularly relevant for organic matter accumulation in soils, as organic matter is stabilised in soil through its interaction with the finest mineral particles. Indeed, organic matter associated with the finest particles (i.e. fine silt and clay) is usually older (Eusterhues et al. 2003) and has longer residence times (Balesdent 1996) than the organic matter in the coarser fractions. In the Pyrenees, although C and N availability were more important explaining topsoil basal respiration, soils developed on granites showed high rates of basal respiration (Garcia-Pausas et al. 2008), suggesting that they may contain a higher proportion of non-stabilised organic matter.

9.3.2 Climate

Carbon stocks are the result of the net balance between C inputs through primary production and C outputs through microbial mineralisation as well as leaching and erosion (Fig. 9.1).

In alpine areas, both primary production and microbial mineralisation are constrained by low temperatures, particularly during wintertime. Given that soils in the mountain areas have a relatively high amount of organic C, it can be suspected that microbial mineralisation might be more strongly limited by climate than primary production. However, there is some evidence that the maximum soil C stocks are found in the subalpine belt and that from that point upwards the soil organic C stocks tend to reduce with the elevation (Djukic et al. 2010), reaching close-to-zero levels at unvegetated substrates of extreme altitudes (Körner 2003). This reduction of C stocks is due to the reduced plant cover and productivity, reduced rooting depth, and also because soils are generally younger at high altitudes (Fig. 9.2). This general trend is expected to differ between the north- and south-facing slopes. Indeed, in the Pyrenees Garcia-Pausas et al. (2007) observed that the reduction in C stocks with altitude was sharper at the north-facing slopes, probably because at high altitudes the environmental conditions on the south-facing slopes are more favourable for plant growth (Fig. 9.3).

The microclimate environment also determines the characteristics of the soil organic matter and thus its turnover. There is an indirect effect mediated by climate-driven changes in the plant community composition and structure (see below), but also a direct effect of climatic conditions on organic matter quality. Soils developed on high altitudes are usually rich in labile and particulate organic C (Leifeld et al. 2009; Budge et al. 2011). As occurs with altitude, the severe conditions at the north-facing soils also cause a higher accumulation of poorly degraded organic matter than at south-facing slopes (Egli et al. 2015). These C pools appear to have long residence times, as shown by radiocarbon dating (Leifeld et al. 2009; Budge et al. 2011), which

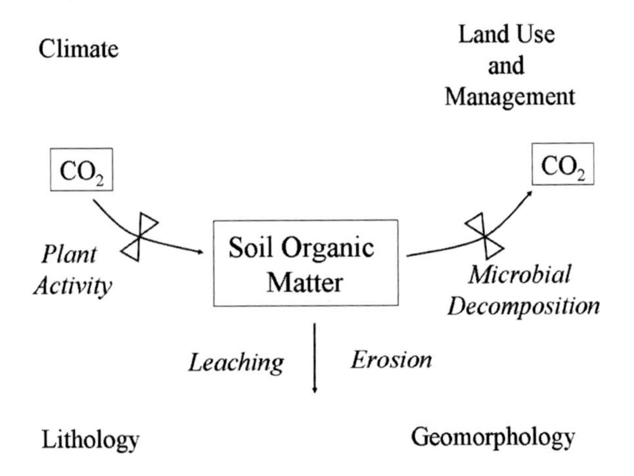

Fig. 9.1 Driving variables and processes involved in soil organic matter dynamics and stocks

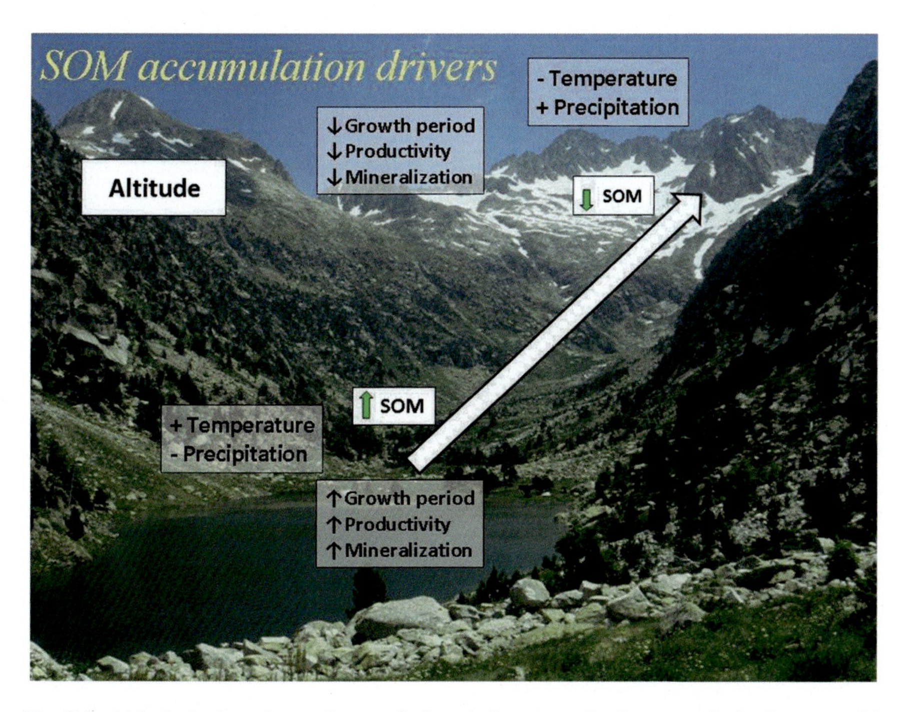

Fig. 9.2 Altitude is the primary factor of climatic heterogeneity in mountain landscapes, with high-altitude areas having generally low (−) temperatures and high (+) precipitation compared to low-altitude areas. It causes short growing seasons, low plant productivity and low soil C mineralisation rates (*downwards arrows*) in high-altitude sites compared to the bottom of the valleys. Photo: J. Garcia-Pausas

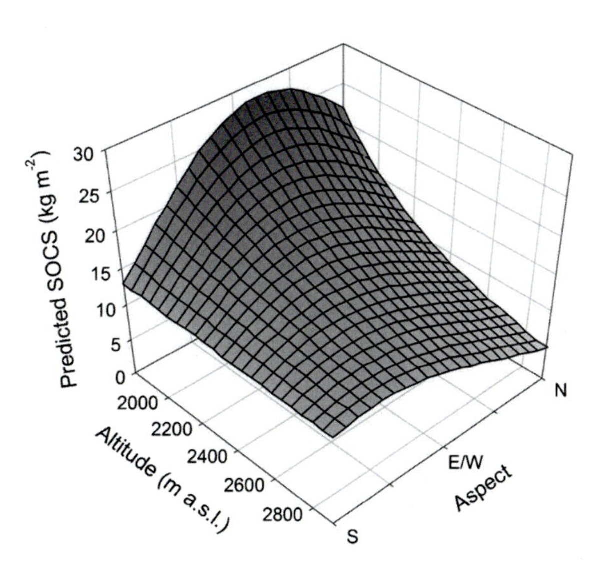

Fig. 9.3 Predicted soil organic C stocks (SOCS) as a function of altitude and aspect in the Pyrenean subalpine and alpine grasslands. From Garcia-Pausas et al. (2007) with permission of Springer

is attributed to the harsh conditions for residue decomposition, the low soil pH and nutrient limitations. However, in areas where environmental conditions are unfavourable for decomposition, the degree of physicochemical stabilisation of organic matter, as measured by incubation under standard conditions, is lower (Garcia-Pausas et al. 2008). This feature could make these C pools particularly vulnerable to future climate and land-use changes.

The temperature variations in altitude and aspect, as well as the microtopography and the predominant wind direction largely determine the distribution and duration of the snowpack cover. The duration of the snowpack cover has important implications for the soil organic C dynamics, as it determines not only the temperature and moisture of the underlying soil but also the length of the plant growing season, the plant community composition, the microbial activity and nutrient dynamics. Indeed, snow cover maintains soil temperature relatively high compared to the air temperatures during winter, allowing the maintenance of unfrozen conditions (Edwards et al. 2007). This isolation is because of the low thermal conductivity of the snow, particularly when it is fresh and non-compacted (Körner 2003). Consequently, topsoil temperature under the snow is usually stable around 0 °C, even when air temperatures are far below zero. This allows the microbial processes to continue in winter (Schmidt and Lipson 2004), causing an increase in the winter CO_2 efflux (Walker et al. 1999) and also a faster decomposition of the leaf litter (Baptist et al. 2010; Saccone et al. 2013) under the snow than in non-covered soils. However, when the snowpack melts in late winter and before the snowfall in late autumn, soils are usually exposed to temperatures well below 0 °C, undergoing frequent episodes of freezing and thawing.

Soil frost does not allow the belowground plant production, but an earlier peak in fine root production during the subsequent growing season has been observed by Tierney et al. (2001) after an experimental snow removal in forest ecosystems. They also reported significant increases in fine root mortality, resulting in an increased root turnover. In grasslands, Kreyling et al. (2008) indicated that recurrent freeze–thaw events reduced root length during the subsequent growing season, but also increased aboveground productivity.

Freeze–thaw events alter C and N dynamics, affecting root production and turnover, soil microbial activity, soil C and N availability and its mineralisation. It has been observed that repeated freeze–thaw cycles as well as prolonged frost increase C and nutrient concentration in the soil solution, which can eventually be lost by leaching (Fitzhugh et al. 2001; Freppaz et al. 2007; Wipf et al. 2015). Also, a burst of CO_2 and N_2O emissions from thawing soils has often been observed (Nielsen et al. 2001; Teepe et al. 2001; Matzner and Borken 2008), as well as higher emissions during the subsequent growing season (Blankinship and Hart 2012). This response is partially explained by the decomposition of the microbial necromass (Herrmann and Witter 2002), but a reduction of microbial biomass in thawing episodes has not been detected in alpine soils (Lipson et al. 2002; Freppaz et al. 2007). Another source of the CO_2 flush when soil thaws is the death of fine roots due to the soil frost. The decomposition of the fine root litter increases the

CO_2 efflux from these soils, but the release of soluble cell constituents from dead fine roots is the main factor that contributes to the observed short-term CO_2 and N_2O pulse after thawing (Matzner and Borken 2008), and the higher concentrations of N and P in the soil solution (Fitzhugh et al. 2001; Freppaz et al. 2007; Wipf et al. 2015) after freeze–thaw events. The third source of the CO_2 flushed upon a freeze–thaw event is the change in the soil structure. Soil freezing causes a disruption of soil aggregates because the ice crystals expand breaking the bonds between mineral particles. This breakdown of the aggregates makes the organic matter protected within the aggregates available for microorganisms. Macroaggregates are more susceptible to these disruptive forces than microaggregates, and their vulnerability is enhanced with increasing soil water content, while high clay, organic matter and Fe-oxide concentrations reduce the disruptive effects of freezing (Six et al. 2004).

9.3.3 *Plant Community Composition*

The effect of abiotic environment on soil C content and its stability is in part mediated by its effect on plant communities. It is well known that abiotic factors determine the composition and characteristics of the plant community, which in mountain ecosystems is also strongly related to topography (Sebastià 2004) and bedrock type. The composition of plant communities determines some functional characteristics that can be relevant for the organic matter production and allocation. For instance, although most of the root characteristics are species-specific, Pohl et al. (2011) showed that in alpine ecosystems graminoids usually have a large proportion of fine roots compared to forbs or shrubs. This feature may enhance topsoil aggregate stability under grasses (Pohl et al. 2009), which could be relevant for stabilising C in soils. Other characteristics of the vegetation such as above- and belowground productivity and allocation, rooting depth, horizontal root expansion may influence the C inputs and its persistence in the soil.

Plant community composition is in turn related to the quality of soil organic matter and, consequently to its decomposition rates. The low quality for decomposition of the organic matter produced by woody shrubs, with high lignin and polyphenol content, greatly differs from that produced by herbaceous plants, which is much more readily decomposable. But also among herbaceous plants, there can be significant differences, as occurs with the N-enriched organic matter produced by legumes. Thus plant communities differing in the biochemical characteristics of the biomass can lead to differences in the biochemical quality of soil organic matter. Indeed, Eskelinen et al. (2009) indicated that high proportion of forbs in an alpine tundra of northern Europe were related to low soil C/N ratios and high soluble N/phenolics ratios, causing in turn differences in the associated microbial communities.

9.4 Effects of Land-Use and Management Changes on Soil C Dynamics and Stocks

Although at the global scale remote areas still exist in mountain regions, in temperate European mountains the diversity of habitats resulting from the complex topography and multiple microclimates has been shaped, similarly to the lowland, by the human activities for centuries (Montserrat and Fillat 1994; Gassiot Ballbè et al. 2017 in the present book). Over the past centuries, low-intensity agriculture, farming and forestry have created and structured semi-natural habitats that constitute the contemporary landscapes in temperate mountains (Fig. 9.4). For some of these habitats, the sustainability of ecosystem services, at least at short- and mid-term, is linked to the continuity of human involvement and may be threatened by changes in the historical use of the mountain ecosystems (MacDonald et al. 2000; Regato and Salman 2008).

Due to the inherent physical constraints of the vast majority of mountain valleys that prevented the modernization of agriculture, traditional and sustainable low-input farming systems had mainly remained invariable until the last decades (Plieninger et al. 2006). In recent times, however, the long agropastoral tradition was altered by the integration of local economies into the global market and the

Fig. 9.4 Low-intensity agriculture and farming for centuries have shaped subalpine landscapes. Alinyà mountain, eastern Pre-pyrenees. Photo: Pere Casals

Table 9.2 Change (%) in the number of farms and livestock heads during 1989–2009 in five counties of the Catalan Pyrenees

	Val d'Aran	Alta Ribagorça	Pallars Sobirà	Cerdanya	Ripollès
Farms					
Bovine	−77	−56	−33	−22	−30
Ovine	−39	−44	−58	−29	−35
Livestock heads					
Bovine	−9	6	135	81	32
Ovine	21	−35	−8	−10	−43

Elaborated using data from the Ministry of Agriculture, Livestock, Fisheries and Food (Government of Catalonia)

emergence of new values and interests. As a consequence, traditional and sustainable multifunctional activities were abandoned and replaced by more purely production-oriented ones. Pastures located on steep slopes and at higher altitudes, requiring intensive labour, were abandoned while, at the same time, agriculture and livestock raising in accessible fertile lowland fields and productive mountain grasslands has intensified (Tasser and Tappeiner 2002; Bartolomé et al. 2005; Hopkins and Holz 2006). Changes in traditional farming practices have been observed across the European mountains (MacDonald et al. 2000). In the Pyrenees, traditional pastoral systems were characterised by an extensive management of the herd and the local transhumance to the communal alpine pastures in summer. Also, herds of sheep moved each year from the Ebro basin to summer mountain grasslands (Montserrat and Fillat 1994). In the last decades, in the Pyrenees, but also in most of the temperate European mountains, the redirection of the labour force to other employments, mainly related to the tertiary sector, caused changes in the farming management practices. As a consequence, some farms were abandoned while other intensified the management to accommodate socio-economic changes and labour resources. Together with a decrease in the number of farms, an increase of stocking number per farm and a shift of the stockbreeding to forms with low labour requirements (i.e. ovine to bovine) are common trends that allowed becoming more cost-efficient (Table 9.2).

9.4.1 Grazing Intensification

The adjustment of traditional farm households to a more intensive production and pluriactivity had entailed spatial changes in the grazing practices and land management. In the Pyrenees and Alps, the decline in shepherding has led to localised concentrations of stock around more easy-to-reach alpine grasslands, frequently resulting in overgrazing of high-quality pastures, while grazing intensity on steep slopes will likely decline (MacDonald et al. 2000).

SOC is a function of the balance between inputs from primary production and outputs through decomposition (Fig. 9.1). As a result of overgrazing, the quantity of the inputs to the soil may be reduced due to the aboveground biomass removal by animals. Also, the reduced plant biomass (i.e. less photosynthetic tissue) in heavily grazed grasslands causes a reduction of the aboveground productivity in comparison to the non-grazed grasslands (Ferraro and Oesterheld 2002). Although grazing can stimulate aboveground plant productivity under light or moderate grazing intensities through the so-called compensatory growth, it has been shown that heavy defoliation may lead to substantial reduction of the aboveground production (Chen et al. 2006; Zhao et al. 2008). In this case, plants respond to defoliation by allocating more C aboveground and thus reducing root biomass and productivity. The magnitude of this effect increases with the intensity of defoliation, the nutrient availability and water availability (Zhao et al. 2008; Klumpp et al. 2009). So there are site-specific sustainable grazing regimes that allow the conservation of C stocks, but when that grazing pressures are exceeded, inputs into the soil may be reduced (Georgaidis et al. 1989; Ferraro and Oesterheld 2002; Gao et al. 2008).

Overgrazing causes an alteration in soil physical, chemical and biological properties, resulting in changes in vegetation cover, a degradation of soil and a loss of soil C stocks. A typical feature of grazing activity is spatial heterogeneity. Animals tend to graze on areas with the most nutritious plants whereas select particular landscapes features for resting and ruminating. As a consequence, different types of vegetation develop which, in turn, influences the subsequent behaviour of the animals. Plant nutrient contents and soil nutrient availability increase from grazed to resting areas (Badia et al. 2008). In highly grazed areas, in comparison with only lightly grazed ones, the availability of P increases as a consequence of cattle grazing and defecation, which may accelerate the P cycling (Güsewell et al. 2005). An increase in fresh organic C (i.e. faeces) and nutrient availabilities as a consequence of animal frequentation may increase the microbial decomposition of native soil organic C. In addition, an excess of trampling and continuous overgrazing increases the area of bare soil and the risk of soil erosion. In the Tibetan plateau, the degradation of grasslands due to land-use change and overgrazing caused relevant losses of soil organic C in the last 30 years (Xie et al. 2007). In the Alps, erosion rates can be considerably higher (4.4–20 Mg ha^{-1} $year^{-1}$) on grasslands with clear signs of degradation of the vegetation cover (Meusburger and Alewell 2014).

9.4.2 *Grazing Abandonment*

Abandonment of pasturelands and traditional farming practices is a widespread phenomenon in the mountain areas of Europe (MacDonald et al. 2000). While the impacts on several environmental and landscape values are evident (Tasser et al. 2007), the effects on C dynamics and soil C stocks are less apparent. The net effects of land-use changes on C stocks are the result of the changes in the inputs and output drivers in the short- and long term after abandonment. Land-use changes may also affect the biochemical quality of litter inputs, which is a major factor influencing the organic C accumulation in soils (Liao and Boutton 2008). In ecosystems with high belowground allocation, such as grasslands, root dynamics represent the primary source for building up soil organic matter (Rasse et al. 2005; Piñeiro et al. 2006). This situation mostly applies to grazed grasslands, where a substantial amount of aboveground production is removed by grazing animals. When grass species are not palatable, both above- and belowground productions may be of the same order of magnitude. For example, in subalpine *Festuca eskia* grasslands in the Pyrenees, Montané et al. (2010) estimated an aboveground production of about 200 g m^{-2} $year^{-1}$ while the root production estimated by 15 cm-depth ingrowth cores was about 150 g m^{-2} $year^{-1}$.

Belowground biomass production and turnover have been related to microclimate as well as to land-use management (Guo et al. 2007; Leifeld et al. 2015). In the short term, grazing abandonment of subalpine grasslands allows higher aboveground biomass and accumulation of substantial amounts of necromass, but the effects on belowground biomass remain controversial. In general, belowground productivity increases in response to grazing removal (Ruess et al. 1998; Johnson and Matchett 2001; Smit and Kooijman 2001) but some studies did not find significant effects (McNaughton et al. 1998; Bazot et al. 2005) or even negative effects (Frank et al. 2002; Pucheta et al. 2004). Controversial findings may be partly explained by the physiological responses of plants to defoliation, but also by changes in plant species composition that may translate to differences in productivity, C allocation patterns and rooting depths at the ecosystem level. Indeed, Lanta et al. (2009) reported changes in plant species composition and richness in a 3-year field experiment with grazing and abandonment, and also showed a reduction of belowground biomass in non-grazed grasslands of the White Carpathians in the Czech Republic. In a 2-year grazing exclusion experiment in the Pyrenees, root production decreased in grazed grasslands in comparison with non-grazed ones. This response occurred right after the grazing event, and no apparent effects on yearly belowground C input were found (Garcia-Pausas et al. 2011).

9.4.3 Soil C Stocks in Grazed and Ungrazed Mountain Grasslands

The effect of grazing abandonment on soil C stocks has mostly been studied through the comparison of grazed and ungrazed areas, but the short-term effects of grazing on SOC is inconsistent to date, with both increases and decreases reported in response to increased grazing pressure. Although the effect of herbivory on plant productivity and C allocation is still under debate, abandonment of light, extensive grazing management might reduce soil stocks (Schuman et al. 1999; Pucheta et al. 2004).

Thus in a survey of grasslands in the Pyrenees, we found that abandoned grasslands had lower SOC stocks in the uppermost 20 cm of soil than grazed ones (Casals et al. 2004). However, this result may just reflect that the less productive grasslands were abandoned. In summary, changes in soil C stocks of mountain grasslands as a consequence of grazing abandonment are, at least in the short term, small and no clear trends may be stated.

9.4.4 Effects of Shrub Encroachment on Soil C Dynamics and Stocks

In the long term, grazing abandonment may involve a change in the dominant functional groups and often leads to shrub encroachment (Fig. 9.5). This shift is often observed in many mountain areas (MacDonald et al. 2000) and can lead to long-term expansion of forests (Gehrig-Fasel et al. 2007; Améztegui et al. 2010). Shrub encroachment into grasslands has been documented in the Pyrenees (Molinillo et al. 1997; Roura-Pascual et al. 2005) and the Central System ranges of the Iberian Peninsula (Sanz-Elorza et al. 2003). However, depending on the site characteristics, this can be a slow process. Indeed, Pardo et al. (2015) did not

Fig. 9.5 Shrub encroachment (*Cytisus balansae* ssp. *europaeus*) into mountain grasslands. Durro mountains (Alta Ribagorça, Central Pyrenees). Photo: Pere Casals

observe great changes in vegetation richness and composition after two decades of grazing exclusion in subalpine grasslands in the central Pyrenees. In Collada de Montalto (Central Pyrenees), shrub encroachment occurred mostly in grass patches inside the shrublands while woody proliferation into open grassland was less evident. Comparing the border between grassland and shrubland, we estimated a mean shrub expansion of the border into the mesic grassland of 2.0 ± 1.4 m (n = 263) in the period between 1997 and 2014 (unpublished data).

When woody plant invasion occurs, the shift from grass- to a shrub-dominated ecosystem entails significant changes in the production and placement of the inputs of litter (i.e. aboveground or belowground) and on factors that regulate soil organic matter mineralisation such as microclimate, biochemical quality of organic matter and the structure of the microbial community.

9.4.4.1 Litter Inputs

After shrub encroachment, the pattern of litter inputs changes from a belowground predominance in grasslands to an increase of aboveground deposition in shrublands. This shift is due to the differential allocation patterns between grasses and shrubs (Lett et al. 2004). For instance, in *Cytisus balansae* shrublands that had invaded subalpine grasslands of the Pyrenees, aboveground litter input was estimated as high as fourfold the root litter inputs in the top 15 cm of soil (Montané et al. 2010).

Surface litter is partially decomposed to CO_2, but a fraction is incorporated into the mineral horizons as a dissolved or particulate organic matter where it is mineralised or stabilised. Indeed, litter layers under shrubs may be an important source of dissolved organic C into the mineral soil, with a flux that may represent up to 35% of the annual litterfall C (Kalbitz and Kaiser 2008). Also, free particulate organic matter plays a significant role in the increase of soil organic C in the uppermost layers after woody plant encroachment in grasslands (Liao et al. 2006). Consequently, shrub encroachment into grasslands modifies the amount and placement of organic matter into the soil, but the effect on C sequestration also depends on the C loss from decomposing litter and soil organic matter.

9.4.4.2 Microclimate

Changes in the plant cover after pasture abandonment and shrub encroachment modify incoming solar radiation and precipitation to the soil. After grazing removal, the accumulation of standing necromass and litter reduces soil temperature and may increase soil water content (Rosset et al. 2001). In the Pyrenees, cumulative degree-days above 0 °C from May to November at 5 cm-depth soil were about 20% lower beneath woody canopies than under grasses (Montané et al. 2010). Lower temperatures under shrub canopies likely reduces above- and/or belowground litter decomposition which may be the predominant mechanism behind higher SOC after shrub encroachment (Smith and Johnson 2004). For instance, a reduction of soil CO_2 efflux after grazing exclusion in the Tibetan Plateau has been attributed in part to its lower soil temperature (Chen et al. 2016). In the Pyrenees, *Festuca eskia* roots incubated for 1 year in buried litterbags in a subalpine soil decomposed slightly slower under shrubs than in paired grasslands (20.1 ± 0.42% and 22.4 ± 1.44% mass loss, respectively) (Casals et al. 2010). In addition, using buried labelled wheat roots mixed with soil, Casals et al. (2010) showed that ^{13}C loss was about four percent units lower in root bags incubated for 1 year in non-grazed grassland plots and seven percent units lower under shrubs than in paired grazed grasslands. As these results derived from the incubation of standard labelled material, they mainly reflect a change to a less favourable soil environment for root decomposition due to either grazing exclusion or shrub encroachment. Therefore, a decrease in soil temperature may contribute to explain lower root decomposition rates after grassland abandonment and shrub encroachment.

9.4.4.3 Biochemical Quality and Microbial Activity

It is widely known that litter nutrient concentration and organic matter quality (e.g. lignin content) are the main factors determining litter decomposition rates (Cornwell et al. 2008). Grasslands typically have a high density of fine roots that are poorly lignified and with high turnover rates, thus providing a relatively labile C substrate for microbial activity. In contrast, the proliferation of shrubs may increase the presence of lignified roots with lower turnover rates. After shrub encroachment, low quality of litter inputs, with large amounts of secondary compounds such as lignin or polyphenolic substances, may hinder decomposition and promote C accumulation (Pérez-Harguindeguy et al. 2000; Shaw and Harte 2001; McCulley et al. 2004; Liao and Boutton 2008).

The biochemical quality of litter may differ between species. In the Pyrenees, an aboveground litter of grasses showed marked differences in the chemical composition from that of the two main invading shrubs of that area (*Cytisus balansae* and *Juniperus communis*). The litter of both shrubs had higher concentrations of recalcitrant compounds (e.g. lignin, lipids, suberin) and a low concentration of either N (conifer) or P (legume) relative to grass litter (Montané et al. 2010). Consequently, the higher organic C found in the upper mineral soil layer under shrubs compared to the grassland was mainly attributed to the slower decomposition of shrub litter and the transfer of litter-derived C into the soil. However, the presence of grass litter, with high N and P concentrations, may enhance microbial activity and prime the decomposition of recalcitrant shrub litter. As a result, at least in the short term after shrub proliferation when both shrub and grass litters coexist, the shrub litter accumulation pattern is altered (Montané et al. 2013).

Defoliation induces an increase of root exudation (Paterson et al. 2005). Therefore when grassland is abandoned a reduction of labile C inputs into the soil can be expected. Also, when shrubs proliferate, their lower fine root density may cause further reduction of C inputs by exudation. This decrease of labile C release from roots may cause a significant reduction of microbial activity (Hamilton and Frank 2001) and also a lower stimulation of soil organic matter mineralisation (i.e. priming) that usually occurs in the presence of labile C (Kuzyakov et al. 2000). Priming effect on soil organic matter mineralisation is particularly relevant in the rhizosphere and, although its magnitude is variable, it increases with the rate of rhizospheric C inputs (Paterson and Sim 2013) and may account for a substantial fraction of the SOM-derived CO_2 efflux (Cheng and Kuzyakov 2005). In addition, this reduction of priming effect may cause in turn a reduction of the nutrient availability for plants (Hamilton and Frank 2001).

The change of root exudates, as well as the fate of particulate organic matter also promotes a change in the microbial community composition (Grayston et al. 2004). Indeed, fungal growth and activity seems to be generally favoured in surface horizons after grazing or agricultural abandonment (Zornoza et al. 2009; Lopez-Sangil et al. 2011) and a higher fungal-to-bacterial activity ratio seems to promote a conservative cycling of nutrients in soil and C accumulation (Wardle et al. 2004; Gordon et al. 2008). Therefore a reduction in soil organic C decomposition in abandoned sites is expected.

9.4.4.4 Soil C Stocks in Shrub-Encroached Grasslands

In summary, shrub encroachment into mountain grasslands increases soil organic carbon in the upper soil mineral profile compared to the grassland soil (Montané et al. 2007). This net C increase may be explained by lower aboveground and belowground litter decomposition after shrub proliferation due to lower soil temperatures and lower biochemical quality of shrub organic matter. Lower litter quality may promote a shift in the composition of the microbial community to a slow-growth strategy, typical of a fungal-dominated microbial community (Bardgett et al. 2005), which may contribute to explain lower decomposition. The reduction of fine root density with the proliferation of shrubs may also reduce the rates of root exudation, which may decrease the priming effect on soil organic matter mineralisation thus contributing to the conservation of soil C stocks.

9.4.4.5 Shrubland Management and Soil C Stocks

Shrub encroachment into grasslands involves the replacement of one dominant growth form by another one, and it is likely to impact on ecosystem structure and functions (Lett and Knapp 2005). In the Pyrenees, a decrease in diversity

(Anthelme et al. 2007) and increases in soil C storage (Montané et al. 2007) have been reported after shrub proliferation into grasslands. Woody encroachment increases the risk of fire propagation by incrementing both fuel load and fuel continuity. In these encroachment-prone communities, managers may have to decide between reducing shrub proliferation to maintain biodiversity and grazing potential or allowing the shrub proliferation to increase C sequestration.

In the Pyrenees, shepherds have traditionally used fire as a management tool to improve grass productivity and transform encroached land into grassland. Today prescribed burning is usually carried out by fire brigades or foresters in winter when snowy or wet conditions limit the impact of the fire on soils and herbaceous plants (Rigolot et al. 2002). Mechanical thinning is also applied to revert encroached grasslands. How these management options affect organic matter dynamics and soil C stocks remains an issue.

9.5 Conclusions and Further Research Needs

Agricultural land-use changes in the European mountains show antagonist trends, intensification at the bottom of valleys and other productive grasslands, whereas the less productive grasslands located on steep slopes and at higher altitudes are being increasingly abandoned. A mechanistic understanding of how these changes affect relevant ecological processes, such as biodiversity or C stocks, is necessary to predict the effects of global change on ecosystem function and deliver appropriate management recommendations.

The decline of agropastoral activities is especially pervasive in high mountain grasslands. As a consequence, pasture abandonment, especially of marginal and less productive lands, and shrub encroachment into grasslands have become the most significant trends in land use, which may be observed all around European mountains, to a greater or lesser extent. Short-term changes in soil C stocks as a consequence of grazing abandonment are difficult to detect due to the large size of the organic matter pool as compared to the small changes in the C inputs, and also to the high spatial variability of soil C stocks (Conant and Paustian 2002; Smith 2004). Smith (2004), using a modelling approach, demonstrated that a change in SOC may not be detectable until about 7–10 year after the experiment, assuming an increase in soil C input of 20–25% (Smith 2004). Therefore, well monitored long-term exclusion experiments would be very useful to measure changes in the C stocks caused by management changes.

Higher soil organic matter after shrub encroachment into grasslands may be explained by the high content of recalcitrant compounds such as lignin and polyphenols in the plant-derived organic matter inputs, which slows the decomposition of soil organic matter and delays its incorporation into the protected pools in the mineral soil.

Due to high fine root density and rhizosphere exudation rates, grassland soils show higher priming effect on C mineralisation than in woodland soils (Waldrop and Firestone 2004). The effect of woody plant invasion on rhizosphere priming is still unknown and could have a significant impact on C balance.

Acknowledgements This study summarises the work done in different research projects funded by the *Ministerio de Economia y Competitividad*, Spain (Carbopas, REN2002-04300-C02-02; VULCA, CGL2005-08133-CO3; GRACCIE Consolider Program, CSD2007-00067) and by the European Commission (GHG-Europe project, FP7-ENV-2009-1, project No. 244122). J.G.P. and P.C. are financially supported by the Spanish *Ministerio de Economía y Competitividad*, through *Juan de la Cierva* and *Ramón y Cajal* contracts, respectively.

References

Améztegui A, Brotons L, Coll L (2010) Land-use changes as major drivers of mountain pine (*Pinus uncinata* Ram.) expansion in the Pyrenees. Glob Ecol Biogeogr 19:632–641

Anthelme F, Villaret JC, Brun JJ (2007) Shrub encroachment in the Alps gives rise to the convergence of sub-alpine communities on a regional scale. J Veg Sci 18:355–362

Badia D, Martí C, Sánchez JR, Fillat F, Aguirre J, Gómez D (2008) Influence of livestock soil eutrophication on floral composition in the Pyrenees mountains. J Mt Sci 5:63–72

Balesdent J (1996) The significance of organic separates to carbon dynamics and its modelling in some cultivated soils. Eur J Soil Sci 47:485–493

Baptist F, Yoccoz NG, Choler P (2010) Direct and indirect control by snow cover over decomposition in alpine tundra along a snowmelt gradient. Plant Soil 328:397–410

Bardgett RD, Bowman WD, Kaufmann R, Schmidt SK (2005) A temporal approach to linking aboveground and belowground ecology. Trends Ecol Evol 20:534–641

Baritz R, Seufert G, Montanarella L, Van Ranst E (2010) Carbon concentrations and stocks in forest soils of Europe. For Ecol Manag 260:262–277

Bartolomé J, Lopez ZG, Broncano MJ, Plaixats J (2005) Grassland colonization by *Erica scoparia* (L.) in the Montseny Biosphere Reserve (Spain) after land-use changes. Agr Ecosyst Environ 111:253–260

Bazot S, Mikola J, Nguyen C, Robin C (2005) Defoliation-induced changes in carbon allocation and root soluble carbon concentration in field-grown *Lolium perenne* plants: do they affect carbon availability, microbes and animal trophic groups in soil? Funt Ecol 19:886–896

Berninger F, Susiluoto S, Gianelle D, Bahn M, Wohlfahrt G, Sutton M, Garcia-Pausas J, Gimeno C, Sanz MJ, Dore S, Rogiers N, Furger M, Eugster W, Balzarolo M, Sebastià MT, Tenhunen J, Staszewski T, Cernusca A (2015) Management and site effects on carbon balances of European mountain meadows and rangelands. Boreal Environ Res 20:748–760

Blankinship JC, Hart SC (2012) Consequences of manipulated snow cover on soil gaseous emission and N retention in the growing season: a meta-analysis. Ecosphere 3, art 1

Budge K, Leifeld J, Hiltbrunner E, Fuhrer J (2011) Alpine grassland soils contain large proportion of labile carbon but indicate long turnover times. Biogeosciences 8:1911–1923

Casals P, Garcia-Pausas J, Romanyà J, Camarero L, Sanz MJ, Sebastià MT (2004) Effects of livestock management on carbon stocks and fluxes in grassland ecosystems in the Pyrenees. In: Lüscher et al A (eds) in land use systems in grassland dominated regions, Grassland science in Europe, vol 9. Swiss Grassland Society (AGFF), Zürich, pp 136–138

Casals P, Garcia-Pausas J, Montané F, Romanya J, Rovira P (2010) Root decomposition in grazed and abandoned dry Mediterranean dehesa and mesic mountain grasslands estimated by standard labelled roots. Agr Ecosyst Environ 139:759–765

Chen J, Zhuo X, Wang J, Hruska T, Shi W, Cao J, Zhang B, Xu G, Chen Y, Luo Y (2016) Grazing exclusion reduced soil respiration but increased its temperature sensitivity in a Meadow Grassland on the Tibetan Plateau. Ecol Evol 6:675–687

Chen Y, Lee P, Lee G, Mariko S, Oikawa T (2006) Simulating root responses to grazing of a Mongolian grassland ecosystem. Plant Ecol 183:265–275

Cheng W, Kuzyakov Y (2005) Root effects on soil organic matter decomposition. In: Zobel RW, Wright SF (eds) Roots and soil management: interactions between roots and the soil. Agronomy monograph no 48. ASA, CSSA and SSSA, Madison, Wisconsin, pp 119–143

Conant RT, Paustian K (2002) Spatial variability of soil organic carbon in grasslands: implications for detecting change at different scales. Environ Pollut 116:127–135

Cornwell WK, Cornelissen JHC, Amatangelo K, Dorrepaal E, Eviner VT, Godoy O, Hobbie SE, Hoorens B, Kurokawa H, Pérez-Harguindeguy N, Quested HM, Santiago LS, Wardle DA, Wright IJ, Aerts R, Allison SD, van Bodegom P, Brovkin V, Chatain A, Callaghan TV, Díaz S, Garnier E, Gurvich DE, Kazakou E, Klein JAL, Read J, Reich P, Soudzilovskaia NA, Vaieretti MV, Westoby M (2008) Plant species traits are the predominant control on litter decomposition rates within biomes worldwide. Ecol Lett 11:1065–1071

Djukic I, Zehetner F, Tatzber M, Gerzabek MH (2010) Soil organic-matter stocks and characteristics along an Alpine elevation gradient. J Plant Nutr Soil Sci 173:30–38

Doblas-Miranda E, Rovira P, Brotons L, Martínez-Vilalta J, Retana J, Pla M, Vayreda J (2013) Soil carbon stocks and their variability across the forests, shrublands and grasslands of peninsular Spain. Biogeosciences 10:8353–8361

Edwards AC, Scalenghe R, Freppaz M (2007) Changes in the seasonal snow cover of alpine regions and its effect on soil processes: a review. Quat Int 172:162–163

Egli M, Lessovaia SN, Chistyakov K, Inozemzev S, Polekhovsky Y, Ganyushkin D (2015) Microclimate affects soil chemical and mineralogical properties of cold alpine soils of the Altai Mountains (Russia). J Soil Sediment 15:1420–1436

Eskelinen A, Stark S, Mänistö M (2009) Links between plant community composition, soil organic matter quality and microbial communities in contrasting tundra habitats. Oecologia 161:113–123

Eusterhues K, Rumpel C, Kleber M, Kögel-Knabner I (2003) Stabilisation of soil organic matter by interactions with minerals as revealed by mineral dissolution and oxidative degradation. Org Geochem 34:1591–1600

Ferraro DO, Oesterheld M (2002) Effect of defoliation on grass growth: a quantitative review. Oikos 98:602–606

Fitzhugh RD, Driscoll CT, Groffman PM, Tierney GL, Fahey TJ, Hardy JP (2001) Effects of soil freezing, disturbance on soil solution nitrogen, phosphorus, and carbon chemistry in a northern hardwood ecosystem. Biogeochemistry 56:215–238

Frank DA, Kuns MM, Guido DR (2002) Consumer control of grassland plant production. Ecology 83:602–606

Freppaz M, Williams BL, Edwards AC, Scalenghe R, Zanini E (2007) Simulating soil freeze/thaw cycles typical of winter alpine conditions: Implications for N and P availability. Appl Soil Ecol 35:247–255

Gao YZ, Giese M, Lin S, Sattelmacher B, Zhao Y, Brueck H (2008) Belowground net primary productivity and biomass allocation of grassland in Inner Mongolia is affected by grazing intensity. Plant Soil 307:41–50

Garcia-Pausas J, Casals P, Camarero L, Huguet C, Sebastià MT, Thompson R, Romanyà J (2007) Soil organic carbon storage in mountain grasslands of the Pyrenees: effects of climate and topography. Biogeochemistry 82:279–289

Garcia-Pausas J, Casals P, Camarero L, Huguet C, Thompson R, Sebastià MT, Romanyà J (2008) Factors regulating carbon mineralization in the surface and subsurface soils of Pyrenean mountain grasslands. Soil Biol Biochem 40:2803–2810

Garcia-Pausas J, Casals P, Romanyà J, Vallecillo S, Sebastià MT (2011) Seasonal patterns of belowground biomass and productivity in mountain grasslands in the Pyrenees. Plant Soil 340:315–326

Gassiot Ballbè E, Mazzucco N, Clemente Conte I, Rodríguez Antón D, Obea Gómez L, Quesada Carrasco M, Díaz Bonilla S (2017) The beginning of high mountain occupations in the Pyrenees. Human settlements and mobility from 10,500 cal BP to 4500 cal BP. In: Catalan J,

Ninot JM, Aniz MM (eds) Challenges for high mountain conservation in a changing world. Springer, pp 75–105
Gehrig-Fasel J, Guisan A, Zimmermann NE (2007) Tree line shifts in the Swiss Alps: climate change or land abandonment? J Veg Sci 18:571–582
Georgaidis NJ, Ruess RW, McNaughtib SJ (1989) Ecological conditions that determine when grazing stimulates grass production. Oecologia 96:157–161
Gordon H, Haygarth PM, Bardgett RD (2008) Drying and rewetting effects on soil microbial community composition and nutrient leaching. Soil Biol Biochem 40:302–311
Grayston SJ, Campbell CD, Bardgett RD, Mawdsley JL, Clegg CD, Ritz K, Griffiths BS, Rodwell JS, Edwards SJ, Davies WJ, Elston DJ, Millard P (2004) Assessing shifts in microbial community structure across a range of grasslands of differing management intensity using CLPP, PLFA and community DNA techniques. Appl Soil Ecol 25:63–84
Guo LBB, Wang MB, Gifford RM (2007) The change of soil carbon stocks and fine root dynamics after land use change from native pasture to a pine plantation. Plant Soil 299:251–262
Güsewell S, Jewell PL, Edwards PJ (2005) Effects of heterogeneous habitat use by cattle on nutrient availability and litter decomposition in soils of an Alpine pasture. Plant Soil 268:135–149
Hamilton EW III, Frank DA (2001) Can plants stimulate soil microbes and their own nutrient supply? Evidence from a grazing tolerant grass. Ecology 82:2397–2402
Herrmann A, Witter E (2002) Sources of C and N contributing to the flush in mineralization upon freeze-thaw cycles in soils. Soil Biol Biochem 34:1495–1505
Hopkins A, Holz B (2006) Grassland for agriculture and nature conservation: production, quality and multi-functionality. Agron Res 4:3–20
IUSS Working Group WRB (2015) World reference base for soil resources 2014, update 2015 international soil classification system for naming soils and creating legends for soil maps. World Soil Resources Reports No 106. FAO, Rome
Jobbágy EG, Jackson RB (2000) The vertical distribution of soil organic carbon and its relation to climate and vegetation. Ecol Appl 10:423–436
Johnson LC, Matchett JR (2001) Fire and grazing regulate belowground processes in tallgrass prairie. Ecology 82:3377–3389
Kalbitz K, Kaiser K (2008) Contribution of dissolved organic matter to carbon storage in forest mineral soils. J Plant Nutr Soil Sci 171:52–60
Kirschbaum MUF (2000) Will changes in soil organic carbon act as a positive or negative feedback on global warming? Biogeochemistry 48:21–51
Klumpp K, Fontaine S, Attard E, Le Roux X, Gleixner G, Soussana JF (2009) Grazing triggers soil carbon loss by altering plant roots and their control on soil microbial community. J Ecol 97:876–885
Kopáček J, Kaňa J, Šantrůčková H (2006) Pools and composition of soils in the alpine zone of the Tatra Mountains. Biologia, Bratislava 61(Suppl 18):S35–S49
Körner C (2003) Alpine plant life. Functional plant ecology of high mountain ecosystems, 2nd edn. Springer
Kreyling J, Baierkuhnlein C, Pritsch K, Schloter M, Jentsch A (2008) Recurrent soil freeze-thaw cycles enhance grassland productivity. New Phytol 177:938–945
Kuzyakov Y, Friedel JK, Stahr K (2000) Review of mechanisms and quantification of priming effects. Soil Biol Biochem 32:1485–1498
Lal R (2008) Carbon sequestration. Philos T R Soc B 363:815–830
Lanta V, Doležal J, Lantová P, Kelíšek J, Mudrák O (2009) Effects of pasture management and fertilizer regimes on botanical changes in species-rich mountain calcareous grassland in Central Europe. Grass Forage Sci 64:443–453
Leifeld J, Zimmermann M, Fuhrer J, Conen F (2009) Storage and turnover of carbon in grassland soils along an elevation gradient in the Swiss Alps. Glob Change Biol 15:668–679
Leifeld J, Meyer S, Budge K, Sebastià MT, Zimmermann M, Fuhrer J (2015) Turnover of grassland roots in mountain ecosystems revealed by their radiocarbon signature: role of temperature and management. PLoS ONE 10(3):e0119184. doi:10.1371/journal.pone.0119184

Lett MS, Knapp AK (2005) Woody plant encroachment and removal in mesic grassland: production and composition responses of herbaceous vegetation. Am Midl Nat 153:217–231
Lett MS, Knapp AK, Briggs JM, Blair JM (2004) Influence of shrub encroachment on aboveground net primary productivity and carbon and nitrogen pools in a mesic grassland. Can J Bot 82:1363–1370
Liao JD, Boutton TW (2008) Soil microbial biomass response to woody plant invasion of grassland. Soil Biol Biochem 40:1207–1216
Liao JD, Boutton TW, Jastrow JD (2006) Storage and dynamics of carbon and nitrogen in soil physical fractions following woody plant invasion of grassland. Soil Biol Biochem 38:3184–3196
Lipson DA, Schadt CW, Schmidt SK (2002) Changes in soil microbial community structure and function in an alpine dry meadow following spring snow melt. Microb Ecol 43:307–314
Lopez-Sangil L, Rousk J, Wallander H, Casals P (2011) Microbial growth rate measurement reveal that land-use abandonment promotes a fungal dominance of SOM decomposition in grazed Mediterranean ecosystems. Biol Fert Soils 47:129–138
MacDonald D, Crabtree JR, Wiesinger G, Dax T, Stamou N, Fleury P, Gutierrez Lazpita J, Gibon A (2000) Agricultural abandonment in mountain areas of Europe: environmental consequences and policy response. J Environ Manage 59:47–69
Matzner E, Borken W (2008) Do freeze-thaw event enhance C and N losses from soils of different ecosystems? A review. Eur J Soil Sci 59:274–284
McCulley RL, Archer SR, Boutton TW, Hons FM, Zuberer DA (2004) Soil respiration and nutrient cycling in wooded communities developing in grasslands. Ecology 85:2804–2817
McNaughton SJ, Banyikwa FF, McNaughton MM (1998) Root biomass and productivity in a grazing ecosystem: the Serengeti. Ecology 79:587–592
Meusburger K, Alewell C (2014) Soil erosion in the Alps. Experience gained from case studies (2006–2013). Federal Office for the Environment, Bern. Environmental Studies No 1408, 116 pp
Molinillo M, Lasanta T, García Ruíz JM (1997) Managing mountainous degraded landscapes after farmland abandonment in the central Spanish Pyrenees. Environ Manag 21:587–598
Montané F, Rovira P, Casals P (2007) Shrub encroachment into mesic mountain grasslands in the Iberian Peninsula: effects of plant quality and temperature on soil C and N stocks. Glob Biogeochem Cy 21:GB4016. doi:10.1029/2006GB002853
Montané F, Romanyà J, Rovira P, Casals P (2010) Aboveground litter quality changes may drive soil organic carbon increase after shrub encroachment into mountain grasslands. Plant Soil 337:151–165
Montané F, Romanyà J, Rovira P, Casals P (2013) Mixtures with grass litter may hasten shrub litter decomposition after shrub encroachment into mountain grasslands. Plant Soil 368:459–469
Montserrat P, Fillat F (1994) The systems of grassland management in Spain. In: Breymeyer A (ed) Managed grasslands. Elsevier, B.V., Amsterdam
Nielsen CB, Groffman PM, Hamburg SP, Driscoll CT, Fayey TJ, Hardy JP (2001) Freezing effects on carbon and nitrogen cycling in northern hardwood forest soils. Soil Sci Soc Am J 65:1723–1730
Pardo I, Doak DF, García-González R, Gómez D, García MB (2015) Long-term response of plant communities to herbivore exclusion at high elevation grasslands. Biodivers Conserv 24:3033–3047
Paterson E, Sim A (2013) Soil-specific response functions of organic matter mineralization to the availability of labile carbon. Glob Change Biol 19:1562–1571
Paterson E, Thornton B, Midwood AJ, Sim A (2005) Defoliation alters the relative contributions of recent and non-recent assimilate to root exudation from *Festuca rubra*. Plant Cell Environ 28:1525–1533
Pérez-Harguindeguy N, Díaz S, Cornelissen JHC, Vendramini F, Cabido M, Castellanos A (2000) Chemistry and toughness predict leaf litter decomposition rates over a wide spectrum of functional types and taxa in central Argentina. Plant Soil 218:21–30

Piñeiro G, Paruelo JM, Oesterheld M (2006) Potential longterm impacts of livestock introduction on carbon and nitrogen cycling in grasslands of Southern South America. Glob Change Biol 12:1267–1284
Plieninger T, Höchtl F, Spek T (2006) Traditional land-use and nature conservation in European rural landscapes. Environ Sci Policy 9:317–321
Pohl M, Alig D, Körner C, Rixen C (2009) Higher plant diversity enhances soil stability in disturbed alpine ecosystems. Plant Soil 324:91–102
Pohl M, Stroude R, Buttler A, Rixen C (2011) Functional traits and root morphology of alpine plants. Ann Bot-London 108:537–548
Pucheta E, Bonamici I, Cabido M, Díaz S (2004) Below-ground biomass and productivity of a grazed site and a neighbouring ungrazed exclosure in a grassland in central Argentina. Austral Ecol 29:201–208
Rasse DP, Rumpel C, Dignac MF (2005) Is soil carbon mostly root carbon? Mechanisms for a specific stabilisation. Plant Soil 269:341–356
Regato P, Salman R (2008) Mediterranean mountains in a changing world: guidelines for developing actions plans. IUCN
Rigolot E, Lambert B, Pons P, Prodon R (2002) Management of a mountain rangeland combining periodic prescribed burnings with grazing: impact on vegetation. In: Trabaud L, Prodon R (eds) Fire and biological processes. Backhuys Publishers, Leiden, pp 325–337
Rosset M, Montani M, Tanner M, Fuhrer J (2001) Effects of abandonment on the energy balance and evapotranspiration of wet subalpine grassland. Agr Ecosyst Environ 86:277–286
Roura-Pascual N, Pons P, Etienne M, Lambert B (2005) Transformation of a rural landscape in the Eastern Pyrenees between 1953 and 2000. Mt Res Dev 25:252–261
Ruess RW, Hendrick RL, Bryant JP (1998) Regulation of fine root dynamics by mammalian browsers in early successional Alaskan taiga forests. Ecology 79:2706–2720
Saccone P, Morin S, Baptist F, Bonneville JM, Colace MP, Domine F, Faure M, Geremia R, Lochet J, Poly F, Lavorel S, Clément JC (2013) The effect of snowpack properties and plant strategies on litter decomposition during winter in subalpine meadows. Plant Soil 363:215–229
Sanz-Elorza M, Dana ED, Gonzalez A, Sobrino E (2003) Changes in the high-mountain vegetation of the central Iberian Peninsula as a probable sign of global warming. Ann Bot-London 92:273–280
Schmidt SK, Lipson DA (2004) Microbial growth under the snow: implications for nutrient and allelochemical availability in temperate soils. Plant Soil 259:1–7
Schuman GE, Reeder JD, Manley JT, Hart RH, Manley WA (1999) Impact of grazing management on the carbon and nitrogen balance of a mixed-grass rangeland. Ecol Appl 9:65–71
Sebastià MT (2004) Role of topography and soils in grassland structuring at the landscape and community scales. Basic Appl Ecol 5:331–346
Shaw MR, Harte J (2001) Control of litter decomposition in subalpine meadow-sagebrush steppe ecotone under climate change. Ecol Appl 11:1206–1223
Six J, Bossuyt H, Degryze S, Denef K (2004) A history of research on the link between (micro) aggregates, soil biota, and soil organic matter dynamics. Soil Till Res 79:7–31
Smit A, Kooijman AM (2001) Impact of grazing on the input of organic matter and nutrients on the soil in a grass-encroached Scots pine forest. For Ecol Manag 142:99–107
Smith DL, Johnson L (2004) Vegetation-mediated changes in microclimate reduce soil respiration as woodlands expand into grasslands. Ecology 85:3348–3361
Smith P (2004) How long before a change in soil organic carbon can be detected? Glob Change Biol 10:1–6
Stöhr D (2007) Soils—heterogeneous at a Microscale. In: Wieser G, Tausz M (eds) Trees at the upper limits. Springer, Dordrecht, pp 37–56
Tasser E, Tappeiner U (2002) Impact of land-use changes on mountain vegetation. Appl Veg Sci 5:173–184
Tasser E, Walde J, Tappeiner U, Teutsch A, Noggler W (2007) Land-use changes and natural reforestation in the Eastern Central Alps. Agr Ecosyst Environ 118:115–129

Teepe R, Brumme R, Beese F (2001) Nitrous oxide emissions from soil during freezing and thawing periods. Soil Biol Biochem 33:1269–1275

Tierney GL, Fahey TJ, Groffman PM, Hardy JP, Fitzhugh RD, Driscoll CT (2001) Soil freezing alters root dynamics in a northern hardwood forest. Biogeochemistry 56:175–190

Waldrop MP, Firestone MK (2004) Microbial community utilization of recalcitrant and simple carbon compounds: impact of oak-woodland plant communities. Oecologia 138:275–284

Walker MD, Walker DA, Welker JM, Arft AM, Bardsley T, Brooks PD, Fahnestock JT, Jones MH, Losleben M, Parsons AN, Seastedt TR, Turner PL (1999) Long-term experimental manipulation of winter snow regime and summer temperature in arctic and alpine tundra. Hydrol Process 13:2315–23130

Wardle DA, Bardgett RD, Klironomos JN, Setälä H, van der Putten WH, Wall DH (2004) Ecological linkages between aboveground and belowground biota. Science 304:1629–1633

Wipf S, Sommerkorn M, Stutter MI, Jasper Wubs ER, van der Wal R (2015) Snow cover, freeze-thaw, and the retention of nutrients in an oceanic mountain ecosystem. Ecosphere 6, art. 207

Xie ZB, Zhu JG, Liu G, Cadisch G, Hasegawa T, Chen CM, Sun HF, Tang HY, Zheng Q (2007) Soil organic carbon stocks in China and changes from 1980s to 2000s. Glob Change Biol 13:1989–2007

Zhao W, Chen SP, Lin GH (2008) Compensatory growth responses to clipping defoliation in *Leymus chinensis* (Poaceae) under nutrient addition and water deficiency conditions. Plant Ecol 196:85–99

Zollinger B, Alewell C, Kneisel C, Meusburger K, Gärtner H, Brandová D, Ivy-Ochs S, Schmidt MWI, Egli M (2013) Effect of permafrost on the formation of soil organic carbon pools and their physical-chemical properties in the Eastern Swiss Alps. CATENA 110:70–85

Zornoza R, Guerrero C, Mataix-Solera J, Scow KM, Arcenegui V, Mataix- Beneyto J (2009) Changes in soil microbial community structure following the abandonment of agricultural terraces in mountainous areas of Eastern Spain. Appl Soil Ecol 42:315–323

Chapter 10
The Importance of Reintroducing Large Carnivores: The Brown Bear in the Pyrenees

Santiago Palazón

Abstract Large carnivores are keystone species in the ecosystems where they inhabit. Their loss may provoke an imbalance at several levels of the ecosystem. Conservation strategies for existing populations of large carnivores and restoration programmes of disappeared populations can help at maintaining the ecosystem balance and foster the perception links of humans with nature. The case of the restoration of the brown bear population in the Pyrenees during the last 20 years is a successful example of conservation measures carried out to assure the coexistence between this species and the local society, which economy is based on extensive livestock, beekeeping and tourism. In this chapter, I describe the role of large carnivores in mountain ecosystems and the context and development of the of the brown bear in the Catalan Pyrenees as an example of the challenges of large carnivore conservation in a rural context but a high influence from nearby urban areas.

Keywords Large Carnivores · Bear · Wolf · Population restauration · Mountains

10.1 Introduction

Large carnivores are characterised by their large size and for being apex predators, i.e. predators at the top of the food web (Edwards 2014). Populations of large carnivores are typically at low densities and with low reproductive potential partly because of their large spatial requirements and their vulnerability to habitat destruction (Rosenblatt et al. 2013; Noss et al. 1996). This feature makes populations of large carnivores especially vulnerable to catastrophic events or continued declines, from which they recover slowly (Edwards 2014). Doing research on large carnivores is also difficult because of the time scale, geographical scope and high costs involved (Estes 1996). Historically in Europe, they have competed for food

S. Palazón (✉)
Fauna and Flora Service, Ministry of Territory and Sustainability,
Government of Catalonia, Dr. Roux, 80, 08017 Barcelona, Spain
e-mail: santiago.palazon@gencat.cat

J. Catalan et al. (eds.), *High Mountain Conservation in a Changing World*,
Advances in Global Change Research 62, DOI 10.1007/978-3-319-55982-7_10

resources and space with humans and continue to do so. Some can be dangerous to humans and are considered fearsome. Human attitudes are an important component of large carnivore management and conservation (Piédallu et al. 2016). Attitudes towards large carnivores of humans co-inhabiting places with them are not always of acceptance.

There are five large carnivores in Europe: the brown bear (Fig. 10.1), the wolf, the European lynx, the Iberian lynx and the wolverine. The first three are distributed in different locations throughout Europe, whereas the last two live at the furthest distance from each other but still within the geographical limits of Europe: the Iberian lynx in the Southern Iberian Peninsula and the wolverine in Northern Scandinavia and Russia.

The conservation of these five large European carnivores is a priority for the European Union and Nature2000. In the words of the Large Carnivore Initiative for Europe's (LCIE; www.lcie.org) member John Linnell, "*People familiar with the presence of large carnivores know how to adapt to them, but because these large carnivores were gone for so long, we have forgotten how to share our living space with large, hairy and potentially dangerous animals. When large carnivores reappear, we have to adapt our way of taking care of livestock again*". Therefore, with the slogan "Working together across the European Union to conserve and manage large carnivores", LCIE works to conserve Europe's large carnivores through the protection afforded by the Habitat Directive, based on the premises of human safety, the protection of livestock with means and measures, and awareness campaigns for local human populations.

Fig. 10.1 Male and female brown bears together in May 2016, Pyrenees

Human populations have for long displaced large carnivores populations thought, e.g. hunting, urban development and habitat loss, being displaced to unmanaged and protected natural environments. However, the opposite is now happening in Europe: populations of large carnivores are increasing the occupied territory, including places with human populations are dense with almost 100 people per square kilometre (Chapron et al. 2014; Boitani et al. 2015). Except for the Iberian lynx, large European carnivores are making a comeback in Europe. One-third of the European territory is now home to at least one large carnivore. Compared to figures from the mid-twentieth century, current populations remain stable or are on the rise. For example, Europe now boasts a population of 18,000 bears and 12,000 wolves (Naves et al. 2015; Chapron et al. 2014). The recovery of these populations of large carnivores is partly due to the change in the social to more positive perceptions and attitudes towards large carnivores.

The processes regulating the population and community dynamics of the large carnivores include bottom-up forces such as primary production (soil, CO_2 and H_2O), nutrient dynamics (nitrogen, phosphorus, potassium and others) and energy cycles; top-down forces such as predation, risk effects and trophic cascades; non-predatory interactions, including facilitation and inter- and intra-specific competition for resources; and pulse and press disturbance and perturbation events such as fires, storms, wave action, floods and drought (Rosenblatt et al. 2013; Estes et al. 2001). Alternative stable states occur when perturbations of sufficient magnitude and direction push ecosystems from one basin of attraction to another (Estes et al. 2011). Trophic cascades are defined by Estes et al. (2011) and Paine (1980) as "*the propagation of impacts by consumers on their prey downward through food webs*" and by Edwards (2014) as "*a series of interactions at more than one trophic level where apex predators suppress mesopredators or prey, leading to an increase in number and/or diversity of primary producers*".

Large predators are keystone species for the ecosystems, i.e. they are at relatively low abundance and biomass, but have a large ecological effect on its community or the ecosystem (Power et al. 1996; Miller et al. 2001). The disappearance or reduction of these keystone species may lead to increases in the (now uncontrolled by top predation) mesopredators and preys and therefore a reduction of primary resources.

10.2 The Benefits of Large Predators

What role do large carnivores play in an ecosystem? This is not an easy question to answer. Apex predators ensure biodiversity. They are the engineers of the ecosystem because they can change its dynamics and increase habitat heterogeneity and biodiversity (Ritchie et al. 2012). They prevent the spread of diseases and invasive species. They also maintain the physical and chemical conditions of the soil and water (Estes et al. 2011). They can connect different, often distant, habitats and their behaviour affects biogeochemical and nutrient dynamics. How do large carnivores affect ecosystems? Mainly in two ways: through prey control and mesopredator control. Because of the effects apex predators have on an ecosystem, their presence benefits the ecosystem as a whole. To revisit the previous example, when populations of sea otters began to increase in the Aleutian archipelago, sea urchin numbers dropped, the kelp forest grew and the number of fish increased.

10.3 Impacts on Preys

Large carnivores control prey populations by direct and indirect means. Carnivores directly reduce the number of preys through predation (Terborgh 1988; Terborgh et al. 1997; Estes et al. 1998; Miller et al. 2001). Their impact on populations of prey species is a form of biological control. Apex predators limit the number of herbivores and reduce pressure on plants (first and third trophic levels). For example, the wolf (*Canis lupus*) and the lynx (*Lynx* sp.) prey on ungulate populations (Jedrzejewski et al. 1993; Okarma 1995; Valdmann et al. 2005). The brown bear preys on ungulates to a lesser degree (Vulla et al. 2009), but in Northern Europe, it preys heavily on the calves of reindeer (*Rangifer tarandus*) and moose (*Alces alces*) (Swenson et al. 2007) and in North America, the brown bear and black bear are major predators of neonatal ungulates (Zager and Beecham 2006). In Bialowieza, Poland, the wolf and lynx prey significantly on small ungulates, but less on wild boar (*Sus scrofa*), moose and bison (*Bison bison*) since their numbers are limited by food availability and weather conditions such as snow (Jedrzejewska and Jedrzejewski 2005). In Norwegian forests, roe deer (*Capreolus capreolus*) density declined predators were present in the habitat (Melis et al. 2009). The population went from 1485 roe deer individuals per 100 km^2 when there were no predators to 605 with one predator (European lynx or wolf) and 167 with both predators.

Carnivores' indirect control has effects on resources, preys behaviour and diseases. It causes prey species to alter their behaviour to reduce their vulnerability (McLaren and Peterson 1994; Fitzgibbon and Lazarus 1995; Schmitz 1998; Brown 1999; Miller et al. 2001). The prey chooses different habitats, food sources, group sizes and activity times, and reduces the amount of time used for feeding (Miller et al. 2001). If a predator selects from a wide range of prey species, the predator's presence may cause all prey species to reduce their respective niches, thus reducing competition among these species. Prey species compete for limited resources and decrease the diversity through competitive exclusion (Paine 1966; Terborgh et al. 1997). The control on preys also becomes a sanitary control because it prevents the proliferation of infectious diseases among prey populations (Miller et al. 2001). Predation risk by wolves affects the number and behaviour of moose on Isle Royale (McLaren and Peterson 1994), as well as elk (*Cervus elaphus*) behaviour in Yellowstone National Park (Ripple and Beschta 2006). This herbivore control, in turn, mainly affects the balsam fir forest by regulating seedling establishment, sapling recruitment, sapling growth rates, litter production in the forest and soil nutrient dynamics (Pastor et al. 1988; Post et al. 1999, 2000; Miller et al. 2001). In Shark Bay, Australia, a prey species of the tiger shark (*Galeocerdo cuvier*) altered its behaviour and habitat use in the presence of the shark to balance the risk of predation and its foraging (Heithaus et al. 2012). These behavioural changes cascade to the seagrass community by altering its biomass, structure, composition and nutrient dynamics (Burkholder et al. 2013).

10.4 Impact on Mesopredators

Besides impacting on preys, large predators also have a bearing on mesopredators (i.e., medium-sized predators) by decreasing their densities and reducing competition among both (Palomares and Caro 1999). In Norway, the European lynx preys on red foxes (Sunde et al. 1999), whereas in Australia the dingo (*Canis dingo*) preys on introduced red foxes and feral domestic cats (an invasive species in Australia), and has positive effects on native medium-sized marsupials (Letnic et al. 2012). Therefore, in Australia, introducing the dingo to reduce invasive mesopredators is better than using lethal control with poison (Letnic et al. 2012). Large predators also have an indirect impact by modifying the behaviour of other predators through the creation of "fear landscapes", i.e. areas of the territory that are avoided by mesopredators due to the presence of large predators (Sergio et al. 2008), as the use of trails and paths (Hayward and Marlow 2014).

However, large predators can also produce the opposite effect on mesopredators, by having a positive effect on them. Eagles, vultures, nocturnal raptors, foxes, martens, wild boars, raccoons, dogs and other small predators and omnivores take advantage of the increase in food left by large predators in the form of carrion (especially during heavy winters) (Fig. 10.2).

10.5 Others Impacts on Ecosystems

Individual specialisation of large top predator behaviours also affects ecosystems (Rosenblatt et al. 2013). Direct impact includes transporting nutrients between habitats (Holtgrieve and Schindler 2011). Predators can also act as vectors between different habitats, potentially affecting nutrient and biogeochemical dynamics through localised behaviours (Rosenblatt et al. 2013). Predators have an indirect impact by affecting nutrient transport between ecosystems by acting on prey species that transport nutrients (Croll et al. 2005; Maron et al. 2006; Rosenblatt et al. 2013). Large predators may contribute to the creation of heterogeneous nutrient patterns in ecosystems through their own nutrient recycling (nitrogen and phosphorus), thus inducing behavioural modifications in prey species that affect the distribution of nutrients (Schmitz et al. 2000). This interaction may enable large top predators to strongly influence the structure, composition and spatial patterns of local areas through their participation in bottom-up processes (Rosenblatt et al. 2013). This behaviour could create hotspots of foraging and nutrient recycling.

Fig. 10.2 Male brown bear born in the Pyrenees (March 2004)

Some examples of this can be found in terrestrial and marine ecosystems. By relocating salmon carcases from streams to riparian areas and through the consumption and excretion of salmon-derived nutrients, the Alaskan brown bear is responsible for up to 24% of riparian nitrogen budgets (Helfield and Naiman 2006). Arctic foxes in the Aleutian archipelago can reduce seabird-mediated nutrient inputs from the ocean to terrestrial areas, thus cause grassland habitats to shift to dwarf shrub/forb-dominated ecosystems (Croll et al. 2005; Maron et al. 2006). Bull sharks (*Carcharinus leucas*) move between the different habitats of the Shark River Estuary in Florida and link these habitats through trophic interactions (Matich et al. 2011). Such behaviours by these large mobile consumers may lead to transport of limiting nutrients from relatively nutrient-rich marine habitats to oligotrophic freshwater/estuarine habitats, which could affect community composition of primary producers (Rosenblatt et al. 2013). The reintroduction of wolves in Yellowstone Park has reduced the positive indirect effects of ungulates on soil nitrogen mineralisation and potentially the nitrogen supply for plant growth (Frank 2008). The introduction of rats and arctic foxes has reduced soil fertility and plant nutrition on high-latitude islands by disrupting seabirds and their sea-to-land nutrient subsidies, and this has had striking effects on plant community composition (Wardle et al. 2007). A linked process of consumption and excretion can recycle nutrients such as nitrogen and phosphorus to similar ratios of other sources of nutrients and contribute to heterogeneity in nutrient dynamics, thus increasing process diversity.

10.6 What Happens When Large Predators Disappear?

Since the Late Pleistocene, over-exploitation, hunting, agriculture, livestock, habitat loss and destruction have had a major impact on the populations of large carnivores living in terrestrial and marine habitats. As human civilisation has progressed, human density has massively increased, and the populations of large carnivores have decreased or been eliminated. As a result, large predators have been entirely removed from the system in many areas, or their numbers have been severely reduced (Weber and Rabinowitz 1996; Terborgh et al. 1999; Woodroffe 2000; Miller et al. 2001).

The loss of large carnivores has long-term effects on ecosystem stability (Miller et al. 2001). Because they have disappeared from large territories where they once lived, top-down processes previously regulated by them have been considerably altered (Rosenblatt et al. 2013). Trophic chains and cascades shorten, and animal and plant populations suffer alterations in their numbers and dynamics. Without top predators, populations of prey species increase; the increase of herbivores leads to greater pressure on plants, reducing vegetation cover and increasing the prevalence of diseases. Changes in animal and plant populations, also alter nutrient and biogeochemical cycles (soil and water) (Estes et al. 2011, Ritchie and Johnson 2009). In the absence of top predators, mesopredator populations increase, thus leading to so-called "mesopredator release" (Soulé et al. 1988) and a decline in populations of smaller prey species. Summarising, the loss of predators limits the ecosystem's ability to restore the process of top-down control.

Apex predators have disappeared from different ecosystems. The extirpation of large top predators in Northern Arizona in the early twentieth century led to the irruption (sudden and rapid increase) of the mule deer (*Odocoileus hemionus*) population, which caused over-browsing, a reduction in woody browse cover and eventually famine among the deer (Rosenblatt et al. 2013; Binkley et al. 2006). In Yellowstone National Park, the extirpation of wolves in the early twentieth century led to an increase in the elk (*Cervus elaphus*) population, a concomitant decrease in the recruitment of deciduous tree species and related effects on ecosystem structure and function (Ripple and Beschta 2012; Rosenblatt et al. 2013). In the Aleutian archipelago, the arctic fox (*Alopex* sp.) limits the number of sea birds, thus decreasing the amount of nutrients transported from sea to land and maintaining the herbaceous structure of the tundra (Croll et al. 2005). When sea otters (*Enhydra lutris*) were overexploited by the fur trade in the North Pacific, marine invertebrate herbivores increased in number, especially sea urchins (*Strongylocentrotus* spp.) and devastated kelp forests. This fact produced a cascade of indirect effects that reduced diversity in fish, shorebirds, invertebrates and raptors (Estes 1996; Estes and Duggins 1995; Miller et al. 2001; Estes et al. 2003). After 7 years of isolation in the recently formed Lago Guri reservoir in Venezuela, nearly 75% of the vertebrate species have disappeared from the scattered islands that are too small to hold the jaguar (*Panthera onca*) or the puma (*Puma concolor*). The few species that remain are hyper-abundant and have gross effects on the plant community. There is also little regeneration of the canopy trees (Terborgh et al. 1997).

10.7 The Situation of the Brown Bear, Wolf and European Lynx in the Pyrenees

The native brown bear population in the Pyrenees disappeared during the last 10 years, although the population located in the Central Pyrenees was wiped out earlier in the 1990s (Camarra and Parde 1990; Parellada et al. 1995; Alonso et al. 1993; Casanova 2005). The last female bear died in 2004, along with the oldest male, and the last male died in 2010 (DARP 2015). Currently, there is only one male born from a released Slovenian male and the last Pyrenees female. The current bear population (Table 10.1) was reintroduced from a set of individuals captured in the Balkans (Slovenia) and released in the Pyrenees in 1996–1997 (two females and one male) in 2006 (four females and one male) (Quenette et al. 2000) and 2016 (one male). In 2016, the bear population was divided into two areas: the Central Pyrenees subpopulation, with more than 35 bears identified, and the Atlantic Pyrenees subpopulation, with only two males identified.

The wolf was once distributed throughout Catalonia (NE Spain) from the Mediterranean coast to the Pyrenees. At the end of the eighteenth century, 200–250 wolves were being hunted per year (0.8 wolves per 100 km^2), but by the beginning of nineteenth century, only 10–50 wolves per year were killed (0.02 wolves per 100 km^2) (Ruiz-Olmo 1995). These data indicate the decline in the wolf population in Catalonia and the Pyrenees. Population declines and extinctions spread progressively from east to west and north to south (Ruiz-Olmo 1995). Two animals killed in 1924 and one in 1935 in the Tortosa Mountains (Southwest Catalonia) were the last wolves hunted in Catalonia (Ruiz-Olmo 1995; Aguilar-Amat 1924). The wolf had therefore disappeared from the Pyrenees by the beginning of twentieth century. However, thanks to the natural dispersal of individuals from two European populations (Italian and Cantabrian) the wolf has been living in the Pyrenees for the last 20 years. Individuals from remaining populations of wolves in northern Spain have sporadically dispersed into the western Pyrenees, but most individuals are immediately eliminated. Italian wolves can also sporadically cross the south-east France and reach the Eastern Pyrenees (France and Catalonia). France and Catalonia are now protecting this population (Table 10.2), though it is currently considered a group of lone individuals rather than a proper population.

Table 10.1 How many bears inhabit the Pyrenees Mountains?

Year	Minimum determinate number	Minimum revised number[a]
2010	19	20
2011	22	23
2012	22	24
2013	25	28
2014	31	31
2015	29	32
2016	39	-

[a]The revision of last year can change the bear number of former years

Table 10.2 How many wolves inhabit the Pyrenees Mountains?

Year	France	Catalonia	Eastern Pyrenees
2010	2	2	3
2011	2	0	2
2012	2	1	3
2013	2	1	2
2014	2–3?	1	2–3?
2015	2–3?	1	2–3?

The European lynx once lived in the Pyrenees (Beaufort 1965; Arribas 2004; Rodríguez-Varela et al. 2015), but probably disappeared in the second half of twentieth century (Ruiz-Olmo 2001). Today, there is no lynx in the Pyrenees, but a European lynx release pilot project is planned in Catalonian in coming years (Ruiz-Olmo et al. 2006; Magrama 2015a).

10.8 Problems Generated by the Presence of the Brown Bear and Wolf in Catalonia and the Pyrenees

The brown bear and wolf created problems in the Pyrenees through their attacks on livestock (sheep and goats) and beehives such as social and political conflict, fear among the locals and tourists, and their competition with hunters. The objectives of the Catalan government is: (1) to preserve large carnivore populations; (2) to conserve mountain livestock (sheep, goats, cows and horses) since the maintenance of the mosaic landscape (pastures and forests) has an important input in the economy of the mountain communities and (3) to protect mountain beekeeping activities (apiculture) because it is important for the economy of Catalan agricultural activity and because cross-pollination of mountain flora is essential for the ecosystem.

10.9 Damage to Livestock and Beehives

Brown bears prey on beehives and livestock (sheep and goats). They can sometimes chase away horses and cows and even wound them. Wolves prey on livestock (sheep, cows and calves, foals and shepherd dogs) and lynx also prey on livestock, mainly sheep. Between 1996 and 2015, 320 bear attacks in Catalonia caused 629 victims (369 dead sheep and goats and 249 damaged beehives) (Fig. 10.3). The annual average was 16 attacks (SD = 6.44) and 31.45 victims (SD = 18.57). The Catalonian government and other public administrations paid an annual average of €5990 (SD = 3045.04) in compensation. This figure only includes compensation for certified bear attacks, but not compensation for political or social reasons.

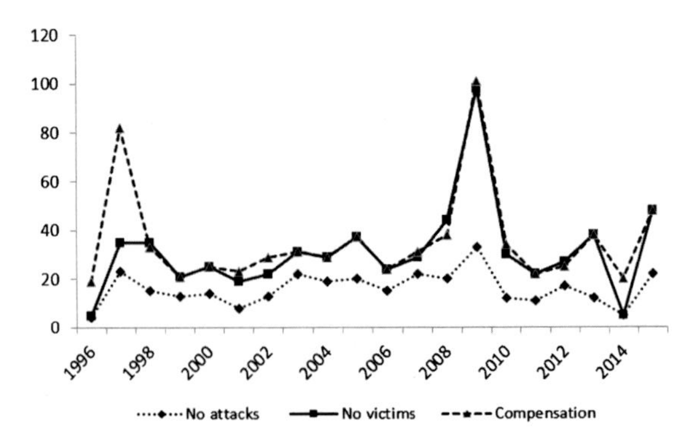

Fig. 10.3 Evolution by year of brown bear attacks (number of attacks and victims) in Catalonia: 1996–2015

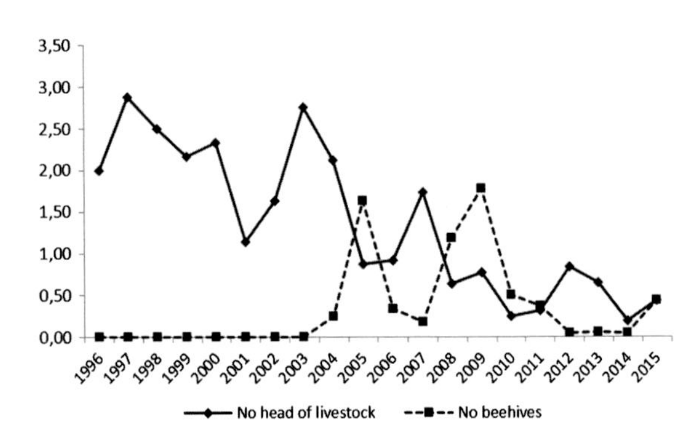

Fig. 10.4 Evolution of attacks (number of victims) on domestic livestock and beehives by one brown bear in Catalonia by year: 1996–2015

Over the years, the bear population has increased while damage has decreased or remained at similar levels. The damage caused by each bear has therefore decreased significantly (Fig. 10.4). In France, the number of attacks, victims and the cost is higher than in Catalonia (Bautista et al. 2016).

10.10 Protective Measures Implemented in Catalonia

The goal of the Catalan government is to reduce the damage caused by these animals effectively. Small flocks were first grouped into big ones (>1000 head), and shepherds and assistants were hired to manage them. Flocks and beehives were enclosed in electrified enclosures. Protection dogs were donated to livestock owners and shepherds. The living conditions of shepherds in the mountains were improved through the construction and remodelling of cabins and shelters. The Bear Support and Conservation Teams carry out monitoring of the brown bear population. They also help in managing livestock and apiculture and provide information to stockbreeders, shepherds, hunters, tourists and town councils. All these actions entailed considerable expenditure, as well as a great deal of time and human and material resources.

Fig. 10.5 A female bear with two cubs of the year in the Pyrenees of Catalonia (September 2016)

Although prevention is more expensive than paying compensation, it is better than paying for the damage caused by bears because it prevents social conflict. Keeping attacks and damage to a minimum is required to avoid social conflicts in regards to the presence of brown bears. Moreover, the measures applied for protection against the brown bear also set a positive precedent for the current reintroduction of the wolf and the possible future reintroduction of the lynx. The flock protection measures applied can be useful for brown bears, wolves, lynx and even feral dogs. In Catalonia, the Decree 176 of 31 July 2007 and Order AAM/147 of 8 May 2014 provide legislation on the payment of compensation for damage caused by protected species: the brown bear, wolf and vulture. The aim is to increase the density of brown bears (i.e. the number of animals in the geographic area occupied) and the area of distribution (Fig. 10.5); consequently, the number of attacks will be likely to increase. Therefore, protection measures must be implemented after the first attack to avoid more damages.

10.11 Encounters Between Humans and Large Carnivores

Another associated problem is the encounters between humans and large carnivores, most of them occurring with the brown bear. Due to the bear's size and power, attacks can occasionally cause severe injuries to people or even death.

However, in Europe, attacks are scarce (Naves et al. 2015): one hunter was killed in Sweden in 2004, one person in Finland in 2006 (Ordiz 2014) and three attacks with three injured people have occurred in the Cantabrian Mountains (N Spain) during the last 25 years. There are extreme cases of wolf attacks in Russia. In close encounters with bears, people were perceived as a danger by the bear. People have to avoid dangerous situations by applying simple measures when walk or work in a forest with bears. In the Pyrenees, several encounters have occurred in last 20 years between humans and bears. In most of them, the bear slowly ran away to get away from people after it was warned of their presence. These encounters have mainly happened with bear teams and photographers. In 1997, the female Melba encountered a French hunter to defend her 3-year-old cubs. She was shot to death. In 2004, the female Cannelle faced a French hunter to protect her 1-year-old cub (Cannellito) from hunting dogs. She was also shot to death. In 2004, when the male Papillon was found dead (due to natural causes), the subsequent autopsy revealed the presence of 50 pellets inside his body. Hunters have also shot the male bears Boutxy and Kouki in France several times. In 2007, the female Francka was hit and killed by a military van on a French highway. In 2008, the female Hvala attacked a hunter when she was running away from a wild boar being hunted with dogs; she was pregnant. One hunter, who was over age 70, suffered superficial wounds in his arm, a small bite in his boot and fell and broke his leg. The government paid a compensation after the trial.

10.12 Potential Economic Benefits Generated by the Presence of Large Carnivores

One of the benefits that could be produced by large carnivores in the Pyrenees is tourism. There are two forms of potential tourism involving large carnivores: hunting tourism and nature tourism or ecotourism. Hunting is only possible when the species boasts a large, self-sustainable population and when studies indicate that an annual culling will not affect long-term species population. Ecotourism can be implemented fully, but clear roles must be defined to ensure the bear population is affected as little as possible.

10.12.1 Hunting Tourism

According to Annex II of the Directive Habitats (EU), the brown bear is a protected species. According to Annexes IV and V, the wolf and the European lynx can be hunted in accordance with the laws of each European country. In Spain, wolf hunting is permitted only under certain circumstances. However, there is currently considerable controversy among hunters, stockbreeders, the public administration and ecologists about the number of wolves that can be hunted and the methods that can be used. When large predators prey on ungulates, the ungulate population density drops, fewer damages are caused to agriculture and conflicts between farmers and hunters are avoided. Moreover, ungulate populations are healthier, so the spread of diseases reduces. Finally, hunting trophies are larger and more attractive to hunters.

10.12.2 Nature Tourism

Nature tourism mainly consists of hiking, mountaineering and animal watching. These activities have increased in recent years in several European countries and are expected to offer many opportunities in the future through the creation of services, businesses and jobs. However, it is first necessary to change the local mindset. People need to take a positive approach to the presence of large carnivores and see them as an opportunity. It will be necessary to take advantage of the positive aspects and reduce the negative effects. In the United States, this mindset shift took place 20 years ago (Mech 1996). Nature tourism is also an opportunity for environmental education. In the Pyrenees, wildlife tourism started with the brown bear in France and now in Catalonia, and can be expanded to include the wolf and the lynx. There are already a few bear observation spots in the Pyrenees, but observers must stay away from the bears to avoid interferences with their behaviour. In the future, feeding spots can be built with supplementary food, as well as observation hides for watching, photo and filming. Nature business has already taken off in the Pyrenees and tourists can now visit the areas inhabited by bears. This activity makes necessary to draft regulations to protect the animals and the habitat. It is crucial to strike a balance between conservation and business. In nature tourism, excessive disturbance of the habitat should be avoided; predators must learn to adapt to additional food and the presence of humans (family bears can be dangerous, and the final solution may be to kill these bears). Also relevant, among other factors, is the alteration of wolf behaviour and interference from young wolves (howling imitations) (Magrama 2015b). Nature tourism has an impact on brown bear behaviour (Nevin et al. 2005), and contact with humans increases the risk of death of these large carnivores (Ordiz 2014).

Nature tourism is an alternative to hunting and hunters. Unlike hunting, where an animal can only be "sold" once, nature tourism enables an animal to be "sold" several times. Associated products can also be sold. Other locations on the Iberian Peninsula have an advantage over the Pyrenees. In the Culebra Mountains, nature tourism focuses on wolf watching because it is relatively easy to watch these animals. Nature tourism in the area has generated ten times more income than hunting tourism; thus, society will have to decide if both kinds of tourism are compatible. The Cantabrian and Pyrenees Mountains are comparable in many ways, and I believe that a modified Cantabrian model could be transferred to the Pyrenees.

10.13 Do the Pyrenees Have Enough Room and the Right Habitats for Large Predators?

Sainz de la Maza and Nunes (2001) studied the habitat preferences of the brown bear in Catalonia using a Geographic Information System (GIS) based on radio tracking data. The results show that the bear selected an altitude between 800 and 1800 m a.s.l., a north to north-east orientation, a slope of 30°–60°, beech and fir and humid oak forests, a distance from villages between 1 and 2.5 km, a distance from roads between 0.5 m and 2 km, and a distance from forest trails of between 0.1 and 1 km. The results showed that trophic and refuge interest areas accounted for 60–70% of the territory. Martin et al. (2012) compared habitat suitability between the Cantabrian and Pyrenees Mountains and found good transposition of the Cantabrian model to the Pyrenees. However, there was a lack of balance between food resources for bears (scarce at high altitudes) and human presence (higher at low altitudes) when comparing both areas. Ruiz-Olmo et al. (in press) compared habitat selection between native bears (disappeared) and the reintroduced bears and made a model of distribution and habitat quality. The brown bear population has increased by between 9 and 15% in recent years, which means that bears have adapted to a new habitat, with high quality. The fragmented habitat has large non-available areas. The reintroduced bear population uses the same habitat as the extinct native bears. The current situation is that the number of bears is low and the maximum capacity of the environment in regard to holding brown bears has not been reached.

10.14 Conclusion

Large carnivores play a significant role in the ecosystem. However, to defend the presence of large carnivores, coexistence between nature and society must be achieved by providing information, raising awareness and implementing effective measures. We must understand that the first response of stockbreeders, livestock owners and shepherds might be “no bears”, but it is important to keep the conversation open.

Stockbreeders have to change their mindset and the way they work. They must become true stockbreeders. Coexistence will only be possible if preventive measures are in place and attacks and damages are very scarce. These measures will avoid social unrest and keep the calm. We must apply common sense and go back to traditional methods (for instance, sheep keepers and protection dogs) focusing on today’s technology. As a result of many years of livestock breeding without predators, the habit of not protecting herds has become a common practice.

The last question is why should the brown bear, wolf and lynx be reintroduced into the Pyrenees? The answer has many different components. Axe predators are endangered and protected species in Catalonia, Spain and the European Union (legal component). They are evolutionary responses to adaptation to different environments (scientific component). They perform a crucial ecological function and are keystone species in the ecosystem (ecological component). They are umbrella species in the sense that the results of protecting the brown bear and wolf include habitat protection and conservation of smaller species and invertebrates (conservation component). They are unrepeatable emblematic species of the Pyrenees (aesthetic component). They can revitalise the Pyrenees economy enhancing the nature tourism (economic component). These species also have the right to live in the ecosystems where they have always lived. Their disappearance was caused by human pressure from hunting, poaching and poison, and we are collectively responsible for that (ethical component).

The current situation in the Pyrenees is very different than it was 50–100 years ago. The damage caused by large carnivores does not represent an economic disaster, and compensation has been paid to cover all actual, proven cases of damage. The presence of the brown bear and wolf has not limited traditional activities in the Pyrenees and is not expected to do so in the future. There is enough room in the Pyrenees for the local human population and self-sufficient large predator populations to live together. These answers confirm that conserving, protecting and promoting large carnivore populations are positive for ecosystems and society.

Bears are similar to humans. They are generalist omnivores, intelligent and boast binocular sight (Rockwell 1991). They are similar in height when they stand up. Wolves have many of the features of human hunter-harvesters. They are members of a pack, hunt cooperatively and defend their territory (López 1978). The intrinsic recognition of the value of wildlife is on the rise, included the importance of large predators.

Acknowledgements I would like to express my thanks to the institutions and public administrations in France, Andorra, Navarra, Aragon and Catalonia on the reintroduction of the brown bear in the Pyrenees. I have worked with them since 1996. I would also like to thank the institutions and public administrations on wolf monitoring in France and Catalonia, with which I have worked since 2000.

References

Aguilar-Amat JB (1924) Dades per un catàleg de mamífers de Catalunya. Trab Mus Cienc Nat Barcelona 7(4):3–51

Alonso M, Pando A, Toldrà L (1993) El oso pardo en Cataluña. In: Naves J, Palomero G (eds) El oso pardo en España. Colección Técnica, Icona. Madrid

Arribas O (2004) Fauna y paisaje de los Pirineos en la era glaciar. Lynx Edicions

Bautista C, Naves J, Revilla E, Fernández N, Albrecht J, Schart AK, Rigg R, Karamanlidis AA, Jerina K, Huber D, Palazón S, Kont R, Ciucci P, Groff C, Dutsov A, Seijas J, Quenette PY, Olszanska A, Shkvyria M, Adamec M, Ozolins J, Jonozovic M, Selva N (2016) Patterns and correlates of claims for brown bear damages on a continental scale. J Appl Ecology. doi:10.1111/1365-2664.12708

Beaufort F (1965) Lynx des Pyrénées, Felis. L. Lynx lynx. Mammalia 29:589–601

Binkley D, Moore MM, Romme WH, Brown PM (2006) Was Aldo Leopold right about the Kaibab deer herd? Ecosystems 9:227–241. doi:10.1007/s10021-005-0100-z

Boitani L, Alvarez F, Anders O, Andren H, Avanzinelli E, BalysV, Blanco JC, Breitenmoser U, Chapron G, Ciucci P, Dutsov A, Groff C, Huber D, Ionescu O, Knauer F, KojolaI, Kubala J, Kutal M, Linnell J, Majic A, MannilP, Manz R, Marucco F, Melovski D, Molinari A, Norberg H, Nowak S, Ozolins J, Palazon S, Potocnik H, Quenette PY, Reinhardt I, Rigg R, Selva N, Sergiel A, Shkvyria M, Swenson J, Trajce A, Von Arx M, Wolfl M, Wotschikowsky U, Zlatanova D (2015) Key actions for Large Carnivore populations in Europe. Institute of Applied Ecology (Rome, Italy). Report to DG Environment, European Commission, Bruxelles. Contract no. 07.0307/2013/654446/SER/B3

Brown JS (1999) Vigilance, patch use, and habitat selection: foraging under predation risk. Evol Ecol Res 49–71

Burkholder DA, Heithaus MR, Fourqurean JW, Wirsing A, Dill LM (2013) Patterns of top-down control in a sea grass ecosystem: could a roving apex predator induce a behavior-mediated trophic cascade? J Anim Ecol 82:1192–1202. doi:10.1111/1365-2656.12097

Camarra JJ, Parde JM (1990) The brown bear in France, status and management in 1985. Aquilo Ser Zool Tome 27:93–96

Casanova E (2005) L'ós del Pirineu. Crònica d'un extermini. Col·lecció Guímet. Pagès Editors, Lleida

Chapron G, Kaczensky P, Linnell JD, Von Arx M, Huber D, Andrén H et al (2014) Recovery of large carnivores in Europe's modern human-dominated landscapes. Science 346:1517–1519

Croll DA, Maron JL, Estes JA, Danner EM, Byrd GV (2005) Introduced predators transform subarctic islands from grassland to tundra. Science 307:1959–1961

DARP (2015) Status of the brown bear in Catalonia: 2014 (In Catalan). Non published Report

Edwards M (2014) A review of management problems arising from reintroductions of large carnivores. J Young Investig 27:11–16

Estes JA (1996) Carnivores and ecosystem management. Wildlife Soc Bul 24:390–396

Estes JA, Duggins DO (1995) Sea otters and kelp forests in Alaska: generality and variation in a community ecological paradigm. Ecol Monogr 65:75–100

Estes JA, Tinker MT, Williams TM, Doak DF (1998) Killer whale predation on sea otters linking oceanic and near shore ecosystems. Science 282:473–476

Estes JA, Crooks K, Holt R (2001) Predators, the ecological role of. In: Levin S (ed) Encyclopedia of Biodiversity, vol 4. Academic Press, San Diego, CA, pp 857–878

Estes JA, Riedman ML, Staedler MM, Tinker MY, Lyon BE (2003) Individual variation in prey selection by sea otters: Patterns, causes and implications. J Anim Ecol 72:144–155

Estes JA, Terborgh J, Brashares JS, Power ME, Berger J, Bond WJ, Carpenter SR, Essington TE, Holt RD, Jackson JBC et al (2011) Trophic downgrading of planet earth. Science 333:301–306

FitzGibbon CD, Lazarus J (1995) Antipredator behavior of Serengeti ungulates: Individual differences and population consequences. In: Sinclair ARE, Arcese P (eds) Serengeti II: dynamics, management, and conservation of an ecosystem. University of ChicagoPress, Chicago IL, pp 274–296

Frank DA (2008) Evidence for top predator control of a grazing ecosystem. Oikos 117:1718–1724

Hayward MW, Marlow N (2014) Will dingoes really conserve wildlife and can our methods tell? J Appl Ecol 51:835–838

Heithaus M, Wirsing AJ, Dill LM (2012) The ecological importance of intact top predator populations: a synthesis of 15 years of research in a sea grass ecosystem. Marine Freshw Res 63:1039–1050

Helfield J, Naiman RJ (2006) Keystone interactions: Salmon and bear in riparian forests of Alaska. Ecosystems 9:167–180

Holtgrieve G, Schindler DE (2011) Marine derived nutrients, bioturbation, and ecosystem metabolism: reconsidering the role of salmon in streams. Ecology 92:373–385

Jedrzejewska B, Jedrzejewski W (2005) Large carnivores and ungulates in European temperate forest ecosystems: bottom-up and top-down control. In: Ray JC, Redford KH, Steneck RS, Berger J (eds) Large carnivores and the conservation of biodiversity. Island Press, Washington, DC, pp 230–246

Jedrzejewski W, Schmidt K, Milkowski L, Jedrzejewska B, Okarma H (1993) Foraging by lynx and its role in ungulate mortality: the local (Białowieża Forest) and the Palaearctic viewpoints. Acta Theriol 38:385–403

Letnic M, Ritchie EG, Dickman CR (2012) Top predators as biodiversity regulators: the dingo *Canis lupus dingo* as a case study. Biol Rev 87:390–413

López BH (1978) Of Wolves And Men. Charles Scribner's and Sons, New York, New York, USA

Magrama (2015a) Estudio de viabilidad de la reintroducción de lince europeo *Lynx lynx* en los Pirineos Centrales. Unpublished document

Magrama (2015b) Buenas prácticas para la observación de oso, lobo y lince en España. Ministerio de Agricultura, Alimentación y Medio Ambiente. Secretaría General Técnica. Centro de Publicaciones. Madrid

Maron J, Estes JA, Croll DA, Danner EM, Elmendorf SC, Buckelew SL (2006) An introduced predator alters Aleutian island plant communities by thwarting nutrient subsidies. Ecol Monogr 76:3–24

Martin J, Revilla E, Quenette PY, Naves J, Allaine D, Swenson E (2012) Brown bear habitat suitability in the Pyrenees: transferability across sites and linking scales to make the most of scarce data. J Appl Ecol 49:621–631

Matich P, Heithaus MR, Layman CA (2011) Contrasting patterns of individual specialization and trophic coupling in two marine apex predators. J Anim Ecol 80:294–305

McLaren BE, Peterson RO (1994) Wolves, moose, and tree rings on Isle Royale. Science 266:5190

Mech LD (1996) A new era for carnivore conservation. Wildlife Soc Bull 24:397–401

Melis C et al (2009) Predation has a greater impact in less productive environments: variation in roe deer, *Capreolus capreolus*, population density across Europe. Glob Ecol Biogeogr 18:724–734

Miller B, Dugelby B, Fopreman D, Martínez del Río C, Noss R, Phillips M, Reading R, Soulé ME, Terborgh J, Willcox L (2001) The importance of large carnivores to healthy ecosystems. Endanger Spec Updat 18:202–210

Naves et al (2015) Brown bear attacks on humans in Europe: an overview for the period 2000–2015

Nevin OT, Gilbert BK (2005) Perceived risk, displacement and refuging in brown bears: positive impacts of ecotourism? Biol Conser 121:611–622

Noss RF, Quigley HB, Hornocker MG, Merrill T, Paquet PC (1996) Conservation biology and carnivore conservation in the rocky mountains. Conser Biol 10:949–963

Okarma H (1995) The trophic ecology of wolves and their predatory role in ungulate communites of forest ecosystems in Europe. Acta Theriol 40:335–386

Ordiz A (2014) Altera el turismo de naturaleza el comportamiento de los grandes carnívoros? Quercus 341:14–21

Paine R (1966) Food web complexity and species diversity. Am Nat 100:65–75

Paine RT (1980) Food webs: linkage, interaction strength and community infrastructure. J Anim Ecol 49:666–685

Palomares F, Caro TM (1999) Interspecific killing amongmammalian carnivores. Am Nat 153:492–508

Parellada X, Alonso M, Toldrà L (1995) Ós bru. In: Ruiz-Olmo J, Aguilar A (eds) Atlas dels Grans Mamífers de Catalunya. Lynx Ed, pp 124–129

Pastor J, Naiman RJ, Dewey B (1988) Moose, microbes and boreal forests. Bioscience 38:770–777

Piédallu B, Quenette PY, Mounet C, Lescureux N, Borelli-Massines M, Dubarry E, Camarra JJ, Gimenez O (2016) Spatial variation in public attitudes towards brown bears in the French Pyrenees. Biol Conser 197:90–97

Post E, Peterson RO, Stenseth NC, McLaren BE (1999) Ecosystem consequences of wolf behavioural response to climate. Nature 401:905–907

Post DM, Pace ML, Hairston NG (2000) Ecosystem size determines food-chain length in lakes. Nature 405:1047–1049

Power ME, Tilman D, Estes JA, Menge BA, Bond WJ, Mills LS, Daily G, Castilla JC, Lubchenco J, Paine RT (1996) Challenges in the quest for keystones. Bioscience 46:609–620
Quenette PY, Alonso M, Chayron L, Cluzel P, Dubarry E, Dubreuil D, Palazón S, Pomarol M (2000) Preliminary results of the first transplantation of brown bears in the French Pyrenees. Ursus 12:115–120
Ripple WJ, Beschta RL (2006) Linking wolves to willows via risk-sensitive foraging by ungulates in the northern Yellowstone ecosystem. For Ecol Manage 230(1–3):96–106. doi:10.1016/j.foreco.2006.04.023
Ripple WJ, Beschta RL (2012) Trophic cascades in Yellowstone: the first 15 years after wolf reintroduction. Biol Conser 145:205–213
Ritchie EG, Johnson CN (2009) Predator interactions, mesopredator release and biodiversity conservation. Ecol Lett 12:982–998
Ritchie EG, Elmhagen B, Glen AS, Letnic M, Ludwig G, McDonald RA (2012) Ecosystem restoration with teeth: what role for predators? Trends Ecol Evol 27:265–271
Rockwell D (1991) Giving voice to the bear, North American Indian myths, rituals and images of the bear. Roberts Rinehart Publishers, Niwot, Colorado, USA
Rodríguez-Varela R, Ureña I, García N, Cregut-Bonmoure E, Mannino MA, Arsuaga JL, Valdiosera C (2015) Ancient DNA evidence of Iberian lynx palaeoendemism. Quat Sci Rev 112:172–180
Rosenblatt AE, Heithaus MR, Mather ME, Matich P, Nifong JC, Ripple WJ, Silliman BR (2013) The roles of large top predators in coastal ecosystems: new insights from long term ecological research. Oceanography 26:156–167
Ruiz-Olmo J (1995) Llop. In: Ruiz-Olmo J, Aguilar A (eds) Atlas dels Grans Mamífers de Catalunya. Lynx Ed, pp 136–139
Ruiz-Olmo J (2001) El misterioso lince de los Pirineos: muchos indicios pero sin evidencias en los últimos cincuenta años. Quercus 182:12–19
Ruiz-Olmo J, Pomarol M, Macias M, Palazón S, Batet T, Lampreave G, López-Martín JM, Such-Sanz À, Camps D, Mañas F, Parellada X, RodríguezI, Alàs J (2006) Informe tècnic per a la valoració de la reintroducció del linx europeo *Lynx lynx* a Catalunya. Generalitat de Catalunya-Forestal Catalana
Ruiz-Olmo J, Palazón S, Camps D, Quenette PY, Batet A, Antona X, Pla M, Villero D, Decaluwe F, Brotons L, Melero Y (in press) Do reintroduced bears (*Ursus arctos*) present the same habitat patterns as extinct native bears in the Pyrenees? Modeling distribution and habitat quality
Sainz de la Maza P, Nunes J (2001) Los sistemas de información geográfica aplicados al estudio, la gestión y la conservación del hábitat del oso pardo (*Ursus arctos*) en un área del Pirineo catalán. Treballs de la Societat Catalana de Geografia 52:245–264
Schmitz OJ (1998) Direct and indirect effects of predation and predation risk in old-field interaction webs. Am Nat 151:327–340
Schmitz OJ, Hamback PA, Beckerman AP (2000) Trophic cascades in terrestrial systems: a review of the effects of carnivore removal on plants. Am Nat 155:141–153
Sergio F, Caro T, Brown D, Clucas B, Hunter J, Ketchum J, McHugh K, Hiraldo F (2008) Top predators as conservation tools: ecological rationale, assumptions and efficacy. Annu Rev Ecol Evol Syst 39:1–19
Soulé ME, Bolger ET, Alberts AC, Wright J, Sorice M, Hill S (1988) Reconstructed dynamics of rapid extinctions of chaparral-requiring birds in urban habitat islands. Conserv Biol 2:75–92
Sunde P, Overskaug K, Kvam T (1999) Intraguild predation of lynxes on foxes: evidence of interference competition? Ecography 22:521–523
Swenson JE, Dahle B, Busk H, Opseth O, Johansen T, Söderberg A, Wallin K, Cederlund G (2007) Predation on moose calves by European Brown bears. J Wild Manag 71:1993–1997
Terborgh J (1988) The big things that run the world—a sequel to E. O. Wilson. Conser Biol 2:402–403
Terborgh J, Lopez L, Tello J, Yu D, Bruni AR (1997) Transitory states in relaxing land bridge islands. In: Laurance WF, Bierregaard RO Jr (eds) Tropical forest remnants: ecology,

management, and conservation of fragmented communities. University of Chicago Press, Chicago IL, pp 256–274
Terborgh J, Estes JA, Paquet P, Ralls K, Boyd D, Miller B, Noss R (1999) Role of top carnivores in regulating terrestrial ecosystems. In: Soulé M, Terborgh J (eds) Continental conservation: scientific foundations of regional reserve networks. Island Press, Covelo, CA, pp 39–64
Valdmann H, Andersone-Lilley Z, Koppa O, Ozolins J, Bagrade G (2005) Winter diets of wolf Canis lupus and lynx Lynx lynx in Estonia and Latvia. Acta Theriol 50:521–527
Vulla E, Hobson KA, Korsten M, Leht M, Martin AJ, Lind A, Männil P, Valdmann H, Saarma U (2009) Carnivory is positively correlated with latitude among omnivorous mammals: evidence from brown bears, badgers and pine martens. Ann Zool Fen 46:395–415
Wardle DA, Bellingham PJ, Fukami T, Mulder CPH (2007) Promotion of ecosystem carbon sequestration by invasive predators. Biol Lett 3:479–482
Weber W, Rabinowitz A (1996) A global perspective on large carnivore conservation. Conser Biol 10:1046–1055
Woodroffe R (2000) Predators and people: using human densities to interpret declines of large carnivores. Animal Conser 3:165–173
Zager P, Beecham J (2006) The role of American black bears and brown bears as predators on ungulates in North America. Ursus 17:95–108

Part IV
Global Change and High Mountain Conservation

Chapter 11
Life-History Responses to the Altitudinal Gradient

Paola Laiolo and José Ramón Obeso

Abstract We review life-history variation along elevation in animals and plants and illustrate its drivers, mechanisms and constraints. Elevation shapes life histories into suites of correlated traits that are often remarkably convergent among organisms facing the same environmental challenges. Much of the variation observed along elevation is the result of direct physiological sensitivity to temperature and nutrient supply. As a general rule, alpine populations adopt 'slow' life cycles, involving long lifespan, delayed maturity, slow reproductive rates and strong inversions in parental care to enhance the chance of recruitment. Exceptions in both animals and plants are often rooted in evolutionary legacies (e.g. constraints to prolonging cycles in obligatory univoltine taxa) or biogeographic history (e.g. location near trailing or leading edges). Predicting evolutionary trajectories into the future must take into account genetic variability, gene flow and selection strength, which define the potential for local adaptation, as well as the rate of anthropogenic environmental change and species' idiosyncratic reaction norms. Shifts up and down elevation in the past helped maintain genetic differentiation in alpine populations, with slow life cycles contributing to the accumulation of genetic diversity during upward migrations. Gene flow is facilitated by the proximity of neighbouring populations, and global warming is likely to move fast genotypes upwards and reduce some of those constraints dominating alpine life. Demographic buffering or compensation may protect local alpine populations against trends in environmental conditions, but such mechanisms may not last indefinitely if evolutionary trajectories cannot keep pace with rapid changes.

Keywords Bet hedging · Centre-periphery hypothesis · Local adaptation · Ontogeny · Phenotypic plasticity · Reproductive allocation · Slow-fast life-history continuum · Survival

---

P. Laiolo (✉) · J.R. Obeso
Research Unit of Biodiversity, Spanish National Research Council,
Principado de Asturias, Oviedo University, Campus de Mieres, 33600 Mieres, Spain
e-mail: paola.laiolo@csic.es

J. Catalan et al. (eds.), *High Mountain Conservation in a Changing World*,
Advances in Global Change Research 62, DOI 10.1007/978-3-319-55982-7_11

11.1 Introduction

Mountain environments present a challenge to any living organism, and elevation gradients, with their sharp physical and ecological transitions, have been a favourite scenario for approaching general questions about adaptive change in life histories. Growth, development, maturation, reproduction and survival patterns of organisms are remarkably diversified along elevation, but also tend to converge in similar environmental conditions. Literature on the subject is abundant, but also widely dispersed and poorly integrated with respect to the plant and animal realms, and also between endotherms (animals that primarily produce their own heat) and ectotherms (in which body temperature tends to match environmental temperature, or requires behavioural thermoregulation). Undeniably, focal organisms differ in substantial ways, but the similarities in scope, objective, and often in findings among studies along elevation gradients provide an opportunity for a synergic appraisal with insights from both vascular plants and animals. Here, we first highlight those abiotic and biotic factors that are major determinants of life-history variation in elevation. We then illustrate the primary mechanisms of evolutionary adaptation to such variation, involving environmental and genotypic variation, their interaction and covariation. We discuss the limitations to environmental fit within and among species, like those imposed by intrinsic or evolutionary constraints. We review published literature on elevation patterns in growth, reproduction and survival in populations and species of plants and animals, and discuss the elevation life-history continuum in these taxonomic contexts. Our intention is neither to provide an extensive survey of literature on specific taxonomic groups, which in some cases are adequately covered (e.g. insects and birds: Hodkinson 2005; Hille and Cooper 2015; Boyle et al. 2015), nor to perform a meta-analysis given the literature bias towards specific taxa. Rather, we aim to provide a broad and synthetic appraisal of life-history variation along elevation. We discuss whether the climatic and ecological shifts occurring along elevation select some strategy within the life-history continuum and whether responses are, to a certain degree, comparable within and between unitary animals and modular plants. In spite of obvious connections with life histories, we intentionally omit to canter on seasonality of plant and animal life-cycle events, i.e. phenology per se, as this would require different spatial and temporal perspectives from those adopted here.

11.2 Environmental Variation in Elevation

11.2.1 Temperature

Among the abiotic factors that vary among mountainsides, temperature is the most critical for both plants and animals given the magnitude of variation it displays, and its profound effect on biochemical or physiological processes (Sibly and Calow 1986).

It decreases at a rate of 0.54–0.65 °C per 100 m of ascent, but significant variation is introduced by meteorology, local topography and height above the ground (Barry and Chorley 1987; Körner 2003; Rolland 2003). At cold temperatures, physical and chemical reactions slow down, as do the assimilation of energy and metabolic activity (Schmidt-Nielsen 1997). This process has the most remarkable consequences for plants and ectotherms, and either leads to reduced activity or triggers costly homeostatic responses to offset the passive reaction to reduced temperature and freezing (Sakai and Larker 1987; Gillooly et al. 2001). Endotherms may be affected in a similar way when they are outside their thermoneutral zone (Angilletta et al. 2010), and lower critical temperatures are reached more frequently at high elevations. Dormancy (including diapause and hibernation) is a mechanism to escape cold weather or resourcc shortage over the winter and corresponds to a period when growth, development and physical activity are temporarily arrested. Through its effect on metabolism, temperature limits the rates of production throughout ontogeny and reproduction, thus directly influencing phenology, growth and reproductive patterns, and life-history correlates such as body size (Atkinson 1996; Atkinson and Sibly 1996; Angilletta et al. 2004). Temperature is also tightly associated with seasonality, and in turn, with productivity, and thus also controls the above processes indirectly via these variables.

11.2.2 Atmospheric Pressure

Atmospheric pressure, the moisture content of air and the partial pressures of biologically relevant gases, such as oxygen and carbon dioxide, decrease relatively uniformly with increasing elevation, and impact gas exchanges in plants and respiration in animals. Pressure effects on photosynthesis are, however, smaller than predicted from the decline in the ambient pCO_2 alone because the increased rate of molecular diffusivity, induced by thinner air, is counteracted by the descent of atmospheric temperature, decelerating diffusion. Moreover, up to 80% of the total CO_2 transfer resistance between air and the chloroplasts is in the liquid phase, which is not influenced by pressure (Körner 2007). An improvement in the rate of oxygen intake with elevation has been observed in animals through ventilation or changes in blood composition (Rourke 2000), as well as increased porosity of bird eggshells to facilitate O_2 diffusion to the embryo (Körner 2007).

11.2.3 Precipitation

Unlike temperature and atmospheric pressure, precipitation exhibits non-linear relationships with elevation, and regional, rather than global, patterns. Precipitation tends to increase with altitude at low elevations, but exhibits no pattern at very high elevations, or declines above the cloud zone in tropical areas (Nagy and Grabherr 2009). Orographic precipitation in the form of snow typically characterises high elevations and dictates, together with seasonal changes in temperature and photoperiod, the duration of the growing and/or breeding seasons of alpine organisms, apart from providing thermally protective snow blankets (Hodkinson 2005). Since evapotranspiration declines with elevation, the balance for plant water supplies is rarely critical. However, in periods of high evaporative demand or prolonged soil moisture depletion, the need to reduce transpiration may affect the rates of CO_2 photosynthetic uptake and nutrient assimilation (Schulze and Chapin 1987). This trade-off has direct consequences for alpine plant recruitment, survival and growth patterns, and ultimately shapes their phenology and morphology (Körner 2003), as well those of the fauna they host (Hodkinson 2005).

11.2.4 Primary Productivity

Mountain environments display sharp gradients in soil fertility, largely dependent on climate, bedrock, soil structure and age, micro-topography and soil fertilisation by primary consumers. At high elevations, low temperatures (or a short duration of mild temperatures) negatively affect soil enzymatic activities, the rate of nutrient mineralisation and turnover, thus reducing nutrient availability for primary producers with consequences for upper trophic levels (Laiolo et al. 2015a). In some mountain areas, shorter grazing seasons and reduced nutrient inputs from herbivores also contribute to lower productivity, and the permanence of nutrients shortens in steep and shallow soils (Mariotti et al. 1980; Huber et al. 2007). Alpine plants may respond by reducing their size and enhancing mineral nutrient concentration (Körner 2003), although there are exceptions (Laiolo et al. 2015a). It is also worth mentioning that soil nutrient concentration does not always reflect availability for organisms, as nutrients may be in a form that cannot be absorbed. Nutrient availability unquestionably affects the overall performance of plants up to higher trophic levels and plays a key role in life-history evolution (Stearns 1992).

11.2.5 Biotic Interactions

The above factors, together with land area and history, are evoked as the primary drivers of the decline in species richness observed in many taxa on mountains, as well as of the decrease of biotic interactions (McCain and Grytnes 2010). Under the Stress Gradient Hypothesis, negative interactions should decline with environmental harshness while positive ones should increase (Brooker 2006). Negative trends of competition, predation and parasitism with elevation have indeed been documented in a large number of studies (Hodkinson 2005; Boyle 2008; Meléndez et al. 2014), but some positive relationships have as well (e.g. Abbate and Antonovics 2014).

Greater agreement exists on the enhancement of plant facilitation with elevation, documented in floral communities across the globe (Callaway et al. 2002). Together with productivity, variation in extrinsic mortality related to interactions is a major driver of organisms' life strategies, influencing patterns of growth, development, age and size at maturity, allocation in self-maintenance and parental care, as detailed further in the text.

11.3 The Process of Life-History Evolution

11.3.1 Mechanisms

Life-history variation among mountain populations can be explained by both phenotypic plasticity and local adaptation. The former, by means of which organisms with the same genetic constitution adjust their development to the current conditions, is generally considered poorly efficient for coping with extreme environments (Grime 1974; DeWitt et al. 1998). Some studies have shown that plastic genotypes bear a cost of low performance in unfavourable alpine habitats when compared to locally adapted genotypes (Emery et al. 1994; Stöcklin et al. 2009; Fischer and Karl 2010). However, purely plastic life-history responses to changes in resource availability among elevations have been described (Dobson and Murie 1987; Blanckenhorn 1997; Sears and Angilletta 2003; Yeh and Price 2004).

More frequently, selection for stress tolerance induces ecotypic differentiation in the form of local adaptations, exemplified by significant non-additive gene–environment interactions and populations that show genetic differences, and performances, corresponding to the conditions met along the gradient (Törang et al. 2015; Muir et al. 2014). Since environmental variation occurs at small spatial scales across elevations, strong local selection and limited gene flow are required to promote local adaptation sensu *strictu* (i.e. demonstrated by comparing performances after reciprocal transplants; Kawecki and Ebert 2004), and population size may also matter (Leimu and Fischer 2008). Local adaptation also results in gene–environment covariation and non-random distributions of genotypes along the gradient, an evolutionary pathway fairly well documented in mountain species. If, for instance, slow-growing genotypes are favoured at high elevations, where low temperatures also act to slow organisms' growth rates, then genetic and environmental influences on phenotypic expression covary positively. This process of co-gradient selection explains size reduction in alpine plants, with genotypes for small size found primarily on uplands where the environment also hampers somatic growth (Aarssen and Clauss 1992; Byars et al. 2007). The same environmental context may however select for genetically rapid growth and development to compensate for environmental conditions that slow down these processes (Conover and Schultz 1995), a countergradient pattern often adopted by ectotherms in cool environments.

Across populations or species, life histories can be interpreted as the result of the optimisation of individual phenotypes, or the development of evolutionarily stable strategies, with respect to the environment (Stearns 1992). This process should lead to similar elevation trends within and among species. The greatest variability among species in organismal design, evolutionary history and ecological niche enhances the opportunity for differentiation as compared to variation within species, but these factors also constrain environmental fit proportionally more among species, as detailed below.

11.3.2 Constraints

Body size is one of the most crucial intrinsic constraints to differentiation in life histories and should always be accounted for when analysing the fit of life histories to the environment. It engenders a continuum with, at one extreme, large species that grow slowly, mature late, and live long, and at the other extreme, small species adopting the opposite strategy (Sibly and Brown 2007). Biological rates, such as growth rates or reproductive biomass production, tend to obey simple allometric scaling laws regardless of the living conditions or taxonomic group (Enquist et al. 1999; Laiolo et al. 2015b). Trade-offs in resource allocation among competing functions foster a second dimension of correlated traits, also known as the 'slow-fast life-history continuum', rooted in the concept of *r*/*K*-selection. The cost of reproduction, i.e. the reduction in future reproduction resulting from current investments in reproduction, is the most prominent trade-off, describing the constraining relationships between growth, survival and reproduction (Reznick 1985; Obeso 2002). Great allocation in reproduction is associated with fast and short lives even when the effect of body mass or environment is controlled for, a correlation also defined as the 'pace-of-life' syndrome in comparative animal studies (Ricklefs and Wikelski 2002). In plants, vegetative growth is strongly hampered by investment in sexual reproduction, and is a crucial component of this continuum. Within species, this trade-off lies beneath the population process of 'demographic compensation', or negative correlations of fitness components across environmental gradients and towards species' range margins (Villellas et al. 2015). At margins, vital rates tend to decline because of poorer conditions of the environment ('centre-periphery' hypothesis: Lawton 1993; Vucetich and Waite 2003; Angert and Schemske 2005).

Evolutionary history, relatively unimportant at the intraspecific level, dictates the options available to selection, with traits of more closely related species responding more similarly to environmental factors (Harvey and Clutton-Brock 1985).

11.3.3 Drivers

Beyond these constraints, natural selection shapes life histories into suites of correlated traits, often remarkably convergent among alpine organisms facing the same environmental challenges. Life-history theory predicts that components of the environment, such as resources, predation, herbivory, competition, disease or physical stresses, favour different combinations of life histories and yield general patterns in their variation. Altogether, these factors can be grouped in two major extrinsic drivers of variation, associated broadly with resource availability and disturbance, respectively. Abundant resources, such as food and light, and competitive environments foster fast growth, short life cycles and high levels of reproduction, thus a 'fast' strategy (Clark and Clark 1992; Ghalambor and Martin 2001). This 'fast' strategy is also favoured when a disturbance, for instance, disease and predation, increases juvenile extrinsic mortality (Franco and Silvertown 1996). The investment in parental care reflects responses to perceived risks and environmental stresses. High juvenile (extrinsic) mortality or reduced recruitment may yield greater allocation in offspring quality instead of quantity to enhance (intrinsic) juvenile survival and recruitment (Clutton-Brock 1991; Armstrong and Westoby 1993). Life-history theory also predicts that where environmental conditions are not constant across years, individuals should favour a bet-hedging strategy. This strategy involves a reduction in annual breeding performance to reduce the probability of investing too much in reproduction during poor years, and an increase in self-maintenance so that reproduction can be attempted over multiple years (Stearns 1976).

As we detailed above, mortality risk (especially for juveniles) associated with disease, competition and predation tends to decline, while those associated with abiotic stress or a paucity of resources increases in alpine environments. These factors are expected to tilt the life-history continuum towards slower life cycles and enhanced offspring quality vs. quantity, thus a 'slow' strategy. Moreover, the environmental variability of alpine regions, consisting both of predictable (seasonality) and less predictable components (e.g. between-year variability), should also favour the evolution of bet-hedging strategies and longer lifespan. This combination of longevity and limited reproductive effort reduces the deleterious effects of environmental stochasticity on population growth and persistence.

In the next section, we present an assessment of the above predictions, reviewing literature that measured responses of life histories along elevation clines in major taxonomic groups, as summarised in Table 11.1.

Table 11.1 Summary of the main life-history characteristics of upland populations and species of different taxonomic groups

Taxonomic group	Responses in uplands (or at the cold extreme of the gradient)	References
Insects and other arthropods	Prolonged generation time	Hodkinson (2005), Schmoller (1970)
	Short larval stage	Tanaka and Brookes (1983), Zettel (2000)
	Fast growth	Dingle et al. (1990), Berner et al. (2004), Laiolo and Obeso (2015), Chown and Klok (2003)
	Small adult size	Laiolo et al. (2013)
	Low fecundity	Reviewed by Hönek (1993)
	Fewer but larger eggs	Fischer et al. (2003), Hayashi and Hamano (1984), Mashiko (1990), Hancock et al. (1998), Wilhelm and Schindler (2000)
	Parthenogenesis	Wachter et al. (2012)
Fishes	Increased longevity	Pauly (1980), Beverton (1987)
	Fewer but larger eggs	Sternberg and Kennard (2013)
	Egg guarding	Sternberg and Kennard (2013)
Amphibians	Increased longevity	Zhang and Lu (2012)
	Larger eggs and reduced offspring size	Liao et al. (2014)
	Fast growth in common garden experiments	Berven et al. (1979), Berven (1982)
Reptiles	Increased adult survival	Adolph and Porter (1993)
	Viviparity	Braña et al. (1991), Tinkle and Gibbons (1977)
	Reduced eggshell thickness	Mathies and Andrews (1995)
	Increased offspring size (oviparous species)	Sinervo et al. (1992)
	Infrequent reproduction	Shine (2005)
Birds	Reduced fecundity	Badyaev (1997a), Badyaev and Ghalambor (2001), Boyle et al. (2015), Laiolo et al. (2015b)
	Prolonged parental care	Badyaev (1997a), Badyaev and Ghalambor (2001), Boyle et al. (2015)
	Shift from sexual to parental behaviour in males	Badyaev (1997b), Snell-Rood and Badyaev (2008), Apfelbeck and Goymann (2011), Bastianelli et al. (2015)
Mammals	Longer lifespan of adults, older age at reproduction	Bronson (1979), Zammuto and Millar (1985), Yoccoz and Ims (1999)
	Lower litter size	Dunmire (1960), Fleming and Rauscher (1978), Smith and McGinnis (1968)
	Increased parental care	Festa-Bianchet et al. (1994)

(continued)

Table 11.1 (continued)

Taxonomic group	Responses in uplands (or at the cold extreme of the gradient)	References
Vascular plants	More investment in maintenance and less in reproduction	Jónsdóttir (2011)
	Clonal growth and vegetative reproduction	Stöcklin (1992), Klimes et al. (1997)
	Iteroparity; high adult survival	Bliss (1971), Hautier et al. (2009), Milla et al. (2009), García and Zamora (2003), Arx et al. (2006), Kim and Donohue (2011)
	Reduction in seed bank size	Molau and Larsson (2000)
	Low seedling recruitment, high seedling mortality	Bliss (1971), Hautier et al. (2009), Milla et al. (2009)
	Pseudovivipary	Sarapult'tsev (2001)

11.4 Empirical Evidence in Animals

11.4.1 Insects and Other Arthropods

Two major obstacles are faced by ectotherms in uplands or with broad altitudinal distribution: low or decreasing ambient temperature and short or decreasing growing/breeding seasons. Insect growth, development, reproduction, dormancy and diapause are timed in relation to these constraints through alternative strategies. The most common are the reduction of the length of the larval stage and the acceleration of growth. Insects from collembolans to orthopterans have been shown to reduce the number of instars (i.e. the number of moults to achieve the adult stage) or the timing of diapause to complete their annual cycle earlier in uplands (Tanaka and Brookes 1983; Zettel 2000). This response is typically associated with thermal conditions close to a species' tolerance range at either low or high extremes (Esperk et al. 2007). Growing faster at high elevations is a widespread alternative (Dingle et al. 1990; Berner et al. 2004; Laiolo and Obeso 2015), and represents one of the best examples of countergradient genotypic variation opposing physiological responses to temperature.

The reduction of instar number or development time lead to smaller adult body sizes, a pattern commonly observed in upland insect populations and species (Laiolo et al. 2013). Body size constrains fecundity, thus these strategies may carry direct fecundity costs (Hönek 1993). These costs are obviated when prolonging development or generation time over the years rather than restraining them within a single year. This strategy has been described in alpine populations of both holo- and hemimetabolous insects (Hodkinson 2005) and is more commonly associated with seasonal but non-resource-limited environments. Meanwhile, resource limitation together with high seasonality tends to favour fast growth at the expense of body size and overall reproductive output (Chown and Klok 2003). Prolonged development is achieved either by increasing instar numbers or by extending instar growth over 2 years, which involves passing from pluri- or univoltinism in lowlands, to semivoltinism in highlands, with individuals overwintering at different instar stages over the years (Miles et al. 1997), or entering diapause at different stages (Dingle et al. 1990).

The allometric reduction in egg number with elevation is often accompanied by increased egg size to enhance embryo viability at low temperatures, whereas, at higher temperatures, it pays off to produce more and smaller eggs since offspring mortality is lower (Fischer et al. 2003). Increased adult survival with elevation has been observed in fruit flies (Duyck et al. 2010) and is explained by trade-offs between fecundity and longevity (Norry et al. 2006) or increasing rates of damage from by-products of metabolism in hot temperatures (Leiser et al. 2011). In annual species, however, selection on reproductive schedules may induce the opposite patterns, with accelerated senescence at the completion of reproduction (Tatar et al. 1997).

Asexual, or parthenogenetic, populations frequently appear in high-altitude habitats. This strategy, favoured in areas with few sexual competitors (Peck et al. 1998), permitted persistence in isolated ice-free summits surrounded by glaciers ('nunataks') during the Pleistocene glaciation periods (Wachter et al. 2012).

The optimal life strategies of crustaceans also vary along elevation, and females from stream head or alpine waters, for instance, lay larger eggs but smaller clutches than those from lowlands (decapods: Hayashi and Hamano 1984; Mashiko 1990; Hancock et al. 1998; amphipods: Wilhelm and Schindler 2000). Among arachnids, alpine tundra *Pardosa* wolf spiders display no interpopulation variation in egg number per cocoon, but generation time is twofold than that in lowlands (Schmoller 1970).

11.4.2 Fishes

Despite the high thermal conductivity of water buffers thermal fluctuations over time and space, and thus reduces the opportunities for sharp temperature-driven selection, temperature has a pervasive influence on developmental traits of fishes and adaptation to local thermal regimes as well as plastic responses are well documented (Haugen and Vøllestad 2000). Factors such as water flow or predation risk also affect fish life-history decisions and, in particular, recruitment is a key trait in determining fish allocation to contrasting life-history traits. As a general rule, larger sizes, later maturity and long reproductive lifespans are selected for when recruitment is low (Kennedy et al. 2003; Parra et al. 2014). Sternberg and Kennard (2013) found that among Australian freshwater fishes, egg guarding species that reach maturity at a small size were more frequent in environments with perennial flow and low mean annual temperatures typical of uplands. Conversely, larger-bodied, non-egg guarding, highly fecund fish with small eggs and late maturity were more frequent in environments with high mean annual temperature and temporary flow. This pattern indicates that it may be advantageous to increase parental care or produce fewer but larger eggs in low-temperature stream heads, as also observed in crustaceans. Another consistent intraspecific pattern with temperature is that of longevity, which declines at increasing temperatures (Pauly 1980; Beverton 1987).

Many fish species continue somatic growth after sexual maturation and growth is typically highly plastic (Gotthard 2001). However, there is evidence that developmental decisions and growth patterns of populations are locally adapted (Nicieza et al. 1994; Haugen and Vøllestad 2000) and respond to climate, for instance, along latitudes, in a countergradient fashion as for other ectotherms (Conover and Present 1990).

11.4.3 Amphibians

Juvenile development to metamorphosis or sexual maturity strongly influences amphibian adult fitness and has been the target of a large number of studies along climate gradients (Berven and Gill 1983; Altwegg and Reyer 2003; Laugen et al. 2003; Muir et al. 2014). These have shown that ontogenetic traits tend to adjust to a countergradient variation pattern across elevations. Berven et al. (1979) and Berven (1982) showed, for instance, that the mountain larvae of the green frog (*Rana clamitans*) and wood frog (*R. sylvatica*) complete metamorphosis faster and at a larger size than their lowland counterparts in a common environment. Despite their higher genetic growth capacity, however, in nature they metamorphose later because of strong climatic constraints.

Again, similar to other taxonomic groups, anurans and urodeles from high-altitudes invest in larger eggs but reduce offspring number (Liao et al. 2014) and increase mean and maximum age, as well as maturation age (Zhang and Lu 2012). Apart from intrinsic trade-offs, hypoxia has been evoked as a possible cause of increased longevity in high-altitude regions (Zhang and Lu 2012).

11.4.4 Reptiles

Viviparity in squamate reptiles has been explained in terms of climate selection for longer periods of egg retention where juvenile mortality is high because of extended cold exposure (Tinkle and Gibbons 1977). However, this does not apply to the independent evolution of viviparity in freshwater fishes (Pollux et al. 2009) and amphibians (Vitt and Caldwell 2013). In viviparous lizards and snakes, gravid females actively thermoregulate and provide embryos of higher temperatures for development, a behaviour that reduces juvenile mortality as compared to conditions in nest regimes (Braña et al. 1991). Viviparity, however, requires modifications to other reproductive features, such as eggshell thickness and clutch size, both decreasing with the degree of viviparity and elevation (Mathies and Andrews 1995). Moreover, pregnant females pay metabolic costs for maintaining higher body temperatures, as visibly appreciable in postpartum body condition. However, the ultimate consequences for survival depend upon a combination of factors, from stored reserves and thermal conditions to the capability of using current food intake during reproduction (Lourdais et al. 2002; Cox et al. 2010). Many reptiles are in fact predominant 'capital breeders', i.e. they use reserves gathered over long periods prior to the year of reproduction, but nevertheless are able to optionally integrate energy from current feeding ('income') (Shine 2005).

In oviparous species, a decrease in clutch size translates into an increase in offspring size (Sinervo et al. 1992), infrequent reproduction (Shine 2005) and enhanced annual survival rate (Adolph and Porter 1993). Lizards exhibit indeterminate and fully plastic growth (Sears and Angilletta 2003) with a few documented cases of countergradient variation (Sinervo 1990). Although no comparative analyses have addressed interspecific variation along elevation gradients, a trend for 'slow' life strategies in cold-environment or slow-metabolism taxa, in contrast to 'fast' strategies in hot-environment or fast-metabolism taxa, has been highlighted (Bauwens and Díaz-Uriarte 1997; Shine 2005).

11.4.5 Birds

Several comparative reviews on avian intra- and interspecific patterns highlight a strategy of reduced annual fecundity, e.g. reduced clutch size or reproductive attempts per year, with elevation. As an opposite pattern, the duration of the incubation and nesting phases increase with elevation (Krementz and Handford1984; Badyaev 1997a; Badyaev and Ghalambor 2001; Boyle et al. 2015; Hille and Cooper 2015; Laiolo et al. 2015b). Ruling out the effect of body size, low predation pressures and poor food availability likely contribute to prolonged parental care, as a result of bird parents spending increasingly longer periods outside the nest (Boersma 1982). This fact inevitably slows the development of young, but protects parental survival (Martin 2002) and improves food provisioning where resources are scarce and scattered and where chicks have increased metabolic demands because of cold weather.

At the intraspecific level, survival has been shown to increase with elevation in some study cases (e.g. Bears et al. 2009), but pairwise comparisons of closely related species, subspecies or populations are not conclusive in this respect (Badyaev and Ghalambor 2001; Boyle et al. 2015). It is possible that large variations in extrinsic (environmental) mortality, essentially independent of the choices made by individuals, override slight differences in intrinsic mortality between entities with a great degree of shared history. Tests with phylogenetically and functionally diverse bird assemblages indeed suggest remarkable variation in survival patterns (Laiolo et al. 2015b; Bastianelli et al. 2017).

Elevation clines in reproductive allocation and parental care have crucial consequences for the expression of costly sexual characters in passerines with bi-parental care (Badyaev and Ghalambor 2001). Males from upland grounds have to shift rapidly from sexual to parental behaviours, which requires that testosterone be maintained at low levels, or to rapidly decline, to avoid an undermining of reproductive success by testosterone-driven aggressiveness (Apfelbeck and Goymann 2011). These conditions should tilt the balance between parental and mating effort of males towards the former, considering that opportunities for additional mating and extra-pair fertilizations may decline with elevation, because of low densities or synchronic reproduction. As a matter of evidence, Badyaev (1997b) and Snell-Rood and Badyaev (2008) found reduced plumage dimorphism and shorter and simpler songs in high elevation Cardueline species, while Apfelbeck and Goymann (2011) and Bastianelli et al. (2015) highlighted weaker male territorial aggressiveness in *Phoenicurus* and *Anthus* species.

11.4.6 Mammals

Literature is limited in mammals compared to other taxonomic groups. One possible explanation is that life-history strategies are constrained by aspects of the ecological niche (e.g. aquatic, aerial or terrestrial life; diurnal vs. nocturnal habits) with a poor relationship with elevation (Fisher et al. 2001; Bielby et al. 2007; Sibly and Brown 2007). Mammal growth and reproduction tend to be highly plastic (Hansen and Boonstra 2000), and temporal patterns are often more divergent than spatial ones. Seasonality is a strong driver of life-history diversification. Thus, elevational clines should be envisioned, but high latitudes have instead been the favourite scenario for analysing major temperature and photoperiod influences. Polyestrous rodents represent one of the best examples of seasonal diversification: spring-born young grow fast, mature early and reproduce in the year of birth, while those born later grow slowly, overwinter as immature and reproduce the next year (Bronson 1989).

When analysing variation across elevation, Bronson (1979), Zammuto and Millar (1985) and Yoccoz and Ims (1999) found that highland populations of ground squirrel (*Spermophilus columbianus*) and voles (*Chionomys nivalis*) have longer lifespans, lower litter sizes and later ages at reproduction than those from lowlands. Lower litter size in ground squirrels depends on reduced ovulation rates because embryonic mortality is low and decreases with elevation (Bronson 1979). A reduction in litter size with altitude has also been recorded in the deer mice *Peromyscus maniculalus* (Dunmire 1960; Fleming and Rauscher 1978) and holds at the interspecific level among species of this genus (Smith and McGinnis 1968). In the alpine collared pika *Ochotona collaris*, adult survival is the trait that contributes most to population growth rate, and fecundity is less variable than in other lagomorphs (Morrison and Hik 2007). As previously mentioned, buffering of survival and bet hedging are thought to secure persistence in alpine environments.

Among mountain goats (*Oreamnos americanus*), females augment parental care to enhance juvenile survival in high elevations (Festa-Bianchet et al. 1994). Male parental care, although rare in mammals, emerges as a facultative behaviour at very low population densities (Barash 1975) and in extreme cold, arid or seasonal environments (Kleiman and Malcolm1981). In the dwarf hamster *Phodopus campbelli*, for instance, male presence is essential to guarantee pup survival and growth, because it alleviates female thermoregulatory stress and thus water loss due to maternal hyperthermia, which compromises milking (Wynne-Edwards 1995, 1998). Similarly, pups of the alpine marmot *Marmota marmota*, due to small sizes, have reduced thermal inertia and take advantage of the energy spent by all family members during hibernation ('social thermoregulation'; Arnold 1988). Young survival, in particular, is positively associated with the number of subordinate males, which also participate in the surveillance of the family's territory (Arnold 1993; Allainé and Theuriau 2004).

11.5 Empirical Evidence in Plants

11.5.1 Interspecific Variation

In mountain and alpine environments, life histories are often characterised by long-lived iteroparous perennial life cycles. A trade-off between allocation to vegetative growth and sexual reproduction is expected as a consequence of nutrient limitation. Thus, the increased allocation to vegetative growth should reduce the availability of resources for reproduction (Obeso 2002). In general terms, alpine and arctic plants invest more in maintenance and less in reproduction (Jónsdóttir 2011).

In the harsh climatic environment of high altitudes, new plant establishment is a particularly risky (unsuccessful) mode of reproduction because of the high nutrient demand of seed production (Watson 1984), infrequent germination and low seedling survival (Bliss 1971; Scherff et al. 1994). Accordingly, there may be a reduction in seed rain and seed bank size as elevation increases (Molau and Larsson 2000). The demography of alpine plant populations is often characterised by low seedling recruitment and high seedling mortality at early developmental stages compared with lower-elevation populations (Bliss 1971; Hautier et al. 2009; Milla et al. 2009). In general terms, this implies that the successful establishment ex-novo of new genets (independent physiological units, or clonal colonies, sensu Watson and Casper 1984) is infrequent. However, these paradigms regarding alpine plants are currently changing and seedling establishment may be more common and successful than previously thought (Jolls and Bock 1983; Chambers et al. 1990; Forbis 2003, Forbis and Doak 2004; Giménez-Benavides et al. 2007; Venn and Morgan 2009; Kim and Donohue 2011).

The main evidence of the rarity of seedling establishment in alpine plants is the fact that the size-class distributions within the populations are often characterised by the absence of the smaller size-classes (Philipp 1997; Jónsdóttir 2011). These distributions may also be a consequence of longer intervals between 'windows' of regeneration by seeds in the extremely variable alpine conditions (Eriksson 1997). Additionally, taking into account that long-lived alpine plants may reach ages of one thousand years or more, the selective pressures that conditioned the establishment of the parent plant were likely not the same that seeds and seedlings currently face.

Persistence of established genets, through somatic maintenance, clonal growth and vegetative reproduction is thought to be one of the most remarkable adaptations to the conditions of high mountain habitats and its importance tends to increase with altitude. Survival of adult plants has been suggested to be a key demographic parameter for maintaining alpine plant populations, and their demography is often characterised by high adult survival compared with lower-elevation populations (Bliss 1971; Hautier et al. 2009; Milla et al. 2009; García and Zamora 2003; Kim and Donouhe 2011). As a consequence, the decline of annual species with increasing altitude is remarkable, as it is the number of long-lived species that relies on clonal reproduction for population maintenance (Stöcklin 1992; Klimes et al. 1997).

Despite the above generalisations, life histories of alpine plants are highly diverse due to a great variety of growth and multiplication models. This diversity may be associated with different growth forms and varying degrees of physiological integration within genets. Most alpine plants are clonal perennials, the lifespan of which is one order of magnitude longer than that of non-clonal perennials (de Witte and Stöcklin 2010). Clonal perennials range from 'splitters' to 'extensive integrators'. In the former case, the new clonal individuals (ramets) split from the parental genet shortly after their development (seeds produced by agamospermy, bulbils, plantlets, and some bulbs and tubers) and in the latter, the offspring ramets (normally rhizomes) remain physiologically integrated with the parent genet throughout their lifetime. There is an intermediate situation ('intermediate integrators') in which the offspring ramets remain connected to the parental plant for a time, as is the case of stolons, rosettes, rhizomes and root shoots (Jónsdóttir 2011).

Arctic and alpine non-clonal perennial lifespans from several decades to more than one hundred years are common (Callaghan and Emanuelsson 1985) and genet age of 'extensive integrator' clonal perennials may reach over one thousand years or more. As an expected consequence of a trade-off between longevity and sexual reproduction, the allocation to sexual reproduction is generally lower in clonal than in non-clonal plants (Jónsdóttir 1995; Stenström 1999; Stenström and Jónsdóttir 2006).

Taking into account the reduction in reproductive allocation at high elevations, we can expect that plants have developed some adaptations in their life histories to reduce the risk of costly reproductive investment. In this sense, we can expect alpine plants to increase offspring survival throughout life-history variables related to parent care: larger seed size to produce larger seedlings, pseudovivipary (Lee and Harmer 1980) and nursing of seedlings to increase their survival. Established cushion plants (such as *Silene acaulis*) can act as nurses of seedlings increasing their survival (Bliss 1971). This nursing effect has been mostly observed in an interspecific context. However, we can predict that this may be an important phenomenon from an intraspecific perspective, as has been proposed in environmental contexts others than alpine ones (Fajardo and McIntire 2011).

Pseudoviviparity consists of the formation of vegetative diaspores in inflorescences, with the already developed flower parts undergoing proliferation and transformation into leaf-like structures (Pijl 1972). Species with pseudovivipary are mostly found in arctic, alpine and arid environments. In the local high-altitude floras, the proportion of pseudoviviparous species reaches 10% and, in exceptional cases, even up to 25% (Sarapul'tsev 2001). These habitats may favour pseudovivipary because they are extraordinarily coarse-grained for seedling establishment and the probability of an offspring being dispersed to a suitable patch is very low. The success of pseudovivipary may also be related to the problems of establishment and growth in the short, cold growing seasons of these regions (Lee and Harmer 1980; Elmqvist and Cox 1996). Furthermore, parental care is not restricted to seedling establishment, as the survival of daughter ramets may be greatly enhanced by translocation of resources from the parental plant through the vascular connections. This extended parental care depends on the degree of physiological integration or independence and is prolonged in the case of 'extensive integrators' (Callaghan 1984; Jónsdóttir 2011). As a rather general trend, parental care to seeds is substituted by parental care to daughter ramets, which are much more costly to produce but exhibit much higher survival. Seedling survival is probably the most critical stage in the life histories of long-lived perennial alpine plants, determining species' distribution and range shifts (Kitajima and Fenner 2000).

Seed weight should be affected by altitude because heavier seeds are more likely to produce larger seedlings that successfully establish in harsh conditions (Westoby et al. 1992), which is in accordance with the 'stress-tolerance' hypothesis (survival depends on plant stress resistance). However, despite the fact that elevation gradients in seed mass have repeatedly been reported (Baker 1972; Blionis and Vokou 2005), findings were often conflicting and had not revealed any consistent pattern thus far. Although an increase in seed mass with elevation was reported by Pluess et al. (2005), there is also evidence of negative relationships between seed mass and elevation supporting the 'energy constraints' hypothesis, which states that lower temperatures and shorter growing seasons at higher elevations may reduce resource acquisition and the energy available for seed development and seed provisioning (Baker 1972; Körner 2003; Bu et al. 2007). Additionally, seed size is subjected to allometric constraints and thus determined by plant size variation with altitude.

In detail, Pluess et al. (2005) tested the hypotheses that between related species-pairs and among populations of single species a similar trend for increasing seed weight with increasing altitude should be present. These authors determined seed weights from 29 species-pairs, with each pair consisting of one species occurring in a lowland area and a congeneric species from a high altitude area. Compared to the related lowland species, 55% of the alpine species had heavier seeds, 3% (one species) had lighter seeds and 41% had seeds of approximately equal weight. However, Wu and Du (2009), who examined the hypothesis of a positive effect of altitude on both interspecific and intraspecific variation in seed mass, found that in 50% of the 44 species that occurred in both low and high altitudes, seed mass increased with altitude, but decreased in the other 50%. Moreover, Wang et al. (2014) examined seed mass variation in 42 species of *Rhododendron* along an altitudinal gradient from a few hundred metres to 5500 m above sea level on the Tibetan Plateau. They found that seed length, width, surface area and wing length were negatively correlated with altitude, and positively with plant height. Conversely, Qi et al. (2014), using a large database involving 1355 species from the Tibetan Plateau, found a non-significant seed mass-elevation relationship across all species after controlling for phylogeny and plant height. These authors also found a mass-dependent response to the elevation gradient: smaller seeds tended to increase in mass with elevation but large seeds tended to decrease.

11.5.2 Intraspecific Variation

When the same plant species occurs along a mountainside, within-species variation in life histories is expected since a suitability gradient is found within each mountain range (Körner 2003). Depending on the biogeographic origin of the species, plants occurring at the highest or lowest altitudinal limits should face especially harsh constraints on reproduction and establishment via seeds (Hampe and Petit 2005; Arrieta and Suárez 2006; Giménez-Benavides et al. 2007). In this sense, the 'centre–periphery' hypothesis proposes that conditions for the regeneration of plant populations are less suitable in the boundaries than in the centre of the distribution area, and at the same time, life cycles should slow down at high altitudes (Lawton 1993; Vucetich and Waite 2003; Angert and Schemske 2005).

Arx et al. (2006) used the width of annual rings in roots to study plant demography along an altitude gradient after determining plant age and lifetime growth in three perennial forbs. For all three species, the plants from the highest altitudes tended to be considerably older and produced more flowering shoots than lowland plants. Highland plant growth, estimated by ring width, was approximately half that of lowland plants. However, ring width of the high-altitude plants increased during the first years and later decreased. These results highlight the importance of investing resources in plant growth during the first years to ensure plant establishment. This initial investment in growth is a characteristic behaviour of life cycles in which mortality decreases considerably with the age of the individual.

When comparing demography and life-history traits of populations of *Erysimun capitatum* from alpine and low-elevation populations, Kim and Donohue (2011) found that mortality of all life stages was higher at lower elevations than at an alpine site. At the same time, they found that low-elevation plants reproduced more quickly and were more frequently semelparous than alpine plants.

Thus, low-elevation semelparous populations depended primarily on seedling recruitment and precocious reproduction, whereas alpine plants tended to be iteroparous and to produce more vegetative rosettes. These results showed an altitudinal variation in parity (number of reproductive events), and its demographic consequences, indicating that plastic or evolutionary changes in this trait have a clear influence on population performance along altitudinal gradients.

As the allocation of resources to reproduction results in a reduction of allocation to vegetative growth and, therefore, an impact on future reproductive success, the trade-off between allocation to reproduction and vegetative growth is also a determinant of iteroparous perennial cycles within species. Hautier et al. (2009) conducted a transplant experiment to assess the influence of both the altitudinal origin of populations and the altitude of the growing site on vegetative growth and reproductive investment in *Poa alpina*. According to the general trend in plants, the variation in reproductive investment was mainly explained by plant size. However, the vegetative growth and the relative reproductive allocation decreased in populations originating from higher altitudes compared to populations originating from lower altitudes. They also found that the importance of plasticity was scarce in relation to genetic effects and interpreted these results as a consequence of local adaptations.

Gao-Lin et al. (2011) tested the hypothesis that seed mass was positively correlated with altitude within species in four congeneric Saussurea (Asteraceae) that occur in the Tibetan Plateau. They found a general trend of a significant increase in seed mass with altitude. Contrarily, Meng et al. (2014) showed that along an altitudinal gradient in the Hengduan Mountains, mean seed weight of *Sinopodophyllum hexandrum* decreased significantly. Pluess et al. (2005) compared seed weights among populations of four species from different habitats and with different life histories along an altitude gradient (*Scabiosa lucida, Saxifraga oppositifolia, Epilobium fleischeri and Carex flacca*). In all the four species, they found no indication for heavier seeds at higher altitudes. Similarly, in the cactus *Gymnocalycium monvillei* seedling height increased with altitude, whereas seed mass was not related to this variable (Bauk et al. 2015).

Assessing adaptive differentiation of plant populations along altitude gradients is useful for predicting how they may respond to climatic change. Local adaptation along altitudinal gradients has been demonstrated in several alpine plant species after reciprocal transplant experiments (Byars et al. 2007; Kim and Donohue 2013; Toräng et al. 2015) or transplants to a common garden (Stenström et al. 2002). However, information about local adaptation in traits related directly to life history is still scarce. Leimu and Fischer (2008) reviewed the information about local adaptations and found that although local plants performed better than foreign plants in 71% of the studies, local adaptation, sensu stricto, was demonstrated in approximately 40% of the case studies.

Surprisingly, genetic diversity of alpine plant populations is not as depleted as predicted from small population sizes and repeated vegetative multiplication, a fact that suggests that gene flow and repeated seedling recruitment during succession might be more frequent than commonly thought (Diggle et al. 1998; Pluess and Stöcklin 2004; Reisch et al. 2007).

11.6 Discussion

11.6.1 Current Patterns

This review shows that much of life-history variation in elevation is the result of direct physiological sensitivity to temperature and nutrient supply, which is then modified secondarily by evolutionary responses that refine the relationship with the environment. Generally, organisms as diverse as animal ectotherms, endotherms and plants inhabit mountaintops by adopting 'slow' life cycles, involving longer lifespan, delayed maturity, slow reproductive rates, including clonal or parthenogenetic spreading and strong inversions in parental care to augment juvenile survival where recruitment is limited. There are however exceptions, for instance, slow life cycles are precluded to obligatory annual organisms in seasonal environments. Moreover, traits may not reflect optimality at lower and upper margins of species ranges, also depending on the position of the mountain ridge with respect to species' overall geographic distribution (Fig. 11.1). Jiménez-Alfaro et al. (2014) showed, for instance, that plant species from different geographic regions are filtered in different ways by altitude, and that constraints on reproduction and establishment via seeds may vary, being generally strong for lowland species at their highest elevation or arctic and alpine species at their lowest limits (Hampe and Petit 2005; Arrieta and Suárez 2006; Giménez-Benavides et al. 2007).

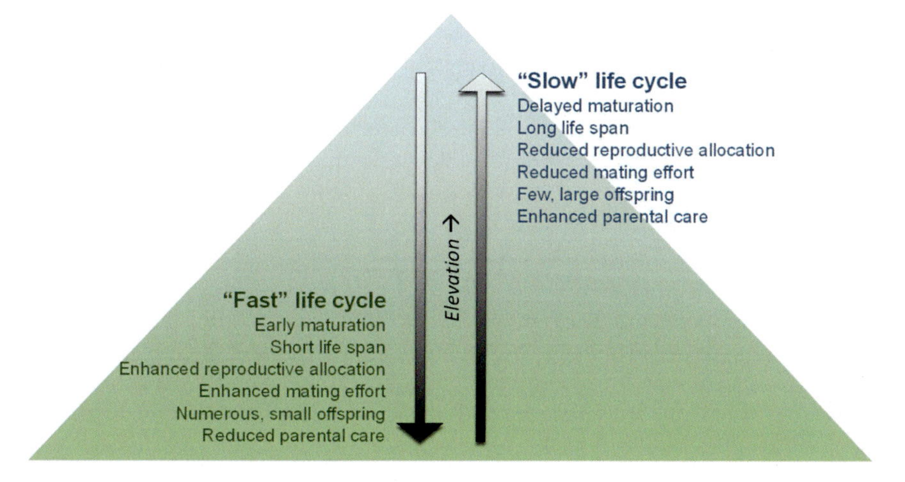

Fig. 11.1 Main trends of life-history variation observed along the elevation gradients

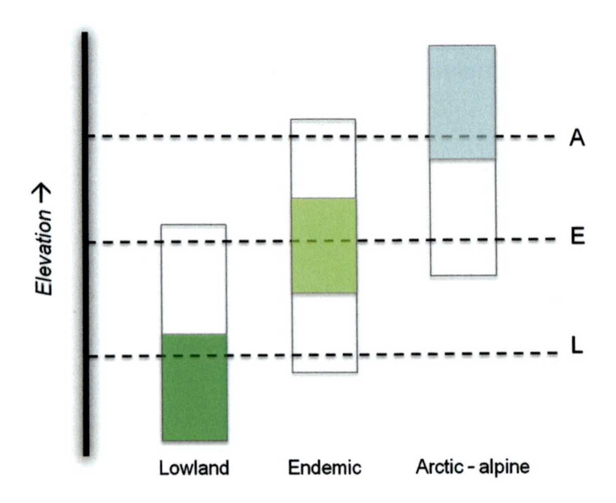

Fig. 11.2 Rectangles showing the distribution area of different species depending on their biogeographic origin. The centre of the distribution is depicted in *black* and the periphery in *white*. Letters *A*, *E*, *L* indicate locations with optimal conditions for regeneration of Arctic-alpine, Endemic and Lowland species, respectively, but non-optimal conditions for the other groups

Contrarily, species at the centre of their distribution areas (e.g. endemic species) should face optimal conditions for regeneration (Fig. 11.2).

11.6.2 Evolutionary and Plastic Responses to Environmental Change

A frequent assumption is that populations negatively affected by ongoing climate change, when lacking the plasticity to tolerate it, will migrate more readily than undergoing evolutionary change to produce new phenotypes. Implicit in this view is the observation that modern populations of many species that shifted ranges in the past display life-history adaptations to current climatic conditions. However, the tolerance ranges of migrating species are not 'static' during distribution shifts: differential survival of migrating individuals, or their propagules, sieve out genotypes that do not tolerate local conditions, and differential growth and reproduction further promote adaptation of physiological characteristics (Davis and Shaw 2001). In this scenario, negative genetic correlations among life-history traits, such as those between survival and reproduction, may slow (or impede) the responses to selection of single traits, as compared, for instance, to the responses of unrelated traits, as those conferring physiological tolerance (Davis and Shaw 2001).

Apart from dispersal and selection, gene flow and recombination are essential elements in evolutionary change during range shifts. When dealing with altitudinal migration, gene flow is facilitated by the proximity of neighbouring populations. In the case of species with broad elevational distribution, if the upper elevation limit is the leading edge of the migrating front, adaptations to the novel environment may be enhanced by gene transfer from the centre of the range. This feature would likely lead to the spread of 'fast' genotypes that may allow persistence in warmer conditions.

Although a 'fast' strategy may enhance the speed of range expansion (Burton et al. 2010), 'slow' cycles protect species from the dramatic loss of genetic variability during upward migration: when a few founders colonise a remote patch, delayed maturation allows genetic diversity to accumulate through recruitment of additional individuals (Austerlitz et al. 2000). It is also often argued that alpine specialists inhabiting narrow elevation bands may lack substantial genetic variability for traits under selection, but evidence of this phenomenon is not strong, because the distribution of these species is the result of past climate-driven shifts up and down elevation gradients that helped maintain genetic differentiation (Galbreath et al. 2009; Wachter et al. 2012). Hence, predicting evolutionary trajectories into the future must take into account the past persistence of many relict species (Hampe and Petit 2005).

The interactions between climate shifts with phenotypic plasticity or heritable variation in reaction norms are also crucial to envisage species responses to changing environmental conditions (Winkler et al. 2002; Both and Visser 2005; Jensen et al. 2008; Williams et al. 2015). Temperature increases may not necessarily have the expected worst impact on ectotherm metabolism because they may adjust thermoregulation and activity to prevailing temperatures (Aguado and Braña 2014), as alpine plants do with respiration (Larigauderie and Körner 1995). It is also worth stressing that warming reduces some of those constraints dominating alpine life, as evidenced by improved survival or reproductive output of a number of species (e.g. Day et al. 1999; Erschbamer 2007; Ozgul et al. 2010). Similarly, Barrett et al. (2015), using data from a long-term research project in the Arctic, demonstrated enhanced reproductive allocation in connection with improved air and soil temperature. Differences in species' responses are, however, huge, making generalisations of responses and predictions of effects very weak. Experiments and observations within the same community or environmental context often highlight highly idiosyncratic responses in growth and reproduction to changing temperature and resource availability (Wookey et al. 1993; Arft et al. 1999; Dorman and Woodin 2002; Wipf et al. 2009).

11.6.3 Demographic Responses to Environmental Change

Patterns of life history determine the dynamics of populations when facing environmental variation, and life-history traits have a differential influence in this process. Perturbations (either cyclic or stochastic) can trigger substantial fluctuations in population size when reproductive parameters have the greatest influence on the finite rate of population growth, corresponding to a 'fast' life strategy. In contrast, perturbations are buffered when survival parameters have the largest relative influence on growth rate, i.e. a 'slow' strategy (Sæther and Bakke 2000; Oli and Dobson 2003). Therefore, when facing environmental change, a 'slow' life strategy is expected to confer more stable dynamics, high resistance and low resilience, as opposed to a 'fast' life strategy, which induces more cyclic or chaotic population dynamics and low resistance, but a greater chance of recovery. Processes like demographic buffering (temporal stability) and demographic compensation (spatial compensation) may buffer local alpine populations against trends in environmental conditions (such as climate warming) (Villellas et al. 2015), although such compensatory responses may not last indefinitely (Doak and Morris 2010).

11.6.4 Future Research

In spite of widespread evidence of adaptations to climate in the past, there is a need to assess whether these changes will occur as readily during the present period of climate change since the range shifts documented in the past are below the rates required to track climate in the future (Davis and Shaw 2001). More experiments coupled with quantitative genetics are required to appreciate the magnitude of genetic constraints and genetic variation for traits critical to survival and reproduction, as well as molecular and demographic studies assessing the potential for dispersal and gene flow.

This overview highlights the scarcity of information and the gaps in our knowledge about life-history variation along elevation gradients. There is a need to expand the taxonomic focus because there has been a disproportionate effort on northern-latitude cold environments (such as the Arctic) compared to mountain and alpine systems in many groups. Life-history knowledge should also be improved: reproductive variables such as seed, clutch or litter size or number have been a favourite target of research, but lifespan and age at first reproduction are virtually unknown for the majority of alpine species. Ultimately, processes within and among species should be integrated, such that their changes can be linked to community-wide processes. This integration will improve our capability for predicting the response of alpine flora and fauna to the combinations of current, novel environmental drivers.

References

Aarssen LW, Clauss MJ (1992) Genotypic variation in fecundity allocation in *Arabidopsis thaliana*. J Ecol 80:109–114

Abbate JL, Antonovics J (2014) Elevational disease distribution in a natural plant–pathogen system: insights from changes across host populations and climate. Oikos 123:1126–1136

Adolph SC, Porter WP (1993) Temperature, activity, and lizard life histories. Am Nat 142:273–295

Aguado S, Braña F (2014) Thermoregulation in a cold-adapted species (Cyren's Rock Lizard, *Iberolacerta cyreni*): influence of thermal environment and associated costs. Can J Zool 92:955–964

Allainé D, Theuriau F (2004) Is there an optimal number of helpers in alpine marmot family groups? Behav Ecol 15:916–924
Altwegg R, Reyer H-U (2003) Patterns of natural selection on size at metamorphosis in water frogs. Evolution 57:872–882
Angert AL, Schemske DW (2005) The evolution of species' distributions: reciprocal transplants across the elevation ranges of *Mimulus cardinalis* and *M. lewisii*. Evolution 59:1671–1684
Angilletta MJ, Steury TD, Sears MW (2004) Temperature, growth rate, and body size in ectotherms: fitting pieces of a life-history puzzle. Integr Comp Biol 44:498–509
Angilletta MJ, Cooper BS, Schuler MS, Boyles JG (2010) The evolution of thermal physiology in endotherms. Front Biosci 2:861–881
Apfelbeck B, Goymann W (2011) Ignoring the challenge? Male black redstarts (*Phoenicurus ochruros*) do not increase testosterone levels during territorial conflicts but they do so in response to gonadotropin-releasing hormone. Proc R Soc Lond B 278:3233–3242
Arft AM et al (1999) Response patterns of tundra plant species to experimental warming: a meta-analysis of the International Tundra Experiment. Ecol Monogr 16:491–511
Armstrong DP, Westoby M (1993) Seedlings from large seeds tolerated defoliation better: a test using phylogeneticaly independent contrasts. Ecology 74:1092–1100
Arnold W (1988) Social thermoregulation during hibernation in alpine marmots (*Marmota marmota*). J Comp Physiol B 158:151–156
Arnold W (1993) Social evolution in marmots and the adaptive value of joint hibernation. Verh Dtsch Zool Ges 86:79–93
Arrieta S, Suarez F (2006) Marginal holly (*Ilex aquifolium* L.) populations in Mediterranean central Spain are constrained by a low-seedling recruitment. Flora 201:152–160
Arx GV, Edwards PJ, Dietz H (2006) Evidence for life history changes in high-altitude populations of three perennial forbs. Ecology 87:665–674
Atkinson D (1996) Ectotherm life history responses to developmental temperature. In: Johnston IA, Bennett AF (eds) Animals and temperature: phenotypic and evolutionary adaptation. Cambridge University Press, Cambridge, pp 183–204
Atkinson D, Sibly RM (1996) On the solution to a major life history puzzle. Oikos 77:359–364
Austerlitz F, Mariette S, Machon N, Gouyon PH, Godelle B (2000) Effects of colonization processes on genetic diversity: differences between annual plants and tree species. Genetics 154:1309–1321
Badyaev AV (1997a) Avian life history variation along altitudinal gradients: an example with cardueline finches. Oecologia 111:365–374
Badyaev AV (1997b) Altitudinal variation in sexual dimorphism: a new pattern and alternative hypotheses. Behav Ecol 8:675–690
Badyaev A, Ghalambor CK (2001) Avian life-history strategies in relation to elevation: evidence for a trade-off between fecundity and parental care. Ecology 82:2948–2960
Baker G (1972) Seed weight in relation to environmental conditions in California. Ecology 53:997–1010
Barash DP (1975) Ecology of paternal behavior in the hoary marmot (*Marmota caligata*): an evolutionary interpretation. J Mammal 56:613–618
Barrett RT, Hollister RD, Oberbauer SF, Tweedie CE (2015) Arctic plant responses to changing abiotic factors in northern Alaska. Am J Bot 102:2020–2031
Barry RG, Chorley RJ (1987) Atmosphere, weather and climate, 5th edn. Routledge, 448 pp
Bastianelli G, Seoane J, Álvarez-Blanco P, Laiolo P (2015) The intensity of male-male interactions declines in highland songbird populations. Behav Ecol Sociobiol 69:1493–1500
Bastianelli G, Tavecchia G, Meléndez L, Seoane J, Obeso JR, Laiolo P (2017) Surviving at high elevations: an inter- and intraspecific analysis in a mountain bird community. Oecologia doi:10.1007/s00442-017-3852-1
Bauk K, Pérez-Sánchez R, Zeballos SR, Las Peñas ML, Flores J, Gurvich DE (2015) Are seed mass and seedling size and shape related to altitude? Evidence in *Gymnocalycium monvillei* (Cactaceae). Botany 93:529–533
Bauwens D, Díaz-Uriarte R (1997) Covariation of life-history traits in lacertid lizards: a comparative study. Am Nat 149:91–111

Bears H, Martin K, White GC (2009) Breeding in high-elevation habitat results in shift to slower life-history strategy within a single species. J Anim Ecol 78:365–375

Berner D, Körner C, Blanckenhorn WU (2004) Grasshopper populations across 2000 m of altitude: is there life history adaptation? Ecography 27:733–740

Berven KA (1982) The genetic basis of altitudinal variation in the wood frog *Rana sylvatica*. I. An experimental analysis of life history traits. Evolution 36:962–983

Berven KA, Gill DE (1983) Interpreting geographic variation in life-history traits. Am Zool 23:85–97

Berven KA, Gill DE, Smith-Gill SJ (1979) Counter gradient selection in the green frog, *Rana clamitans*. Evolution 33:609–623

Beverton RJ (1987) Longevity in fish: some ecological and evolutionary considerations. Evolution of longevity in animals. Springer, New York, pp 161–185

Bielby J, Mace GM, Bininda-Emonds ORP, Cardillo M, Gittleman JL, Jones KE, Purvis A (2007) The fast-slow continuum in mammalian life history: an empirical reevaluation. Am Nat 169:748–757

Blanckenhorn WU (1997) Altitudinal life history variation in the dung flies *Scathophaga stercoraria* and *Sepsis cynipsea*. Oecologia 109:342–352

Blionis GJ, Vokou D (2005) Reproductive attributes of Campanula populations from Mt Olympos, Greece. Plant Ecol 178:77–88

Bliss LC (1971) Arctic and alpine plant life cycles. Annu Rev Ecol Syst 2:405–438

Boersma PD (1982) Why some birds take so long to hatch. Am Nat 120:733–750

Both C, Visser ME (2005) The effect of climate change on the correlation between avian life-history traits. Glob Change Biol 11:1606–1613

Boyle WA (2008) Can variation in risk of nest predation explain altitudinal migration in tropical birds? Oecologia 155:397–403

Boyle WA, Sandercock BK, Martin K (2015) Patterns and drivers of intraspecific variation in avian life history along elevational gradients: a meta-analysis. Biol Rev. doi:10.1111/brv.12180

Braña F, Bea A, Arrayago MJ (1991) Egg retention in lacertid lizards: relationships with reproductive ecology and the evolution of viviparity. Herpetologica 47:218–226

Bronson FH (1989) Mammalian reproductive biology. University of Chicago Press

Bronson MT (1979) Altitudinal variation in the life history of the golden-mantled ground squirrel (*Spermophilus lateralis*). Ecology 60:272–279

Brooker RW (2006) Plant–plant interactions and environmental change. New Phytol 171:271–284

Bu HY, Chen XL, Xu XL, Liu K, Jia P, Du GZ (2007) Seed mass and germination in an alpine meadow on the eastern Tsinghai-Tibet plateau. Plant Ecol 191:127–149

Burton OJ, Phillips BL, Travis JM (2010) Trade-offs and the evolution of life-histories during range expansion. Ecol Lett 13:1210–1220

Byars SG, Papst W, Hoffmann AA (2007) Local adaptation and cogradient selection in the alpine plant, *Poa hiemata*, along a narrow altitudinal gradient. Evolution 61:2925–2941

Callaghan TV (1984) Growth and translocation in a clonal southern hemisphere sedge: *Uncinia meridensis*. J Ecol 72:529–546

Callaghan TV, Emanuelsson U (1985) Population structure and processes of tundra plants and vegetation. In: White J (ed) The population structure of vegetation. Junk Publishing, Dordrecht, pp 399–439

Callaway RM et al (2002) Positive interactions among alpine plants increase with stress. Nature 417:844–848

Chambers JC, MacMahon JA, Brown RW (1990) Alpine seedling establishment: the influence of disturbance type. Ecology 1:1323–1341

Chown SL, Klok CJ (2003) Altitudinal body size clines: latitudinal effects associated with changing seasonality. Ecography 26:445–455

Clark DA, Clark DB (1992) Life history diversity of canopy and emergent trees in a neotropical rain forest. Ecol Monogr 62:315–344

Clutton-Brock TH (1991) The evolution of parental care. Princeton University Press
Conover DO, Present TM (1990) Countergradient variation in growth rate: compensation for length of the growing season among Atlantic silversides from different latitudes. Oecologia 83:316–324
Conover DO, Schultz ET (1995) Phenotypic similarity and the evolutionary significance of countergradient variation. Trends Ecol Evol 10:248–252
Cox RM, Parker EU, Cheney DM, Liebl AL, Martin B, Calsbeek R (2010) Experimental evidence for physiological costs underlying the trade-off between reproduction and survival. Funct Ecol 24:1262–1269
Davis MB, Shaw RG (2001) Range shifts and adaptive responses to Quaternary climate change. Science 292:673–679
Day TA, Ruhland CT, Grobe CW, Xiong F (1999) Growth and reproduction of Antarctic vascular plants in response to warming and UV radiation reductions in the field. Oecologia 119:24–35
de Witte LC, Stöcklin J (2010) Longevity of clonal plants: why it matters and how to measure it. Ann Bot 106:859–870
DeWitt TJ, Sih A, Wilson DS (1998) Costs and limits of phenotypic plasticity. Trends Ecol Evol 13:77–81
Diggle PK, Lower S, Ranker TA (1998) Clonal diversity in alpine populations of *Polygonum viviparum* (Polygonaceae). Int J Plant Sci 159:606–615
Dingle H, Mousseau TA, Scott SM (1990) Altitudinal variation in life cycle syndromes of California populations of the grasshopper, *Melanoplus sanguinipes* (F.). Oecologia 84:199–206
Doak DF, Morris WF (2010) Demographic compensation and tipping points in climate-induced range shifts. Nature 467:959–996
Dobson FS, Murie JO (1987) Interpretation of intraspecific life history patterns: evidence from Columbian ground squirrels. Am Nat 129:382–397
Dorman CF, Woodin SJ (2002) Climate change in the Arctic: using plant functional types in a meta-analyzes of field experiments. Funct Ecol 16:4–17
Dunmire WW (1960) An altitudinal survey of reproduction in Peromyscus maniculatus. Ecology 41:174–182
Duyck PF, Kouloussis N, Papadopoulos N, Quilici S, Wang JL, Jiang CR, Müller HG, Carey JR (2010) Lifespan of a *Ceratitis* fruit fly increases with higher altitude. Biol J Linn Soc 101:345–350
Elmqvist T, Cox PA (1996) The evolution of vivipary in flowering plants. Oikos 1:3–9
Emery RJN, Chinnappa CC, Chmielewski JG (1994) Specialization, plant strategies, and phenotypic plasticity in populations of *Stellaria longipes* along an elevational gradient. Int J Plant Sci 155:203–219
Enquist BJ, West GB, Charnov EL, Brown JH (1999) Allometric scaling of production and life-history variation in vascular plants. Nature 401:907–911
Eriksson O (1997) Clonal life histories and the evolution of seed recruitment. In: de Kroon H, van Groenendael J (eds) The ecology and evolution of clonal plants. Backhuys Publishers, Leyden, pp 221–226
Erschbamer B (2007) Winners and losers of climate change in a central alpine glacier foreland. Arct Antarct Alp Res 39:237–244
Esperk T, Tammaru T, Nylin S (2007) Intraspecific variability in number of larval instars in insects. J Econ Entomol 100:627–645
Fajardo A, McIntire EJB (2011) Under strong niche overlap conspecifics do not compete but help each other to survive: facilitation at the intraspecific level. J Ecol 99:642–650
Festa-Bianchet M, Urquhart M, Smith KG (1994) Mountain goat recruitment: kid production and survival to breeding age. Can J Zool 72:22–27
Fischer K, Karl I (2010) Exploring plastic and genetic responses to temperature variation using copper butterflies. Clim Res 43:17–30

Fischer K, Brakefield PM, Zwaan BJ (2003) Plasticity in butterfly egg size: why larger offspring at lower temperatures? Ecology 84:3138–3147

Fisher DO, Owens IP, Johnson CN (2001) The ecological basis of life history variation in marsupials. Ecology 82:3531–3540

Fleming TH, Rauscher RJ (1978) On the evolution of litter size in *Peromyscus leucopus*. Evolution 32:45–55

Forbis TA (2003) Seedling demography in an alpine ecosystem. Am J Bot 90:1197–1206

Forbis TA, Doak DF (2004) Seedling establishment and life history trade-offs in alpine plants. Am J Bot 91:1147–1153

Franco M, Silvertown J (1996) Life history variation in plants: an exploration of the fast-slow continuum hypothesis. Philos Trans R Soc B Biol Sci 351:1341–1348

Galbreath KE, Hafner DJ, Zamudio KR (2009) When cold is better: climate-driven elevation shifts yield complex patterns of diversification and demography in an alpine specialist (American pika, Ochotona princeps). Evolution 63:2848–2863

Gao-Lin W, Fu-Ping T, Wei L, Zhen-heng L (2011) Seed mass increase along altitude within four Saussurea species in Tibetan Plateau. Pol J Ecol 59:381–389

García D, Zamora R (2003) Persistence, multiple demographic strategies and conservation in long-lived Mediterranean plants. J Veg Sci 14:921–926

Ghalambor CK, Martin TE (2001) Fecundity-survival trade-offs and parental risk-taking in birds. Science 292:494–497

Gillooly JF, Brown JH, West GB, Savage VM, Charnov EL (2001) Effects of size and temperature on metabolic rate. Science 293:2248–2251

Giménez-Benavides L, Escudero A, Iriondo JM (2007) Local adaptation enhances seedling recruitment along an altitudinal gradient in a high mountain Mediterranean plant. Ann Bot 99:723–734

Gotthard K (2001) Growth strategies of ectothermic animals in temperate environments. Environment and animal development. BIOS Scientific, Oxford, pp 287–304

Grime JP (1974) Vegetation classification by reference to strategies. Nature 250:26–31

Hampe A, Petit RJ (2005) Conserving biodiversity under climate change: the rear edge matters. Ecol Lett 8:461–467

Hancock MA, Hughes JM, Bunn SE (1998) Influence of genetic and environmental factors on egg and clutch sizes among populations of *Paratya australiensis* Kemp (Decapoda: Atyidae) in upland rainforest streams, south-east Queensland. Oecologia 115:483–491

Hansen TF, Boonstra R (2000) The best in all possible worlds? A quantitative genetic study of geographic variation in the meadow vole, *Microtus pennsylvanicus*. Oikos 89:81–94

Harvey PH, Clutton-Brock TH (1985) Life history variation in primates. Evolution 39:559–581

Haugen TO, Vøllestad LA (2000) Population differences in early life-history traits in grayling. J Evol Biol 13:897–905

Hautier Y, Randin CF, Stocklin J, Guisan A (2009) Changes in reproductive investment with altitude in an alpine plant. J Plant Ecol 2:125–134

Hayashi KI, Hamano T (1984) The complete larval development of *Caridina japónica* De Man (Decapoda, Caridea, Atyidae) reared in the laboratory. Zool Sci 1:571–589

Hille SM, Cooper CB (2015) Elevational trends in life histories: revising the pace-of-life framework. Biol Rev 90:204–213

Hodkinson ID (2005) Terrestrial insects along elevation gradients: species and community response to altitude. Biol Rev 80:489–513

Honek A (1993) Intraspecific variation in body size and fecundity in insects: a general relationship. Oikos 66:483–492

Huber E et al (2007) Shift in soil–plant nitrogen dynamics of an alpine–nival ecotone. Plant Soil 301:65–76

Jensen LF, Hansen MM, Pertoldi C, Holdensgaard G, Mensberg KL, Loeschcke V (2008) Local adaptation in brown trout early life-history traits: implications for climate change adaptability. Proc R Soc B 275:2859–2868

Jiménez-Alfaro B, Marcenó C, Bueno A, Gavilán R, Obeso JR (2014) Biogeographic deconstruction of alpine plant communities along altitudinal and topographic gradients. J Veg Sci 25:160–171
Jolls CL, Bock JH (1983) Seedling density and mortality patterns among elevations in *Sedum lanceolatum*. Arct Alp Res 1:119–126
Jónsdóttir IS (1995) Importance of sexual reproduction in arctic clonal plants and their evolutionary potential. In: Callaghan TV, Oechel WC, Gilmanow T, Molau U, Maxwell B, Tyson M, Sveinbjörnsson B, Holten JI (eds) Global change and arctic terrestrial ecosystems. European Commission, Luxemburg, pp 81–88
Jónsdóttir IS (2011) Diversity of plant life histories in the Arctic. Preslia 83:281–300
Kawecki TJ, Ebert D (2004) Conceptual issues in local adaptation. Ecol Lett 7:1225–1241
Kennedy BM, Peterson DP, Fausch KD (2003) Different life histories of brook trout populations invading mid-elevation and high-elevation cutthroat trout streams in Colorado. West N Am Naturalist 63:215–223
Kim E, Donohue K (2011) Demographic, developmental and life-history variation across altitude in *Erysimum capitatum*. J Ecol 99:1237–1249
Kim E, Donohue K (2013) Local adaptation and plasticity of *Erysimum capitatum* to altitude: its implications for responses to climate change. J Ecol 101:796–805
Kitajima K, Fenner M (2000) Ecology of seedling regeneration. In: Fenner M (ed) Seeds: the ecology of regeneration in plant communities. CAB International, Wallingford, UK, pp 331–359
Kleiman DG, Malcolm JR (1981) The evolution of male parental investment in mammals. Parental care in mammals. Springer, US, pp 347–387
Klimes L, Klimesova J, Hendriks R, van Groenendael J (1997) Clonal plant architecture: a comparative analysis of form and function. In: de Kroon H, van Groenendael J (eds) The ecology and evolution of clonal plants. Backhuys Publishers, Leiden, pp 1–29
Körner C (2003) Alpine plant life: functional plant ecology of high mountain ecosystems. Springer, Berlin
Körner C (2007) The use of 'altitude' in ecological research. Trends Ecol Evol 22:569–574
Krementz DG, Handford P (1984) Does avian clutch size increase with altitude? Oikos 256–259
Laiolo P, Obeso JR (2015) Plastic responses to temperature vs. local adaptation at the cold extreme of the climate gradient. Evol Biol 42:473–482
Laiolo P, Illera JC, Obeso JR (2013) Local climate determines intra- and interspecific variation in sexual size dimorphism in mountain grasshopper communities. J Evol Biol 26:2171–2183
Laiolo P, Illera JC, Meléndez L, Segura A, Obeso JR (2015a) Abiotic, biotic and evolutionary control of the distribution of C and N isotopes in food webs. Am Nat 185:169–182
Laiolo P, Seoane J, Illera JC, Bastianelli G, Carrascal LM, Obeso JR (2015b) The evolutionary convergence of avian lifestyles and their constrained coevolution with species' ecological niche. Proc R Soc B 282:20151808
Larigauderie A, Körner C (1995) Acclimation of leaf dark respiration to temperature in alpine and lowland plant species. Ann Bot 76:245–252
Laugen AT, Laurila A, Räsänen K, Merilä J (2003) Latitudinal countergradient variation in the common frog (*Rana temporaria*) development rates–evidence for local adaptation. J Evol Biol 16:996–1005
Lawton JH (1993) Range, population abundance and conservation. Trends Ecol Evol 8:409–413
Lee JA, Harmer R (1980) Vivipary, a reproductive strategy in response to environmental stress? Oikos 35:254–265
Leimu R, Fischer M (2008) A meta-analysis of local adaptation in plants. PLoS ONE 3(12): e4010–e4010
Leiser SF, Begun A, Kaeberlein M (2011) HIF-1 modulates longevity and healthspan in a temperature-dependent manner. Aging Cell 10:318–326
Liao WB, Lu X, Jehle R (2014) Altitudinal variation in maternal investment and trade-offs between egg size and clutch size in the Andrew's toad. J Zool 293:84–91

Lourdais O, Bonnet X, Shine R, DeNardo D, Naulleau G, Guillon M (2002) Capital-breeding and reproductive effort in a variable environment: a longitudinal study of a viviparous snake. J Anim Ecol 71:470–479

Mariotti A, Pierre D, Vedy JC, Bruckert S, Guillemot J (1980) The abundance of natural nitrogen-15 in the organic matter of soils along an altitudinal gradient. CATENA 7:293–300

Martin TE (2002) A new view of avian life-history evolution tested on an incubation paradox. Proc R Soc B 269:309–316

Mashiko K (1990) Diversified egg and clutch sizes among local populations of the fresh-water prawn *Macrobrachium nipponense* (De Haan). J Crust Biol 10:306–314

Mathies T, Andrews RM (1995) Thermal and reproductive biology of high and low elevation populations of the lizard *Sceloporus scalaris*: implications for the evolution of viviparity. Oecologia 104:101–111

McCain CM, Grytnes JA (2010) Elevational gradients in species richness. In: Encyclopedia of life sciences (ELS). Wiley, Chichester

Meléndez L, Laiolo P, Mironov S, García M, Magaña O, Jovani R (2014) Climate-driven variation in the intensity of a host-symbiont animal interaction along a broad elevation gradient. PLoS ONE 9:e101942

Meng LH, Wang Y, Luo J, Yang YP, Duan YD (2014) The Trade-Off and Altitudinal Variations in Seed Weight-Number in *Sinopodophyllum hexandrum* (Royle) Ying (Berberidaceae) Populations from the Hengduan Mountains. Pol J Ecol 62:413–419

Miles JE, Bale JS, Hodkinson ID (1997) Effects of temperature elevation on the population dynamics of the upland heather psyllid *Strophingia ericae* (Curtis) (Homoptera: Psylloidea). Glob Change Biol 3:291–297

Milla R, Giménez-Benavides L, Escudero A, Reich PB (2009) Intra-and interspecific performance in growth and reproduction increase with altitude: a case study with two Saxifraga species from northern Spain. Funct Ecol 23:111–118

Molau U, Larsson E-L (2000) Seed rain and seed bank along an alpine altitudinal gradient in Swedish Lapland. Can J Bot 78:728–747

Morrison SF, Hik DS (2007) Demographic analysis of a declining pika Ochotona collaris population: linking survival to broad-scale climate patterns via spring snowmelt patterns. J Anim Ecol 76:899–907

Muir AP, Biek R, Thomas R, Mable BK (2014) Local adaptation with high gene flow: temperature parameters drive adaptation to altitude in the common frog (*Rana temporaria*). Mol Ecol 23:561–574

Nagy L, Grabherr G (2009) The biology of Alpine habitats. Oxford University Press, New York

Nicieza AG, Reyes-Gavilán FG, Braña F (1994) Differentiation in juvenile growth and bimodality patterns between northern and southern populations of Atlantic salmon (*Salmo salar* L.). Can J Zool 72:1603–1610

Norry FM, Sambucetti P, Scannapieco AC, Loeschcke V (2006) Altitudinal patterns for longevity, fecundity and senescence in *Drosophila buzzatii*. Genetica 128:81–93

Obeso JR (2002) The costs of reproduction in plants. New Phytol 155:321–348

Oli MK, Dobson FS (2003) The relative importance of life-history variables to population growth rate in mammals: Cole's prediction revisited. Am Nat 161:422–440

Ozgul A et al (2010) Coupled dynamics of body mass and population growth in response to environmental change. Nature 466:482–485

Parra I, Nicola GG, Vøllestad LA, Elvira B, Almodóvar A (2014) Latitude and altitude differentially shape life history trajectories between the sexes in non-anadromous brown trout. Evol Ecol 28:707–720

Pauly D (1980) On the interrelationships between natural mortality, growth parameters, and mean environmental temperature in 175 fish stocks. ICES J Mar Sci 39:175–192

Peck JR, Yearsley JM, Waxman D (1998) Explaining the geographic distributions of sexual and asexual populations. Nature 391:889–892

Philipp M (1997) Genetic diversity, breeding system, and population structure in *Silene acaulis* (Caryophyllaceae) in West Greenland. Opera Bot 132:89–100

Pijl L (1972) Principles of dispersal in higher plants. Springer

Pluess AR, Stöcklin J (2004) Population genetic diversity of the clonal plant *Geum reptans* (Rosaceae) in the Swiss Alps. Am J Bot 91:2013–2021

Pluess AR, Schutz W, Stöcklin J (2005) Seed weight increases with elevation in the Swiss Alps between related species but not among populations of individual species. Oecologia 144:55–61

Pollux BJA, Pires MN, Banet AI, Reznick DN (2009) Evolution of placentas in the fish family Poeciliidae: an empirical study of macroevolution. Annu Rev Ecol Evol Syst 40:271–289

Qi W, Guo S, Chen X, Cornelissen JHC, Bu H, Du G, Cui X, Li W, Liu K (2014) Disentangling ecological, allometric and evolutionary determinants of the relationship between seed mass and elevation: insights from multiple analyses of 1355 angiosperm species on the Eastern Tibetan Plateau. Oikos 123:23–32

Reisch C, Schurm S, Poschlod P (2007) Spatial genetic structure and clonal diversity in an alpine population of *Salix herbacea* (Salicaceae). Ann Bot 99:647–651

Reznick D (1985) Cost of reproduction: an evaluation of the empirical evidence. Oikos 44:257–267

Ricklefs RE, Wikelski M (2002) The physiology/life-history nexus. Trends Ecol Evol 17:462–468

Rolland C (2003) Spatial and seasonal variations of air temperature lapse rates in Alpine regions. J Climate 16:1032–1046

Rourke BC (2000) Geographic and altitudinal variation in water balance and metabolic rate in a California grasshopper, *Melanoplus sanguinipes*. J Exp Biol 203:2699–2712

Sæther BE, Bakke Ø (2000) Avian life history variation and contribution of demographic traits to the population growth rate. Ecology 81:642–653

Sakai A, Larcher W (1987) Frost survival of plants responses and adaptation to freezing stress. Springer, Berlin

Sarapult'tsev IE (2001) The phenomenon of pseudovivipারity in Alpine and Arctomontane Grasses (*Deschampsia* Beauv., *Festuca* L., and *Poa* L.). Russ J Ecol 32:170–178

Scherff EJ, Galen C, Stanton ML (1994) Seed dispersal, seedling survival and habitat affinity in a snowbed plant: limits to the distribution of the snow buttercup, *Ranunculus adoneus*. Oikos 3:405–413

Schmidt-Nielsen K (1997) Animal physiology: adaptation and environment. Cambridge University Press

Schmoller R (1970) Life histories of alpine tundra Arachnida in Colorado. Am Midl Nat 83:119–133

Schulze ED, Chapin FS III (1987) Plant specialization to environments of different resource availability. Potentials and limitations of ecosystem analysis. Springer, Berlin, pp 120–148

Sears MW, Angilletta MJ Jr (2003) Life history variation in the sagebrush lizard: phenotypic plasticity or local adaptation? Ecology 84:1624–1634

Shine R (2005) Life-history evolution in reptiles. Annu Rev Ecol Evol S:23–46

Sibly RM, Brown JH (2007) Effects of body size and lifestyle on evolution of mammal life histories. Proc Natl Acad Sci USA 104:17707–17712

Sibly RM, Calow P (1986) Physiological ecology of animals. Blackwell Scientific Publications

Sinervo B (1990) Evolution of thermal physiology and growth rate between populations of the western fence lizard (*Sceloporus occidentalis*). Oecologia 83:228–237

Sinervo B, Doughty P, Huey RB, Zamudio K (1992) Allometric engineering: a causal analysis of natural selection on offspring size. Science 258:1927

Smith MH, McGinnis JT (1968) Relationships of latitude, altitude, and body size to litter size and mean annual production of offspring in *Peromyscus*. Res Popul Ecol 10:115–126

Snell-Rood E, Badyaev AV (2008) Ecological gradient of sexual selection: elevation and song elaboration in finches. Oecologia 157:545–551

Stearns SC (1976) Life-history tactics: a review of the ideas. Q Rev Biol 51:3–47

Stearns SC (1992) The evolution of life histories, vol 249. Oxford University Press, Oxford

Stenström A (1999) Sexual reproductive ecology of *Carex bigelowii*, an arctic-alpine sedge. Ecography 22:305–313

Stenström A, Jónsdóttir IS (2006) Effects of simulated climate change on phenology and life history traits in *Carex bigelowii*. Nord J Bot 24:355–371

Stenström A, Jónsdóttir IS, Augner M (2002) Genetic and environmental effects on morphology in clonal sedges in the Eurasian Arctic. Am J Bot 89:1410–1421

Sternberg D, Kennard MJ (2013) Environmental, spatial and phylogenetic determinants of fish life-history traits and functional composition of Australian rivers. Freshwater Biol 58:1767–1778

Stöcklin J (1992) Umwelt, Morphologie und Wachstumsmuster klonaler Pflanzen—eine Übersicht. Bot Helv 102:3–21

Stöcklin J, Kuss P, Pluess AR (2009) Genetic diversity, phenotypic variation and local adaptation in the alpine landscape: case studies with alpine plant species. Bot Helv 119:125–133

Tanaka S, Brookes VJ (1983) Altitudinal adaptation of the life cycle in *Allonemobius fasciatus* DeGeer (Orthoptera: Gryllidae). Can J Zool 61:1986–1990

Tatar M, Gray DW, Carey JR (1997) Altitudinal variation for senescence in *Melanoplus* grasshoppers. Oecologia 111:357–364

Tinkle DW, Gibbons JW (1977) The distribution and evolution of viviparity in reptiles. Misc Publ Mus Zool Univ Mich 154:1–55

Toräng P, Wunder J, Obeso JR, Herzog M, Coupland G, Ågren J (2015) Large-scale adaptive differentiation in the alpine perennial herb *Arabis alpina*. New Phyt 206:459–470

Venn SE, Morgan JW (2009) Patterns in alpine seedling emergence and establishment across a stress gradient of mountain summits in south-eastern Australia. Plant Ecol Diver 2:5–16

Villellas J, Doak DF, García MB, Morris WF (2015) Demographic compensation among populations: what is it, how does it arise and what are its implications? Ecol Lett 18:1139–1152

Vitt LJ, Caldwell JP (2013) Herpetology: an introductory biology of amphibians and reptiles. Academic Press

Vucetich JA, Waite TA (2003) Spatial patterns of demography and genetic processes across the species' range: null hypotheses for landscape conservation genetics. Conserv Genet 4:639–645

Wachter GA, Arthofer W, DejacoT Rinnhofer LJ, Steiner FM, Schlick-Steiner BC (2012) Pleistocene survival on central Alpine nunataks: genetic evidence from the jumping bristletail *Machilis pallida*. Mol Ecol 21:4983–4995

Wang Y, Wang JJ, Lai LM, Jiang LH, Zhuang P, Zhang LH, Zheng YR, Baskin JM, Baskin CC (2014) Geographic variation in seed traits within and among forty-two species of *Rhododendron* (Ericaceae) on the Tibetan plateau: relationships with altitude, habitat, plant height, and phylogeny. Ecol Evol 4:1913–1923

Watson MA (1984) Developmental constraints—effect on population growth and patterns of resource allocation in a clonal plant. Am Nat 123:411–426

Watson MA, Casper BB (1984) Morphogenetic constraints on patterns of carbon distribution in plants. Ann Rev Ecol Syst 15:233–258

Westoby J, Jurado E, Leishmann M (1992) Comparative evolutionary ecology of seed size. Trends Ecol Evol 7:368–372

Wilhelm FM, Schindler DW (2000) Reproductive strategies of *Gammarus lacustris* (Crustacea: Amphipoda) along an elevation gradient. Funct Ecol 14:413–422

Williams CM, Henry HA, Sinclair BJ (2015) Cold truths: how winter drives responses of terrestrial organisms to climate change. Biol Rev 90:214–235

Winkler DW, Dunn PO, McCulloch CE (2002) Predicting the effects of climate change on avian life-history traits. Proc Natl Acad Sci USA 99:13595–13599

Wipf S, Stoeckli V, Bebi P (2009) Winter climate change in alpine tundra: plant responses to changes in snow depth and snowmelt timing. Clim Change 94:105–121

Wookey PA, Parsons AN, Welker JM, Potter JA, Callaghan TV, Lee JA, Press MC (1993) Comparative responses of phenology and reproductive development to simulated environmental change in sub-arctic and high arctic plants. Oikos 67:490–502

Wu GL, Du GZ (2009) Seed mass in *Kobresia*-dominated communities in alpine meadows at two different elevations. Isr J Ecol Evol 55:31–40

Wynne-Edwards KE (1995) Biparental care in Djungarian but not Siberian dwarf hamsters (*Phodopus*). Anim Behav 50:1571–1585

Wynne-Edwards KE (1998) Evolution of parental care in *Phodopus*: conflict between adaptations for survival and adaptations for rapid reproduction. Am Zool 38:238–250

Yeh PJ, Price TD (2004) Adaptive phenotypic plasticity and the successful colonization of a novel environment. Am Nat 164:531–542
Yoccoz NG, Ims RA (1999) Demography of small mammals in cold regions: the importance of environmental variability. Ecol Bull 47:137–144
Zammuto RM, Millar JS (1985) Environmental predictability, variability, and *Spermophilus columbianus* life history over an elevational gradient. Ecology 66:1784–1794
Zettel J (2000) Alpine Collembola—adaptations and strategies for survival in harsh environments. Zool Anal Complex Syst 102:73–89
Zhang L, Lu XIN (2012) Amphibians live longer at higher altitudes but not at higher latitudes. Biol J Linn Soc 106:623–632

Chapter 12
Non-equilibrium in Alpine Plant Assemblages: Shifts in Europe's Summit Floras

Christian Rixen and Sonja Wipf

Abstract Climate warming has been more pronounced in Arctic and alpine areas, and changes in the mountain flora can be expected as the temperature envelope moves upslope. On the one hand, alpine habitats will shrink due to upward migration of species from lower areas, such as trees and tall plants. On the other hand, extinctions of summit plants may be slowed down considerably by the high diversity of microhabitats, the longevity of alpine plants and positive plant–plant interactions in extreme environments. This review chapter attempts to document and monitor vegetation changes on mountain summits. Vegetation surveys that repeat century-old historical vegetation records show considerable upward migration and subsequent increases in species on summits. This trend apparently has accelerated in recent decades. Detailed monitoring of the last decade in European mountain ranges, however, shows that this vegetation change may be at the cost of rare endemic species and alpine specialists in drier Mediterranean regions. This chapter furthermore reviews other factors than temperature influencing alpine vegetation, namely precipitation and snow, nutrients, atmospheric CO_2 concentrations and land use. A subsequent question is how threatened mountain flora is by the ongoing environmental changes. Finally, this chapter discusses options for conservation and land use in high-alpine areas.

Keywords Climate change · Alpine plants · Long-term monitoring · Warming · Snow · High mountain conservation

---

C. Rixen (✉) · S. Wipf
WSL Institute for Snow and Avalanche Research SLF, Flüelastrasse 11, CH-7260 Davos Dorf, Switzerland
e-mail: rixen@slf.ch

J. Catalan et al. (eds.), *High Mountain Conservation in a Changing World*, Advances in Global Change Research 62, DOI 10.1007/978-3-319-55982-7_12

12.1 Introduction

Mountain plant species are already showing strong responses to climate change, for instance through upwards shifts in distribution limits (Grabherr et al. 1994; Walther et al. 2002; Lenoir et al. 2008). Species distribution models predict that this will lead to a contraction or total loss of high-alpine species' distribution ranges in the longer term (Engler et al. 2011), as their potential new habitat decreases in area at higher altitude (Körner 2007), while they might become out-competed and replaced by species from lower elevations (Engler et al. 2011). Through these mechanisms, species distribution models predict losses of over one-third of all species of the alpine vegetation belt for some regions of the Alps, and even higher extinction rates in other European mountain ranges (Engler et al. 2011).

This chapter will give an overview of our current knowledge of vegetation change in alpine regions with a particular focus on mountain summits in the Swiss Alps and across Europe. First, results from different monitoring approaches will be reviewed. One way to study vegetation changes is to repeat historical surveys, as many historical species lists from mountain summits are available from about a century ago, in some cases even from 170 years ago. Another suitable approach is standardised monitoring that was initiated relatively recently but capture shorter term vegetation changes in great detail (e.g. Roth et al. 2014). The Global Observation Research Initiative in Alpine Environments (GLORIA) for instance was initiated in 2001 across many European mountains and has now research sites on summits all over the world (Grabherr et al. 2000). The first analyses from GLORIA have demonstrated rapid vegetation changes on European summits (Pauli et al. 2012; Gottfried et al. 2012).

This chapter will then focus on different factors influencing mountain vegetation. The most discussed cause for vegetation changes is climate warming, but it is important to take also other factors of global change into consideration, such as atmospheric CO_2 concentrations, nutrient availability, land use, etc. Although temperature is, without a doubt, an important climatic driver of alpine plant distribution, it probably strongly interacts with precipitation and soil moisture (Elmendorf et al. 2012a, b) especially in the form of snow (Grytnes et al. 2014).

Given ongoing climate and vegetation change, the question arises how threatened mountain flora actually is. On the one hand, habitat for high-alpine specialists will most likely shrink in a warming climate. On the other hand, the high diversity of microhabitats on mountains (Scherrer and Körner 2011) and the longevity of many mountain plants may prevent extinctions or at least result in a delayed extinction debt (Dullinger et al. 2012). This book chapter will outline our current knowledge about the extinction risk of alpine plants.

Finally, the key question remains whether humans can contribute to the preservation of alpine plants or to prevent their local extinctions. Hence, the final section of this chapter will outline opportunities for conservation, appropriate forms of land-use, conservation and restoration measures in high-alpine environments.

12.2 Rapid Climate Change in Arctic and Alpine Areas

Mountain ecosystems are projected to experience more dramatic climate warming than most other regions of the world (Pepin et al. 2015; IPCC 2014). The Swiss Alps, for instance, have already experienced a warming of 1.8 °C since the Little Ice Age in the mid-nineteenth century (Begert et al. 2005) (Fig. 12.1), and the warming during the past 30 years was twice as high compared with the Northern Hemisphere (Böhm et al. 2001; Rebetez and Reinhard 2008). Since then the duration of snow cover has decreased in many regions of the world (IPCC 2007), and glaciers in the Alps have lost about 35% of their surface area (Hoelzle et al. 2007). Migration of plants and animals to higher elevations are impressive indicators for these profound changes in climate (Walther et al. 2002; Seimon et al. 2007).

12.3 Re-surveys of Historical Vegetation Records on Summits

Several studies have used re-surveys of historical data of summits floras to study long-term vegetation changes in high-alpine regions (Grabherr et al. 1994; Klanderud and Birks 2003; Walther et al. 2005). Summits are easy to relocate, which makes them equivalent to permanent plots. Moreover, summits are particularly important in the context of climate change-driven upward shifts, as they represent the last resort before species go extinct due to the absence of suitable habitats at even higher altitudes. Summits might thus provide one of the most exact,

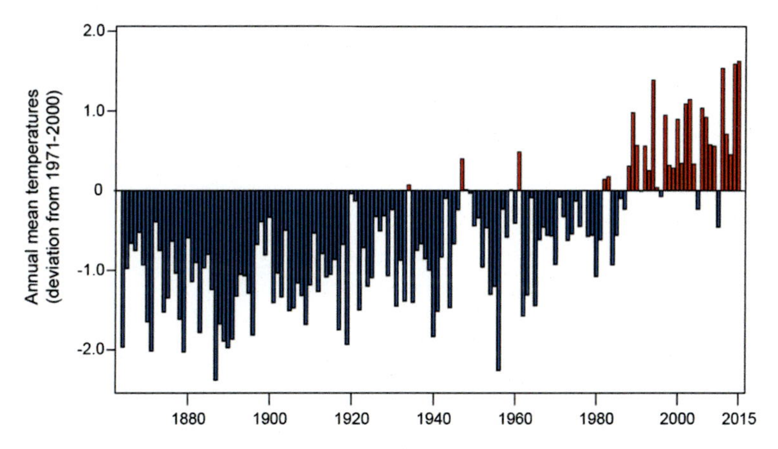

Fig. 12.1 Temperature anomalies (annual deviation from long-term mean) since the Little Ice Age at five climate stations (mean values) in Switzerland above 1000 m a.s.l. (Grand St. Bernard, Sils Maria, Davos, Engelberg, Säntis). Based on data from Begert et al. (2005)

most sensitive, and longest term indicators for floristic change and species loss due to climate change in the world. As a baseline for these re-survey studies serves plant species lists of mountain summits collected in the late nineteenth and early twentieth century by some of the most renowned botanists of their time, who were aiming to explore the elevation limits of vascular plant life (Stöckli et al. 2011). Europe is unique in harbouring a large number of such historical datasets, and Swiss botanists had a leading role. Almost 200 historical summit records of high quality (Stöckli et al. 2011) exist from Switzerland alone, and dozens more from the French and Italian Alps, the Pyrenees, the Scottish Highlands and the Scandes (e.g. Moen and Lagerstrom 2008; Odland et al. 2010; Grytnes et al. 2014; Klanderud and Birks 2003).

In general, previous re-survey studies on summits found an enrichment of the plant community and that species from lower elevation had been colonising higher elevations over the past century (see Fig. 12.2; Hofer 1992; Grabherr et al. 1994, 2001; Camenisch 2002; Walther et al. 2005; Holzinger et al. 2008; Kullman 2010; Wipf et al. 2013a). The rate of upward migration of plant species varied between studies, ranging from 4 (Grabherr et al. 2001) to 28 m per decade (Walther et al. 2005). However, these results were based on studies with relatively few samples (approx. 30 summits). While most studies suggest climate warming as a main driver of these changes, changes in winter precipitation might be an additional factor that fosters high-alpine community change (Grytnes et al. 2014).

Analyses of species traits indicated that species with seeds adapted to long-distance dispersal (i.e. with wings or similar) were particularly successful new

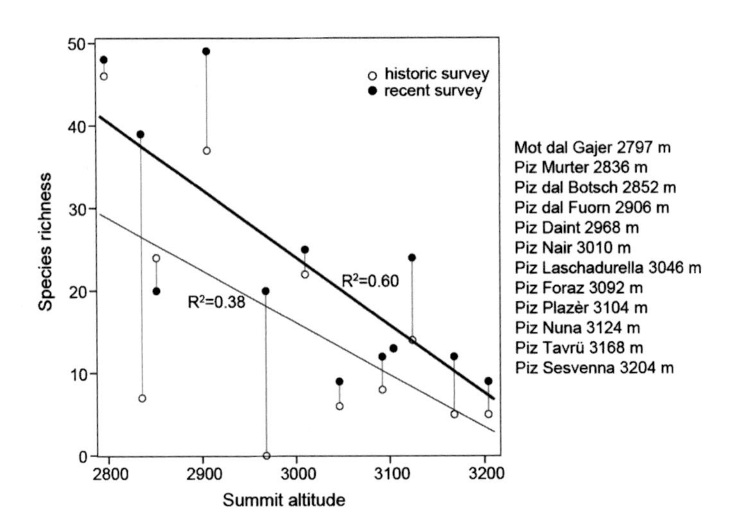

Fig. 12.2 Species numbers on 12 summits in the Swiss National Park region along a gradient in elevation as recorded in historical times by Josias Braun-Blanquet (1911–1927, *thin line*) (Braun-Blanquet 1958) and in recent times (2010–2012, *bold line*). Based on Wipf et al. (2013a)

colonisers (Holzinger et al. 2008; Vittoz et al. 2009; Matteodo et al. 2013). Generally, trait characteristics of new colonisers on summits were similar to those of lowland communities (Matteodo et al. 2013), further illustrating the general upwards trend of plants in mountain areas. Also, there is evidence that the biodiversity change has accelerated during the past 30 years comparable to recent temperature increase (Wipf et al. 2013b; Walther et al. 2005).

The single most prominent example for long-term vegetation change on a mountain summit is certainly Piz Linard in South East Switzerland (Wipf et al. 2013b; see Fig. 12.3). This mountain was first visited and botanized by Oswald Heer in 1835. He then recorded only a single plant species at the summit (*Androsace alpina*, Alpine Rock-Jasmine, Primulaceae) but noted many other plant species at a lower elevation of the same mountain (Fig. 12.3). Piz Linard was then re-visited and re-botanized eight times, the last time in 2011, making a total of nine botanical records in 176 years (Table 12.1). Species numbers had increased due to

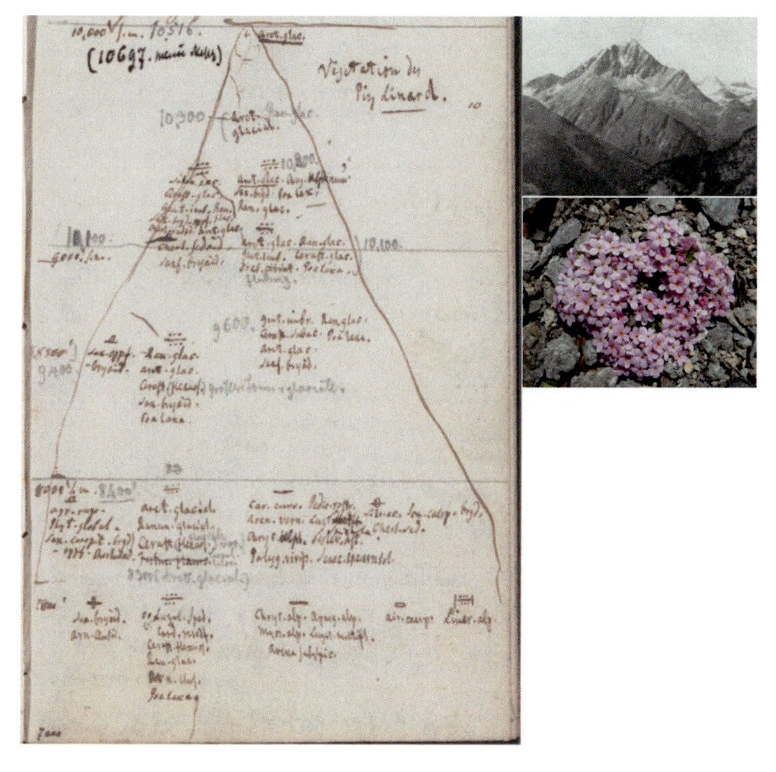

Fig. 12.3 Excerpt from Oswald Heer's 1835 notebook (Heer 1835). Diagram of species occurrences on Piz Linard with *Aretia glacialis* = *Androsace alpina* (small photo; C. Rixen) at the summit. A transcript of this figure with modern species names is available in Wipf et al. (2013b). The photo of Piz Linard was taken by Josias Braun-Blanquet (Braun-Blanquet 1957), one of the re-surveys in the 1930s and 1940s. Reprinted by permission of the publisher (Taylor & Francis Ltd, http://www.tandfonline.com)

Table 12.1 Species occurrences and abundances on Piz Linard summit (3410 m a.s.l.; uppermost 30 m) from 1835–2011. Abundances are indicated by colour: light grey, low (<5 individuals); intermediate grey, intermediate (<10 individuals); dark grey, high (>10 individuals). No abundances are available for the 1864 and 1895 records. Species present in the uppermost 10 m are indicated with *double asterisks*. Highest observations of the species elsewhere in south-eastern Switzerland up to 1911 are listed for comparison. See details in Wipf et al. (2013b). In 2014, Doronicum clusii was rediscovered on the summit, but no full species record was taken (Wipf et al. personal communication)

	Presence on Piz Linard summit									Highest historical observation elsewhere		
Species	1835	1864	1895	1911	1937	1947	1992	2003	2011	Altitude (m)	Location	Year
Androsace alpina (L.) Lam.	**	**	**	**	**	**	**	**	**	3400[+]	Piz Kesch[1]	1894
Leucanthemopsis alpina (L.) Heywood		**		**	**	**			*	3400[+]	Piz Tschierva[2]	1906
Ranunculus glacialis L.		**	**	**	**	**	**	**	**	3500	Piz Tschierva[2]	1906
Saxifraga bryoides L.			**	**	**	**	**	**	**	3418	Piz Kesch[1]	1894
Saxifraga oppositifolia L.			**	**	**	**	**	**	**	3465	Piz Palü[3]	1835
Poa laxa Haenke				**	**	**	**	**	**	3400[+]	Piz Kesch[1]	1894
Draba fladnizensis Wulfen				**	**	**	*	*	*	3400[+]	Piz Kesch[1]	1894
Gentiana bavarica L.				*	*	*	*	*	*	3400	Piz Tschierva[2]	1906
Cerastium uniflorum Clairv.					*	*	*	*	**	3400	Piz Kesch[1]	1894
Saxifraga exarata Vill.					*	*		*	**	3380	Piz Kesch[1]	1894
Luzula spicata (L.) DC.							*	*	*	3262[+]	Piz Languard[4]	1905
Cardamine resedifolia L.							*		*	3280	Piz Julier[4]	1910
Sedum alpestre Vill.								*	*	3250	Piz Languard[4]	1905
Doronicum clusii (All.) Tausch								*		3260[+]	Piz Languard[4]	1905
Cerastium pedunculatum Gaudin									**	3100[+]	Munt Cotschen[3]	1835
Erigeron uniflorus L.									*	3262[+]	Piz Languard[4]	1905
Gnaphalium supinum L.									*	3262[+]	Piz Languard[4]	1905
Total species number	1	3	4	8	10	10	10	12	16			

[+]found at higher altitude on Piz Linard in 1911.
Source: [1]Schibler (1897, 1929); [2]Rübel (1912); [3]Heer (1885); [4] Braun (1913).

upward migration to a total of 16 species in 2011. Interestingly, the species number stagnated (at ten species) during three records between 1937 and 1992 but then increased considerably to 16 species in the past decades, which matches well the recent temperature increase. Most already present species increased in abundance and colonised new areas of the summit, while new arrivals mainly established at sites with already high species richness (Wipf et al. 2013b). Species that appeared after 1992 differed from species already present previously by having had a 200 m lower maximum altitude in the region during the early twentieth century. Although the conclusions that can be drawn from one single mountain are limited, the example of Piz Linard is nevertheless highly illustrative of ongoing vegetation changes on summits that are also supported by larger studies.

12.4 Extensive Monitoring of Recent Changes in Summit Plants

Re-sampling of historical vegetation surveys proved to be a very useful tool to study long-term vegetation changes on summits, but has the downside of some methodological uncertainties (Stöckli et al. 2011; Burg et al. 2015). This problem

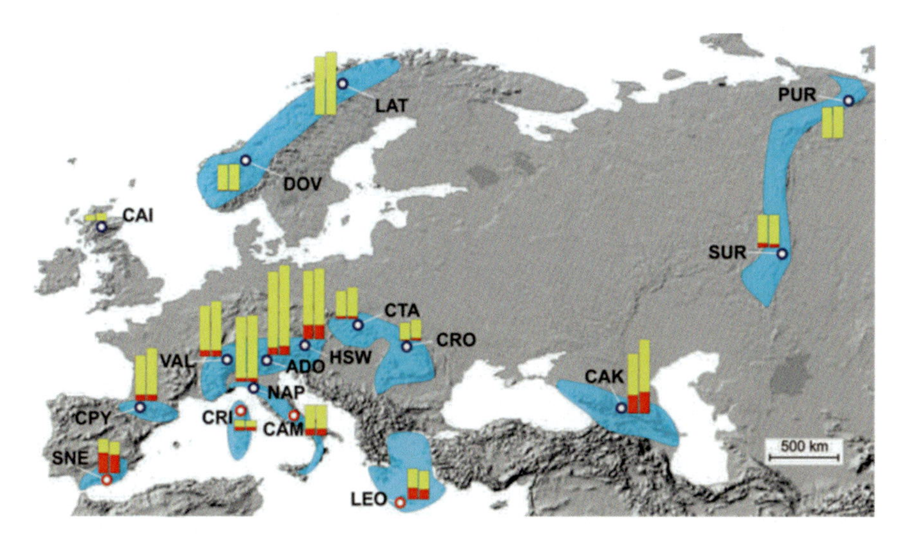

Fig. 12.4 GLORIA network (Pauli et al. 2012). Vascular plant species numbers in 17 European study regions. *Blue circles* indicate boreal and temperate, *red circles* indicate Mediterranean regions. Bars show the number of species found in 2001 (*left bar*) and 2008 (*right bar*); the proportion of endemic species is shown in *red*. Species number (endemic number) per region in 2001/in 2008: LAT (N-Scandes/Sweden, 109(0)/118(0); PUR (Polar Urals/Russia), 58(0)/60(0); DOV (S-Scandes/Norway), 49(1)/50(1); CAI (Cairngorms/UK), 10(0)/14(0); SUR (S-Urals/Russia), 62(9)/62(7); CTA (High Tatra/Slovakia), 53(5)/60(5); HSW (NE-Alps/Austria), 130(27)/134(27); CRO (E-Carpathians/Romania), 33(2)/40(5); ADO (S-Alps/Italy), 158(14)/170(17); VAL (W-Alps/Switzerland), 96(12)/105(12); NAP (N-Apennines/Italy), 123(7)/126(7); CPY (Central Pyrenees/Spain), 87(12)/101(12); CAK (Central Caucasus/Georgia), 113(35)/140(41); CRI (Corsica/France), 20(7)/19(7); CAM (Central Apennines/Italy), 57(13)/57(13); SNE (Sierra Nevada/Spain), 65(39)/60(35); LEO (Lefka Ori-Crete/Greece), 58(22)/54(19). *Blue-shaded areas* indicate the respective maximum distribution of species defined as endemic (12); most endemics have a far more narrow distribution area. From Pauli et al. (2012). Reprinted with permission from AAAS

was addressed in the Global Observation Research Initiative in Alpine Environments (GLORIA, http://www.gloria.ac.at), which provides a detailed protocol to record summit vegetation (Grabherr et al. 2000). Within less than a decade, significant changes in vegetation were already found in monitoring sites across Europe that were set up in 2001 and repeated in 2008 (Pauli et al. 2012; Gottfried et al. 2012) (and very recently in 2015). In the boreal-temperate mountain regions of Europe, species number had increased by nearly four species on average (Fig. 12.4; Pauli et al. 2012). In Mediterranean mountain regions, however, species number had decreased by ca. 1.5 species, possibly because recent climatic trends have decreased the availability of water in the European south.

Another interesting analysis of the same GLORIA data set looked at how much the vegetation change indicated warmer conditions, i.e. if species migrating upwards reported a warmer environment than before (so-called thermophilisation, Gottfried et al. 2012). Across the entire data set, the vegetation indicated

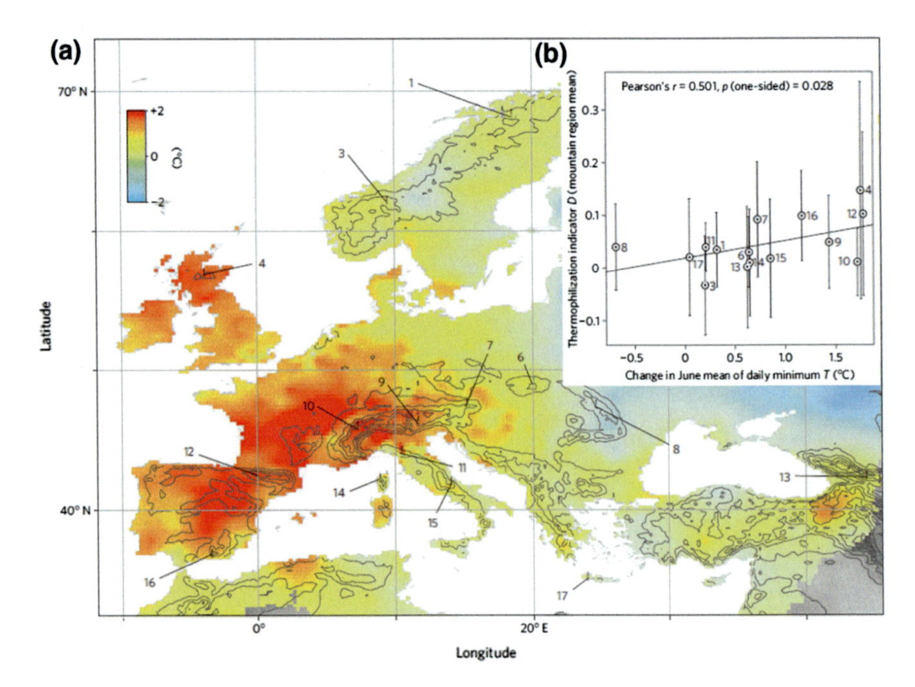

Fig. 12.5 Summit thermophilisation. The thermophilisation indicator D of mountain regions is correlated with temperature change. **a** Change in June mean of daily minimum temperature (map prepared from data provided by E-OBS (Haylock et al. 2008), resolution 0:25°), calculated as the difference between the averages of two time periods that precede plant data recording: prior 2008 (2003–2007)–prior 2001 (1996–2000). The numbers indicate the mountain regions and are referenced in (Gottfried et al. 2012). **b** Correlation of D with the change in June mean of daily minimum temperature (prior 2008–prior 2001) in the study regions (data derived from the map in a), using a one-sided test following the null hypothesis of no positive correlation. Vertical lines are 95% confidence intervals of D for the mountain regions, and a linear regression line is shown. Reprinted by permission from Macmillan Publishers Ltd: Nature, Gottfried et al. (2012)

thermophilisation, however, differences between mountain ranges were considerable. Most importantly, regions with most pronounced warming during the study period showed the highest thermophilisation (Fig. 12.5). These results illustrate how relatively rapid vegetation changes to climate warming can be.

12.5 Global Change, Not Only Climate Change: Snow Versus Temperature, Impacts of Nutrients, CO_2 Concentration, Land Use, Grazing

Temperature is one of the most important factors influencing high-alpine vegetation (Körner 2003). However, it always needs to be considered in combination with precipitation and, more specifically, snow. Temperature and light are responsible for most physiological processes in alpine plants (Körner 2003), however, while snow is lying on the ground, warm atmospheric temperatures are of little effect on the plants below the snow (but see Starr and Oberbauer 2003; Palacio et al. 2015). Furthermore, there is evidence that with climate warming precipitation in the form of snow can actually increase and subsequently delay the timing of snowmelt and hence shorten the vegetation period (Bjorkman et al. 2015). Hence, to understand climate effects on alpine vegetation, we need to know summer temperatures and the timing of snowmelt.

Unfortunately, detailed snow information with high temporal and spatial resolution is often not easy to obtain. One possible approach is to use information of climate stations that not only record temperature but also snow cover below the stations. This approach has been used in the Swiss Alps where more than 100 metro stations have been employed since 1998 (Jonas et al. 2008; Rammig et al. 2010; Fontana et al. 2008). Plant phenology and productivity were analysed between stations and between years, and variables related to precipitation and snowmelt explained as much variance or more than temperature variables (Jonas et al. 2008). Also, vegetation change on Piz Linard (see above, Wipf et al. 2013b) and on Scandinavian mountains (Grytnes et al. 2014) seemed to be partly driven by the snow distribution on the summit. Furthermore, the small-scale distribution of snow in complex alpine terrain is extremely important for the distribution of plants: within the same elevation, the date of snowmelt can differ by more than a month within a few metres depending on topography (Rixen et al. 2010), which influences vegetation and plant populations considerably (for studies on the snowbed species *Salix herbacea* see Wheeler et al. 2015; Sedlacek et al. 2015; Cortes et al. 2014). Hence, future efforts should clearly focus on not only explaining vegetation changes by temperature but by a combination of temperature and precipitation/snow cover.

Apart from temperatures and precipitation, also factors such as nutrient input, elevated atmospheric CO_2, extreme events, land use, grazing, etc. need to be taken into consideration as drivers of vegetation change. Although nitrogen input is usually smaller (Hiltbrunner et al. 2005) and land-use less intensive at high elevation compared to lowlands, they are by no means negligible (Boutin et al. 2015). There is evidence that nitrogen deposition could affect alpine plants more than climate warming (Bobbink et al. 2010). Grazing by sheep can be observed up to the highest alpine grasslands e.g. in the Alps or the Pyrenees, and might over the long term have changed alpine vegetation composition profoundly. Abandonment of remote or steep areas, which is common e.g. in the Pyrenees and the Alps, is hence likely to change vegetation again, but in combination with climate change it is unlikely that vegetation will change back to its previous composition. Elevated atmospheric CO_2 concentrations did not enhance plant growth in alpine grasslands (Inauen et al. 2012; Korner et al. 1997) but in shrub communities at treeline (Anadon-Rosell et al. 2014; Dawes et al. 2013, 2014) where the bilberry (*Vaccinium myrtillus*) showed more growth, possibly at the cost of smaller or less responsive plant species.

12.6 Alpine Plants on the Verge to Extinction or Safe in Cold Microhabitats?

Although most studies demonstrate an impressive increase in species numbers and local colonisations, only relatively little local extinction, and no particular traits or species groups that were mainly affected by local extinctions, were found in long-term vegetation surveys in European alpine ecosystems (Hofer 1992; Walther et al. 2005; Wipf et al. 2013b; Grabherr et al. 2001; Matteodo et al. 2013). Moreover, the summit flora has been found to become more similar in composition over time, and there is evidence that many high-alpine species that were already present on few summits have meanwhile also colonised further summits (Kammer et al. 2007; Jurasinski and Kreyling 2007). Thus, up to now, we see many winners, but few losers on Europe's mountain summits even after several decades of ongoing climate warming.

Species distribution models recently predicted mountain flora to be threatened unequally across Europe in the twenty-first century (Engler et al. 2011). Specifically, temperature increase and precipitation decrease are expected to be more pronounced in e.g. the Alps and the Pyrenees than, e.g. in the Norwegian Scandes (Engler et al. 2011), which can be seen, in part, already in the temperature changes occurred in the recent past (Gottfried et al. 2012). Short-term floristic changes on European summits analysed by the GLORIA initiative indeed indicate a signal towards an increased prevalence of species with higher temperature preferences over 8 years that correlates with the magnitude of recent warming (see above, Gottfried et al. 2012).

While most studies agree that species upwards shifts are already happening, there is little consensus on potential losses of alpine biota due to future climate change. At first glance, the modelled projections of massive extinction rates in high-alpine species, and the observational findings of strong increase in summits species numbers even contradict each other. However, as the expected local extinctions are thought to be driven by competition through species rising to higher altitudes, it could be expected that an initial enrichment with new colonisers will be followed by an extinction of the formerly local species after a certain time lag (Dullinger et al. 2012; Engler et al. 2009). On the other hand, evidence for competitive replacement of high-alpine species is, at best, weak, even after decades of ongoing climate warming. Also, species in cold habitats are assumed to be less affected by competition (Pellissier et al. 2013) and their niches to be more closely related to their physiological limits (Normand et al. 2009). However, these studies do not take into account that some alpine species are true cold species, i.e. that they are not able to adapt their physiology (dark respiration) to a warmer temperature (Larigauderie and Körner 1995).

In contrast to the massive range contractions and high extinction rates among high-alpine plants predicted by species distribution models, recent micrometeorological studies show that due to the large variety of different microhabitats on a small spatial scale, the alpine belt offers a large number of small-scale "refugia" that

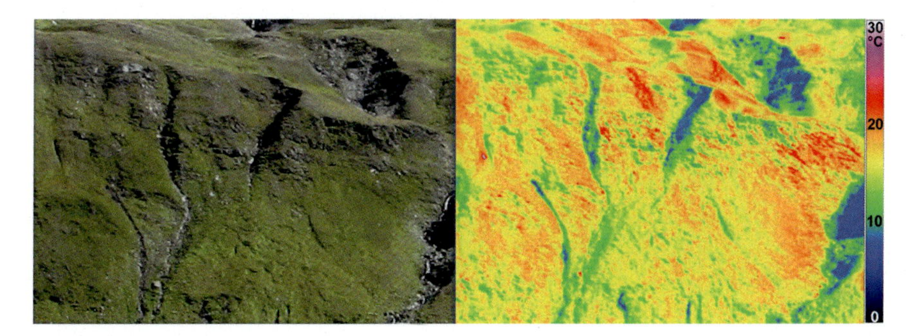

Fig. 12.6 Topography and surface temperatures on an NNW exposed slope at the Furka Pass in the Swiss Alps (elevation gradient of c. 100 m at c. 2450 m asl) on 29 August 2008, under full direct solar radiation (12–18 h). Topography, slope and aspect create a mosaic of habitats with very different temperatures. During one growing season temperature means of different microhabitats can differ by more than 10 °C

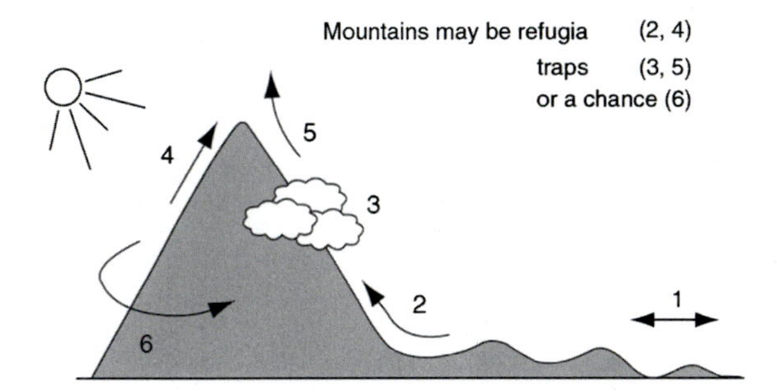

Fig. 12.7 Where to go in a warmer climate? Species from the lowlands may have difficulties to find suitable habitat as migration distances are long (1). Mountains can be refugia (2, 4) or traps (3, 5 if cloud forests shrink or mountains are to low). But often appropriate habitat can be nearby due to the mosaic of microhabitats on mountains (6). Reprinted from Körner (2013), with permission from Elsevier

could potentially meet the climatic requirements of high-altitude plants under warmer climate (Fig. 12.6, Scherrer and Körner 2011). Warmer "current micro-sites" and colder "future refugia" can persist at the same altitude, thus within a small distance of each other (Fig. 12.7). According to these studies, high-alpine plants should be well buffered against climate change, as they will only have to disperse over small distances to reach microsites that correspond with their climatic niche, rather than over large altitudinal distances as assumed by species distribution models. The point, however, is not so much about dispersion but about finding favourable sites to install and to grow when the place is already occupied. Many alpine species are in fact pioneer plants.

Support for the hypothesis of co-existence in separate microsites also arises from species distribution models themselves: if they operate with a spatial resolution too large to reflect small-scale microclimatic variability in the terrain, their predictions of extinction risks will be too high, and models with finer scales end up with strikingly lower losses of high-alpine species (Randin et al. 2009). Thus, in alpine terrain with its high microsite diversity, coexistence between new colonisers and persisting high-alpine species may be possible if they do not show any niche overlap nor occupy the same microsites.

Even if species occupy the same microsite, they do not automatically out-compete each other. Neighbour facilitation, whereby plant individuals benefit from the presence of their interspecific neighbours, is a widespread phenomenon, especially under harsh environmental conditions (Brooker et al. 2008; Callaway et al. 2002; Choler et al. 2001; Wipf et al. 2006). Positive neighbour interactions can affect alpine plant diversity as much as climate (Cavieres et al. 2014). Neighbours can, e.g. ameliorate the microclimatic, environmental, and soil conditions while competing for the same resources at the same time (space, light, nutrients). If this facilitative force outbalances the competition, then facilitation fosters the coexistence of plant species on a small spatial scale (Kikvidze et al. 2001; Rixen and Mulder 2009) and could also play a major role in the colonisation of new sites through species from lower altitudes. There is even evidence that positive species interactions can extend species distributions into otherwise unfavourable habitats (le Roux et al. 2012). Hence, it is conceivable that facilitative neighbour interactions enable the coexistence of high-alpine species and new colonisers on mountain summits, which could counterbalance projected extinctions. Nevertheless, shifts in net interactions with environmental severity may differ among indicators of severity, growth forms and scales (Dullinger et al. 2007). Ongoing and future research will need to target at understanding if upward migration of plant species will lead to a loss of high-alpine specialists, or if the mosaic of microhabitats within one elevation range will provide enough buffer to prevent species loss, or if facilitation between neighbours enables the coexistence in the same microhabitat.

12.7 From Knowledge to Action? Towards Conservation of High Mountain Flora

Facing ongoing climate and vegetation change, the question remains if plants can adapt to new conditions and if humans can preserve alpine plants and prevent extinctions. We have seen above that the small-scale heterogeneity of the alpine landscape may provide habitat for alpine plants in a changing climate (Scherrer and Körner 2011). Adaptation of alpine plants through gene flow may also provide mechanisms to withstand changing environmental conditions (Cortes et al. 2014).

Nevertheless, upward migration of trees and plants from lower elevation will reduce the area with high-alpine habitat, and, in mountain ranges with human land use, measures for conservation and restoration need to be considered. On the one hand, moderate grazing can prevent or slow down tall competitive plants from outcompeting small alpine plants. On the other hand, if grazing pressure increases because alpine habitat decreases, erosion in steep areas might be the consequence. Also, pressures

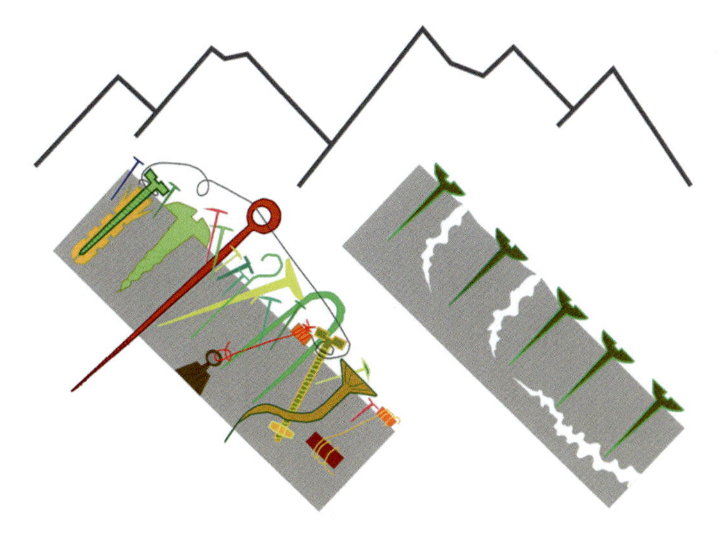

Fig. 12.8 Plants, with their diverse root systems, can be seen as the screws and nails of mountain ecosystems. From Körner and Spehn (2002) and Körner (2004) with permission of Springer Nature

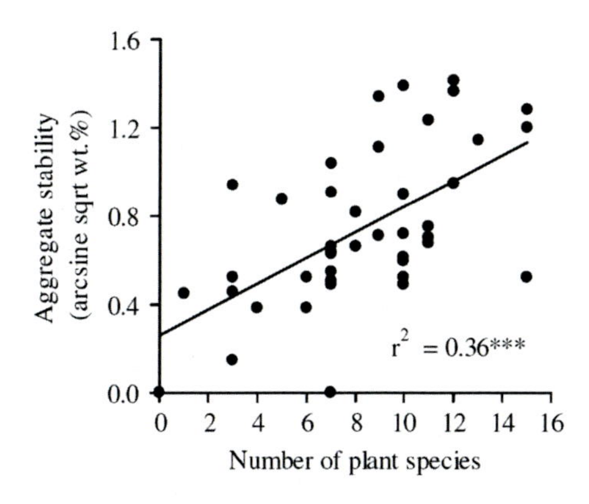

Fig. 12.9 Relationship between aggregate stability (weight percent, arcsine square root transformed) and number of plant species on ski slopes (Pohl et al. 2009). Reprinted with permission of Springer

related to tourism (trampling, skiing, etc.) might become more severe if the alpine area shrinks (Pickering et al. 2003; Rixen and Rolando 2013; Rixen et al. 2011).

An important aspect to prevent erosion in alpine areas is that biodiversity provides an ecosystem function that is particularly relevant in steep terrain, namely soil aggregate stability. Figure 12.8 illustrates the idea that a high number of species and growth forms might be more likely to stabilise the uppermost soil horizons than a monoculture (Körner and Spehn 2002; Körner 2004). Although intuitive, this concept and the hypothesis have not often been tested in alpine vegetation. On disturbed ski slopes in the Swiss Alps, however, it was indeed shown that plant diversity explained soil aggregate stability better than all another measured plant, root and soil parameters (Fig. 12.9, Pohl et al. 2009). Hence, it is important to avoid severe disturbance in sensitive alpine areas and, if necessary, to restore disturbed areas with seeds of adapted plants from high altitudes and with a large number of plant species to provide high plant diversity (Locher Oberholzer et al. 2008).

It clearly remains a challenge for future research to fully understand and appreciate if and how humans can provide conservation measures, appropriate intensities of grazing, etc. in alpine areas to prevent or reduce extinctions of alpine plant species. The established monitoring initiatives to document changes in alpine vegetation (see above) clearly need to be continued to improve our understanding of risks for alpine flora and possibly provide solutions for the future. Mountains are biodiversity hotspots, which provide numerous ecosystem services also for the lowlands, and hence we have a responsibility to maintain their precious habitats and flora.

References

Anadon-Rosell A, Rixen C, Cherubini P, Wipf S, Hagedorn F, Dawes MA (2014) Growth and phenology of three dwarf shrub species in a six-year soil warming experiment at the alpine treeline. PLoS ONE 9:e100577–e100577

Begert M, Schlegel T, Kirchhofer W (2005) Homogeneous temperature and precipitation series of Switzerland from 1864 to 2000. Int J Climatol 25:65–80

Bjorkman AD, Elmendorf SC, Beamish AL, Vellend M, Henry GHR (2015) Contrasting effects of warming and increased snowfall on Arctic tundra plant phenology over the past two decades. Global Change Biol 21:4651–4661

Bobbink R, Hicks K, Galloway J, Spranger T, Alkemade R, Ashmore M, Bustamante M, Cinderby S, Davidson E, Dentener F, Emmett B, Erisman JW, Fenn M, Gilliam F, Nordin A, Pardo L, De Vries W (2010) Global assessment of nitrogen deposition effects on terrestrial plant diversity: a synthesis. Ecol Appl 20:30–59

Böhm R, Auer I, Brunetti M, Maugeri M, Nanni T, Schoner W (2001) Regional temperature variability in the European Alps: 1760–1998 from homogenized instrumental time series. Int J Climatol 21:1779–1801

Boutin M, Lamaze T, Couvidat F, Pornon A (2015) Subalpine Pyrenees received higher nitrogen deposition than predicted by EMEP and CHIMERE chemistry-transport models. Sci Rep 5:12942

Braun-Blanquet J (1957) Ein Jahrhundert Florenwandel am Piz Linard (3414 m). Bulletin du Jardin Botanique de l'Etat Bruxelles Volume jubilaire Walter Robyns:221–232

Braun-Blanquet J (1958) Über die obersten Grenzen pflanzlichen Lebens im Gipfelbereich des Schweizerischen Nationalparks Ergebnisse der wissenschaftlichen Untersuchungen im Schweizerischen Nationalpark

Brooker RW, Maestre FT, Callaway RM, Lortie CL, Cavieres LA, Kunstler G, Liancourt P, Tielbörger K, Travis JMJ, Anthelme F, Armas C, Coll L, Corcket E, Delzon S, Forey E, Kikvidze Z, Olofsson J, Pugnaire F, Quiroz CL, Saccone P, Schiffers K, Seifan M, Touzard B, Michalet R (2008) Facilitation in plant communities: the past, the present, and the future. J Ecol 96:18–34

Burg S, Rixen C, Stoeckli V, Wipf S (2015) Observation bias and its causes in botanical surveys on high-alpine summits. J Veg Sci 26:191–200

Callaway RM, Brooker RW, Choler P, Kikvidze Z, Lortie CJ, Michalet R, Paolini L, Pugnaire FL, Newingham B, Aschehoug ET, Armas C, Kikodze D, Cook BJ (2002) Positive interactions among alpine plants increase with stress. Nature 417:844–848

Camenisch M (2002) Veränderungen der Gipfelflora im Bereich des Schweizerischen Nationalparks: Ein Vergleich über die letzten 80 Jahre. Jahresbericht der Naturforschenden Gesellschaft Graubünden 111:27–37

Cavieres LA, Brooker RW, Butterfield BJ, Cook BJ, Kikvidze Z, Lortie CJ, Michalet R, Pugnaire FI, Schoeb C, Xiao S, Anthelme F, Bjoerk RG, Dickinson KJM, Cranston BH, Gavilan R, Gutierrez-Giron A, Kanka R, Maalouf J-P, Mark AF, Noroozi J, Parajuli R, Phoenix GK, Reid AM, Ridenour WM, Rixen C, Wipf S, Zhao L, Escudero A, Zaitchik BF, Lingua E, Aschehoug ET, Callaway RM (2014) Facilitative plant interactions and climate simultaneously drive alpine plant diversity. Ecol Lett 17:193–202

Choler P, Michalet R, Callaway RM (2001) Facilitation and competition on gradients in alpine plant communities. Ecology 82:3295–3308

Cortes AJ, Waeber S, Lexer C, Sedlacek J, Wheeler JA, van Kleunen M, Bossdorf O, Hoch G, Rixen C, Wipf S, Karrenberg S (2014) Small-scale patterns in snowmelt timing affect gene flow and the distribution of genetic diversity in the alpine dwarf shrub *Salix herbacea*. Heredity 113:233–239

Dawes MA, Hagedorn F, Handa IT, Streit K, Ekblad A, Rixen C, Koerner C, Haettenschwiler S (2013) An alpine treeline in a carbon dioxide-rich world: synthesis of a nine-year free-air carbon dioxide enrichment study. Oecologia 171:623–637

Dawes MA, Zweifel R, Dawes N, Rixen C, Hagedorn F (2014) CO_2 enrichment alters diurnal stem radius fluctuations of 36-yr-old Larix decidua growing at the alpine tree line. New Phytol 202:1237–1248

Dullinger S, Kleinbauer I, Pauli H, Gottfried M, Brooker R, Nagy L, Theurillat JP, Holten JI, Abdaladze O, Benito JL, Borel JL, Coldea G, Ghosn D, Kanka R, Merzouki A, Klettner C, Moiseev P, Molau U, Reiter K, Rossi G, Stanisci A, Tomaselli M, Unterlugauer P, Vittoz P, Grabherr G (2007) Weak and variable relationships between environmental severity and small-scale co-occurrence in alpine plant communities. J Ecol 95:1284–1295

Dullinger S, Gattringer A, Thuiller W, Moser D, Zimmermann NE, Guisan A, Willner W, Plutzar C, Leitner M, Mang T, Caccianiga M, Dirnboeck T, Ertl S, Fischer A, Lenoir J, Svenning J-C, Psomas A, Schmatz DR, Silc U, Vittoz P, Huelber K (2012) Extinction debt of high-mountain plants under twenty-first-century climate change. Nat Clim Change 2:619–622

Elmendorf SC, Henry GHR, Hollister RD, Bjork RG, Bjorkman AJ, Callaghan TV, Cooper E, Cornelissen JHC, Day TA, Fosaa AM, Gould WA, Grétarsdóttir J, Harte J, Hermanutz L, Hik DA, Hofgaard A, Jarrad F, Jonsdottir IS, Keuper F, Klanderud K, Klein JA, Koh S, Kudo G, Lang S, Lowen V, May JL, Mercado J, Michelsen A, Molau U, Pieper S, Robinson CH, Siegart L, Myers-Smith I, Oberbauer SF, Post E, Rixen C, Schmidt NM, Shaver GR, Tolvanen A, Totland O, Troxler T, Wahren CH, Webber PJ, Welker JM, Wookey PA (2012a) Global assessment of simulated climate warming on tundra vegetation: Heterogeneity over space and time. Ecol Lett 15:164–175

Elmendorf SC, Henry GHR, Hollister RD, Bjork RG, Boulanger-Lapointe N, Cooper EJ, Cornelissen JHC, Day TA, Dorrepaal E, Elumeeva TG, Gill M, Gould WA, Harte J, Hik DS, Hofgaard A, Johnson DR, Johnstone JF, Jonsdottir IS, Jorgenson JC, Klanderud K, Klein JA, Koh S, Kudo G, Lara M, Levesque E, Magnusson B, May JL, Mercado-Diaz JA, Michelsen A, Molau U, Myers-Smith IH, Oberbauer SF, Onipchenko VG, Rixen C, Schmidt NM, Shaver GR, Spasojevic MJ, Porhallsdottir PE, Tolvanen A, Troxler T, Tweedie CE, Villareal S, Wahren CH, Walker X, Webber PJ, Welker JM, Wipf S (2012b) Plot-scale evidence of tundra vegetation change and links to recent summer warming. Nat Clim Change 2:453–457

Engler R, Randin CF, Vittoz P, Czaka T, Beniston M, Zimmermann NE, Guisan A (2009) Predicting future distributions of mountain plants under climate change: does dispersal capacity matter? Ecography 32:34–45

Engler R, Randin CF, Thuiller W, Dullinger S, Zimmermann NE, Araujo MB, Pearman PB, Le Lay G, Piedallu C, Albert CH, Choler P, Coldea G, De Lamo X, Dirnbock T, Gegout JC, Gomez-Garcia D, Grytnes JA, Heegaard E, Hoistad F, Nogues-Bravo D, Normand S, Puscas M, Sebastia MT, Stanisci A, Theurillat JP, Trivedi MR, Vittoz P, Guisan A (2011) 21st century climate change threatens mountain flora unequally across Europe. Global Change Biol 17:2330–2341

Fontana F, Rixen C, Jonas T, Aberegg G, Wunderle S (2008) Alpine grassland phenology as seen in AVHRR, VEGETATION, and MODIS NDVI time series—a comparison with in situ measurements. Sensors 8:2833–2853

Gottfried M, Pauli H, Futschik A, Akhalkatsi M, Barancok P, Alonso JLB, Coldea G, Dick J, Erschbamer B, Calzado MRF, Kazakis G, Krajci J, Larsson P, Mallaun M, Michelsen O, Moiseev D, Moiseev P, Molau U, Merzouki A, Nagy L, Nakhutsrishvili G, Pedersen B, Pelino G, Puscas M, Rossi G, Stanisci A, Theurillat JP, Tomaselli M, Villar L, Vittoz P, Vogiatzakis I, Grabherr G (2012) Continent-wide response of mountain vegetation to climate change. Nat Clim Change 2:111–115

Grabherr G, Gottfried M, Pauli H (1994) Climate effects on mountain plants. Nature 369:448–448

Grabherr G, Gottfried M, Pauli H (2000) GLORIA: A global observation research initiative in alpine environments. Mt Res Dev 20:190–191

Grabherr G, Gottfried M, Pauli H (2001) Long-term monitoring of mountain peaks in the Alps. In: Burga CA, Kratochwil A (eds) Biomonitoring: general and applied aspects on regional and global scales. Chapter C. Aspects of global change in the Alps and in the high arctic region. Tasks for Vegetation Science 35. Kluwer Academic Publishers, Dordrecht, pp 153–177

Grytnes J-A, Kapfer J, Jurasinski G, Birks HH, Henriksen H, Klanderud K, Odland A, Ohlson M, Wipf S, Birks HJB (2014) Identifying the driving factors behind observed elevational range shifts on European mountains. Glob Ecol Biogeogr 23:876–884

Haylock MR, Hofstra N, Tank AMGK, Klok EJ, Jones PD, New M (2008) A European daily high-resolution gridded data set of surface temperature and precipitation for 1950–2006. J Geophys Res-Atmos 113

Heer O (1835) Nachlass of Prof. Oswald Heer. Zentralbibliothek Zürich 63:55

Hiltbrunner E, Schwikowski M, Körner C (2005) Inorganic nitrogen storage in alpine snow pack in the central Alps (Switzerland). Atmos Environ 39:2249–2259

Hoelzle M, Chinn T, Stumm D, Paul F, Zemp M, Haeberli W (2007) The application of glacier inventory data for estimating past climate change effects on mountain glaciers: a comparison between the European Alps and the Southern Alps of New Zealand. Glob Planet Change 56:69–82

Hofer HR (1992) Veränderungen in der Vegetation von 14 Gipfeln des Berninagebietes zwischen 1905 and 1985. Berichte des Geobotanischen Institutes der Eidg Technischen Hochschule, Stiftung Rübel 58:39–54

Holzinger B, Hulber K, Camenisch M, Grabherr G (2008) Changes in plant species richness over the last century in the eastern Swiss Alps: elevational gradient, bedrock effects and migration rates. Plant Ecol 195:179–196

Inauen N, Koerner C, Hiltbrunner E (2012) No growth stimulation by CO_2 enrichment in alpine glacier forefield plants. Global Change Biol 18:985–999

IPCC (2007) Climate change 2007—the physical science basis. Cambridge University Press

IPCC (2014) Climate change 2014: synthesis report. contribution of working groups I, II and III to the fifth assessment report of the intergovernmental panel on climate change. Geneva, Switzerland

Jonas T, Rixen C, Sturm M, Stoeckli V (2008) How alpine plant growth is linked to snow cover and climate variability. J Geophys Res-Biogeo 113

Jurasinski G, Kreyling J (2007) Upward shift of alpine plants increases floristic similarity of mountain summits. J Veg Sci 18:711–718

Kammer PM, Schob C, Choler P (2007) Increasing species richness on mountain summits: upward migration due to anthropogenic climate change or re-colonisation? J Veg Sci 18:301–306
Kikvidze Z, Khetsuriani L, Kikodze D, Callaway RM (2001) Facilitation and interference in subalpine meadows of the central Caucasus. J Veg Sci 12:833–838
Klanderud K, Birks HJB (2003) Recent increases in species richness and shifts in altitudinal distributions of Norwegian mountain plants. Holocene 13:1–6
Körner C (2003) Alpine plant life, 2nd edn. Springer Verlag, Berlin
Körner C (2004) Mountain biodiversity, its causes and function. Ambio Spec Rep:11–17
Körner C (2007) The use of 'altitude' in ecological research. Trends Ecol Evol 22:569–574
Körner C (2013) Alpine ecosystems. In: SA L (ed) Encyclopedia of biodiversity. 2nd edn. Elsevier pp 148–157
Körner C, Spehn EM (eds) (2002) Mountain biodiversity: a global assessment. CRC Press, New York
Korner C, Diemer M, Schappi B, Niklaus P, Arnone J (1997) The responses of alpine grassland to four seasons of CO_2 enrichment: a synthesis. Acta Oecologica-Int J Ecol 18:165–175
Kullman L (2010) A richer, greener and smaller Alpine world: review and projection of warming-induced plant cover change in the Swedish Scandes. Ambio 39:159–169
Larigauderie A, Körner C (1995) Acclimation of leaf dark respiration to temperature in alpine and lowland plant species. Ann Bot 76:245–252
le Roux PC, Virtanen R, Heikkinen RK, Luoto M (2012) Biotic interactions affect the elevational ranges of high-latitude plant species. Ecography 35:1048–1056
Lenoir J, Gegout JC, Marquet PA, de Ruffray P, Brisse H (2008) A significant upward shift in plant species optimum elevation during the 20th century. Science 320:1768–1771
Locher Oberholzer N, Streit M, Frei M, Andrey C, Blaser R, Meyer J, Müller U, Reidy B, Schutz M, Schwager M, Stoll M, Wyttenbach M, Rixen C (2008) Richtlinien Hochlagenbegrünung. Ingenierbiologie 2:3–33
Matteodo M, Wipf S, Stoeckli V, Rixen C, Vittoz P (2013) Elevation gradient of successful plant traits for colonizing alpine summits under climate change. Environ Res Lett 8
Moen J, Lagerstrom A (2008) High species turnover and decreasing plant species richness on mountain summits in Sweden: Reindeer grazing overrides climate change? Arct Antarct Alp Res 40:382–395
Normand S, Treier UA, Randin C, Vittoz P, Guisan A, Svenning J-C (2009) Importance of abiotic stress as a range-limit determinant for European plants: insights from species responses to climatic gradients. Glob Ecol Biogeogr 18:437–449
Odland A, Hoitomt T, Olsen SL (2010) Increasing vascular plant richness on 13 high mountain summits in Southern Norway since the early 1970s. Arct Antarct Alp Res 42:458–470
Palacio S, Lenz A, Wipf S, Hoch G, Rixen C (2015) Bud freezing resistance in alpine shrubs across snow depth gradients. Environ Exp Bot 118:95–101
Pauli H, Gottfried M, Dullinger S, Abdaladze O, Akhalkatsi M, Alonso JLB, Coldea G, Dick J, Erschbamer B, Calzado RF, Ghosn D, Holten JI, Kanka R, Kazakis G, Kollar J, Larsson P, Moiseev P, Moiseev D, Molau U, Mesa JM, Nagy L, Pelino G, Puscas M, Rossi G, Stanisci A, Syverhuset AO, Theurillat JP, Tomaselli M, Unterluggauer P, Villar L, Vittoz P, Grabherr G (2012) Recent plant diversity changes on Europe's mountain summits. Science 336:353–355
Pellissier L, Brathen KA, Vittoz P, Yoccoz NG, Dubuis A, Meier ES, Zimmermann NE, Randin CF, Thuiller W, Garraud L, Van Es J, Guisan A (2013) Thermal niches are more conserved at cold than warm limits in arctic-alpine plant species. Glob Ecol Biogeogr 22:933–941
Pepin N, Bradley RS, Diaz HF, Baraer M, Caceres EB, Forsythe N, Fowler H, Greenwood G, Hashmi MZ, Liu XD, Miller JR, Ning L, Ohmura A, Palazzi E, Rangwala I, Schoener W, Severskiy I, Shahgedanova M, Wang MB, Williamson SN, Yang DQ, Mt Res Initiative EDWWG (2015) Elevation-dependent warming in mountain regions of the world. Nature Clim Change 5:424–430
Pickering CM, Harrington J, Worboys G (2003) Environmental impacts of tourism on the Australian Alps protected areas—judgments of protected area managers. Mt Res Dev 23:247–254

Pohl M, Alig D, Körner C, Rixen C (2009) Higher plant diversity enhances soil stability in disturbed alpine ecosystems. Plant Soil 324:91–102

Rammig A, Jonas T, Zimmermann NE, Rixen C (2010) Changes in alpine plant growth under future climate conditions. Biogeosciences 7:2013–2024

Randin CF, Engler R, Normand S, Zappa M, Zimmermann NE, Pearman PB, Vittoz P, Thuiller W, Guisan A (2009) Climate change and plant distribution: local models predict high-elevation persistence. Global Change Biol 15:1557–1569

Rebetez M, Reinhard M (2008) Monthly air temperature trends in Switzerland 1901-2000 and 1975-2004. Theor Appl Clim 91:27–34

Rixen C, Mulder CPH (2009) Species removal and experimental warming in a subarctic tundra plant community. Oecologia 161:173–186

Rixen C, Rolando A (eds) (2013) The impacts of skiing and related winter recreational activities on mountain environments. Bentham. doi:10.2174/97816080548861130101

Rixen C, Schwoerer C, Wipf S (2010) Winter climate change at different temporal scales in *Vaccinium myrtillus*, an Arctic and alpine dwarf shrub. Polar Res 29:85–94

Rixen C, Teich M, Lardelli C, Gallati D, Pohl M, Putz M, Bebi P (2011) Winter tourism and climate change in the Alps: an assessment of resource consumption, snow reliability, and future snowmaking potential. Mt Res Dev 31:229–236

Roth T, Plattner M, Amrhein V (2014) Plants, birds and butterflies: short-term responses of species communities to climate warming vary by taxon and with altitude. PloS one 9

Scherrer D, Körner C (2010) Infra-red thermometry of alpine landscapes challenges climatic warming projections. Global Change Biol 16:2602–2613

Scherrer D, Körner C (2011) Topographically controlled thermal-habitat differentiation buffers alpine plant diversity against climate warming. J Biogeogr 38:406–416

Sedlacek J, Wheeler JA, Cortes AJ, Bossdorf O, Hoch G, Lexer C, Wipf S, Karrenberg S, van Kleunen M, Rixen C (2015) The response of the alpine dwarf shrub *Salix herbacea* to altered snowmelt timing: lessons from a multi-site transplant experiment. PloS one 10

Seimon TA, Seimon A, Daszak P, Halloy SRP, Schloegel LM, Aguilar CA, Sowell P, Hyatt AD, Konecky B, Simmons JE (2007) Upward range extension of Andean anurans and chytridiomycosis to extreme elevations in response to tropical deglaciation. Global Change Biol 13:288–299

Starr G, Oberbauer SF (2003) Photosynthesis of arctic evergreens under snow: implications for tundra ecosystem carbon balance. Ecology 84:1415–1420

Stöckli V, Wipf S, Nilsson C, Rixen C (2011) Using historical plant surveys to track biodiversity on mountain summits. Plant Ecol Divers 4:415–425

Vittoz P, Dussex N, Wassef J, Guisan A (2009) Diaspore traits discriminate good from weak colonisers on high-elevation summits. Basic Appl Ecol 10:508–515

Walther GR, Post E, Convey P, Menzel A, Parmesan C, Beebee TJC, Fromentin JM, Hoegh-Guldberg O, Bairlein F (2002) Ecological responses to recent climate change. Nature 416:389–395

Walther GR, Beissner S, Burga CA (2005) Trends in the upward shift of alpine plants. J Veg Sci 16:541–548

Wheeler JA, Schnider F, Sedlacek J, Cortes AJ, Wipf S, Hoch G, Rixen C (2015) With a little help from my friends: community facilitation increases performance in the dwarf shrub Salix herbacea. Basic Appl Ecol 16:202–209

Wipf S, Rixen C, Mulder CPH (2006) Advanced snowmelt causes shift towards positive neighbour interactions in a subarctic tundra community. Global Change Biol 12:1496–1506

Wipf S, Rixen C, Stöckli V (2013a) Veränderung der Gipfelfloren in der Nationalparkregion. Cratschla:12–13

Wipf S, Stoeckli V, Herz K, Rixen C (2013b) The oldest monitoring site of the Alps revisited: accelerated increase in plant species richness on Piz Linard summit since 1835. Plant Ecol Divers 6:447–455

Chapter 13
Changes in Climate, Snow and Water Resources in the Spanish Pyrenees: Observations and Projections in a Warming Climate

Enrique Morán-Tejeda, Juan Ignacio López-Moreno and Alba Sanmiguel-Vallelado

Abstract The Pyrenees constitute one of the greatest sources of freshwater in the Spanish territory, but, like many other mountain systems in the world, they are subject to environmental changes that ultimately affect the availability of water resources in areas downstream. In this study, we offer an assessment of hydrological changes in the Pyrenees, from a warming climate perspective, including climate and snow cover trends, changes in the timing of river flows, and future changes under climate change scenarios. Overall, we found that increasing temperatures are responsible for a lesser accumulation of snow over time, although with spatial differences. As a consequence, the occurrence of spring flows (that largely depend on snowmelt) on the studied rivers, has shifted earlier by approximately one month (from mid-June to mid-May). Future projections, which are made by coupling regional climate models outputs and hydrological modelling, indicate that observed decrease in snow accumulation and shifts in streamflow timing will exacerbate in a warmer short-term future (2050). The amount of water yields will not change significantly, only will suffer a slight decrease due to increased evapotranspiration. Observed and projected hydrological changes must be considered by water managers and environmental technicians if a sustainable management of the water resource and the mountain territory is to be done.

Keywords Mountain hydrology · Hydrological changes · Snow cover · River flow · Mountain climate projections

E. Morán-Tejeda (✉)
Department of Geography, Universitat de Les Illes Balears,
Cra. de Valldemossa, Km 7.5, 07122 Palma (Illes Balears), Spain
e-mail: e.moran@uib.eu

J.I. López-Moreno · A. Sanmiguel-Vallelado
Pyrenean Institute of Ecology, Consejo Superior de Investigaciones Científicas,
Avda. Montañana 1005, 50192 Zaragoza, Spain

J. Catalan et al. (eds.), *High Mountain Conservation in a Changing World*,
Advances in Global Change Research 62, DOI 10.1007/978-3-319-55982-7_13

13.1 Introduction

Amongst many other ecological functions, mountains play a major role as freshwater reservoirs, this being especially relevant in arid and semi-arid regions (Viviroli et al. 2007). Either in the form of aquifers, lakes, snow or ice cover, the water stored by mountains often represents the principal source of water for most populated areas that are usually located downstream. Thus any process involving changes in the hydrological cycle in the mountains, e.g. in climate variables, land cover, soil properties, or snow cover (Arnell 1999; Foley et al. 2005; Stewart et al. 2005; Bormann et al. 2007; López-Moreno et al. 2011), will have an impact on water availability downstream. Recent observed worldwide warming is inducing environmental changes in high mountains related to their capacity for storing water, including a decrease in the accumulation and accelerated melting of snow, the consequent reduction in the duration of the snowpack, the receding of glaciers and the loss of permafrost (Barnett et al. 2005; Lemke et al. 2007; Adam et al. 2009). Mountains and the process of snow accumulation-melting are hotspots for climate change impacts (Beniston 2003), due to the high sensitivity of the snow cover to seasonal temperatures, especially in low-to-middle elevation sites (Morán-Tejeda et al. 2013). Streamflow, as the ultimate output of mountain systems, will reflect these environmental signals by changing its timing and magnitude, and thus becomes not only a vital resource for populations, but also an important indicator of environmental changes, including global warming, for scientific research.

Climate, snow cover and streamflow are closely connected, and their interactions in high mountain terrains will impact the availability of water for downstream populations. In arid or semi-arid environment, such the Mediterranean, societies have adapted to an irregular availability of water (as a consequence of the high inter-and intra-annual variability of precipitation and temperature) mainly through the construction of dams in mountain valleys for storing water during the periods of water excess, and its release during the high demand and/or low water availability seasons (which are usually coincident). The rapid changes that mountain flows are experiencing, not only due to climate change/variability but also due to land use shifts through afforestation/deforestation and natural land cover expansion (Poyatos et al. 2003), makes it necessary to adapt the water consumption strategies in areas downstream including the management of dams located in mountain valleys (López-Moreno et al. 2004). The assessment of processes involving climate, snow, and hydrology by scientists, at high elevations, must be therefore a key step in planning conservation strategies for mountain territories.

The Pyrenees, located in the transition of Atlantic and Mediterranean climate influences, constitute a paradigmatic example of mountains undergoing rapid changes in environmental conditions, with a potential impact on the availability of water resources for downstream populations. Water availability downstream is highly related to the timing of snow accumulation and snowmelt (López-Moreno and García-Ruiz 2004), which is in turn related to the climate variability. The winter snowpack acts as a natural water reservoir that further provides sustained water during spring and summer. The south slopes of the Pyrenees drain to the Ebro River, which flows through a region with semi-arid climatic conditions (low values of precipitation, $\sim$350 mm yr^{-1} and large rates of potential evapotranspiration, $\sim$1200 mm yr^{-1}) but with large irrigation area (9800 km^2) and agriculture production, and high populated cities including Zaragoza or Lleida. Thus, the melting waters draining from the Pyrenees are a strategic resource for the economy and development of the downstream region. Moreover, the social development and ecological value of the Pyrenees are highly related to water in solid form. The permanent snowpack during winter months is a major source of economic income for Pyrenean peoples through the touristic industry derived from the practice of ski and other winter sports (Lasanta et al. 2007; Pons et al. 2015).

This work aims to present a general description of the interactions of climate variability/trends, snow cover, and water resources during the last decades in the Pyrenees, as well as to provide a simple but reliable projection in the availability of water resources in the short-term future. For this, we focus our analysis in (1) trends in climate indices that summarize far-from-normal conditions in temperature and precipitation over the mountain chain; (2) temporal trends and spatial variability in snow cover depth over the last decades; (3) changes in the timing of Pyrenean river flows, related to the observed trends in climate indices and snow cover; (4) projection of future changes in river flows by means of hydrological modelling and regional climate models output coupling.

13.2 Study Area

The Pyrenees is a mountain range located in SW Europe at the northern edge of the Iberian Peninsula and constitutes the natural border between Spain and France (Fig. 13.1). Altitude within the area ranges from 500 m to a maximum elevation of 3404 m a.s.l, and its extension is more than 50,000 km^2. The relief is firmly split by the river network due to the disposition of the main valleys (north–south), which are perpendicular to the Pyrenean structures (west–east) (Peña and Lozano 2004).

The climate of the Pyrenees is influenced by its location between the Atlantic Ocean to the west and the Mediterranean Sea to the east. The Central Pyrenees shows a greater Continental influence. Moreover, topographic heterogeneity introduces a noticeable variability to the distributions of precipitation and temperature (Del Barrio et al. 1990; García-Ruiz et al. 2001).

The temperature gradient proposed for the entire Pyrenean range varies according to the author: 0.6 (García-Ruiz et al. 1986), 0.68 (Del Barrio et al. 1990) and 0.63 °C/100 m (López-Moreno 2006). Based on these gradients, the annual 0 °C isotherm is located around 2900 m (Chueca-Cía et al. 2003). However,

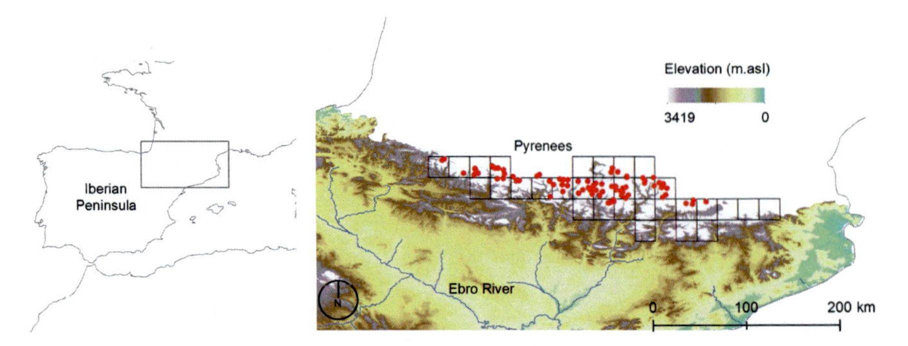

Fig. 13.1 Location of the Pyrenees, with *red points* indicating the snow poles for snow depth measurements, and *square grid* indicating the spatial resolution of the climatic database Spain2.0 in the Pyrenean territory (elevation > 1200 m a.s.l.)

between November and April, it falls at approximately 1600–1700 m a.s.l. (García-Ruiz et al. 1986), representing the lower limit of the stable winter snow cover. In the mountains, annual precipitation exceeds 2000 mm and sometimes reaches more than 2500 mm in the highest divides (García-Ruiz et al. 2001). Most of the annual precipitation falls during the cold season in the western areas, and during spring and autumn in the eastern regions.

Snow depth and duration in the Pyrenees show a marked spatial variability and vertical gradients. There exists a trend towards greater and more prolonged accumulation of snow in areas exposed to Atlantic climatic conditions (westward) compared with areas influenced by Mediterranean conditions (eastward) (López-Moreno et al. 2009). The vertical gradient in accumulated snow is evident, with values ranging between 121 mm (accumulated snow water equivalent) at 1500 m a.s.l., and 908 mm at 3000 m. a.s.l. The duration of snowpack shows a marked increase with altitude, ranging from an average of 50.1 days at 1500 m to 263.7 days at 3000 m a.s.l. (López-Moreno et al. 2009).

In general, the timing of snow accumulation and snowmelt markedly modulates the regimes of Pyrenean rivers, as most of the winter precipitation falls as snow in the Pyrenees (López-Moreno and Garcia-Ruiz 2004). Thus flows during winter (December–February) are low and uniform, and flows during spring/beginning of summer (March–July) are greater and have fluctuations (García-Ruíz et al. 1986; Beguería et al. 2003; López-Moreno and García-Ruiz 2004; Bejarano et al. 2010). Regional discharge contrasts have been assessed in relation to basin elevation and the longitudinal climate gradient established in the mountain range (López-Moreno and García-Ruiz 2004). Where large snow accumulation is possible, at the highest and more continental massifs, a delay in the onset of snow melting (mid-April on average) and higher spring flows is usually observed. Downstream the peak spring discharges decrease in intensity and the low winter discharges increase (Beguería et al. 2003).

13.3 Data and Methods

Different types of datasets have been used to accomplish the objectives of this work:

- For studying climate trends and variability, we used the Spain02 database (Herrera et al. 2012), which is a regular 0.2° latitude/longitude (approximately 20 km) daily gridded dataset containing precipitation, maximum and minimum temperatures covering continental Spain and the Balearic Islands during the period 1950–2007. From this, we extracted only pixels located in the Pyrenees range, with an elevation (>1200 m a.s.l.) that ensured the specific climate characteristic of mountains to be captured. A validation of Spain02 against observations shows that temperatures are positively biased (because grid points are generally at lower elevations than the AEMET (Agencia Estatal de Meteorología—the Spanish Meteorological Agency) stations selected for validation). However, this does not affect the computation of trends since the temporal variability of temperature and precipitation is well represented by the gridded dataset ($R^2 > 0.6$).
- Snow thickness measures were provided by the ERHIN program (Evaluación de los Recursos Hídricos procedentes de la Innivación—assessment of water resources from snow) of the Spanish Ministry of Environment. The ERHIN program has been taking snow measurements in the Spanish mountains since mid-1980 with fixed snow poles; three measurements were carried out for every year: the first in late January or early February, the second in March and the third in mid-April or early May. Quality criteria, including a maximum number of three data gaps were set up to obtain reliable snow data series. The data period was set from 1986 (first year of snow sampling) to 2007 (to make it coincide with the last year of the climatic series). Some of the selected series still had few data gaps, and only those with less than 15% of missing data were filled using the results of linear regressions with the best-correlated series (i.e. those where $R^2 > 0.7$). A total of 84 series of snow depth in April–May, covering most of the Pyrenees area, were finally selected.
- Daily streamflow data was collected from the national water agency of Spain, Centro de Estudios Hidrográficos (CEDEX, http://hercules.cedex.es/anuarioaforos/default.asp). To make sure that snowmelt pulses were present in all river regimes, we selected only rivers located in the foothills of mountain systems whose drainage watersheds had a mean elevation exceeding 800 m.a.s.l., and had no presence of the reservoirs or impoundment systems upstream of the gauge station. Data from seven gauge stations corresponding to six rivers were finally selected.

Some climatic and hydrological indices were computed in order to capture the conditions that can favour snow accumulation and melt and their signal in the hydrographs.

- The daily resolution of Spain02 database enabled the behaviour of far-from-normal events to be investigated. We computed a series of indices that combine moisture and heat conditions based on the tails of the frequency distributions of daily precipitation and mean temperature series. Based on the frequency distributions of precipitation (skewed) and temperature (normal), we selected the 25th and 75th percentiles, and through their combination we obtained four types of days: wet-cold, wet-warm, dry-cold and dry-warm days (Beniston and Goyette 2007). The yearly exceedance series of these joint thresholds were then calculated for winter and spring, and their trends over time were estimated.

- In order to explore changes in the timing of mountain flows, we calculated several indices related to the time of the year when the spring pulse from melting waters is recorded: the day of the maximum spring flows, the day when the 75th percentile of yearly (considering the water year with a start in October 1st) flows is reached, and the day when spring pulse begins (Cayan et al. 2001). By looking at temporal trends in the values of these indices, we can observe if the nival signal of Pyrenean Rivers tended to diminish or augment.

Further, statistical analysis included the computation of correlations using the Pearson's correlation test, the estimation of temporal trends using Mann–Kendall's test and the estimation of patterns in the evolution of hydro-climatic variables using Principal Component Analysis in S mode.

13.4 Climate Evolution

The evolution of climate conditions prone to snow accumulation and melting is depicted in Figs. 13.2 and 13.3. In general, we observe that joint-quantile indices in the Pyrenees show a clear pattern of monotonic trend for the spring season,

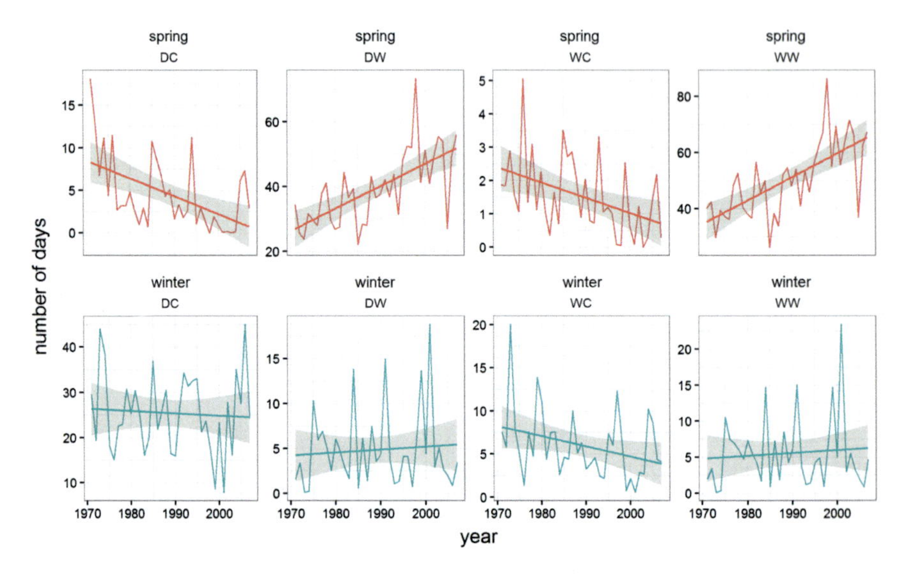

Fig. 13.2 Mean evolution and trend line (*shaded area* 95% confidence interval) for joint-quantiles indices in the Pyrenees. *DW* dry-warm days; *DC* dry-cold days; *WW* wet-warm days; *WC* wet-cold days

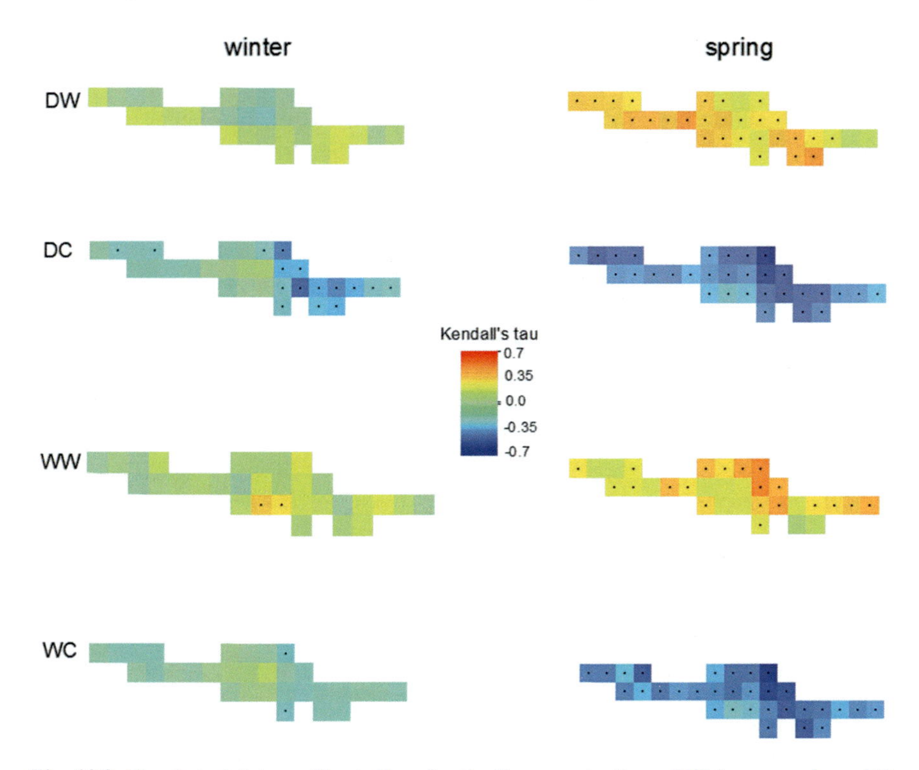

Fig. 13.3 Trends in joint-quantiles indices for the Pyrenean territory. *DW* dry-warm days; *DC* dry-cold days; *WW* wet-warm days; *WC* wet-cold days. Dots indicate a statistically significant trend with 95% of confidence

whereas their evolution seems more stationary for winter. The number of dry-cold and wet-cold days in spring show a steep decrease, and the number of dry-warm and wet-warm days show a clear increase. In winter, only wet-cold days show a decrease. When looking at the spatial distribution of trends (Fig. 13.3), we observe that generally, the trends are more pronounced and significant in the east part of the Pyrenees. These plots reveal that far-from-normal temperatures alone are responsible for increasing/decreasing trends in join-quantile indices, whereas far-from-normal precipitation seems not to have experienced a significant trend during the study period.

We can consider the wet-cold days as the most favourable conditions for snowfalls and the dry-cold days as favourable conditions for snowpack conservation (despite the sublimation that can be caused by intense solar radiation); and their relative warm types as conditions that favour melting or hamper snowpack consolidation. Thus, under such observed trends, it is reasonable to expect a reduction of snowpack in the Pyrenees during the studied period, especially at low elevations, where the zero degrees temperature threshold is reached less often.

13.5 Snow Observations

Figure 13.4 shows the evolution of winter–spring average snow depth (standard-iszed values) in the 84 snow poles located across the Pyrenees and the regional average for all the poles. There is a general decrease in snow accumulation, although the trend is not as steep as that observed for climate indices, and is not statistically significant. We must take into account, however, that the length of snow depth series (1984–2007) is not the same as the length of climate series (1950–2007). Given, the high dependency of snow to climate conditions, and the close relation demonstrated between snow depth and temperature/precipitation (López-Moreno 2005; Morán-Tejeda et al. 2013), we can expect that the slight decrease observed for snow depth in the Pyrenees would be sharper if snow data from previous decades (1950–1980) was available. This point was already noted by López-Moreno (2005) when snow series were statistically simulated from temperature and precipitation, showing a statistically significant negative trend for the period 1950–1999. This tendency is in line with the trends observed in other mountain ranges of southern Europe, such as the Alps, where Scherrer et al. (2004), Marty (2008) or Beniston (2012) have also observed a decreasing trend in snow depth and snow duration, especially at low and middle elevation sites.

Although the majority of the snow poles show a similar evolution in the snow depth, there is some dispersion in the data, especially, during the 1985–1995 decade. To search for a relation between the climate indices and the snow depth evolution, we have thus performed a principal component analysis, in which the

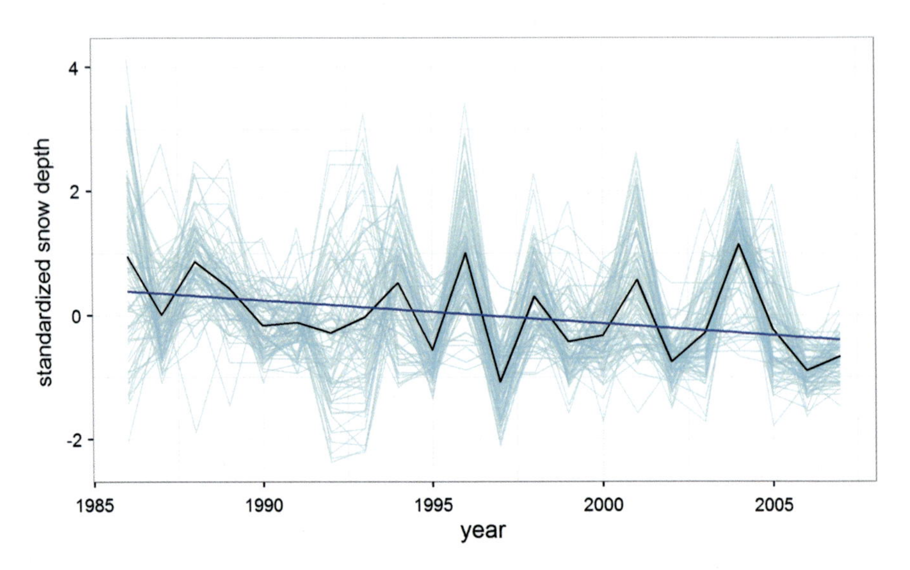

Fig. 13.4 Evolution and trend of winter–spring snow depth across the Pyrenees during the study period. *Light blue lines* indicate the standardized snow depth in each snow pole and *black line* indicates the average

variables to extract the components are the snow poles. Four Principal Components explained 80% of the variance contained in original data. For representation purposes, we only show the evolution of the two main factors, which together explain 60% of the variance. The snow depth evolution for each one is depicted in Fig. 13.5a. Thus, we observe two different patterns of evolution in snow depth, though both are showing a decreasing trend. The extraction of principal components enabled us to establish correlations with the join-quantile indices (Fig. 13.5b). For PC1 we observe that significant coefficients are mainly found with index DW

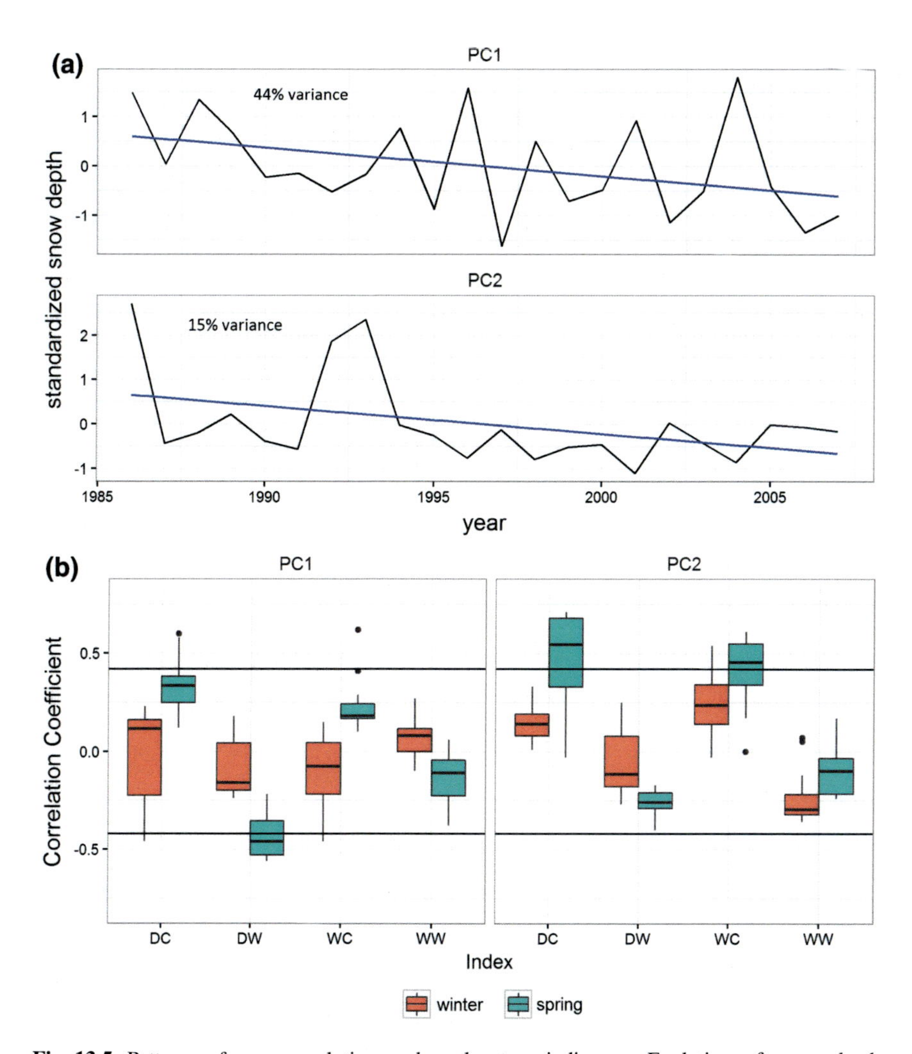

Fig. 13.5 Patterns of snow evolution and explanatory indices. **a** Evolution of snow depth (standardized) for the two principal components that most variance explained. **b** Correlations between the climatic indices in each of the pixel for the Pyrenean region and the two principal components depicted in (**a**)

for spring, correlations being negative, i.e. the less (more) frequent dry-warm days in spring, the higher (lower) yearly snow accumulation. Thus, this pattern of snow evolution, which is the most frequent, responds inversely to the conditions that hamper snow accumulation and favour snow melting. There are also significant correlations (though less in number) with the number of dry-cold days in spring, thus conditions that favour the conservation of the snowpack. In general, years in which cold days (regardless precipitation) prevail over warm days, are years with higher snow accumulation. It may seem strange that conditions that favour snow accumulation (wet-cold days) do not show a significant explanatory capacity for snow evolution (except for some cases, as can be seen by the dots of the boxplot, which correspond to the 95th and 99th percentiles of the distribution); but this shall be interpreted in terms of limiting conditions: wet-cold days is not a limiting factor for inter-annual snow evolution, meaning that there will still be snowfalls in days that are not extremely cold and particularly wet. In contrary dry-warm days is a strong limiting factor for snowpack consolidation.

PC2 which explains less proportion of variance, and thus represents fewer snow poles, shows a different evolution, and different explanatory variables. The index that best correlates with PC2 is DC in spring, with higher correlations than with PC1. Thus, in this case, the conditions favourable for snowpack conservation are a limiting factor. But we also found significant correlations with wet-cold days in spring, i.e. with conditions that favour snowfalls and snow accumulation. Contrary to PC1, where conditions that can favour snowfalls were not a limiting factor, in this case, the occurrence of wet-cold days determines whether the year will by more or less snowy.

To find a geographical explanation for the differentiation of PC1 and PC2 and their differing limiting factors, we have performed a Mann–Whitney–Wilcoxon Test, for the following dependent variables: elevation, latitude and longitude. Previously, we separated the snow poles that best correlated with PC1 from the snow poles that best correlated with PC2, and the test was performed over these two populations. Results of the test inform that elevation of snow poles from PC1 is higher (mean elevation = 2300 m) and differs significantly (p-value = 0.015) from elevation of snow poles from PC2 (mean elevation = 2100). Latitude and longitude did not show significant differences between snow poles of PC1 and PC2 (p values = 0.656 and 0.657 respectively). This is physically consistent with the fact that cold days were not a limiting factor for snow evolution of PC1, but they were for snow evolution of PC2: the higher is the elevation, the less dependent is snow accumulation to conditions of extreme cold, because the zero degrees isotherm (which marks the limit for snowpack to consolidate) will be reached more frequently. Thus at high elevations, the occurrence of warm days, especially in spring is observed to be the most important limiting factor for the consolidation of the snowpack. Our results suggest that precipitation (absence or particularly high values) is not a factor controlling snowpack variability. This result may be, however, a shortcoming of the joint-quantile indices used. Several studies demonstrated that precipitation can be a good predictor of snowpack variability when used as a single covariate. However, its predictive power depends greatly, on the elevation:

at low elevations the main climatic snowpack predictor is temperature, but at elevations at which the zero-degree isotherm remains during most of the winter, the variability of precipitation becomes the most important predictor for snowpack variability (López-Moreno 2005; Morán-Tejeda et al. 2013; Sospedra-Alfonso et al. 2015). Thus, it is advisable to interpret elevation-wise, the trends observed in snow depth and snow duration in the context of climate warming. The lack of significant trends observed at high elevations (as observed in this work) do not mean that warming is not affecting the accumulation of snow, but that warming at those elevations is not yet significant enough as to affect snow accumulation. Further warming projected by climate models will, however, make increase the elevation threshold at which temperatures become a limiting factor for snow accumulation, and negative trends in snowpack depth and duration will be more likely found at increasing elevations (Morán-Tejeda et al. 2013).

13.6 Streamflow Changes

To study for hydrological changes, we selected three Pyrenean Rivers (Ésera, Aragón and Ara Rivers) with natural flow regimes (unaffected by dams or water extractions) and we further defined three indices that represent different moments of the year related to the signal of snow on the hydrograph (Fig. 13.6). SP indicates the day of the year when the streamflow pulse due to snow melting starts (see calculation in Cayan et al. 2001); SM stands for the day of the year when the maximum spring flow is reached; and 75M indicates the day of the year when the 75th percentile of yearly water volume is reached. After these indicators had been computed for every year in each of the three selected rivers, we checked the existence of trends in time using Mann–Kendall and Thiel–Sen's slope estimator tests (Table 13.1). The three indices show a decreasing trend for the three selected

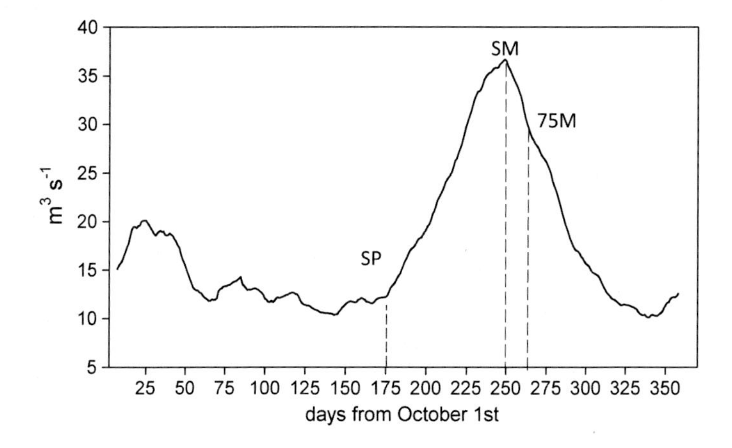

Fig. 13.6 Example of location of hydrological indices across a yearly hydrograph

Table 13.1 Trend statistics for SP, MS and 75M hydrological indices

Index	Trend statistic	Ésera	Aragón	Ara
SP	Slope	−0.65	−0.34	−0.64
	Change	−24.7	−11.2	−21.3
	Tau	−0.15	−0.07	−0.18
	Sig	0.21	0.54	0.16
MS	Slope	−0.48	−1.13	−1.0
	Change	−27.8	−37.6	−33
	Tau	−0.4	−0.38	−0.47
	Sig	0.001	0.002	0.0013
75M	Slope	−0.57	−0.58	−0.6
	Change	−18.3	−18.5	−19.2
	Tau	−0.32	−0.3	−0.36
	Sig	0.01	0.015	0.004

Slope Theil–Sen's slope estimator; *change* change in the number of days over the study period; *tau* = correlation coefficient for the Mann–Kendall's test; *sig* p-value for the test

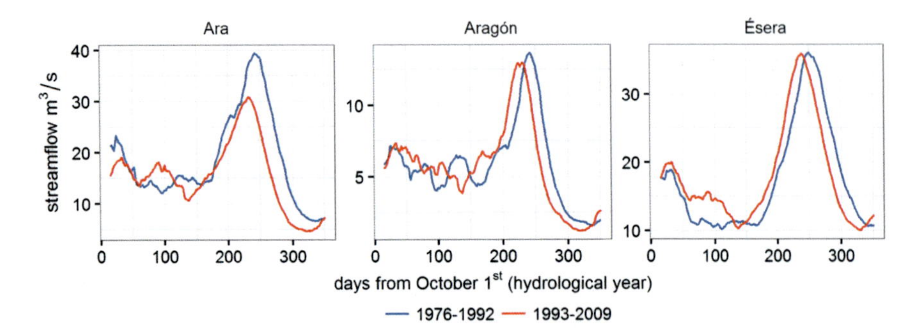

Fig. 13.7 Changes in the hydrograph of Pyrenean Rivers during the study period

rivers, although trends for SP are not significant with a 95% confidence level. Trends in the other two indices are statistically significant: the day when maximum spring flows are reached occurs now a month earlier and the day when the 75th percentile of the yearly streamflow is reached occurs nearly 20 days earlier.

In Fig. 13.7 we can visualize the results of the trends above, in simple plots that represent the average hydrograph of rivers for the first and second halves of the study period. In the three cases, we readily observe that nowadays there is an earlier occurrence of the spring peak, which results from snow melting, than some decades ago; with the correspondent increase in winter flows. In our view, this is the reaction of the hydrological system, to the changes in climate and snow accumulation discussed in previous paragraphs.

With the aim of verifying it, we have computed correlations between the evolution of each hydrological index and the evolution of different seasonal/annual

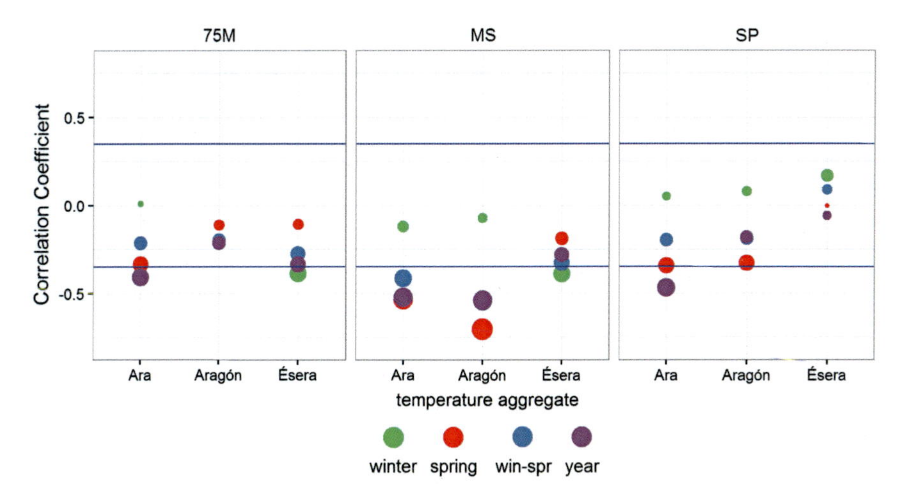

Fig. 13.8 Pearson correlations between hydrological indices and temperature aggregates for the three studied rivers. Circle size is proportional to the correlation level. *Blue horizontal lines* indicate significant level set at 95% of confidence

temperature aggregates (winter, spring, winter–spring and year averages) (Fig. 13.8). Generally, there are negative correlations between the hydrological indices and the temperatures, although some temperature aggregates are a better predictor than others. The hydrological index best explained by temperature evolution is MS, especially in Aragon and Ara Rivers, where average spring temperatures and annual average temperatures show correlation values close to R = −0.7. Indices 75M and SP show lower correlations coefficients, but still significant in Ara River for annual and average spring temperature. These results confirm that, although others factors (e.g. variability in seasonal precipitation) may greatly affect the timing of river flows in snow-fed rivers (Sanmiguel-Vallelado et al. 2017), there is a clear correspondence between the observed increase in temperature and the early occurrence of the spring peak in Pyrenean rivers. This process in not exclusive of the Pyrenees, and has been observed in rivers of snow-dominated regions around the globe, including the Rocky Mountains (Stewart et al. 2005), or New England (Hodgkins et al. 2003) in the United Sates, or the Swiss Alps in Europe (Birsan et al. 2005). The underlying processes behind these trends are the decrease in the snowfall/rainfall ratio in winter and the earlier snowmelt in spring due to higher temperatures.

13.7 Projections for a Warmer Climate

The observed correspondence between trends in temperature and changes in streamflow timing make it logical to expect that further climate warming due to the increase in the atmospheric concentration of greenhouse gases (IPCC 2013), will enhance further hydrological changes. In recent years, a great effort has been made by hydrological scientists to provide reliable projections of future water resources availability in a warmer climate by coupling climate and hydrological models. In this last section, we aim to provide a simple, but feasible, projection of future changes in streamflow of Pyrenean Rivers through a hydrological modelling approach, considering climate warming scenarios. The hydrological simulations were done with SWAT (Soil Water Assessment Tool) model, for the Ésera River, considering the changes in temperature (delta changes) that the Regional Climate Models of the ENSEMBLES project (Hewitt and Griggs 2004), predict for the time window 2035–2065, compared to the control period 1970–2000. We did not consider changes in precipitation due to the high level of uncertainty related to precipitation projections for the future. Details on the model calibration and climate scenarios can be found in Morán-Tejeda et al. (2015). Figure 13.9 shows a summary of results, considering the multi-model 10th percentile, average and 90th percentile temperature projections (Table 13.2). The left plot shows the changes in snow water equivalent (SWE), i.e. the volume of water contained in the snowpack, yearly and monthly (lines). Decrease in SWE is evident in warming conditions with nearly −45% considering the multi-model average. SWE decrease is consistent throughout the entire snow season but is even greater in spring (around Julian day 180–200). As a result of the warming, the duration of the snowpack will be as well reduced an average of 50 days. The hydrological result is depicted in Fig. 13.9 (right plot). We thus observe a very small decrease in annual water yield of −1, 9% for the multi-model average (mainly due to an increase in evapotranspiration), but a significant change in the timing of flows. Spring and summer flow (from May to September) will experience a substantial decrease, but it will be compensated by an increase in winter flows. Overall, the hydrograph would experience a sort of smoothing, with less marked seasonality. This feature would potentially involve changes in water management to satisfy water demands in downstream areas.

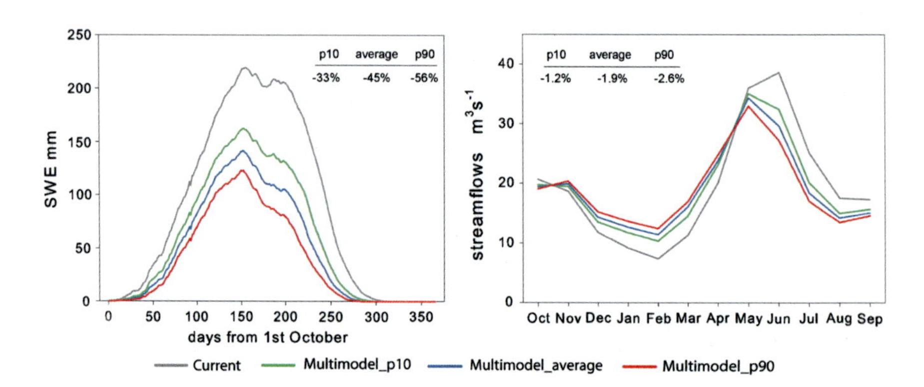

Fig. 13.9 Monthly (*lines*) and yearly (*text*) changes in snow water equivalent (SWE) (*left plot*) and in streamflows (*right plot*), projected by SWAT model for the Ésera river, under climate change conditions

Table 13.2 Temperature changes (°C) projected by RCMs of ENSEMBLES project for 2035–2065 in our study area

	Winter	Spring	Autumn	Summer
p10	+1.2	+1.2	+1.9	+1.8
Average	+1.7	+1.8	+2.7	+2.2
p90	+2.1	+2.5	+3.8	+2.6

As noted above, many reservoirs store water from Pyrenean rivers to supply water during the season of lower availability and higher demand (summer). As observed by López-Moreno et al. (2004), Pyrenean reservoirs have used different patterns of management (basically two) during the last decades, but in a very restricted way, not adapted to the inter-annual variability of river flows. We want to emphasize on the recommendation done by López-Moreno et al. (2004) for a more dynamic management of Pyrenean reservoirs, given the occurred changes in river flows and their possible enhancement during next decades.

Although we did not consider changes in future amounts of precipitation, climate models predict that it is likely to experience a decrease during the 21st century in the Mediterranean region (Giorgi and Lionello 2008). If such predictions are right, we would have to add a decline in streamflow amounts to the observed change in the timing of flows. Another process that was not considered here may involve a further reduction in water availability. The Pyrenees, as many other Spanish mountains are experiencing a recovery of natural vegetation, including shrubs and forests, due to the abandonment of traditional activities that used the mountain slopes for pastures and cropping (Vicente-Serrano et al. 2004; Lasanta-Martínez et al. 2005). It has been experimentally demonstrated that forested catchments yield less amount of water than non-forested ones (e.g. García-Ruiz et al. 2008) due to rainfall interception and evapotranspiration by forest canopy (Llorens and Domingo 2007). Hydrological modelling also shows that future increase of vegetation cover in Pyrenean catchments would reduce river flows (López-Moreno et al. 2014). This aspect involves more complexity and uncertainty for the prediction of future water resources availability, but also gives us key information for improving the management of water in the Pyrenees in a more integrated fashion.

13.8 Conclusions

We show in this work a comprehensive analysis of the evolution of climate, snow cover and water resources in the Spanish Pyrenees, during the last four decades, and the interactions amongst them. This analysis has been possible due to the existence of quality databases of precipitation, temperature, snow depth and streamflows developed and managed by different institutions.

Climatically, we focused on the tails of the climate variables distributions, to find any trend in conditions that could favour/hamper snow accumulation and melting. We found a decrease in the number of dry-cold and wet-cold days, and an increase in the number of warm-dry and warm-wet days, meaning that only far-from-mean temperatures and not precipitation, have experienced significant trends.

Snow depth evolution in the Pyrenees shows spatial variability, but overall we observed a negative trend of snow depth during the study period. Moreover, we found that in the snow depth sites located at higher elevations (2100–2600 m a.s.l.), snow variability correlated inversely with the frequency of dry-warm days in spring, indicating that only in years when exceptionally warm days predominate, snowpack will be reduced, whereas the frequency of cold days is not a limiting factor. On the contrary, in snow measurement sites at lower elevations (1800–2400 m a.s.l.), snow variability correlated with the frequency of dry-cold and wet-cold days in spring, meaning that spring temperature begins to be a limiting factor at this lower sites.

The consequence of the above mentioned trends over water resources has been, so far, a change in the timing of flows, particularly on the occurrence of the spring peak that results from snowmelt. On the studied rivers, there is an earlier occurrence of spring flows during recent years than four decades ago. The increase of spring temperatures triggers an earlier snowmelt, an overall decrease in the duration of the snowpack and an earlier and lower spring peak flow. When projecting future hydrological changes by coupling RCMs outputs and hydrological modelling, we observe that the detected changes are likely to be enhanced if temperatures grow as estimated by climate models: winter flows will augment due to fewer snowfalls and snow accumulation and consequently spring flows will decrease. The amount of water resources available will not suffer big changes, except if a large decrease in precipitation occurs.

Such changes must be acknowledged by water managers and water users from downstream territories and incorporated in long-term water management policies that promote a sustainable use of the water resource in this sensitive territory.

References

Adam JC, Hamlet AF, Lettenmaier DP (2009) Implications of global climate change for snowmelt hydrology in the twenty-first century. Hydrol Process 23:962–972

Arnell NW (1999) The effect of climate change on hydrological regimes in Europe: a continental perspective. Glob Environ Chang-Hum Policy Dimens 9:5–23

Barnett TP, Adam J, Lettenmaier DP (2005) Potential impacts of a warming climate on water availability in snow-dominated regions. Nature 438:303–309

Beguería S, López-Moreno JI, Lorente A, Seeger M, García-Ruiz JM (2003) Assessing the effect of climate oscillations and land-use changes on streamflow in the Central Spanish Pyrenees. Ambio 32:283–286

Bejarano MD, Marchamalo M, de Jalón DG, del Tánago MG (2010) Flow regime patterns and their controlling factors in the Ebro basin (Spain). J Hydrol 385:323–335

Beniston M (2003) Climatic change in mountain regions: a review of possible impacts. Clim Change 59:5–31

Beniston M (2012) Is snow in the Alps receding or disappearing?, WIREs climate change (wiley interdisciplinary reviews/climate change). doi:10.1002/wcc.179

Beniston M, Goyette S (2007) Changes in variability and persistence of climate in Switzerland: exploring 20th century observations and 21st century simulations. Glob Planet Chang 57:1–15

Birsan MV, Molnar P, Burlando P, Pfaundler M (2005) Streamflow trends in Switzerland. J Hydrol 314:312–329

Bormann H, Breuer L, Gräff T, Huisman JA (2007) Analysing the effects of soil properties changes associated with land use changes on the simulated water balance: a comparison of three hydrological catchment models for scenario analysis. Ecol Model 209:29–40

Cayan DR, Dettinger MD, Kammerdiener SA, Caprio JM, Peterson DH (2001) Changes in the onset of spring in the western United States. Bull Am Met Soc 82:399–415

Chueca-Cía J, Julian-Andrés AMA, López-Moreno JI (2003) Variations of Glaciar Coronas, Pyrenees, Spain, during the 20th century. J Glaciol 49:449–455

Del Barrio G, Creus J, Puigdefábregas J (1990) Thermal seasonality of the high mountains belt of the Pyrenees. Mount Res Dev 10:227–233

Foley JA, DeFries R, Asner GP, Barford C, Bonan G, Carpenter SR, Chapin FS, Coe MT, Daily GC, Gibbs HK, Helkowski JH, Holloway T, Howard EA, Kucharik CJ, Monfreda C, Patz JA, Prentice I, Ramankutty N, Snyder PK (2005) Global consequences of land use. Science 309:570–574

García-Ruiz JM, Puigdefábregas TJ, Creus-Novau J (1986) La acumulación de nieve en el Pirineo Central y su influencia hidrológica. Pirineos 127:27–72

García-Ruiz JM, Beguería S, López-Moreno JI, Lorente A, Seeger M (2001) Los recursos hídricos superficiales del Pirineo aragonés y su evolución reciente. Geoforma, Logroño, p 192

García-Ruiz JM, Regüés D, Alvera B, Lana-Renault N, Serrano-Muela P, Nadal-Romero E, Navas A, Latron J, Martí-Bono C, Arnáez J (2008) Flood generation and sediment transport in experimental catchments affected by land use changes in the central Pyrenees. J Hydrol 356:245–260

Giorgi F, Lionello P (2008) Climate change projections for the Mediterranean region. Glob Planet Change 63:90–104

Herrera S, Gutiérrez JM, Ancell R, Pons MR, Frías MD, Fernández J (2012) Development and analysis of a 50-year high-resolution daily gridded precipitation dataset over Spain (Spain02). Int J Climatol 32(1):74–85

Hewitt CD, Griggs DJ (2004) Ensembles-based predictions of climate changes and their impacts (ENSEMBLES). Eos 85:566

Hodgkins GA, Dudley RW, Huntington TG (2003) Changes in the timing of high river flows in New England over the 20th century. J Hydrol 278:244–252

IPCC (2013) Climate change 2013: the physical science basis. Contribution of working group I to the fifth assessment report of the intergovernmental panel on climate change In: Stocker, T.F., D. Qin, G.-K. Plattner, M. Tignor, S.K. Allen, J. Boschung, A. Nauels, Y. Xia, V. Bex and P. M. Midgley (eds). Cambridge University Press, Cambridge, United Kingdom and New York, NY, USA

Lasanta-Martínez T, Vicente-Serrano SM, Cuadrat-Prats JM (2005) Mountain Mediterranean landscape evolution caused by the abandonment of traditional primary activities: a study of the Spanish Central Pyrenees. Appl Geogr 25:47–65

Lasanta T, Laguna M, Vicente-Serrano SM (2007) Do tourism-based ski resorts contribute to the homogeneous development of the Mediterranean mountains? A case study in the Central Spanish Pyrenees. Tourism Manag 28:1326–1339

Lemke P, Ren J, Alley RB, Allison I, Carrasco J, Flato G, Fujii Y, Kaser G, Mote P, Thomas RH, Zhang T (2007) Observations: changes in snow, ice and frozen ground, climate change 2007: The physical science basis. Contribution of working group i to the fourth assessment report of the intergovernmental panel on climate change. In: Solomon S, Qin D, Manning M, Chen Z, Marquis M, Averyt KB, Tignor M, Miller HL (eds) Cambridge University Press. UK, Cambridge, pp 337–383

Llorens P, Domingo F (2007) Rainfall partitioning by vegetation under Mediterranean conditions. A review of studies in Europe. J Hydrol 335:37–54

López-Moreno JI (2005) Recent variations of snowpack depth in the Central Spanish Pyrenees. Arct Antarct Alp Res 37:253–260

López-Moreno JI (2006) Cambio ambiental y gestión de embalses en el Pirineo Central Español. Consejo de Protección de la Naturaleza de Aragón, Zaragoza

López-Moreno JI, García-Ruiz JM (2004) Influence of snow accumulation and snowmelt on streamflow in the central Spanish Pyrenees/Influence de l'accumulation et de la fonte de la neige sur les écoulements dans les Pyrénées centrales espagnoles. Hydrol Sci J 49(5)

López-Moreno JI, Beguería S, García-Ruiz JM (2004) The management of a large Mediterranean reservoir: storage regimens of the Yesa reservoir, upper Aragon river basin, central Spanish Pyrenees. Environ Manag 34:508–515

López-Moreno JI, Goyette S, Beniston M (2009) Impact of climate change on snowpack in the Pyrenees: horizontal spatial variability and vertical gradients. J Hydrol 374:3–4

López-Moreno JI, Vicente-Serrano SM, Morán-Tejeda E, Zabalza J, Lorenzo- Lacruz J, García-Ruiz JM (2011) Impact of climate evolution and land use changes on water yield in the Ebro basin. Hydrol Earth Syst Sci 15:311–322

López-Moreno JI, Zabalza J, Vicente-Serrano SM, Revuelto J, Gilaberte M, Azorin-Molina C, Morán-Tejeda E, García-Ruíz JM, Tague C (2014) Impact of climate and land use change on water availability and reservoir management: scenarios in the Upper Aragón River, Spanish Pyrenees. Sci Total Environ 493:1222–1231

Marty C (2008) Regime shift of snow days in Switzerland. Geophys Res Lett 35:L12501

Morán-Tejeda E, López-Moreno JI, Beniston M (2013) The changing roles of temperature and precipitation on snowpack variability in Switzerland as a function of altitude. Geophys Res Lett 40:2131–2136

Morán-Tejeda E, Zabalza J, Rahman K, Gago-Silva A, López-Moreno JI, Vicente-Serrano S, Lehmann A, Tague C, Beniston M (2015) Hydrological impacts of climate and land-use changes in a mountain watershed: uncertainty estimation based on model comparison. Ecohydrology 8:1396–1416

Peña J, Lozano M (2004) Las unidades del relieve aragonés, Geografía Física de Aragón, Aspectos Generales y Temáticos. Universidad de Zaragoza, Zaragoza

Pons M, López-Moreno JI, Rosas-Casals M, Jover È (2015) The vulnerability of Pyrenean ski resorts to climate-induced changes in the snowpack. Clim Change 131:591–605

Poyatos R, Latron J, Llorens P (2003) Land use and land cover change after agricultural abandonment: the case of a Mediterranean mountain area (Catalan Pre-Pyrenees). Mount Res Dev 23:362–368

Sanmiguel-Vallelado A, Morán-Tejeda E, Alonso-González E, López-Moreno JI (2017) Effect of snow on mountain river regimes: an example from the Pyrenees. Front Earth Sci 1–16

Scherrer SC, Appenzeller C, Laternser M (2004) Trends in Swiss Alpine snow days: the role of local- and large-scale climate variability. Geophys Res Lett 31:L13215

Sospedra-Alfonso R, Melton JR, Merryfield WJ (2015) Effects of temperature and precipitation on snowpack variability in the Central Rocky Mountains as a function of elevation. Geophys Res Lett 42:4429–4438

Stewart IT, Cayan DR, Dettinger MD (2005) Changes toward earlier streamflow timing across western North America. J Clim 18:1136–1155

Vicente-Serrano SM, Lasanta T, Romo A (2004) Analysis of spatial and temporal evolution of vegetation cover in the Spanish Central Pyrenees: role of human management. Environ manag 34:802–818

Viviroli D, Dürr HH, Messerli B, Meybeck M, Weingartner R (2007) Mountains of the world, water towers for humanity: typology, mapping, and global significance. Water Resour Res 43. doi:10.1029/2006WR005653

Chapter 14
Atmospheric Chemical Loadings in the High Mountain: Current Forcing and Legacy Pollution

Lluís Camarero

Abstract Human emissions have changed the chemistry of atmosphere. Potentially toxic chemicals have been spread, and the global cycles of some key elements have been disrupted. Because enhanced atmospheric precipitation and cold trapping caused by elevation, high mountain ecosystems are considered as regional convergence areas of atmospheric pollutants. In this chapter, research on surface waters acidification, pollution by trace elements, and atmospheric nutrient inputs in the Pyrenees is reviewed. Pyrenean lakes have experienced only a moderate acidification, due partly to an also moderate acid load and partly to the neutralising cations carried by dust. Presently, declining concentrations of sulphate in lakes indicate that recovery is proceeding. Pollution by trace elements dates more than two millennia back. The primary accumulation sites are the sediments of lakes. Soils also hold an important burden, and there is evidence that some elements are being currently remobilised. This is causing a delayed pollution, despite deposition of several trace metals is declining. The emissions of artificial reactive nitrogen have caused increased deposition on the Pyrenean catchments, which are thus nitrogen saturated. A parallel increase of phosphorus deposition has occurred, likely caused by climatic reasons. The combined effect of both seems to be an enhanced uptake of nitrogen by phytoplankton causing a lower nitrogen concentration in lakes and a possible shift from phosphorus-to-nitrogen limitation of phytoplankton growth, as well as an incipient eutrophication. All these are examples of impacts in remote natural areas that require a global strategy of conservation beyond the boundaries of the ecosystems affected.

Keywords Atmospheric pollutants · Global biogeochemical cycles · Acidification · Trace elements · Legacy pollution · Airborne nutrients

L. Camarero (✉)
Centre d'Estudis Avançats de Blanes, CSIC, Accés Cala Sant Francesc 14, 17300 Blanes, Girona, Spain
e-mail: camarero@ceab.csic.es

J. Catalan et al. (eds.), *High Mountain Conservation in a Changing World*,
Advances in Global Change Research 62, DOI 10.1007/978-3-319-55982-7_14

14.1 Introduction

One of the main impacts of human activities on the global ecosystem is the change caused on the chemistry of the atmosphere. Such a change is due mainly to the emission of a wide range of chemicals, in the form of both gases and aerosols. Some of them are potentially toxic (such as some trace metals and persistent organic molecules) and are considered as pollutants. In many cases, these pollutants do not exist in nature. Some of the emitted chemicals are present in nature, but the human emissions have caused their levels in the atmosphere to exceed by far the natural ones. This excess causes a disturbance in the planetary biogeochemical cycles, very often with awkward effects. Sometimes there is a biogeochemical bottleneck that causes the accumulation of a compound. An example of this is the rising atmospheric CO_2 (and other greenhouse gases) level, now very widely accepted to be responsible of the human contribution to the global warming. In some other cases, the addition of an element causes the acceleration of the natural cycling. For instance, the introduction of large amounts of some compounds of nitrogen and phosphorus (both are major nutrients for the living organisms) in the environment has caused a nutrient imbalance in the ecosystems. This is of special importance in nutrient-poor systems, such as remote oceanic areas and high mountains, where the atmospheric inputs may have a fertilising effect that leads to an unnatural growth of primary producers, basically plants and algae.

In addition to the emissions, human activities have also the effect of modifying the natural atmospheric transport of substances. A direct way to do so is by changes in the land use; for instance, agricultural practices and deforestation may enhance production of dust and aeolian transport from land; changes in the wild fires regime affect the emission of ashes and gases from burning biomass. But there are also indirect ways that have to do with climate change, although in this case both natural and human-induced causes play a role: droughts and losses of snow cover that enhance dust production; melting of organic permafrost that increases CO_2 and methane emissions from soil; changes in the prevailing winds and patterns of circulation of air masses that carry airborne substances. All these are examples of effects of climate change on the natural atmospheric fluxes of chemicals.

The long-range transport of atmospheric chemicals takes place in the free atmosphere, that is, above the mixing boundary layer which generally has a depth of 1,000–1,500 m. In this sense, the chemistry of the free atmosphere can be considered to reflect the composition of the global atmosphere better than the boundary layer below it. High mountains protrude above the boundary layer on the lowlands around them, and thus intercept the "global" flux of chemicals. Such interception is enhanced by altitude. Orographic precipitation is higher in the mountains causing greater scavenging of airborne substances than in the lowlands. Also, temperature is lower in the summits than in the piedmont and the deposition of volatile compounds by condensation (cold trapping) is favoured. In summary, mountains act as regional (or even global) convergence areas of atmospheric chemical fluxes.

Alpine ecosystems are sensitive to the atmospheric inputs for several reasons. They are headwater catchments where chemical weathering of the bedrock is limited, specially in those on highly insoluble crystalline rocks, so lake and stream waters are very diluted. Consequently their capacity of buffering any acidity coming with precipitation is rather poor. Nutrients coming from the substrate are also scarce. Therefore, the atmospheric supply is an important (if not the larger) fraction of nutrient input to alpine ecosystems. Catchment soils and lake sediments are generally highly organic. This gives them a large capacity for binding pollutants such as trace metals and organic chemicals. In the long run, this great affinity causes a build-up of the pollutant burden stored in soils and sediments. These are some of the main reasons for the sensitivity of alpine systems to atmospheric chemical loadings. They are sensitive in two senses. First, they are actually threatened by certain atmospherically deposited chemicals that may reach levels that have eco-toxic effects or alter the natural ecosystem functioning, and constitute a conservation issue. But second, even in the cases where impacts are not severe, the signs of the chemical inputs are easily noticeable. Inputs that, remember, respond to a global forcing. This makes alpine ecosystems an excellent early warning system of global environmental change.

In this chapter, I am mostly presenting a summary of the results regarding some of the topics mentioned above obtained during almost 30 years of research in the Pyrenees. But beyond the local interest, the results presented here are an example of how atmospheric chemical loadings affect high mountains that also applies to other high mountain ranges.

14.2 Surface Waters Acidification

Since the onset of the Industrial Revolution, the massive burning of fossil fuels has caused the emission of sulphur and nitrogen oxides to the atmosphere. These oxides acidify the atmospheric precipitation, as it was already noticed in the mid-nineteenth century when the term "acid rain" was coined (Smith 1872). Acid rain in its turn acidifies soils and surface waters when these receptor systems cannot buffer the incoming acidity. In that case, deleterious effects occur. The first cases of environmental damage caused by acid deposition were reported at the beginning of the twentieth century. These first cases were attributed to local sources in the vicinities of the impacted sites (Cowling 1982). But in the 1970s the effects of long-range transport of pollutants shifted the focus to a larger scale. There was then widespread evidence of regional acidification of surface water, affecting the most sensitive lacustrine areas in the Northern Hemisphere: the Precambrian Shields of Canada and Scandinavia (Gorham et al. 1986; Odén and Ahl 1979; Likens and Bormann 1974). During the 1980s, on the basis of the scientific studies carried out in the previous years, it was generally accepted that human emissions were producing significant acidification of fresh waters (Schindler et al. 1985) and might be a factor contributing to forest dieback (Pitelka and Raynal 1989) on a regional scale.

Sulphate derived from the sulphur oxide emissions was identified as the main acidifying agent (Neary and Dillon 1988; Sullivan et al. 1988). As a consequence of the overwhelming evidence, serious efforts were made to control the sulphur emissions. Laws aiming to the reduction of emissions came into force in Europe and North America (UNECE 2016a). Recovery of acidified ecosystems did not follow the reduced emissions immediately. There is a hysteresis in the reversal of acidification. However, there is nowadays growing evidence that recovery is in progress, although there are also some uncertainties about how it will be in the next future (Wright et al. 2005). There are for instance confounding factors related to climate change that can mask or counteract the recovery, such as increase of $p\mathrm{CO}_2$ and organic acids in soil and runoff and increased sea salt atmospheric inputs. (Wright et al. 2006). Furthermore, the global emissions of sulphur may rise again as emerging economic powers develop their industry (Streets and Waldhoff 2000; Smith et al. 2011).

As mentioned above, high mountain environments and, in particular, surface waters lying on crystalline bedrocks are sensitive to acidification. The sensitivity of lakes, for instance, can be assessed according to the acid neutralising capacity (ANC) of its waters. ANC is a chemical parameter that measures the difference between dissolved base cations and acid anions. ANC is expressed in terms of equivalents (the usual unit for diluted waters is μeq L^{-1}). ANC is closely related to pH. Negative ANC values indicate acidic lakes, with pH below 5.5. Organisms able to survive below that pH are an exception. In the Pyrenees, c. 2% of lakes have negative ANC. Lakes with ANC between 0 and 50 μeq L^{-1} have a pH in the range 5.5–6.5. Many organisms are affected in some degree by acidity in this range of pH. Furthermore, the buffering capacity is so low that small additions of acid may lead to negative ANC values. Approximately 15% of Pyrenean lakes lie in this range of ANC and pH. pH above 6.5 do not cause particular damage to organisms. Nevertheless, lakes with ANC < 200 μeq L^{-1} are susceptible to reach seasonally low pH values as, for instance, during snowmelt runoff in spring. For this reason, lakes below 200 μeq L^{-1} are considered potentially sensitive to acidification. In the Pyrenees, c. 44% of lakes falls in the range of ANC between 50 and 200 μeq L^{-1}. Altogether, about 60% of Pyrenean lakes are, in higher or lower degree, sensitive to acidification.

How has acidification evolved in the Pyrenean lakes? Modelling on the basis of reconstructed acid deposition and the current chemical features of lakes indicates that the effects of acidification have been moderate in the Pyrenees (Camarero and Catalan 1998). The modelling results indicate that Pyrenean lakes have experienced an ANC loss of c. 35 μeq L^{-1} on average since 1850. This has caused only 2% of lakes to cross the boundary of ANC = 20 μeq L^{-1} (the lower limit for salmonids survival), low in comparison with the 25% in many areas of Central and Northern Europe and the 90% in the most sensitive areas of Scandinavia. The reasons for this are, on the one hand, that the atmospheric acid load has been lower on the Pyrenees than on the more polluted Central and Northern Europe and, on the other hand, that dust inputs from the Iberian Peninsula and Northern Africa are important here. Dust is a large source of base cations able to neutralise acidity in precipitation and to

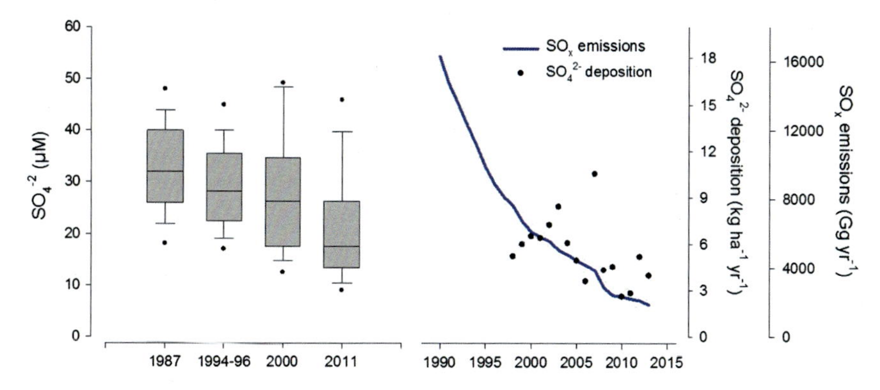

Fig. 14.1 *Left* Box plot showing the distribution of sulphate concentration in Pyrenean lakes on igneous bedrock, where there is no supply of sulphur from rock weathering. Each box corresponds to a synoptic survey carried during the summer of the years indicated in the axis. The *line* within the box represents the median value, limits of the box are the 25th and 75th percentiles, whiskers are the 10th and 90th percentiles and *dots* are the 5th and 95th percentiles. *Right* Trend-line showing the evolution of SO_x emissions to the atmosphere in Western Europe since 1990 (*Source* EMEP http://www.ceip.at/ms/ceip_home1/ceip_home/webdab_emepdatabase/emissions_emepmodels/). The overimposed dots represent yearly sulphate deposition as measured at the Lake Redon sampling station (Central Pyrenees). All three series show a clear decreasing trend

confer ANC to the receiving waters (Psenner 1999). Measurements of the chemical composition of precipitation show that wet air masses coming from southern locations exhibit higher concentration of base cations than those originated in northern areas (Camarero and Catalan 1996).

What is the present-day situation regarding acidification in the Pyrenees? Sulphate deposition has been decreasing since the 1980s, in good agreement with the reduction in sulphur emissions to the atmosphere since then. Lakes are also responding to that, with decreasing sulphate concentrations in lakes on non-sulphur containing bedrocks (Fig. 14.1). The lakes of the Pyrenees are thus useful witnesses of the recovery of surface waters from acidification, even if acidification has not been here as extreme as in other locations.

14.3 Trace Elements Pollution

Acidification was the first case of pollution that raised general concern about the effects of long-range transported pollutants on distant ecosystems. But it is not the oldest case of large scale pollution. Contamination by trace elements started more than two millennia ago.

The term "trace elements" is progressively substituting the formerly used term "heavy metals"; not all elements included in the term are metals, nor heavy. Trace elements include metals, metalloids and organometals that are potentially toxic to biota when found at concentrations higher than the ones usually observed within organisms. In organisms, trace elements are truly found at the level of traces. But they can be present at much higher levels in the environment surrounding those organisms. Lead is one of such trace elements. A widespread contamination by lead over the Northern Hemisphere is the oldest large-scale known pollution caused by human industry (Nriagu 1996). It dates back to the discovery of cupellation (a metallurgical process for refining noble metals) 5,000 years ago, although it increased largely 2,600 years ago with the introduction of coinage during the Ancient Greek and Roman times. Smelting of metal ores in open fires caused the emission of substantial amounts of lead and other metals to the atmosphere. After the decline of the Roman Empire, a second pulse of lead pollution was caused by the exploitation of silver mines in Central Europe during the Middle Age. A third one occurred in modern times, starting with the Industrial Revolution during the nineteenth century and peaking with the great expansion of motor vehicle traffic since the end of the Second World War, in the period in which leaded gasoline was used. Acute lead poisoning is known since the Ancient times, but it was not until the 1960s and early 1970s that studies presented evidence that chronic exposure to relatively low lead levels in the environment were causing neurological impairment in subjects, especially children, with no clinical symptoms of toxicity (Needleman 2004). This finding was crucial to the banning of lead in gasoline (and also in paint for domestic use) in the 1980s. Currently, there are international agreements to further reduce emissions of lead and some other trace elements to the atmosphere (UNECE 2016b).

The Pyrenees have not escaped pollution by trace elements. The sediments from the lakes are a record of the history of pollution in the Pyrenees; changes in the concentration of trace elements in layers of sediment deposited at different moments reveal what was the contamination load at each time. Thus, for instance, the analysis of sediments from Lake Redon (42°38′ N, 0°46′ E; 2240 m a.s.l.) in the Central Pyrenees (Fig. 14.2) showed increasing levels of contaminating lead since the 4th century BCE, a conspicuous peak during the ninth century CE, and a modern peak starting in 1900 (Camarero et al. 1998). Furthermore, lead isotopic composition changed in the peaks from the natural lead composition in the oldest strata towards values closer to that of lead from mining activities. This pattern likely results from the mixing of global, regional and even local influences. The onset of the rise is in good agreement with the accepted global chronology of emissions. But the tallest peaks are more reasonably attributed to mining in the surrounding areas, although in the most recent one the influence of leaded gasoline cannot be distinguished from mining and probably also contributes.

Contamination of lake sediments by lead during the past centuries, if not longer, is common in the Pyrenees. Almost with no exception, all lakes analysed show increased values of lead with an isotopic composition that reveals its polluting origin in the uppermost sediment layers (Bacardit et al. 2012). Other contaminating trace metals such as cadmium, copper, mercury, nickel and zinc are often found accompanying lead. Polluting trace elements are not equally distributed along the whole Pyrenean range. An extensive lake survey showed that modern surface sediments were enriched in lead and mercury with respect to deeper pre-industrial

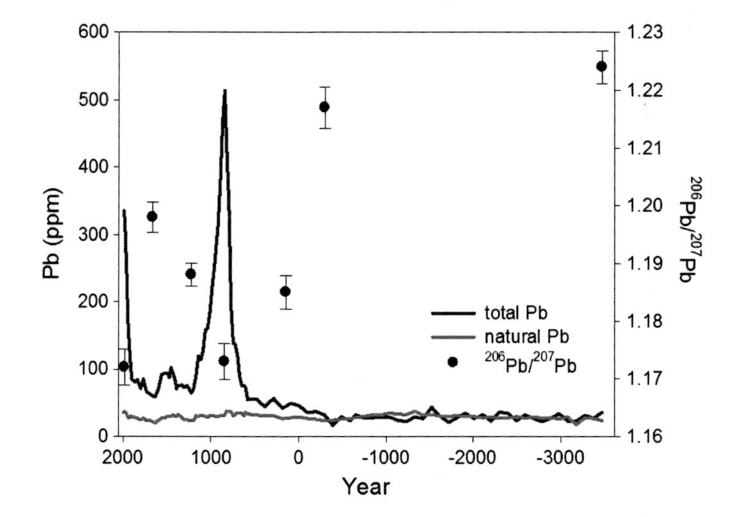

Fig. 14.2 Concentration and isotopic composition of total extractable lead in a sediment core from Lake Redon (Camarero et al. 1998). Natural lead was estimated from the aluminium concentration along the core and the natural lead-to-aluminium ratio. The difference between total and natural lead corresponds to the polluting lead inputs. Total lead consistently diverges from natural lead since c. 300 BCE, and shows two conspicuous peaks in c. 850 CE and present. The isotopic ratio ^{206}Pb/Pb varies accordingly, from the natural value (c. 1.22) to lower values consistent with the mixing with lead coming from the nearby mines of Cierco (c. 1.18) and Bossost (c. 1.15). In the modern peak, also lead from gasoline (c. 1.04) might have some contribution

sediments, and that lakes in the Central and Eastern Pyrenees were more frequently enriched than lakes in the Western side (Camarero 2003). This difference could be hypothetically due to a higher deposition of dust from southern locations carrying those metals adsorbed onto the particles (see below) in the Centre and East.

The concentration of some trace elements in the surface sediments of a number of lakes is surprisingly high for such relatively remote sites where pollutants arrive exclusively through atmospheric deposition (Fig. 14.3). The concentration found in those lakes is comparable to the levels found in sites with intense pollution, such as areas directly affected by urban and industrial sewage discharge or within harbours (Camarero et al. 2009). Some factors likely contribute to these high levels: first, lake sediments are the final sink of many elements deposited in the whole catchment and transported from there to the lake, so it may take place a "focussing" of pollutants towards the sediments; second, sediments of Pyrenean lakes tend to be very organic and have therefore a great affinity and ability to bound metal and metalloid cations; and finally, sedimentary rates in these non-productive lakes are often very low, so the effect of "dilution" of pollutants within the bulk sediment is minimised. As a consequence, a large proportion of the lakes presents concentrations of trace elements in sediments that are above the boundary where the first toxic effects on biota are detected, or even above the level in which severe effects occur.

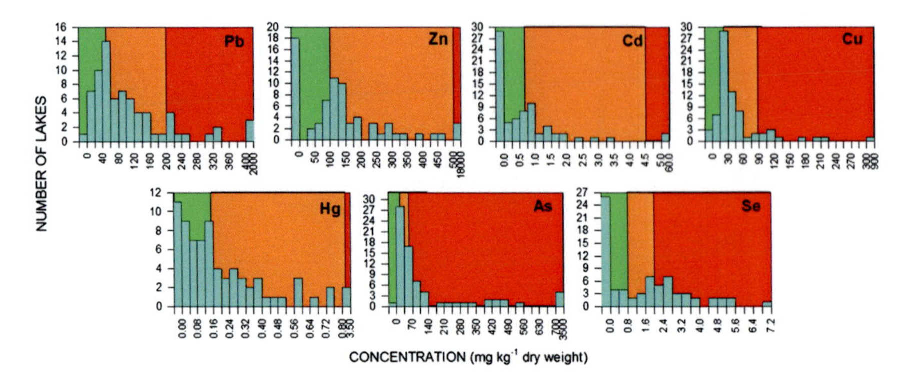

Fig. 14.3 Histograms (*blue bars*) showing the distribution of concentration of several trace elements in sediments from 85 Pyrenean lakes surveyed in 2000 (Camarero 2003). Background colours represent intensity of pollution, based on a worldwide review of aquatic sediments: background (*green*), moderate (*orange*) and intense (*red*), according to Camarero et al. (2009). The boundaries between coloured areas approximately coincide as well with the "lowest effect" (the first effects on organisms are detected) and "severe effect" (lethal effects in >80% of organisms) limits reported from bioassays in sediments

How is presently the atmospheric deposition of trace elements at high altitude in the Pyrenees? Measurements done in the Lake Redon sampling station during the decade 2004–2013 show that the concentration in atmospheric precipitation (volume weighted average) of the most commonly monitored trace elements are in the range of nanogram per liter (ng L^{-1}) but vary by three orders of magnitude, from several thousand for the most abundant to a few tens for the scarcest:

$$\text{Zn } (6500 \text{ ng L}^{-1}) > \text{Pb } (960 \text{ ng L}^{-1}) \sim \text{Cu } (950 \text{ ng L}^{-1}) >$$
$$> \text{Ni } (180 \text{ ng L}^{-1}) > \text{As } (38 \text{ ng L}^{-1}) > \text{Cd } (20 \text{ ng L}^{-1})$$

These values fall within the range of measurements in relatively remote sites in Europe, suggesting that they can be considered as background levels of pollution in atmospheric precipitation, at least in the European context.

The deposition of trace elements varies with time: it fluctuates seasonally, interannually and, in some cases, they show statistically significant decadal trends. Both interannual and seasonal variations are related with the origin of the air masses causing precipitation: air masses coming from the Iberian Peninsula and Northern Africa carry dust, and associated to it there is polluting lead, nickel and cadmium which are transported mainly bound to particles (Bacardit and Camarero 2009). Southern precipitation is more frequent during spring and summer, whereas during winter precipitation comes more often from north. This explains the seasonal fluctuation (higher values in summer) in deposition of those trace elements associated to dust. Similarly, the proportion of both types of precipitation changes from year to year, producing interannual differences in trace elements deposition. In years with a higher frequency of precipitation with a southern origin, lead, nickel and cadmium deposition is higher. In contrast, in years with higher frequency of precipitation with a northern provenance zinc and copper deposition is higher. Beyond this interannual variation, lead, zinc and nickel show a significant decreasing trend between 2004 and 2013, in good agreement with the reduction in the emissions to the atmosphere recorded in Europe during the same period (EEA 2013).

Is the abatement of emissions the end of the story about trace elements pollution? Possibly not. Besides lake sediments, soils are another matrix where trace elements have accumulated during centuries of contamination. The amount of polluting trace elements in high mountain soils (averaged taking into account bare rock areas) in the Pyrenees (Bacardit et al. 2012) can be quantified in milligrams per square metre (mg m^{-2}; Pb and Zn ~ 1000, Ni and Cu ~ 200, Cd ~ 10), whereas yearly atmospheric deposition is measured as micrograms per square metre (µg m^{-2}; Zn ~ 10,000, Pb and Cu ~ 1000, Ni ~ 250, Cd ~ 25). That is, soils hold (on a per unit area) an anthropogenic trace elements burden that is about three orders of magnitude larger than the current yearly deposition, even larger if natural (i.e. coming from the rocks) trace elements are included. So, what is the fate of this pollution legacy? May it be released from soils and contaminate the water courses, for instance? There is evidence for certain elements that this is indeed the case (Bacardit and Camarero 2010). Measurements in some Pyrenean catchments show that the amount of lead that reaches the lakes every year from the terrestrial area is larger than the lead deposited from the atmosphere on the whole catchment (Fig. 14.4). Other trace metals, such as zinc, continue to be stored within the catchments. But as shown above, zinc is deposited at a high rate, so soils could be

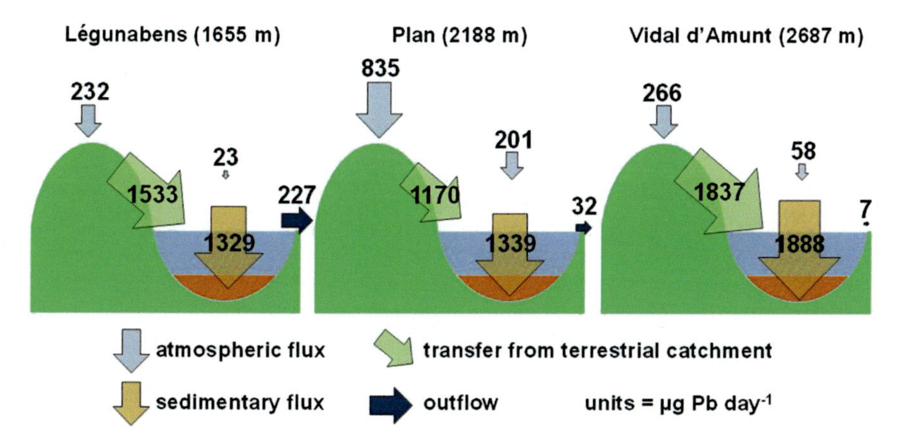

Fig. 14.4 Mass balance modelling (Bacardit and Camarero 2010) of the fluxes of lead during the snow-free season within three catchments and their lakes in the Pyrenees: Légunabens (42°45′ N, 1°26′ E), Plan (42°37′ N, 0°56′ E) and Vidal d'Amunt (42°32′ N, 0°60′ E). The atmospheric flux was measured using precipitation collectors, and it is split between direct deposition onto the lake and the terrestrial catchment according to their respective surface areas. Sedimentary fluxes were measured using sediment traps deployed in the deepest part of the lake. Outflow was calculated from discharge and lead concentration in the epilimnetic water. Transfer of lead from the terrestrial catchment to the lake was estimated so as to balance the mass fluxes. *Arrows* size is proportional to the computed fluxes. The main result is that the transfer of lead from the catchment to the lake is in all cases larger than the atmospheric flux. This indicates that the catchment is a net source releasing lead that was previously stored in soils

eventually saturated and start to release it (Nriagu et al. 1998). And climate change may have a synergistic effect (Klaminder et al. 2010). It has been argued that climate change-related processes, such as intensification of soil carbon mineralisation, melting of permafrost and increased erosion, may accelerate the shift of soils from sink to source.

14.4 Effects of Airborne Nutrients on Ecosystems

Pollutants are not the only substances transported atmospherically. Some airborne substances can provide important elements for life; in this case they are considered nutrients. Nitrogen and phosphorus are of particular interest. Both are key limiting nutrients in ecosystems, and their relative availability determines ecosystem productivity and species composition. Human activities have altered the global cycles of N and P, leading to an increased flux of these elements through the atmosphere.

There is plenty of N in the earth's atmosphere, but it is the form of inert gas generally unavailable for organisms. Organisms need the so called reactive nitrogen, that is, nitrogen-bearing molecules that can be assimilated by primary producers. These molecules are principally nitrate and ammonium. The natural production of reactive nitrogen is low, causing the productivity of many ecosystems to be limited by nitrogen. But since the start of industrialisation, human production of reactive nitrogen has increased dramatically by artificial nitrogen fixation and the use of internal combustion engines (Elser 2011). Artificial nitrogen fixation is an industrial process (Haber–Bosch process) that converts gaseous nitrogen into ammonia, which can be oxidised afterwards to produce nitrite and nitrate. The process has been massively used to produce fertilisers for agriculture, and the increased use of artificial fertilisers has lead in turn to higher emissions of ammonia to the atmosphere from soils and livestock, as well as dust rich in nitrogen from the field crops. In parallel, the use of internal combustion engines in automobiles mainly, but also other industrial combustion processes, have generalised the burning of fossil fuel causing huge emissions of nitrogen oxides to the atmosphere. Nitrogen oxides are precursors of nitrate, which has an acidifying effect but is also a nutrient. As a consequence of all these increased emissions, atmospheric transport and subsequent deposition has become the dominant distribution process of reactive nitrogen on a global basis (Galloway et al. 2008), and nitrogen deposition to ecosystems has increased from $\sim$0.5 kilograms per hectare and year (kg N ha^{-1} year^{-1}) or less in pristine conditions to rates that are nowadays greater than one order of magnitude, exceeding 10 kg N ha^{-1} year^{-1} on average in large areas of the world. Similarly to the case of sulphur oxides, efforts are underway to abate the nitrogen oxides emissions in Europe and North America, but emissions from emerging economies demanding more energy may counteract these reductions on a global balance. Furthermore, the production of reactive nitrogen related with agriculture is about 75% of the total, and it is challenging to reduce it in a world where hundreds of millions of people still suffer from a "fertilizer deficit". And the increased use of biofuels from crops that need to be fertilised with nitrogen adds a new dimension to the problem. Predictions are that reactive nitrogen deposition may double by 2050, with some regions reaching 50 kg N ha^{-1} $year^{-1}$ (Galloway et al. 2004).

Nitrogen deposition has been measured routinely in many places since it was recognised as an acidifying agent. In contrast, phosphorus deposition measurements are much less frequent and, as a matter of fact, the knowledge about the atmospheric phosphorus cycling is limited. In the absence of measurements, most of our present understanding derives from modelling, which is based on the estimation of emissions coupled with models of atmospheric circulation to infer the transport and distribution of atmospheric phosphorus. Phosphorus does not form easily gases that can be emitted, so it has been assumed that it is mainly transported as particles. Under this assumption, the estimations indicate that, on a global basis, most (>80%) of the atmospheric phosphorus is in mineral aerosols, with primary biogenic particles and combustion particles having a weight only in non-dusty regions (Mahowald et al. 2008). They also indicate that atmospheric phosphorus of human origin is only 5% of the total. However, recent work suggests that gaseous phosphorus emissions are important during combustion processes. If gases are also considered, human atmospheric phosphorus should then account for about 50% of the total (Wang et al. 2015). Estimates are therefore still controversial, and this fact highlights the need of a better understanding of the processes of production and speciation of atmospheric phosphorus. In any case, human action may have an strong effect on the atmospheric phosphorus balance, either by enhancing mineral dust production by land-use changes and (indirectly) climate change, or by increasing combustion sources, or both. Although global trends are not clear (mainly because of the lack of data), some research suggests that a significant increase in phosphorus deposition has occurred since pre-industrial times, and that this increase can be causing a change in the biogeochemistry and the trophic status of naturally oligotrophic systems, such as alpine lakes (Brahney et al. 2015). The same can be claimed for nitrogen. Yet phosphorus is present as a constituent of living matter in a lower proportion than nitrogen. Hence, for stoichiometric reasons, smaller changes in phosphorus than nitrogen availability may potentially have a larger effect on primary production. It is therefore necessary to take both nutrients into account in order to assess their true impact.

The nitrate concentrations in streams indicate that the Pyrenean catchments are saturated with nitrogen. The concept of nitrogen saturation means that the catchment is receiving more nitrogen than organisms (plants and microbiota) living in the catchment are able to take up and immobilise, so a part of it is leached out of the catchment, mostly as nitrate in running waters. Nitrate concentrations in the Central Pyrenees streams fluctuate seasonally, with basal values typically between 7 and 10 µeq L^{-1} during the plant growing season and peaks as high as 20–30 µeq L^{-1} during the dormant period. On a scale from 0 to 3 (Stoddard and Traaen 1995), Pyrenean catchments are at saturation stage 2, defined as “saturated, with high N loss”.

The atmospheric nitrogen load from bulk deposition (deposition with rain and the fine fraction of dry deposition) in the central Pyrenees in 2010 was c. 10 kg N ha^{-1}

year^{-1}, matching the global average. This deposition rate has been increasing during the precedent decade, in good agreement with the Spanish inventory of nitrogen emissions; but, surprisingly, the lakes in the Pyrenees did not follow the same trend (Fig. 14.5). In an opposite way, the concentration of dissolved inorganic nitrogen has decreased with time in the lakes (Camarero and Catalan 2012).

Why is nitrogen concentration declining in the lakes? Monitoring data from Lake Redon provide insight into the process that causes this (Fig. 14.6). While nitrogen deposition and nitrogen concentration in Conangles creek (a stream which drains the Lake Redon basin together with a much larger area without lakes) have increased, nitrogen in the lake has been decreasing, in agreement with the trend observed for the whole set of studied lakes. The most plausible reason is that the uptake of nitrogen by primary producers has increased at a higher rate than nitrogen deposition, an increase that might be explained by the fact that phosphorus deposition has increased in parallel to nitrogen deposition. This may be causing a shift from a situation in which phosphorus was the limiting nutrient, so nitrogen in excess of biological needs can accumulate, to a new situation in which this limitation is alleviated by the increased input of atmospheric phosphorus, so more nitrogen is consumed to the point of becoming limiting and, eventually, exhausted. Presumably, limitation by phosphorus is a circumstance that has been induced by the increased deposition of reactive nitrogen of human origin during industrialisation (Goldman 1988). This new change towards nitrogen limitation can be seen as a return to a pristine situation. However, there is a difference. The inputs of both

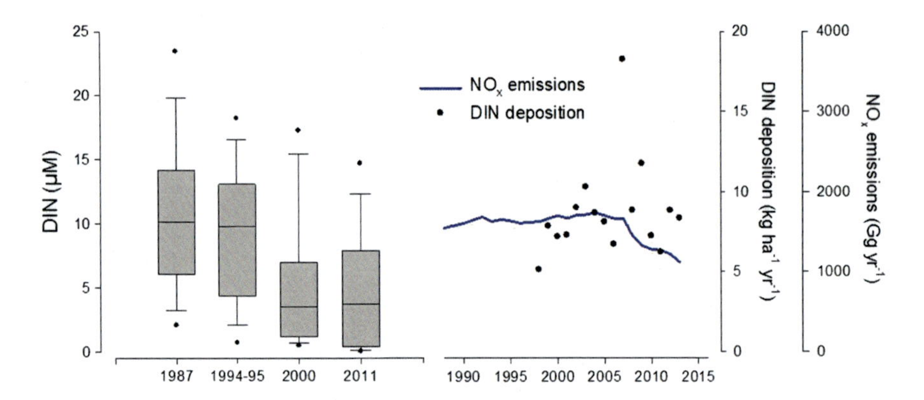

Fig. 14.5 *Left* Box plot showing the distribution of dissolved inorganic nitrogen (DIN) concentration in Pyrenean lakes (percentiles represented as in Fig. 14.1) during four synoptic surveys (Camarero and Catalan 2012). *Right trend-line* showing the evolution of NO_x emissions in Spain (*Source* EMEP, http://www.ceip.at/ms/ceip_home1/ceip_home/webdab_emepdatabase/emissions_emepmodels/) The overimposed *dots* represent yearly DIN deposition as measured at the Lake Redon station. While DIN concentration has decreased since the late 1980s, emission presented an increasing trend up to 2007. From then on, a sharp decrease is reported, likely caused by the application of more stringent measures of reduction and also as an effect of the economic crisis. Deposition of DIN does not show a clear response to that sharp decrease. Up to the year 2009 deposition increased. Only from 2010 on an incipient downward trend could be argued, but it does not seem to be proportional to the reduction in emissions

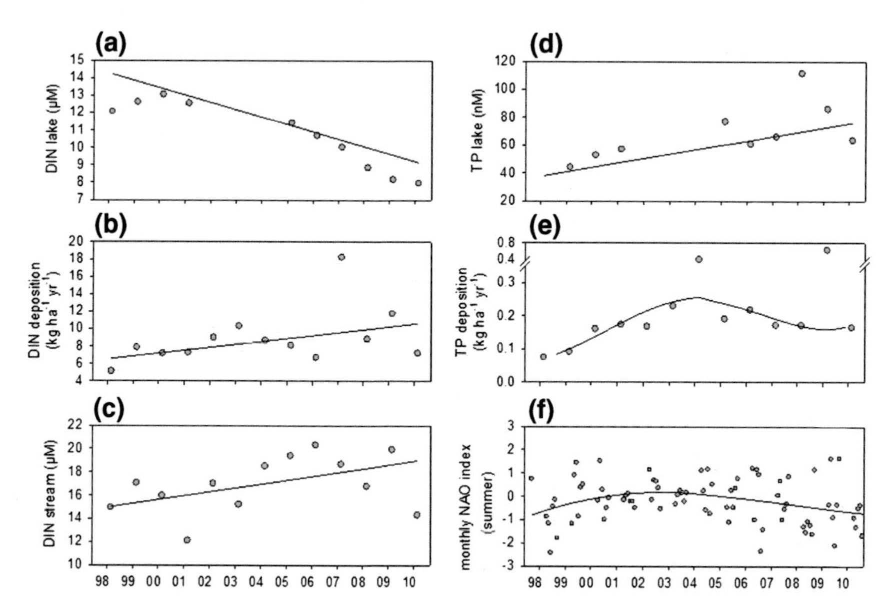

Fig. 14.6 Panels **a–e** show nitrogen and phosphorus in Lake Redon and its catchment (Conangles). **a** DIN concentration in lake water (yearly average); **b** DIN yearly deposition; **c** DIN concentration in stream water (Conangles creek, yearly average); **d** total phosphorus in lake water (yearly average); **e** total phosphorus yearly deposition. Panel **f** shows the values for the monthly NAO index during spring and summer (April–September). While nitrogen deposition has increased, DIN in the lake has decreased. The increase in the stream indicates that the catchment is not a sink for nitrogen. Therefore, an in-lake process should control DIN in the lake. Phosphorus has increased both in deposition and in the lake, suggesting a fertilisation effect that has enhanced nitrogen uptake. The stoichiometry of the lake has shifted towards a hypothetical situation of nitrogen limitation for phytoplankton. Both phosphorus deposition and the NAO index show a *bell-shaped* trend. A higher frequency of positive NAO values in the mid part of the period likely correlates with more frequent dust outbreak episodes carrying phosphorus from Northern Africa to the Pyrenees. This would indicate a certain meteorological control on the transport of atmospheric phosphorus (after Camarero and Catalan 2012)

nutrients are higher now, so the productivity of the ecosystem seems to be enhanced (Fig. 14.7) and we can speak of a change in the trophic status of the Pyrenean lakes towards a relative (in terms of these low-productive waters) “eutrophication”.

Phosphorus comes largely from South, accompanying the dust outbreaks reaching the Pyrenees from Northern Africa and the Iberian Peninsula (Camarero and Catalan 2012). There is a certain correlation between the North Atlantic Oscillation (NAO) index and the production of dust in Northern Africa and subsequent transport towards the Mediterranean (Moulin et al. 1997). Positive NAO values entail dryer meteorological conditions and therefore higher dust inputs to the atmosphere. Changes in the dust and phosphorus inputs may therefore be linked to natural meteorological variations. But in the context of climate change, it can be also argued that human action is contributing indirectly to enlarge the atmospheric dust fluxes as well (Mahowald et al. 2010). The data set presented here could be

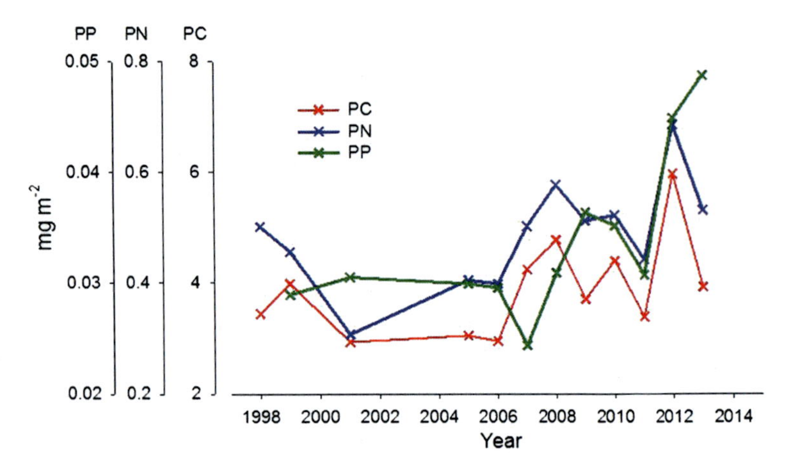

Fig. 14.7 Particulate carbon, nitrogen and phosphorus in the water column of Lake Redon, as indicators of the standing biomass of plankton (excluding crustaceans) in the lake. Values are integrated over the whole water column and are expressed on a per square metre basis. From 2007 on, planktonic biomass has increased. This could be an early indication of a shift of the lake towards a higher trophic status caused by the increase in atmospheric nitrogen and phosphorus deposition

either part of a longer trend or a mere fluctuation. But in any case, what can be deduced from it is that the anthropogenic nitrogen inputs, by causing phosphorus limitation, have made the lake more sensitive to any natural or human-induced phosphorus input than it would be otherwise. It is a combined effect of both nutrients what is causing a response of the system.

14.5 Some Reflections on Conservation

The issues presented here have a characteristic that prevents us from viewing them using a conventional conservation approach: they cannot be treated within the boundaries of the own ecosystems. The systems cannot be isolated from the atmosphere to protect them, and there is no way to "clean" them from the undesirable chemical load already deposited. These facts add a wider dimension to the concept of conservation: actions at a large scale, and not necessarily related or applied in a direct way to the ecosystems to be preserved, are required to protect them. Mountains are not the only example of ecosystems being at risk by a distant threat. But, at least in Europe, they are perceived as the last wilderness left untouched. This perception helps to make a strong point and raise public concern on the message that some kind of global action is required to protect those parts of nature that a priori seemed unaffected by impacts caused by humans.

The results presented here highlight some notions that are of current concern regarding the processes involved in the atmospheric transport and fate of polluting chemicals. The first notion is that they do have a fate. In contrast with an early, simplistic idea that chemicals emitted to the atmosphere were somehow "diluted" down to no-effect levels, i.e. virtually disappearing, research has shown that pollutants may undergo preferential accumulation. By entering the biogeochemical cycles they can be directed to some particular sites where they accumulate to reach significant concentration. This is worsened by the fact that humans have developed a capacity to mobilise chemicals in amounts comparable to (or even much higher than) the global natural fluxes. As a result, accumulated chemicals are not merely a "footprint" of human actions, but they are present in large pools in the environment. Examples of this are the accumulation of trace elements in Pyrenean lakes sediments, and the reactive nitrogen saturation exhibited by streams.

A second important notion is that these pools are a pollutant legacy that may not be the ultimate fate of contaminants. Chemicals stored over time can be remobilised and constitute a secondary source of delayed pollution. There is a legacy of pollution that span not only over space, but also over time. The export of polluting lead accumulated during millennia from catchments in the Pyrenees is an example of that.

Another important emerging idea is that there are synergistic effects caused either by several pollutants or by the interaction between pollutants and climate change. For instance, the combined effect of human nitrogen emissions together with changes in natural emissions of phosphorus-bearing dust (with a possible connection to climate changes) is changing the trophic status of Pyrenean lakes. Such combined effects add complexity to the picture. Knowledge about multi-factor synergistic and confounding effects is still missing, as well as on effects caused by long-term chronic exposure to environmental doses of multi-pollutant mixtures.

And finally, there are also findings that bring some hope. Sulphate decline in surface waters as recorded in the Pyrenees (and other sensitive areas worldwide) suggests that perhaps it is not too naive to think that global actions are still possible to preserve nature. Despite there is a hysteresis in the reversibility of acidification, and that climate change has confounding (or even counteracting) effects on it, it seems that recovery of ecosystems may proceed, provided that polluting emissions are abated and sufficient time is allowed to natural processes to act. In this sense, remote ecosystems such as high mountain catchments have a role not only as warning systems, but also as indicators to assess the effectiveness of policies and efforts for global conservation.

References

Bacardit M, Camarero L (2009) Fluxes of Al, Fe, Ti, Mn, Pb, Cd, Zn, Ni, Cu, and As in monthly bulk deposition over the Pyrenees (SW Europe): the influence of meteorology on the atmospheric component of trace element cycles and its implications for high mountain lakes. J Geophys Res 114:G00D02

Bacardit M, Camarero L (2010) Atmospherically deposited major and trace elements in the winter snowpack along a gradient of altitude in the Central Pyrenees: the seasonal record of long-range fluxes over SW Europe. Atmos Environ 44:582–595

Bacardit M, Krachler M, Camarero L (2012) Whole-catchment inventories of trace metals in soils and sediments in mountain lake catchments in the Central Pyrenees: apportioning the anthropogenic and natural contributions. Geochim Cosmochim Acta 82:52–67

Brahney J, Mahowald N, Ward DS, Ballantyne AP, Neff JCCGB (2015) Is atmospheric phosphorus pollution altering global alpine lake stoichiometry? Glob Biogeochem Cycles 29:1369–1383

Camarero L (2003) Spreading of trace metals and metalloids pollution in lake sediments over the Pyrenees. J Phys IV France 107:249–253

Camarero L, Botev I, Muri G, Psenner R, Rose N, Stuchlik E (2009) Trace elements in alpine and arctic lake sediments as a record of diffuse atmospheric contamination across Europe. Freshw Biol 54:2518–2532

Camarero L, Catalan J (1996) Variability in the chemistry of precipitation in the Pyrenees (northeastern Spain): dominance of storm origin and lack of altitude influence. J Geophys Res 101:29491–29498

Camarero L, Catalan J (1998) A simple model of regional acidification for high mountain lakes: application to the Pyrenean lakes (North-East Spain). Water Res 32:1126–1136

Camarero L, Catalan J (2012) Atmospheric phosphorus deposition may cause lakes to revert from phosphorus limitation back to nitrogen limitation. Nat Commun 3:1118

Camarero L, Masqué P, Devos W, Ani-Ragolta I, Catalan J, Moor HC, Pla S, Sanchez-Cabeza JA (1998) Historical variations in lead fluxes in the Pyrenees (NE Spain) from a dated lake sediment core. Water Air Soil Pollut 105:439–449

Cowling EB (1982) Acid precipitation in historical perspective. Envir Sci Technol 16:A110–A123

EEA (2013) EU emissions of As, Cd, Hg, Ni and Pb, 2002–2011, as a percentage of 2002 emissions. http://www.eea.europa.eu/data-and-maps/figures/eu-emissions-of-as-cd-1

Elser JJ (2011) World Awash Nitrogen Sci 334:1504–1505

Galloway JN, Dentener FJ, Capone DG, Boyer EW, Howarth RW, Seitzinger SP, Asner GP, Cleveland CC, Green PA, Holland EA, Karl DM, Michaels AF, Porter JH, Townsend AR, Vöosmarty CJ (2004) Nitrogen cycles: past, present, and future. Biogeochemistry 70:153–226

Galloway JN, Townsend AR, Erisman JW, Bekunda M, Cai Z, Freney JR, Martinelli LA, Seitzinger SP, Sutton MA (2008) Transformation of the nitrogen cycle: recent trends, questions, and potential solutions. Science 320:889–892

Goldman CR (1988) Primary productivity, nutrients, and transparency during the early onset of eutrophication in ultra-oligotrophic Lake Tahoe, California-Nevada. Limnol Oceanogr 33:1321–1333

Gorham E, Underwood JK, Martin FB, Gordon O (1986) Natural and anthropogenic causes of lake acidification in Nova Scotia. Nature 324:451–453

Klaminder J, Hammarlund D, Kokfelt U, Vonk JE, Bigler C (2010) Lead contamination of subarctic lakes and its response to reduced atmospheric fallout: can the recovery process be counteracted by the ongoing climate change? Environ Sci Technol 44:2335–2340

Likens GE, Bormann FH (1974) Acid rain: a serious regional environmental problem. Science 184:1176–1179

Mahowald N, Jickells TD, Baker AR, Artaxo P, Benitez-Nelson CR, Bergametti G, Bond TC, Chen Y, Cohen DD, Herut B, Kubilay N, Losno R, Luo C, Maenhaut W, McGee KA, Okin GS, Siefert RL, Tsukuda S (2008) Global distribution of atmospheric phosphorus sources, concentrations and deposition rates, and anthropogenic impacts. Glob Biogeochem Cycles 22:GB4026

Mahowald NM, Kloster S, Engelstaedter S, Moore JK, Mukhopadhyay S, McConnell JR, Albani S, Doney SC, Bhattacharya A, Curran MAJ, Flanner MG, Hoffman FM, Lawrence DM, Lindsay K, Mayewski PA, Neff J, Rothenberg D, Thomas E, Thornton PE, Zender CS (2010) Observed 20th century desert dust variability: impact on climate and biogeochemistry. Atmos Chem Phys 10:10875–10893

Moulin C, Lambert CE, Dulac F, Dayan U (1997) Control of atmospheric export of dust from North Africa by the North Atlantic oscillation. Nature 387:691–694

Neary BP, Dillon PJ (1988) Effects of sulphur deposition on lake water chemistry in Ontario. Nature 333:340–343

Needleman H (2004) Lead poisoning. Annu Rev Med 55:209–222
Nriagu JO (1996) A history of global metal pollution. Science 272:223–224
Nriagu JO, Wong HKT, Lawson G, Daniel P (1998) Saturation of ecosystems with toxic metals in Sudbury basin, Ontario, Canada. Sci Total Environ 223:99–117
Odén S, Ahl T (1979) The acidification of Scandinavian lakes and rivers. Ymer, Årsbok
Pitelka LF, Raynal DJ (1989) Forest decline and acidic deposition. Ecology 70:2–10
Psenner R (1999) Living in a dusty world: airborne dust as a key factor for alpine lakes. Water Air Soil Pollut 112:217–227
Schindler DW, Mills KH, Malley DF, Findlay DL, Shearer JA, Davies IJ, Turner MA, Linsey GA, Cruikshank DR (1985) Long-term ecosystem stress: the effects of years of acidification on a small lake. Sci (WashDC) 228:1395–1401
Smith RA (1872) Air and rain: the begginings of a chemical climatology. Longmans Green, London
Smith SJ, van Aardenne J, Klimont Z, Andres RJ, Volke A, Delgado Arias S (2011) Anthropogenic sulfur dioxide emissions: 1850–2005. Atmos Chem Phys 11:1101–1116
Stoddard JL, Traaen T (1995) The stages of nitrogen saturation: Classification of catchments included in 'ICP on Waters'. In: Hornung M, Sutton MA, Wilson RB (eds) Mapping and modelling of critical loads for nitrogen, Proceedings of the grange-over-sands workshop, 24–26 Oct 1994. Institute of Terrestrial Ecology, Edinburg, pp 55–62.I
Streets DG, Waldhoff ST (2000) Present and future emissions of air pollutants in China: SO_2, NOx, and CO. Atmos Environ 34:363–374
Sullivan TJ, Eilers JM, Church MR, Blick DJ, Eshleman KN, Landers DH, DeHaan MS (1988) Atmospheric wet sulphate deposition and lakewater chemistry. Nature 331:607–609
UNECE (2016a) The 1979 Geneva Convention on Long-range Transboundary Air Pollution. http://www.unece.org/env/lrtap/lrtap_h1.html, http://www.unece.org/env/lrtap/lrtap_h1.html. 2016
UNECE (2016b) The 1998 Aarhus Protocol on Heavy Metals. http://www.unece.org/env/lrtap/hm_h1.html
Wang R, Balkanski Y, Boucher O, Ciais P, Penuelas J, Tao S (2015) Significant contribution of combustion-related emissions to the atmospheric phosphorus budget. Nat Geosci 8:48–54
Wright RF, Aherne J, Bishop K, Camarero L, Cosby BJ, Erlandsson M, Evans CD, Forsius M, Hardekopf DW, Helliwell R, Hruska J, Jenkins A, Kopacek J, Moldan F, Posch M, Rogora M (2006) Modelling the effect of climate change on recovery of acidified freshwaters: relative sensitivity of individual processes in the magic model. Sci Total Environ 365:154–166
Wright RF, Larssen T, Camarero L, Cosby BJ, Ferrier RC, Helliwell R, Forsius M, Jenkins A, Kopacek J, Majer V, Moldan F, Posch M, Rogora M, Schöpp W (2005) Recovery of acidified European surface waters. Environ Sci Technol 39:64A–72A

Chapter 15
Importance of Long-Term Studies to Conservation Practice: The Case of the Bearded Vulture in the Pyrenees

Antoni Margalida

Abstract Detailed, long-term scientific studies are necessary for conservation purposes, but with the main handicap to have the continual economic support required for them. Behavioural and conservation biology studies need long-term projects to achieve robust data, but managers, administrations and policy-makers need, in most cases, immediate results. Here I show several examples of the research obtained from a long-term study (1987–2014) in one of the most threatened species in Pyrenean mountains, the bearded vulture (*Gypaetus barbatus*), highlighting the importance of such long-term research. The results show how long-term studies are necessary to identify conservation problems, to understand demographic changes on populations and priorities to apply conservation measures. The study's findings allowed the identification of the negative density-dependent effects on fecundity, the lack of recolonization of new territories outside the current distribution area and the increase in polyandrous trios, suggesting an initial optimal habitat saturation. From a management point of view, the studies show that supplementary feeding sites (SFS) can have detrimental effects on fecundity but increases pre-adult survival. Also, illegal poisoning is increasing, and the demographic simulations suggest a regressive scenario in population dynamics if this factor is not eliminated. More recently, anthropogenic activities through human health regulations that affect habitat quality can suddenly modify demographic parameters. The results obtained about changes in nest-site selection, mating system and demographic parameters can only be achieved through long-term studies,

A. Margalida (✉)
Faculty of Life Sciences and Engineering, Department of Animal Science, University of Lleida, 25198 Lleida, Spain
e-mail: amargalida@ca.udl.cat

A. Margalida
Division of Conservation Biology, Institute of Ecology and Evolution, University of Bern, 3012 Bern, Switzerland

J. Catalan et al. (eds.), *High Mountain Conservation in a Changing World*, Advances in Global Change Research 62, DOI 10.1007/978-3-319-55982-7_15

suggesting the importance of long-term research to provide accurate information to managers and policy-makers to optimise the application of conservation measures.

Keywords Conservation measures · *Gypaetus barbatus* · Long-term studies · Management · Policy-makers · Pyrenees · Threatened species · Vulture

15.1 Introduction

Detailed, long-term scientific studies of long-lived species although necessary are relatively unusual (Clutton-Brock and Sheldon 2010; Mills et al. 2015, 2016), with the main handicap to require a continual economic support (Birkhead 2014). On the other hand, Ph.D. projects, as well managers, administrations and policy-makers need, in most cases, immediate results. As a result, funding agencies usually operate on short-term funding cycles, making it difficult to maintain the continual levels of support required for long-term research.

Nevertheless, such studies are essential to understanding the factors that drive population size and dynamics and to provide crucial information for conservation. Thus, long-term monitoring of wild populations is necessary to obtain the information needed for conservation purposes. For example, some demographic and ecological processes require long-term monitoring. Those relevant for conservation include the need to estimate variation in demographic parameters over time, which affect geometric population growth and establish population trends. For example, the near extinction of some long-lived species as the Condor of California (*Gymnogyps californianus*) related to lead intoxication (Mee and Hall 2007) and the sudden regressive trend on three Asian vulture species during the last decade as consequence of the intoxication by a veterinary drug (Oaks et al. 2004; Green et al. 2004) show the importance of long-term monitoring and the estimation of demographic parameters for conservation practice.

The identification of conservation problems through long-term studies to help threatened species applying most appropriate measures constitutes a valuable tool for managers and practitioners. Accordingly, the discipline of conservation biology needs to be integrated into policy decisions to be effective, assuring that results obtained are immediately applied.

In the case of threatened species, the identification and application of measures to improve the conservation status and population trend are a priority for conservation biologists and policy-makers. This is the case of the bearded vulture (*Gypaetus barbatus*), a long-lived, flashing endangered species that provides critical ecosystem services such as disease and pest control (Margalida et al. 2011a; Margalida and Colomer 2012; Donázar et al. 2016), indirect greenhouse emissions regulation (Morales-Reyes et al. 2015, 2017) or cultural inspiration and recreational activities (Donázar et al. 2016). In Europe, a total of 215 territories have been documented in mountainous ecosystems (del Hoyo et al. 1994; Margalida 2010a), 130 of which are present on the southern face of the Pyrenees (Spain). In this region, several management and conservation actions have been carried out during the last 30 years to improve the status of the species. The historical decline in Europe has been associated with anthropogenic actions, mainly poisoning and direct persecution (Margalida et al. 2008a; Margalida 2010a). Significant human and economic efforts have been dedicated to recovering this threatened species in Europe to reverse its negative population demographic trend (i.e. conservation programmes and reintroduction projects), through funds provided by several life projects, administrations and NGOs (Schaub et al. 2009; Margalida 2010a).

Here, I show several examples of thc research obtained from a long-term study in this threatened species inhabiting mountain biomes, in this case, the Pyrenees (Spain, France and Andorra). The primary goal is to highlight, based on some case study examples (see Carrete et al. 2006a, b; Margalida et al. 2008a, 2014a; Oro et al. 2008), the importance of long-term studies to obtain accurate data for conservation purposes.

15.2 A Long-Term Study Initiated in 90s

In 1992, a search for the SCI papers indexed as "*Gypaetus barbatus*" showed that until this year, only seven papers had been published about the species. The main information was provided on books (Hiraldo et al. 1979; Heredia and Heredia 1990; Mundy et al. 1992) and the unpublished thesis by Chris J. Brown about the subspecies *Gypaetus barbatus meridionalis* in Southern Africa (Brown 1988). General aspects of breeding ecology were provided, but several essential aspects of the ecology and demography were lacking. Under this scenario, the Spanish Ministry of Environment and the autonomous governments of Catalunya, Navarra and Aragón started different conservation actions in this difficult to study species. At the beginning of 90s, under the recovery plan for the bearded vulture applied by the different administrations (Navarra, Catalunya and Aragón) and a Life project, Spain started with the regular monitoring and study of the breeding biology of the species (see Heredia and Heredia 1990; Margalida and Heredia 2005). With this background, the main goals were as follows (see Margalida 2010a):

- Describe the behaviour, demography and ecology of the bearded vulture
- Evaluate the conservation measures already implemented and the future problems that the conservation of the species will have to confront
- Identify the factors causing the species' decline
- Develop management recommendations for its recovery
- Optimise the management of bearded vulture populations from the standpoint of conservation biology.

15.3 Long-Term Changes in Nest-Site Selection and Distribution

15.3.1 Importance of Nest Selection Studies for Conservation

Nest-site selection studies are a valuable tool that provides information about the suitability and limitations for the future population growth and geographic expansion of a species. Explanatory and predictive models are powerful instruments in the study of spatial distributions of populations, and their use is growing exponentially in animal ecology (Guisan and Zimmermann 2000; Guisan et al. 2002, 2013; Rushton et al. 2004; Oppel et al. 2012). Models establish statistical relationships between a response variable, such as the presence or absence of the species, and a set of explanatory variables that usually quantify environmental characteristics such as climate, landscape features and degree of human influence (e.g. Bustamante and Seoane 2004; Guisan and Thuiller 2005; Gavashelishvili and McGrady 2006; Addison et al. 2013; Guisan et al. 2013). This procedure has been regularly employed in wildlife conservation, mainly to determine the probability of future site occupation for expanding populations (Mladenoff et al. 1995; Buckland et al. 1996; Jerina et al. 2003; Hirzel et al. 2004) and to examine habitat suitability for released populations within reintroduction programmes (Bustamante 1998). The accuracy of predictive models is strongly dependent on population equilibrium, biotic interactions and stochastic local events. While such complexity frequently limits the generality of the results (Guisan and Zimmermann 2000; Guisan et al. 2002, 2013), an increasing number of studies take this into account (e.g. Fielding and Haworth 1995; Lindenmayer et al. 1995; Rodriguez and Andrén 1999). Predictive models are a potentially valuable tool to detect long-term trends in those key variables capable of explaining the distribution of organisms. In the case of the bearded vulture, the restricted distribution of the isolated Pyrenean region was a concern from a conservation perspective by making the species more vulnerable to extinction threats. As a consequence, we examined the long-term variation in how individuals select nesting cliffs all over the Spanish Pyrenees (see details in Margalida et al. 2008a). We focussed on variables indicating a high probability of cliff occupancy as determined by a previously published model (Table 15.1, see Donázar et al. 1993; Margalida et al. 2008a).

15.3.2 The Compacting Process on Pyrenean Bearded Vultures

Although the breeding population increased from 53 to 93 territories between 1991 and 2002, the breeding range surprisingly expanded only slightly. New and old nesting cliffs had similar habitat features in relation to topography, altitude and

Table 15.1 Variables used to characterise bearded vulture nesting cliffs and random cliffs. In random cliffs, distances were measured from a point in the centre of the cliff (modified from Donázar et al. 1993 and Margalida et al. 2008a)

Topography
Relief: Topographic irregularity index. A total number of 20 m contour lines, cut by four 1 km lines starting from the nest in directions N, S, E and W.
Cliff local characteristics
Altitude: Altitude of the nest above the sea-level (m).
Cliff: Cliff height, measured as the number of the 20-m contour cuts by a 50-m line perpendicular to the cliff face at nest level.
Orientation: Orientation of the cliff face at the level of the nest. Orientations were scored in increasing shelter from cold, humid winds from the NW which are dominant in the area: 1= NW, 2 = N or W, 3 = NE or SW, 4 = E or S, 5 = SE.
Environmental characteristics of the surrounding area
Forest: Extension (%) of forested areas in a 1 000 m radius around the nest.
Distance Village: Distance to the nearest inhabited village (km).
Inhabitants: Number of inhabitants in the nearest village.
Human disturbance
Kilometres roads: Kilometres of paved and unpaved roads in a 1 000 m radius of the nest.
Distance paved road: Shortest linear distance between the nest and the closest paved road (km).
Distance road: Shortest linear distance between the nest and the closest road, paved or unpaved (km).
Height paved road: Altitudinal difference between the nest and the closest paved road, measured at the point the road is closer to the nest (m). If the nest is lower than the road, a negative value is obtained.
Height road: Altitudinal difference (m) between the nest and the closest road (paved or unpaved). If the nest is lower than the road a negative value was obtained.
Intraspecific relationships
Nearest neighbour: Linear distance between the nest and the closest nest of the nearest neighbour (km).

degree of human influence, but the distance between occupied cliffs was reduced (from 11.1 to 8.9 km). As a result, the probabilities of occupation predicted by the model were lower for newly colonised sites. Interestingly, the cliffs located in peripheral zones, without no previous bearded vulture territories, tended to be smaller (mean height) and were usually located in less steep and lower altitude zones. Also, they were closer in both, distance and altitude, to roads and they had a greater neighbouring human population. The new territories occupied from 1991 to 2002 showed similar habitat characteristics to those occupied before 1991. The only exception was the variable “nearest neighbour” distance. New territories were situated at a significantly smaller distance to other occupied territories than they were before. In relation to randomly selected cliffs, new nest sites were in areas that were more rugged, and in cliffs that were higher than average. In addition, new nesting cliffs were not significantly further from the nearest occupied nest (mean: 8.9 km) than random cliffs (mean: 8.1 km) with similar when comparing old nests to randomly selected cliffs.

The application of models including the variable “nearest neighbour” determined that newly occupied cliffs had intermediate probability values between old nesting cliffs and randomly selected cliffs. Therefore, during the last years, bearded vultures have been selecting a similar kind of cliff as previously chosen in relation to topography and distance from human habitation, the main factors affecting cliff selection. The distance between neighbouring territories decreased in parallel with the increase in population density (from 4.4 to 2.5 territories/1000 km^2 during 1991–2002).

15.3.3 Why Bearded Vultures Do not Extend Their Distribution Range?

Our study suggests that territory compression may occur without serious modification of nesting habitat quality. Our findings show that the selection of nest sites has not changed with the population increase of the Pyrenean bearded vulture. For those variables determining the probability of cliff occupancy (ruggedness, altitude and distance to the nearest human habitation), newly established territories between 1991 and 2002 had similar values to those of traditional territories occupied previously. These new pairs, however, occupied cliffs in between the existing old territories so that the distance between neighbours has reduced by around 20%, which proves territory shrinkage (Both and Visser 2000; Ridley et al. 2004). Cliffs remaining unoccupied by bearded vultures after 1991 in the periphery of the Pyrenees had both lower values in key variables (e.g. relief, altitude, distance to a village) and lower probability of occupancy than those occupied in the Pyrenean range during the same period. These findings seem to contradict what would be expected from a distribution by despotic competition, classically described for birds of prey and other territorial birds (Newton 1998). It could be argued that there is still availability of good-quality breeding sites in the Pyrenees and that, consequently, the population has not reached the saturation threshold necessary to observe the occupation of marginal territories by newly established birds. This argument holds true if the comparison is limited to the examination of the quality of nesting cliffs where no long-term changes were appreciable. A different picture, however, arises when the distance between neighbouring territories is considered. Since this variable was significant in the model fitted in 1991, we can deduce that saturation is taking place and that there is a decrease in territorial quality via a reduction in the distance between breeding pairs. In parallel, it has been demonstrated that the productivity of the Pyrenean population has sharply declined during the last decade as a consequence of the increasing breeding density (Carrete et al. 2006a; Margalida et al. 2014b), a clear symptom of population crowding, as is the appearance of unusual mating systems (Bertran and Margalida 2003; Carrete et al. 2006b; Bertran et al. 2009).

The packing process could also be motivated by the scarcity of suitable breeding places in mountain areas surrounding the Pyrenees. As our analyses demonstrate, potential nesting cliffs in those regions are of lower quality (regarding altitude, topography and human influence) than those existing within the central mountain range. Human presence is known to strongly affect the distribution patterns of bearded vultures and other large raptors, not only respect to the safety of breeding nest sites (Donázar et al. 1993; Margalida et al. 2007a), but also with regard to foraging grounds (Bautista et al. 2004; Gavashelishvili and McGrady 2006). Previous analyses have shown that the breeding success of Pyrenean bearded vultures is affected by human activities and infrastructures (Donázar et al. 1993; Arroyo and Razin 2006) as occurs in other vulture species (Zuberogoitia et al. 2008; Margalida et al. 2011b; but see Oppel et al. 2016). Therefore, it is reasonable to think that the low probability of cliff occupancy outside the Pyrenees is determined, at least partially, by the intense human influence on these peripheral mountain regions. In this sense, the Pyrenees could progressively become more of an ecological island for the bearded vulture surrounded by areas that are inadequate due to the increasingly human influence on the environment. If these factors prevent the population from increasing in numbers and expanding, the risk of extinction in this population will remain high as a consequence of stochastic demographic and environmental phenomena and the limited genetic variability (Godoy et al. 2004).

The relatively low quality of sites in nearby mountains may prevent the expansion of the breeding range, but conspecific attraction and supplementary feeding may also play a role. Our study confirmed that monitoring changes in key variables relevant to habitat selection are useful in determining long-term trends in settlement patterns in heterogeneous environments. The results also suggest that the available nest-site selection model may accurately predict cliff occupancy by bearded vultures in those areas where the distance to the nearest neighbour is not limiting. In particular, the model may be useful in establishing priority areas for reintroduction.

15.4 Changes in Mating System and Population Regulation

15.4.1 Cooperative Breeding

In many bird species, a proportion of mature individuals in the population does not breed (i.e. floaters), mainly because all suitable territories are occupied (Newton 1992). These individuals are a reservoir for the recruitment of new breeders and, therefore, are important for the regulation of the population (López-Sepulcre and Kokko 2005). However, delaying reproduction has a fitness cost, even for a long-lived species, and mature birds may try to enter into the reproductive population by alternative routes: individuals may occupy poor-quality territories (Rodenhouse et al. 1997), or enter into an occupied high-quality territory as a helper (Hatchwell and Komdeur 2000).

Cooperative breeding is a breeding system in which more than a pair of individuals shows parent-like behaviour towards young of a single nest. In its broadest sense, it includes reproductive systems varying from helping by non-breeding offspring to various forms of polygamy, in which more than one male or female share breeding status (Brown 1987; Cockburn 1998; Hatchwell and Komdeur 2000). This is the case of the bearded vulture in the Pyrenees, a species considered monogamous but with several polyandrous territories found since the beginning of the 80s (Heredia and Donázar 1990; Bertran et al. 2009). The copulation behaviour suggests that most of these territories are formed by two males and a female (Bertran and Margalida 2002a, 2003, 2004) but also polyandrous quartets have been described (Margalida et al. 1997).

15.4.2 Contrasting Hypotheses to Explain Changes in Reproductive System

Several hypotheses have been proposed to explain the evolution of this reproductive system. According to the *ecological constraints hypothesis*, cooperative breeding appears when opportunities for independent breeding are limited because of ecological factors, such as low availability of resources or high risk of mortality during dispersal. The *life-history hypothesis*, however, states that cooperative breeding is a consequence of certain life-history traits of a species, such as low adult mortality, that reduce the opportunities for independent breeding. Beyond their differences, both hypotheses stress that direct fitness benefits of looking for independent breeding opportunities do not outweigh the indirect fitness benefits of helping relatives (Pen and Weissing 2000). Although the evolution of cooperative breeding has often been attributed to kin selection (Emlen 1991; Emlen et al. 1991), there is increasing evidence that helpers can be unrelated to the young they are raising (Cockburn 1998; Heg and van Treuren 1998). Other rewards, such as staying in a restricted area to increase their probabilities to acquire a high-quality territory (Heg and van Treuren 1998; Kokko and Sutherland 1998) or, as in polygamous units, to have some chance of breeding (Hartley and Davies 1994) should tempt birds to join an existing breeding pair rather than search for an unlikely opportunity of independent reproduction in a saturated environment.

15.4.3 What Happened with the Pyrenean Population? The Polyandrous Emergence

The analysis of the role of habitat saturation in the mating system of bearded vultures, a territorial and usually monogamous raptor, was assessed by Carrete et al. (2006a). Because the bearded vulture population increased progressively since

1970s (less than 40 breeding territories) until the 90 breeding pairs monitored in 2002 and this increment was within a restricted geographical range (the Pyrenees), so crowding mechanisms severely reduced territory suitability and fitness of territorial birds (Carrete et al. 2006a; Margalida et al. 2008a). In parallel to the increase in population size, some territories became occupied by polyandrous trios (Heredia and Donázar 1990). This fact raised the question of what are the consequences of this behavioural change for dominant breeders as well as for the whole population (Fig. 15.1). Currently, the proportion of polyandrous trios in Spanish Pyrenees is around 35% of the breeding population.

Polyandrous trios tended to appear in traditional territories (i.e. territories occupied before 1988) where productivity tended to be higher when occupied by breeding pairs (see details in Carrete et al. 2006b, Fig. 15.2). However, these high-quality territories became less productive when they received a third individual, suggesting that the appearance of another adult could trigger conflicts that affect reproduction among individuals from the same territory. Accordingly, the addition of a new individual decreased pair's productivity and that this was true even when considering density-dependent effects of the increase of the population over time.

The social organisation of a population is the consequence of the decisions made by individuals to maximise their fitness. Carrete et al. (2006b) found that a substantial proportion of unpaired birds become potential breeders by either entering high-quality territories or by forming polyandrous trios as a strategy to increase their individual performance. Thus annual increases in the number of breeding territories were associated with the proportion of trios, suggesting that the

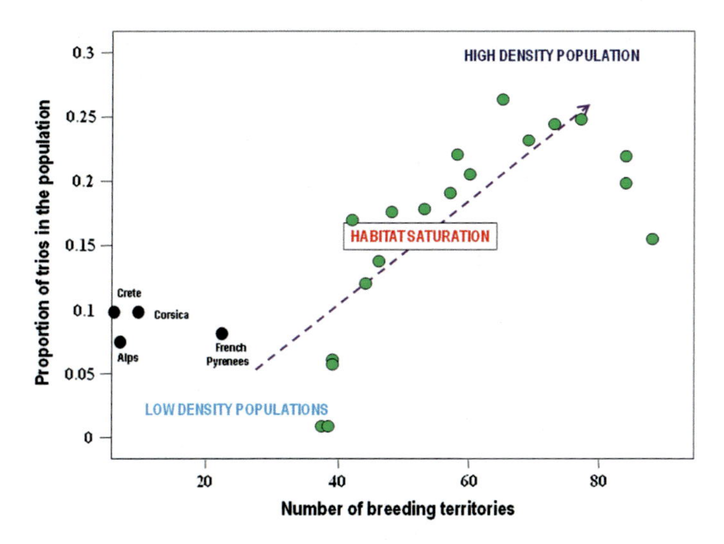

Fig. 15.1 Number of breeding territories and proportion of trios in the Spanish population of bearded vultures (*green dots*) compared with other populations (*black dots*) (modified from Carrete et al. 2006a)

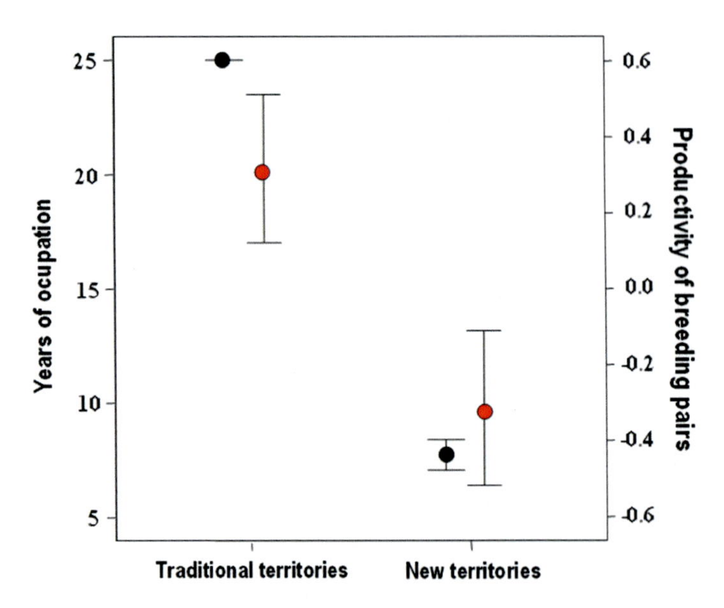

Fig. 15.2 Differences (mean ± s.e.) in time since territory formation (*black dots*) and mean standardised productivity of territories when occupied by breeding pairs (*red dots*) between traditional and new territories (modified from Carrete et al. 2006b)

progressive saturation of the population could be related to their formation. In this sense, the appearance of another adult could trigger conflicts that affect reproduction among individuals from the same territory (Bertran et al. 2009). Accordingly, after fitting territory and year into models to control for the density-dependent decay in the productivity of the population during the study period, we found that the formation of trios in some territories caused productivity depression in the whole population. The decision of some individuals to enter into breeding trios as subordinates also had clear adverse consequences to population demography.

Bearded vultures entered the breeding pool by occupying poor-quality territories (Carrete et al. 2006b) or by queuing in high-quality territories to wait until they are available. Therefore, trio formation may be a strategy for intruders to increase their fitness prospects by mating with a female or by increasing their likelihood of inheriting a high-quality territory when the dominant male dies. Reproductive success of previous owners was compromised when territories changed to trios, suggesting that the third individual was costly. Intruders, however, are not evicted by owners (Bertran and Margalida 2002b), perhaps because trios arise from uncommon ecological constraints and thus the energy invested in an unusually developed agonistic behaviour could be higher than the costs of accepting them (Clutton-Brock and Parker 1995; Hamilton and Taborsky 2005).

Intraspecific variation in mating systems may be determined by ecological features of the environment (Heg and van Treuren 1998) or by conflicts of interest between individuals (Davies and Hartley 1996). The Spanish population of bearded vultures is highly restricted spatially, and both the reproductive and non-reproductive fractions of the population coexist. The decision of some individuals to get into a trio as a subordinate also has clear adverse consequences for the whole population. However, since it appears as a result of ecological constraints, this behaviour could be reversed through management.

15.4.4 Conservation Implications Related with the Apparition of Polyandrous Formations

This unusual mating behaviour is thus compromising the conservation effort directed to this endangered species; management to encourage floaters to settle in other suitable but unoccupied areas may be beneficial. Our findings have direct implications for conservation. To our knowledge, this would be another example of the detrimental effects of food supplementation programmes on the conservation of threatened species (Robertson et al. 2006). Reducing food supply or progressively moving feeding stations outside of the Pyrenees may encourage floaters to settle in other areas where they could find breeding opportunities (see Margalida et al. 2008a, 2013). The translocation of floaters of the endangered Seychelles magpie-robin *Copsychus sechellarum* from a population-saturated island to an unoccupied one was a successful conservation strategy, since individuals become breeders (Kokko and Sutherland 1998). A similar situation could be expected if ecological and behavioural cues exposed in this paper are linked to direct conservation actions for this endangered vulture.

15.5 Density-Dependent Productivity Depression and Supplementary Feeding Management

Two different hypotheses relate density-dependent changes in demographic parameters to population regulation in territorial species. The *interference hypothesis* suggests that reductions in fecundity and/or survival coinciding with an increase in population size are caused by a homogeneous reduction in the quality of available resources due to an increase in agonistic encounters between individuals (Dhondt and Schillemans 1983; Sillett et al. 2004). The *habitat heterogeneity hypothesis*, however, suggests that the progressive occupation of low-quality territories as density increases causes a decline in the average per capita productivity and/or survival of a population even when its variation increases, leading to density-dependent regulation (Rodenhouse et al. 1997; Krüger and Lindström 2001; Sergio and Newton 2003). Here, dominant or early-arriving individuals occupy high-quality areas and, by means of territorial behaviour, relegate subordinate or late-arriving individuals to inferior territories or, when these places are also occupied, to a non-breeding lifestyle (Newton 1998). This pre-emptive settlement pattern, coupled with habitat heterogeneity and density-dependent changes in demography, has been defined as site-dependent population regulation (Rodenhouse et al. 1997). This mechanism, which complements and, in certain kinds of species, may even preclude local crowding mechanisms, can generate negative feedback at all population sizes, sometimes independently of local population densities (Rodenhouse et al. 1997).

The behaviour of floaters may also be a potential regulatory factor (López-Sepulcre and Kokko 2005). These "surplus" individuals that form a buffer against population fluctuations may harm breeding performance through intraspecific conflicts. The establishment of supplementary feeding points within the distribution area of the breeding population has been the most significant management action in terms of time and effort undertaken to help the bearded vulture in the Spanish Pyrenees from the 80s. These sites concentrate a substantial fraction of the floater population. In this sense, the spatial and temporal overlap between the breeding and non-breeding population fractions may affect the fitness of territorial birds by increasing intraspecific interactions.

15.5.1 Density-Dependent Productivity Depression Associated to Supplementary Feeding Sites (SFS)

From 1978 to 2002 the bearded vulture population in the Spanish Pyrenees increased from 38 to 91 breeding pairs. During the same period, the mean annual productivity of the population declined from 0.8 to 0.37 young/territorial pair. Variability among territories was responsible for a large proportion of this decline. However, this variability was not related to habitat heterogeneity per se. Habitat quality index (i.e. an index obtained by considering only relief, altitude and distance to villages) dropped by 13% during the study period, while the same index including distance to the nearest breeding pair declined by 20%. Thus, while the increase in population resulted in some pairs occupying intrinsically poorer territories, the proximity between conspecific breeding pairs seemed to be the most important factor reducing habitat quality and, therefore, productivity.

Productivity in traditional territories (those occupied at least since 1978) was better and more stable than in new ones (those occupied from 1988 onwards, when the population started to increase) initially, from 1988 to 1993. However, these differences lessened and disappeared all together in the final years (1994–2002), when territories became more homogeneous regarding productivity. Accordingly, we found that although the intrinsic quality of traditional territories remained relatively constant through time, a rise in the number of breeding pairs may have increased intraspecific interactions, thereby reducing the other quality index.

Even when all territories are not equally affected, the distances to both the nearest conspecific pair and the nearest supplementary feeding points have a negative effect on productivity. Although there is a certain degree of variability in their responses, territories located near supplementary feeding points that are also near to other bearded vultures breeding territories had lower productivity than territories with less conspecific pressure.

15.5.2 Habitat Heterogeneity and Population Regulation

The above evidence suggests that, in accordance with other studies on territorial raptors (e.g. Krüger and Lindström 2001; Sergio and Newton 2003; Carrete et al. 2006a), habitat heterogeneity plays a key role in the population regulation of bearded vultures. As the Pyrenean bearded vultures are not individually marked, we cannot tell whether habitat heterogeneity is a consequence of sites possessing different suitabilities for reproduction or survival (Breininger and Carter 2003; Lambrechts et al. 2004; Carrete et al. 2006a). However, our findings that productivity declined and its variation increased as bearded vulture populations increased from 38–91 pairs (during 1987–2002) are new and relevant to both basic and applied ecology because they show that population regulation is not merely a result of interference (i.e. ideal free distribution) or pre-emptive use of space. Moreover, traditional discussions on density dependence in territorial systems are mainly based on data obtained from populations in demographic equilibrium, where crowding mechanisms are usually precluded. Our data, collected throughout a period of population growth, show that when high-density situations are encouraged, demographic density dependence in territorial birds can occur because of the combined effects of site quality (ideal despotic distribution) and crowding mechanisms (ideal free distribution). Moreover and no less importantly, we also show that non-breeding birds can make up a significant fraction of the whole population and that their effects on breeding individuals as scramble competitors must be taken into account (López-Sepulcre and Kokko 2005).

Age differences could be proposed as an alternative hypothesis to explain productivity variation between territories (Forslund and Pärt 1995), where inexperienced birds occupying new territories increase their productivity through years and senescence promotes a progressive decay of productivity in traditional sites (Margalida et al. 2008b). Even when we were not able to test age effect on reproduction (Bearded vultures are not individually marked), and therefore we cannot discard it, our data are reliably showing that habitat heterogeneity and interference play a role in productivity depression, explaining a substantial percentage of deviance.

The density of conspecific competitors has been shown to affect territory size in several bird species negatively (see review in Newton 1998), independently of food availability (e.g. Arcese and Smith 1988; Stamps 1990). Although we have no information on either home range size or its change with density, our results suggest that this bearded vulture population may have suffered a process of territorial compression associated with an increase in the number of breeding pairs (nearly 25% reduction in the mean nearest neighbour distance between 1987 and 2002, Margalida et al. 2008a). This fact may be affecting the productivity of the population (see Carrete et al. 2006a). Moreover, Donázar et al. (1993) did not find any relationship between breeding success and distance to conspecific breeding pairs, suggesting that productivity was not limited by any density-dependent mechanism before 1991. Consequences of territory compression have been explored in other species, where increases in density are accompanied by increases in aggressive behaviour among territorial animals and costs associated with territory defence (e.g. Calsbeek and Sinervo 2002; Mougeot et al. 2003; Sillett et al. 2004). In these cases, territory shrinkage and territorial disputes associated with high-density situations affected reproduction and had fitness costs for territorial animals (Gordon 1997; Calsbeek and Sinervo 2002; Ridley et al. 2004), as in our bearded vulture population. However, we found that not all territories were equally affected by increases in the number of breeding pairs. Territories located in high-density situations became less productive and more unpredictable than territories located far away from conspecific pairs, indicating that in the present situation proximity to other breeding pairs could be the main factor promoting territory quality and also, to some extent, habitat heterogeneity in this closed population.

15.5.3 Effects of Supplementary Feeding Sites on Breeding Output

Proximity to supplementary feeding points where non-breeding birds congregate was also detrimental for reproduction (Fig. 15.3). For species with delayed maturity such as many long-lived raptors, spatial segregation between dispersing and breeding birds is a common feature (Newton 1979). This is because preparation for reproduction governs preferences among breeders, while food is the main driving force underlying habitat selection patterns in dispersing birds (e.g. Bustamante et al. 1997; Brown 1997; Mañosa et al. 1998; Hirzel et al. 2004). In our study area, however, the high availability of food resources associated with supplementary feeding points allows a great number of non-breeding bearded vultures—which otherwise would be occupying different areas (Brown 1997; Xirouchakis and Nikolakalis 2002; Hirzel et al. 2004; Margalida et al. 2011a)—to coexist within the spatial distribution of the breeding population. Contrary to the social behaviour observed in other species in which floaters and territorial birds may coexist in areas of high food supply (e.g. Blanco and Tella 1999), bearded vultures are territorial birds that defend exclusive breeding areas against both conspecifics and heterospecific birds (e.g. Bertran and Margalida 2002b; Margalida and Bertran 2000, 2005). Thus, high concentrations of floaters around breeding territories—as

Fig. 15.3 Supplementary feeding sites congregate bearded vultures of different age classes (A. Margalida)

happens near supplementary feeding points (see Sesé et al. 2005; Margalida et al. 2011a)—may increase the time being spent in agonistic encounters and, therefore, reduce breeding success. Moreover, conspecific crowding can be a significant stressor that may alter glucocorticosteroid release, causing both physiological and behavioural changes that may affect population dynamics (Rotllant et al. 1998; Creel 2001; Romero 2004).

In conclusion, these results suggest that vulture populations are regulated as postulated by the site-dependency hypothesis: as the population increases, average productivity decreases because progressively poorer territories are used. The combined effects of the shrinkage of territories and the presence of floaters around supplementary feeding points seem to be the main causes of productivity decline and are therefore the main determinants of territory quality. This has conservation implications, especially concerning the role of supplementary feeding points.

15.5.4 Management Implications for Conservation Plans

The establishment of supplementary feeding points for the management of vulture populations has been used during reintroduction programmes to maintain birds close to release areas (Griffon Vultures, *Gyps fulvus*, in France; Sarrazin et al. 1996), to increase food supply (Piper et al. 1999), or even as a potential solution to reduce poisoning (California Condor, in the United States; Meretsky et al. 2000).

Our results suggest that these management actions aimed at increasing the number of breeding pairs within the present distribution of the species and those attracting nonbreeders within the spatial range of the breeding population of bearded vultures should be reconsidered. In particular, the strategy of food supplementation should be reviewed because it seems to be one of the main potential factors promoting the congregation of non-breeding birds around breeders. Decisions to disperse or to remain in the local population are influenced by local intraspecific competition (Clarke et al. 1997; Perrin and Mazalov 1999; Gandon and Michalakis 2001; Lambin et al. 2001; Serrano et al. 2004). Thus high food availability within the geographical range of the breeding population could be keeping dispersing birds in their natal areas. The expected consequences of reducing food availability would be both an increase in the movements of floaters outside the distribution range of breeding birds, thereby reducing direct interactions in territories located near feeding points, and a geographical expansión of the breeding population to other suitable areas, as is proposed in the Recovery Plan of the species in Spain. Supplementary feeding sites were opened on the basis of their importance in increasing juvenile survival (Heredia and Heredia 1990). In this sense, direct evidence of a causal link between food supplementation and juvenile survival is lacking, although suggested according to the behaviour and survival values obtained (Oro et al. 2008; Margalida et al. 2011a, 2014b). Juvenile populations may have increased as a result of other factors such as an increase in wild ungulate populations (Razin and Bretagnolle 2003) or the reduction in direct human persecution through legislation since early 1980.

15.6 Demography: Effects of Poisoning and Supplementary Feeding

The provision of supplementary food at artificial or supplementary feeding stations or "vulture restaurants" (hereafter SFS), a well-established management tool in the conservation of scavenger populations (Cortés-Avizanda et al. 2010; Moreno-Opo et al. 2015; Cortés-Avizanda et al. 2016), appears to be a potentially useful solution worldwide. SFS has frequently been used to facilitate the recolonization of abandoned areas (Mundy et al. 1992), or to provide safe food sources in areas where carcasses are baited with poisons to control carnivores (e.g. Wilbur et al. 1974) or livestock has been treated with veterinary drugs (Gilbert et al. 2007). To the contrary, adverse effects of artificial feeding stations on demographic parameters of some species (Carrete et al. 2006a) are a concern and prompt us to correctly assess their effect on the demography of target populations.

The relevance of SFS for population viability through increments in the survival rates of non-territorial birds challenges the theoretical low sensitivity of population growth rates to this parameter expected among long-lived species (Saether and Bakke 2000). Conversely, SFS appear to be discouraging population expansion outside the Pyrenees and have been related to habitat saturation processes (Margalida et al. 2008a), which are triggering negative population effects such as reductions in productivity and changes in mating behaviour (Carrete et al. 2006a). Taking into account the life-history characteristics of the species and based on the general hypothesis of positive effects of predictable food supply on survival, we predict that increments in the illegal use of poison during recent decades (Hernández and Margalida 2008, 2009; Margalida 2012; Mateo-Tomás et al. 2012) have decreased survival rates of bearded vultures. However, this effect would have been buffered by the use of SFS, the use of which varies considerably among individuals although a general pattern is associated with age (young birds are more frequently seen at SFS than older ones, Margalida et al. 2011c). Population projections under different management scenarios in which poisoning have been combated with a variable degree of effectiveness are also presented.

15.6.1 Relationship Between Supplementary Feeding and Survival

We assessed the role of supplementary feeding on bearded vulture's survival. The best survival model obtained using the MARK program accounts for 85% of the weight of all models and included the effects of age (young or adults), time and the intensity of use of SFS on survival, and of age on recapture rates. Interestingly, the addition of the individual covariate describing the intensity of use of SFS significantly improved the models. However, the use of SFS decreased with bird age, even after controlling for individual variability (see Oro et al. 2008). Thus, survival was equal for birds of ages up to 4 years old (the age classes more frequently seen at SFS) and remained constant over time (0.944) even after the recent increment in the illegal use of poison. Conversely, survival of older birds (≥ 5 years old) decreased linearly with time (Fig. 15.4), with an average value of 0.878. Other models, including a higher number of age classes (results not shown) or different combinations of age and time, behaved worse. Thus, the results support the hypothesis of improvement of survival rates associated with the use of SFS. Indeed, survival estimates for the two most important age classes at the beginning of the study period (1987) when the use of illegal poison was less marked, but SFS were not available, were 0.787 and 0.961 for young (the age class using SFS) and adults, respectively.

Ecological theory predicts that survival of long-lived species should increase gradually with age because (1) individuals improve their competitive skills or change from vagrant lifestyles to more sedentary ones after territory acquisition

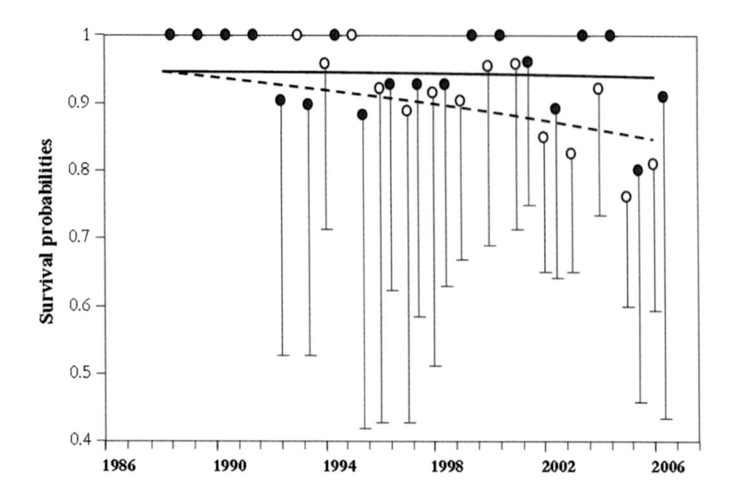

Fig. 15.4 Variation in survival rates of adult (>5 years and older; *white dots*) and young birds (4 years and younger; *red dots*) with time and age using the parameters obtained with the model ($\phi_{(1\_4+5)+T}$, p_A). Mean values and 95% lower confidence intervals are shown, as well as the linear negative trends for adults (*dashed line*) and young (*solid line*). Survival rates estimated as 1 were actually estimable parameters, i.e. years in which all individuals survived. Note that the two trends were parallel in the logit scale (modified from Oro et al. 2008)

and/or (2) natural selection progressively eliminates low-quality individuals (Tavecchia et al. 2001). Even when the underlying mechanisms may be of great interest in understanding the evolution of life histories, it is tough to correctly separate one mechanism from another (Tavecchia et al. 2001; Sanz et al. 2008). All in all, consequences for population dynamics are similar: population growth rates among long-lived species are expected to be highly sensitive to changes in adult survival (Real and Mañosa 1997; Oro et al. 2008; Ortega et al. 2009; Hernández-Matías et al. 2013; Margalida et al. 2015), so natural selection might have minimised variation in this parameter to ensure population stability (Saether and Bakke 2000). However, our results show how survival of bearded vultures changes through a bird's lifespan in an unnatural way, with non-adult birds (<5 years old) having higher and more constant estimates than adults (0.944 and 0.878, respectively, Fig. 15.4). Two main aspects seem to have been directly responsible for this outcome, namely the opening of SFS (directly tested through individual and age-specific frequencies of visits), and the increasing use of illegal poison to control predators (indirectly tested through a temporal trend in survival). Thus, human activities, both through apparently well-intentioned and malicious actions, can perturb evolutionary forces promoting unexpected changes in survival patterns and, therefore, demographic dynamics.

15.6.2 Population Trajectories Under Different Management Scenarios: Which Are the Demographic Effects of SFS?

We built a prospective stochastic age-structured population model (only for females) to explore the functional dependence of λ (the population growth rate) on the demographic rates of the study population. This kind of perturbation analysis allows us to identify potential management targets because variations in demographic parameters with high sensitivity (or elasticity) produce large changes in λ. We also built a retrospective stochastic population model to detect whether parameters used in the prospective model were reliable and described with certainty the behaviour of the population. Here, we used the same structure and number of replications as that of the prospective model but run over the period for which data were collected (20 years), setting the initial population size to that estimated in 1985 (i.e. 37 breeding females).

The current combination of demographic parameters of the Spanish bearded vulture population (scenario 1, Table 15.2) gave a $\lambda = 0.961$. When we explored how this rate changed within a range of different combinations of survival scenarios (all other vital rates remained equal as in further prospective models), we found that the population increased ($\lambda > 1$) only with very high values of adult survival. To the contrary, λ showed low sensitivity to changes in survival of non-adult individuals.

Retrospective analysis showed that simulations performed using actual survival estimates (scenario 1, Table 15.2) fitted quite well with the observed dynamics of the breeding population (mainly at the end of the time series, Fig. 15.5a), despite the uncertainty of the demographic estimates of the model, and the way they

Table 15.2 Survival and reproductive parameters used in Monte Carlo simulations for estimating extinction risk in prospective models for the bearded vulture in the Spanish Pyrenees. In brackets, SE of the estimates (see details in Oro et al. 2008)

Parameters	Supplementary feeding sites			No supplementary feeding sites	
	Poisoning		No poisoning	Poisoning	No poisoning
	Scenario 1	Scenario 2	Scenario 3	Scenario 4	Scenario 5
Young survival	0.944 (0.012)	0.944 (0.012)	0.944 (0.012)	0.787 (0.018)	0.787 (0.018)
Adult survival	0.878 (0.014)	$\phi_{(t)} \cdot 0.264 + rand \cdot 0.241^{\xi}$	0.961 (0.019)	0.878 (0.014)	0.961 (0.019)
Fertility[§]	$0.7359e\text{-}0.0056 \cdot PD^{*}$				

$^{\xi}$ Initial value$_{t=1} = 0.961$

[*]Density-dependence function (PD = population density)

[§]*Source* Carrete et al. (2006a, b)

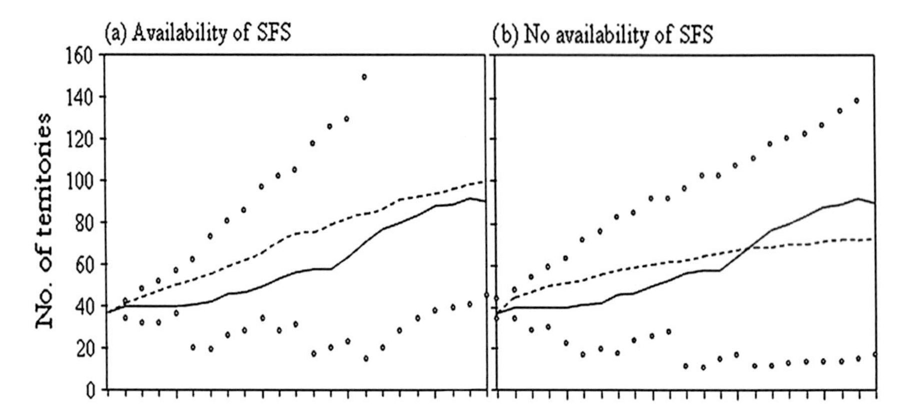

Fig. 15.5 Retrospective analysis of the dynamics of the Spanish population of bearded vultures in the Pyrenees during 1985–2007. Simulations were carried out using estimated values of **a** actual survival rates, i.e. with the effects of poison and SFS; and **b** survival with poison effects but without SFS. *Dashed lines* show the mean value of stochastic trajectories using Monte Carlo simulations, while open dots are the maximum and minimum values of that run. For comparison, we show the observed number of breeding territories through time (*solid lines*) from Oro et al. (2008)

potentially changed with population density (both positively and negatively; author's unpublished data). Retrospective simulations without the positive effects of SFS on young survival but no effects of poison on adult survival rates yielded a lower number of territories than observed in the study population (71 vs. 80 territories, Fig. 15.5). However, population trends were quite similar in all cases (scenario 1: $\lambda = 0.961$, scenario 5: $\lambda = 1.000$, observed: $\lambda = 1.048$) suggesting that population consequences of the progressive reduction in adult survival rates could have been buffered by increments in young survival rates associated with the use of SFS.

15.6.3 Availability of SFS

We envisaged that the illegal use of poison continues affecting survival rates of vultures using the survival rates of pre-adult and adult birds estimated by the best-selected model (see Oro et al. 2008), which were lower than expected likely due to the effect of illegal poison. Thus, we performed a second set of simulations considering that the temporal dynamics of poisoning are dominated by fluctuations, both short and long term, with positive temporal autocorrelation (coloured environmental noise, Inchausti and Halley 2003). Finally, we simulated the behaviour of our population under effective management actions reducing the impact of poisoning, using actual adult survival estimates obtained in our study area but without poison effects.

The three scenarios considered within this group of simulations showed different patterns associated with the effects of poison on survival. When the survival of adults decreases because of the illegal use of poison and survival of young increases due to their use of SFS (scenarios 1 and 2, Fig. 15.6), population trajectories decreased over the years (scenario 1: $\lambda = 0.961$; scenario 2: $\lambda = 0.932$). However, extinction probabilities were not equal in the first and the second scenarios. When the impact of poison was stochastic with no clear trend over time (scenario 1 Table 15.2), probabilities of extinction were nil, although some trajectories attained the quasi-extinction threshold (scenario 1 Table 15.2, Fig. 15.6). Conversely, ARIMA analysis indicated that the way in which illegal poisoning impacted our bearded vulture population (i.e. the colour of the environmental noise) depended on the age class considered. For instance, the decrease of survival among birds older than 4 years during 1986–2007 ($a = 0.264$, $b = 0.241$) showed a long-term temporal autocorrelation (i.e. red noise), whereas for birds up to 4 years old ($a = -0.018$, $b = 0.250$) survival exhibited a short-term temporal autocorrelation

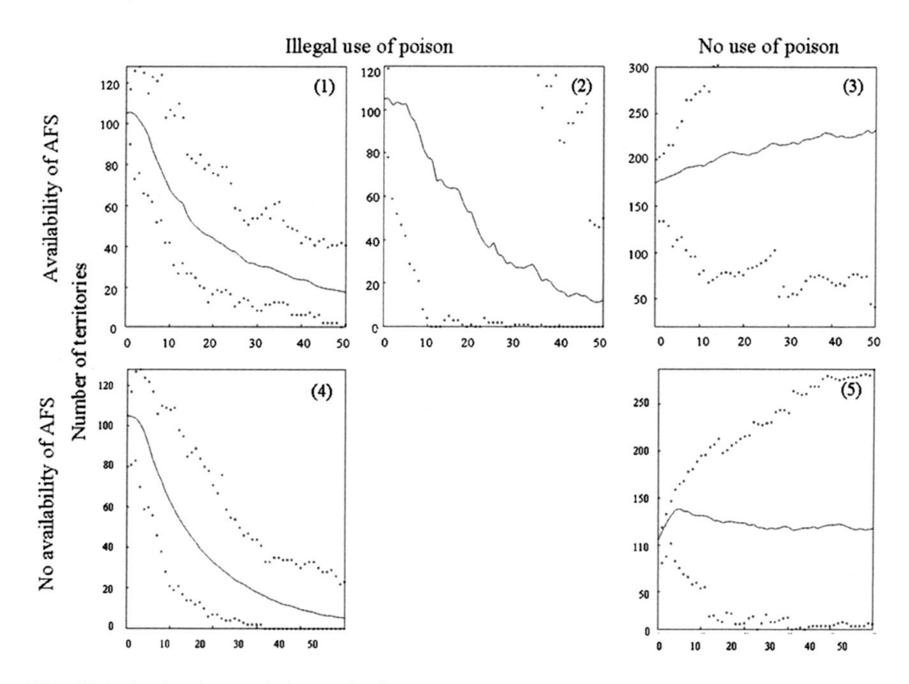

Fig. 15.6 Stochastic population projections estimated through Monte Carlo simulations (50 years of temporal window, 500 runs) for some of the scenarios considered: (1) actual values of adult survival without temporal autocorrelation in the use of poison, (2) and (4) actual values of adult survival with temporal autocorrelation in the use of poison, (3) no impact of poison, assuming that young and adult survival were equal, and (5) no impact of poison, assuming that survival of young, and adults were the same t (see Oro et al. 2008 for details). *Lines* are mean values of the stochastic runs for each time step, while *dots* show the maximum and minimum values of those runs. Projections were carried out on the density of females of any age, which have different scales in each graph

(i.e. blue noise). This temporal autocorrelation in the impact of poison, the most realistic situation (scenario 2 Table 15.2), increased the extinction probabilities of the whole population, with a maximum value over a time horizon of 50 years ($p_e(t = 50) = 1$, mean extinction time: 10.2 years, see scenario 2 Table 15.2, Fig. 15.6). Note that for these two poison-impacted scenarios the mean trajectories were quite similar, except for the largest variability expected in population trajectories when the impact of poison was temporally correlated (scenario 2 Table 15.2, Fig. 15.6). Interestingly, both mean trajectories were relatively stable in early years, probably due to the buffer capacity of recruits resulting from very high young survival.

Increments in the survival rate of adult birds up to that expected when the illegal use of poison was near nil combined with a high survival rate of young (maintained through SFS) predicted a marked increment in the number of breeding territories ($\lambda = 1.014$; Fig. 15.6).

15.6.4 No Availability of SFS

This group of simulations allowed us to disentangle the actual usefulness of SFS to mitigate the effects of illegal poison on population persistence.

Populations suffering from negative effects of illegal poisoning on survival rates but not managed with SFS showed a lower population growth rate and a slightly higher probability of extinction in the time horizon of 50 years than its most conservative counterpart (scenario 1: $\lambda = 0.961$; scenario 4: $\lambda = 0.932$; Fig. 15.6). However, in a scenario of no use of illegal poisoning and no SFS, population numbers are somewhat lower although still rather stable (scenario 5: $\lambda = 1.000$; scenario 3: $\lambda = 1.016$).

Population projections under scenarios of illegal poisoning did not forecast a positive outcome for the most important European core of this species. When the impact of poison on adult survival was stochastic, probabilities of extinction for the population were nil, although some trajectories achieved the quasi-extinction threshold. Of even greater concern, extinction probabilities for the population increased when the effect of illegal poisoning on adult survival followed a temporal autocorrelation, the most probable scenario. Conversely, population projections run with the survival of adults not affected by illegal poisoning predicted better situations, with larger population increments in scenarios of maintenance of SFS, when survival rates of young birds are also improved. The counter-scenario, no use of illegal poisoning and no availability of SFS, predicts slightly lower population sizes but with near stable trajectories. Consequently, an important management action intuitively used to reduce the negative effects of illegal poisoning such as the opening of SFS as appear to be not as effective as expected in saving threatened populations from future negative trends. As in many other long-lived species, survival of adults is the key demographic parameter contributing most to the projected population growth rate (e.g. Saether and Bakke 2000), and management actions should be directed to improve it. However, in the short term, SFS can maintain a large floater surplus that may delay population decline. It is understood that the eradication of the illegal use of poison is neither easy nor time efficient, and hence measures taken to allow more time for the application of proper and more effective management actions can be a useful instrument for conservation. Thus, efforts to determine adequate tools to reduce poisoning risk among adults are important as well. In this sense, experimental work is needed to test the effectiveness of smaller, less predictable SFS located near breeding territories to enhance adult survival while avoiding large aggregations of non-adult birds in their surroundings.

It is worth noting that the reliability of predictive models depends on the robustness of demographic rates and the number of known parameters for each age class. In our case, survival estimates by age classes and fecundity were available and reliable. Nevertheless, some parameters were unknown, mainly the recruitment curve and how it changed with variations in density (Tavecchia et al. 2007). Thus, the structure of the model was a compromise between complexity (due to known demographic patterns such as age-dependent survival or density-dependent fecundity) and simplicity (due to the unknown parameters such as the percentage of breeders at each age class). However, several goals to ensure the maximum reliability at the predictive power of extinction risk were achieved in our modelling by incorporating uncertainty in parameter estimates and stochasticity in population dynamics (Lande et al. 2003). Finally, it is important to note that demographic consequences of artificial increments in survival are limited to those included in our hypothetical scenarios. Complex aspects linked to alterations in natural selection pressures should also be taken into account since a large proportion of young, that in more "natural" situations would have died (low-quality individuals; Tavecchia et al. 2001; Sanz et al. 2008), are now potentially recruited into the breeding population. Moreover, if only SFS-maintained birds are progressively selected, the population can become more dependent on human-supplied food than previously thought.

15.6.5 Usefulness of Supplementary Feeding Sites for the Conservation of Endangered Populations

Supplementary feeding of wild birds is a widespread practice that may alter the natural dynamics of food supply, representing a major intervention in avian ecology. Indeed, supplementary feeding has the potential to change long-term population dynamics and distribution ranges of many species (for a revision see Robb et al. 2008). Therefore, policy-makers and managers have found in supplementary feeding actions a common, straightforward solution to many different conservation challenges of endangered populations, including increasing breeding success (e.g. González et al. 2006; Margalida 2010b; Margalida et al. 2017), providing safe food in areas where carcasses are poisoned or contaminated with veterinary drugs (e.g. Gilbert et al. 2007), promoting the recolonization of abandoned areas (Mundy et al. 1992) and aiding in reintroduction programmes (Chamberlain et al. 2005).

To our knowledge, the Spanish population of bearded vultures represents one of the few cases for which both positive and negative outcomes of artificial feeding sites have been carefully weighed using a population dynamics approach (Fig. 15.7). Recent studies have shown that supplementary feeding can have several adverse effects on this population such as territory compression and coexistence between breeders and floaters (Carrete et al. 2006b; Margalida et al. 2008a), and changes in the mating system reducing territory quality (because of intraspecific interactions) and population productivity (Carrete et al. 2006a). Present results, however, support their usefulness as temporal tools to maintain individuals while more complex objectives such as the eradication of illegal poisoning from the field are achieved. However, the maintenance of SFS should not distract managers from prioritising the long-term viability of this and many other species by eradicating illegal poison use. Taking into account that one of the main threats to this population is its restricted geographic range, SFS can be utilised as a very specific tool for the recovery of the population in peripheral areas, to promote the colonisation of suitable unoccupied areas outside the Pyrenees. Although beyond the scope of this chapter and awaiting scientific support, this latter possibility should be considered a potentially positive aspect of SFS on species conservation.

Fig. 15.7 Adult bearded vulture feeding in a supplementary feeding station in the Catalonian Pyrenees (A. Margalida)

15.7 Effects of Changes in Sanitary Policies in Demographic Parameters

Any factor that disturbs the balance between fecundity and survival will be particularly hazardous for species with slow lifestyles (Owens and Benett 2000; Cardillo et al. 2004). For example, the recent and sudden appearance of non-natural mortality factors related to the ingestion of veterinary drugs (i.e. diclofenac) or the increase in the use of illegal poison baits severely affected populations of Old World vultures (Green et al. 2004, 2006; Virani et al. 2011; Margalida 2012; Margalida et al. 2014c). In parallel, a recent change in European health policy may be exacerbating the precarious status of these large species. Coinciding with the outbreak of bovine spongiform encephalopathy in 2001, scavenger food resources have been reduced because farmers were forbidden from retaining certain dead livestock (Donázar et al. 2009a; Margalida et al. 2010).

Considering the importance of such information for managers and conservationists, our purpose was to test if two human-induced activities (i.e. illegal use of poison baits and regulations in the availability of domestic carcases) can modify life-history traits in a threatened vulture population. Some assessment of food resources and their relation to population dynamics (Colomer et al. 2011; Margalida et al. 2011a; Margalida and Colomer 2012; Martínez-Abraín et al. 2012) and behavioural changes (Zuberogoitia et al. 2010; Margalida et al. 2011d, 2014b) has been documented. In this sense, although food availability can also influence life-history traits and the scientific literature on this topic in birds is abundant (Lack 1968; Western and Ssemakula 1982; Oro et al. 1999, 2004), to our knowledge few studies have attempted to relate the effects of human-induced activities on life-history traits in long-lived raptor species (Martínez-Abraín et al. 2012, Hernández-Matías et al. 2015). The outbreak of bovine spongiform encephalopathy began in 2001, the restrictive legislation forced the closure of 80% of feeding stations (Donázar et al. 2009a; Cortés-Avizanda et al. 2010; Margalida et al. 2014b) and obliged the collection of domestic ungulates to be destroyed. These measures were not implemented effectively until 2005 (Donázar et al. 2009a, 2010; Margalida et al. 2010) when the recovery of dead livestock was collected from farms and most of the feeding stations (80%) closed. Thus, we considered the period 1994–2004 as the “before policy implementation period” (noted as BPI) and 2005–2011 the “after policy implementation period” (noted as API) (Margalida et al. 2012). During the BPI remains of domestic ungulates were present in the field and food availability was considered sufficient to cover the energetic requirements of avian scavengers (Margalida et al. 2011a). However, during API the carcases of domestic animals progressively had to be collected from farms and destroyed in authorised plants (Donázar et al. 2009a; Morales-Reyes et al. 2015). The monitoring of several regions suggested that more than 80% of remains of Ovis/Capra were collected, whereas this proportion reached nearly 100% for the remains of Bos/Equus.

We analysed breeding parameters (during 1994–2011) and survival probabilities (during 1987–2011) to determine if changes in the availability of food supply and/or the effects of illegal poisoning are important anthropogenic mortality factors affecting this population, and whether they can modify life-history traits in a long-lived species. We tested the hypothesis that the reduction in the availability of food resources in the ecosystem (as consequence of a sudden reduction of domestic carcasses in the field from 2005 and the closure of several feeding stations) should decrease survival probabilities, specially that of juveniles and subadults, which are more dependent on this food resource than are older birds (Oro et al. 2008).

15.7.1 Influence of Health Policies on Demographic Parameters

Between 1994 and 2011 we monitored 510 breeding attempts with a total of 298 clutches being laid. The average number of pairs that annually began breeding was 59% (range 42–71%, n = 18) with a similar trend between years, being the proportions of pairs that started breeding before and after policy change similar (57 vs. 61%, Fig. 15.8).

We documented egg-laying dates on 258 occasions, without significant differences between years. However, when we compared egg-laying dates before and after policy change, we found a significant delay following the reduction in food supplies (average 8 January vs. 12 January, $t = 2.44$, $P = 0.016$, Fig. 15.9).

We observed a total of 104 clutches of which 57% consisted of a single egg (remainder held two eggs). The proportion of single-egg clutches increased through time and differed before and after policy change (single-egg clutches: 39.4% vs. 69.6%, Fig. 15.10).

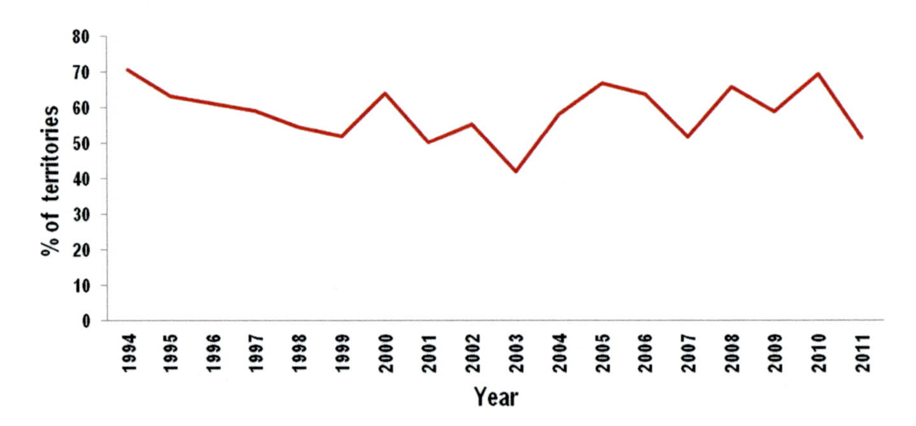

Fig. 15.8 Inter-annual variation in the percentage of bearded vulture territories that had breeding pairs (Margalida et al. 2014a)

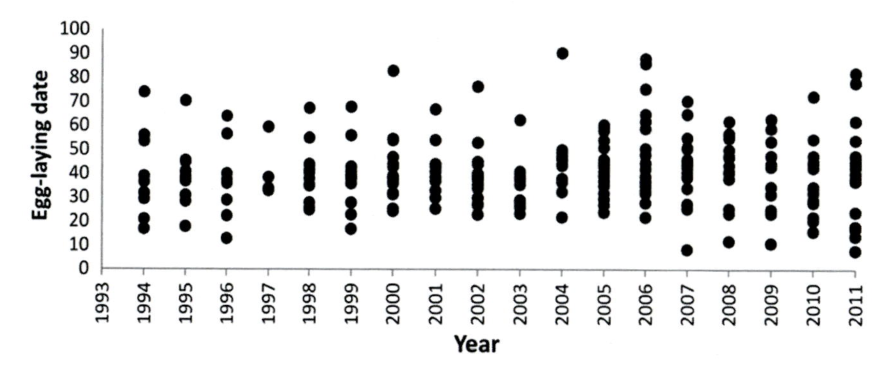

Fig. 15.9 Inter-annual variation in laying dates in the bearded vulture. 0 = 1st December (Margalida et al. 2014a)

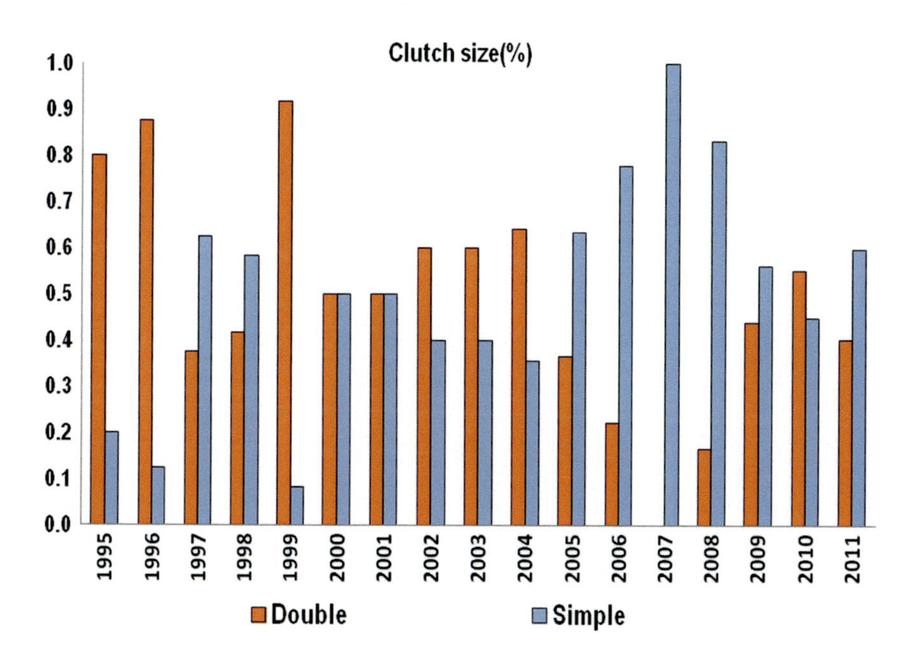

Fig. 15.10 Variation in the clutch size in the bearded vulture in the Pyrenees. *Orange columns* double (2-egg clutches); *blue columns* simple (1-egg clutches) (Margalida et al. 2014a)

Productivity and fledging rate decreased slightly through time, and both values decreased before policy changes, although the differences were not statistically significant (Productivity: 0.37 vs. 0.29; Fledging rate: 0.54 vs. 0.46, Fig. 15.11).

Between 1994 and 2011 we documented a total of 65 cases of mortality. Survival decreased with time in all age classes (adults, >6 years old: $r_s = -0.60$, $P = 0.0091$; subadults, 4–5 years old: $r_s = -0.56$, $P = 0.0147$; juveniles, 1–3 years old: $r_s = -0.48$, $P = 0.042$, $n = 18$). Annual survival differences did not coincide

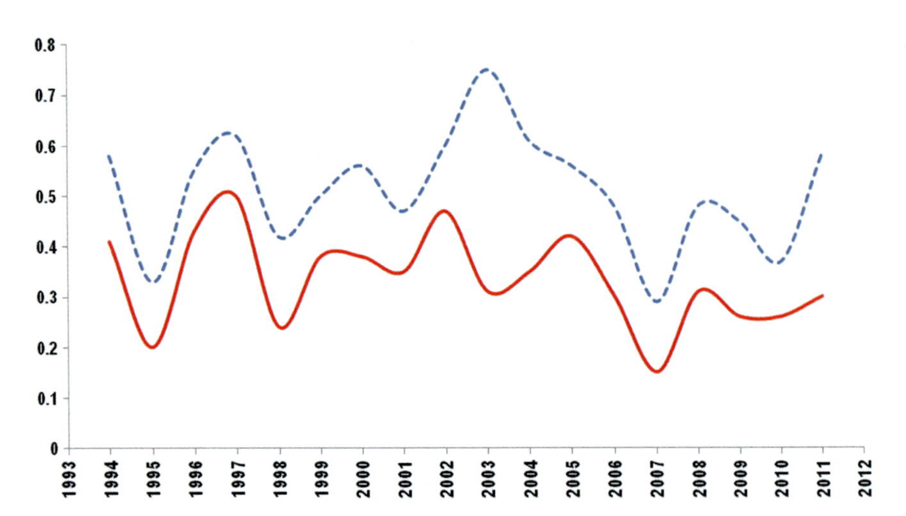

Fig. 15.11 Inter-annual variation in the number of young fledged from all monitored nests (productivity, *red line*) and the number of young fledged from successful nests (fledging rate, *broken line*) in the bearded vulture (Margalida et al. 2014a)

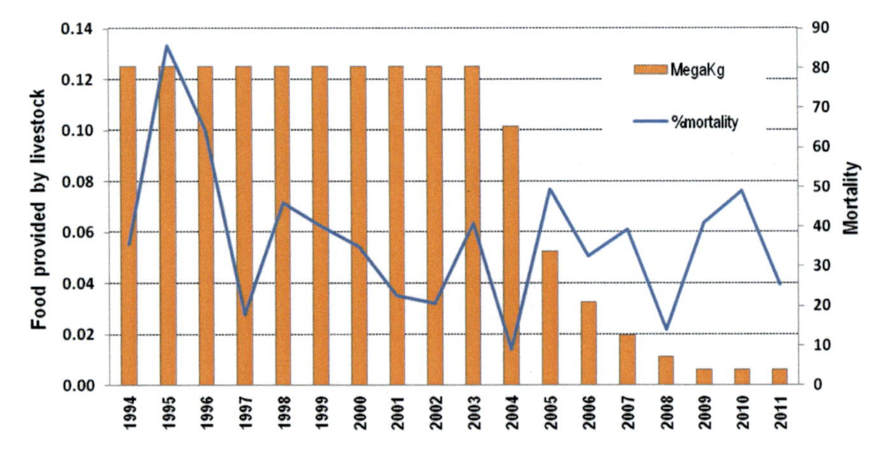

Fig. 15.12 Inter-annual variation in the food biomass (bones) provided by domestic ungulates and adult survival of the bearded vulture (for more details see Margalida et al. 2014a)

with policy change in juvenile and adult age classes (juveniles: 0.975 vs. 0.981; adults: 0.978 vs. 0.946), but did with subadults (0.920 vs. 0.902, P = 0.0059). This coincides with changes in the number of carcasses provided by domestic ungulates that decreased progressively from 2003 ($r_s = -0.91$, $P = 0.0001$, n = 18, Fig. 15.12).

15.7.2 *Effects of Illegal Poisoning and Reduced Food Supplies on Adult and Pre-adult Survival*

The model garnering the greatest support (model 1) showed that the closure of some feeding stations slightly changed survival and did so differently for juveniles (1–2 years old, the age class exploiting feeding stations with higher frequency) and immature (2–3 years old) and subadults (4–5 years old), whereas adults (>6 years old) showed a temporal decline in survival over the study. Survival of juveniles slightly decreased from 0.975 to 0.957, that of immatures and subadults from 0.917 to 0.906, whereas adult survival decreased from 0.975 in 1987–0.901 in 2011 (Table 15.3).

15.7.3 *Contradictions Between Health and Biodiversity Policies*

Recent changes in European health regulations provoked apparent contradictions between the application of sanitary and conservation policies (i.e. to eliminate corpses avoiding the presence of carrion in the field vs. to conserve scavenger species; Donázar et al. 2009b; Margalida et al. 2010). Implementation of the sanitation policy appears to have undermined conservation goals, as evidenced by, for example, in the same area with Eurasian griffon vultures in which an annual decrease in breeding success, reduced population growth, increased mortality on young age classes and changes in diet and behaviour is observed (Donázar et al. 2009b, 2010; Zuberogoitia et al. 2010; Margalida et al. 2011d, 2014c). Our results confirm these observations with empirical, long-term data on another vulture species. After sanitation policies had been enacted, the threatened Pyrenean bearded vulture population had reduced survival, especially of subadult individuals, and adults bred later and less successfully than before policy implementation. In this sense, additional conservation actions, such as supplementary feeding, can buffer the negative effects of policy change on demographic parameters (Oro et al. 2008). However, because supplementary feeding can also have detrimental effects (Carrete et al. 2006b; Robb et al. 2008; Cortés-Avizanda et al. 2016), this tool should only be used as a precautionary measure until the recovery of the previous scenario (i.e. availability of domestic carcases in the field) is achieved.

According to our results, reduction of food supply does not affect the proportion of pairs that start breeding, and so the decision to breed or not was influenced by the change in carcase disposal legislation. Among the European scavenger guild, bearded vultures seem to be the least sensitive to the effects of food limitations (Margalida and Colomer 2012). Their specialised diet based on bone remains implies a foraging resource for which only conspecifics can compete, which lasts

Table 15.3 List of the 15 models explaining bearded vulture survival with lowest QAICc values. Age of individuals was grouped into eight age classes, the last grouping 6-year-old birds and older (see methods). Models referring to 2005 deal with a tipping point in that year, after which API was in short supply probably affecting survival. Models considering different groupings of age classes were noted by their first and age class considered; for instance, a model grouping birds of 2 year and 3 year old was noted as 2 year_3 year. Np = number of identifiable parameters; Δ_i = QAICc difference value between the best model and model i; w_i = weight of model i (Margalida et al. 2014a)

No.	Np	Deviance	QAICc	Δ_i	w_i	Hypothesis tested on survival by time and age
1	28	921.148	979.148	0	0.41	Only two values: before and after 2005 differently for juveniles (1 year) and immature and subadults (2 year_5 year); for adults, trend with time
2	28	922.572	980.636	1.488	0.19	Temporal trend in survival only for adults (>5 year old); the two other age classes considered have constant survival and different for juveniles (1 year) and immatures and subadults (2 year_5 year)
3	28	923.76	981.823	2.675	0.11	Only two values: before and after 2005, but differently for juveniles, immature and subadults (2 year_5 year) and adults
4	28	924.045	982.109	2.961	0.09	Two values: before and after 2005 but only for adults; other age classes had constant survival but different for juveniles (1 year) and immature and subadults (2 year_5 year)
5	28	924.561	982.624	3.476	0.07	Two values: before and after 2005 but only for adults; other age classes had constant survival (1 year_5 year)
6	25	931.319	982.965	3.817	0.06	Constant (all age classes the same survival)
7	27	928.733	984.625	5.477	0.03	Two values: before and after 2005 but only for adults; other age classes had the same constant survival
8	27	928.757	984.675	5.527	0.03	Only two values: before and after 2005 but differently for juveniles, immature and subadults (1 year_5 year) and adults

(continued)

Table 15.3 (continued)

No.	Np	Deviance	QAICc	Δ_i	w_i	Hypothesis tested on survival by time and age
9	31	921.023	985.553	6.405	0.01	Only survival of old subadults and adults (>5 year old) had a temporal trend; other age classes have constant survival and different for each age
10	29	930.044	990.258	11.11	0.00	Only two values: before and after 2005 differently for juveniles (1 year) and immature and subadults (2 year_5 year); for adults, a temporal trend before 2005 and another temporal trend (different slope) after 2005
11	41	904.715	991.165	12.017	0.00	Change only with age
12	29	931.077	991.291	12.143	0.00	Only two values: before and after 2005 but differently for juveniles (1 year) and immature, and subadults (2 year_5 year); for adults, survival had a temporal trend until 2006 and then remained constant
13	30	929.903	992.272	13.124	0.00	Only adult survival changes with time; other age classes had constant survival and grouped differently for juveniles (1 year), immatures (2 year_3 year) and subadults (4 year_5 year)
14	84	810.319	997.854	18.706	0.00	Change with time and age, but recapture equal for all ages except for juveniles
15	94	797.367	1010.411	31.263	0.00	Change with time and age

ten times longer than meat remains and is very nutritious, being an advantage compared to a meat-based diet (Houston and Copsey 1994; Margalida 2008a, b). Thus, although the proportion of pairs that do not start laying annually is important (range 29.4%–58.1), the decision to start breeding, which could be influenced by the physical condition of the individual (Jenouvrier et al. 2005), appears not to compromise the reproduction of bearded vultures. This suggests that, in this isolated population, several buffering mechanisms (e.g. changes in the structure of breeding age, low dispersal movements and/or the availability of natural food) may contribute to maintain stable the proportion of pairs that start laying (Sergio et al. 2011; Oro et al. 2012; Margalida et al. 2013). In this sense, the carcass disposal policy does not seem to affect population movements because pre-adult individuals remained in the study area without greater pre-dispersal movements (Margalida et al. 2013). Thus, a possible explanation is related to the high food availability in the Pyrenees (Margalida et al. 2011a). With respect to non-natural mortality, although mortality factors affect mostly adult age classes (Margalida et al. 2007b; Oro et al. 2008), this seems to play no role in the start of breeding. Thus, buffering capacity through ecological and behavioural processes such as positive feedback from conspecific attraction could explain this regular pattern (Margalida et al. 2011d; Oro et al. 2012).

During the last six years, the proportion of double clutches seemed to decrease in parallel with laying dates. Large avian species often show obligate brood reduction (two eggs laid but only one chick survives) in which siblicide (when more that 90% of last-hatched chicks die, Simmons 1988) may occur, even when food supplies are abundant. The second egg thus serves as insurance against loss of the first egg from infertility, predation, or damage, rather than as a means of rearing two chicks (Stinson 1979; Anderson 1990; Mock et al. 1990). According to Winkler et al. (2002), there is a strong effect of laying date on clutch size, and earlier breeding may lead to larger clutch sizes. An increase of food resources could increase clutch size in fratricide facultative species (Korpimaki and Wiehn 1998), but in obligate fratricide species such as bearded vultures, the second egg seems to be an adaptive mechanism serving as an insurance egg to avoid breeding failure if the first egg does not hatch (Stinson 1979; Margalida et al. 2004). In this respect, in several species, including the bearded vulture, it seems that single-egg clutches are related to old or young females (Frey et al. 1995). According to our results, the increase of single-egg clutches could be related to a reduction in food supply or the stress that changes in food availability could provoke in individuals (i.e. increasing their foraging effort or the agonistic interactions as a consequence of a greater competence for the resources) and the effects of mate loss with the incorporation of less experienced individuals.

15.7.4 The Impact of Policies Decisions on Breeding Ecology and Conservation

The conservation implications of these results suggest that anthropogenic activities modifying habitat quality through human health regulations (i.e. a reduction in food availability) can alter demographic parameters and increase the probability of illegal poison bait consumption thus increasing non-natural mortality. The shifts in demographic parameters can have important conservation consequences, being necessary to adopt preventive mitigation measures on targeted species. As a result, preventive management measures such as supplementary feeding can be effective until more flexible sanitary legislation allows increased trophic availability, minimising the detrimental effects that food shortages and illegal poisoning can cause.

15.8 Concluding Remarks

Scientists and managers generally work at different timescales (Linklater 2003). Managers often need to respond quickly to immediate problems, whereas scientists can sometimes take years to generate appropriate information. This is the case of the bearded vulture in the Pyrenees, a species at which administrations invested substantial economic and human resources in recovering their populations. As our results show, long-term studies are necessary to identify conservation problems, to understand demographic changes on populations and priorities to apply conservation measures. Thanks to the long-term studies, we identified a saturation in the region of optimal habitat for the species, reflected in negative density-dependent effects on fecundity (Carrete et al. 2006a), the lack of recolonization and expansion beyond the current range of new territories (Margalida et al. 2008a) and the increase in polyandrous trios (Carrete et al. 2006b). Also, supplementary feeding sites (SFS) can have detrimental effects on fecundity (Carrete et al. 2006a), despite increasing pre-adult survival (Oro et al. 2008; Margalida et al. 2014b). Thus, as a management action, SFS should be moved into peripheral areas. In this sense, the recent network of protection areas for the feeding of necrophagous species of European interest (PAFs) suggests that these areas may be more efficient for breeders than for floaters, whose home ranges can be considerably larger (Margalida et al. 2016; Morales-Reyes et al. 2017). In the case of bearded vultures, the overlap of core areas (k50%) of breeders with PAFs reached 90.6%, while the overlap was only 64.2% for floaters (Morales-Reyes et al. 2017). The long-term monitoring has revealed how illegal poisoning, the most important non-natural mortality factor (Margalida 2012), is increasing over time and the demographic simulations suggest a regressive scenario in population dynamics if this factor is not eliminated (Oro et al. 2008). Anthropogenic activities through human health regulations that affect habitat quality can also suddenly modify demographic parameters (Margalida et al. 2014c). Our findings suggest a delay in laying dates and a regressive trend in clutch size, breeding success and survival following the policy change. Finally, the recent approved veterinary use of diclofenac in Spain (Margalida et al. 2014c) can threaten the scavenger guild community. This new risk requires a programme of monitoring of non-steroidal anti-inflammatory drug contamination of ungulate carcases available to vultures and of moribund and dead obligate and facultative avian scavengers would be needed to be confident that a damaging level of contamination is not present (Green et al. 2016). The ecosystem services provided by vultures (Moleón et al. 2014; Morales-Reyes et al. 2015) and the conservation of biodiversity implies that managers and policy-makers need to balance the demands of public health protection and the long-term conservation of biodiversity.

Acknowledgements Thanks to the coauthors of different papers in which this chapter has been based: J. Bustamante, M. Carrete, J.A. Donázar, R. Heredia, F.J. Hernández, M.A. Colomer, D. Oro and M. Romero-Pujante. To Generalitat de Catalunya and Ministry of Environment for the continuous support funding this research. Special thanks to J. Bertran, D. García, R. Heredia,

J.A. Sesé, M. Razin, J. Ruiz-Olmo and L.M. González for their help and support during these years. To all the forestry rangers and people involved in the monitoring of the bearded vulture population. The comments of two anonymous reviewers improved the first draft of this chapter. Finally, thanks to J. Catalan and M. Aniz for the invitation to contribute in the "The High Mountain in a changing world: challenges for conservation" workshop.

References

Addison PFE, Rumpff L, Bau SS, Carey JM, Chee YE et al (2013) Practical solutions formaking models indispensable in conservation decision-making. Divers Distrib 19:490–502

Anderson DJ (1990) Evolution of obligate siblicide in boobies. 1. A test of the insurance-egg hypothesis. Am Nat 135:334–350

Arcese P, Smith JNM (1988) Effects of population density and supplemental food on reproduction in song sparrows. J Anim Ecol 57:119–136

Arroyo B, Razin M (2006) Effect of human activities on bearded vulture behaviour and breeding success in the French Pyrenees. Biol Conserv 128:276–284

Bautista LM, García JT, Calmaestra RG, Palacín C, Martín C, Morales MB, Bonal R, Viñuela J (2004) Effect of weekend road traffic on the use of space by raptors. Conserv Biol 18:726–732

Bertran J, Margalida A (2002a) Social organization of a trio of bearded vulture (*Gypaetus barbatus*): sexual and parental roles. J Raptor Res 36:65–69

Bertran J, Margalida A (2002b) Territorial behaviour of the bearded vulture (*Gypaetus barbatus*) in response to the griffon vulture (*Gyps fulvus*). J Field Ornithol 73:86–90

Bertran J, Margalida A (2003) Male-male copulations in polyandrous bearded vultures (*Gypaetus barbatus*): an unusual mating system in raptors. J Avian Biol 34:334–338

Bertran J, Margalida A (2004) Do female control matings in polyandrous bearded vulture (*Gypaetus barbatus*) trios? Ethol Ecol Evol 16:181–186

Bertran J, Margalida A, Arroyo BE (2009) Agonistic behaviour and sexual conflict in atypical reproductive groups: the case of bearded vulture polyandrous trios. Ethology 115:429–438

Birkhead TR (2014) Stormy Outlook for long-term ecology studies. Nature 514:405

Blanco G, Tella JL (1999) Temporal, spatial and social segregation of red-billed choughs between two types of communal roost: a role for mating and territory acquisition. Anim Behav 57:1219–1227

Both C, Visser ME (2000) Breeding territory size affects fitness: an experimental study on competition at the individual level. J Anim Ecol 69:1021–1030

Breininger DR, Carter GM (2003) Territory quality transitions and source–sink dynamics in a Florida Scrub-Jay population. Ecol Appl 13:516–529

Brown JL (1987) Helping and communal breeding in birds: ecology and evolution. Princeton University Press, Princeton, NJ

Brown CJ (1988) A study of the bearded vulture *Gypaetus barbatus* in Southern Africa. PhD Thesis, University of Natal, Pietermaritzburg

Brown CJ (1997) Population dynamics of the bearded vulture *Gypaetus barbatus* in southern Africa. Afr J Ecol 35:53–63

Buckland ST, Elston DA, Beanay SJ (1996) Predicting distributional change, with application to bird distributions in Northeast Scotland. Glob Ecol Biog Lett 5:66–84

Bustamante J (1998) Use of simulation models to plan species reintroductions: the case of the bearded vulture in Southern Spain. Anim Conserv 1:229–238

Bustamante J, Seoane J (2004) Predicting the distribution of four species of raptors (Aves: Accipitridae) in Southern Spain: statistical models work better than existing maps. J Biog 31:295–306

Bustamante J, Donázar JA, Hiraldo F, Ceballos O, Travaini A (1997) Differential habitat selection by immature and adult grey eagle-buzzards *Geranoaetus melanoleucus*. Ibis 139:322–330

Calsbeek R, Sinervo B (2002) An experimental test of the ideal despotic distribution. J Anim Ecol 71:513–523

Cardillo M, Purvis A, Sechrest W, Gittleman JL, Bielby J, Mace GM (2004) Human population density and extinction risk in the world's carnivores. PLoS Biol 2:e197

Carrete M, Donázar JA, Margalida A (2006a) Density-dependent productivity depression in Pyrenean bearded vultures: implications for conservation. Ecol Appl 15:1674–1682

Carrete M, Donázar JA, Margalida A, Bertran J (2006b) Linking ecology, behaviour and conservation: does habitat saturation change mating system in bearded vultures? Biol Lett 2:624–627

Chamberlain CP, Waldbauer JR, Fox-Dobbs K, Newsome SD, Koch PL, Smith DR, Church ME et al (2005) Pleistocene to recent dietary shifts in California condors. Proc Natl Acad Sci USA 102:16707–16711

Clarke AL, Sæther BE, Roskaft E (1997) Sex biases in avian dispersal: a reappraisal. Oikos 79:429–438

Clutton-Brock T, Sheldon BC (2010) The seven ages of *Pan*. Science 327:1207–1208

Clutton-Brock TH, Parker GA (1995) Punishment in animal societies. Nature 373:209–216

Cockburn A (1998) Evolution of helping behaviour in cooperative breeding birds. Annu Rev Ecol Syst 29:141–177

Colomer MA, Margalida A, Sanuy D, Pérez-Jiménez MJ (2011) A bio-inspired computing model as a new tool for modelling ecosystems: the avian scavengers as a case study. Ecol Model 222:34–47

Cortés-Avizanda A, Carrete M, Donázar JA (2010) Managing supplementary feeding for avian scavengers: Guidelines for optimal design using ecological criteria. Biol Conserv 143:1707–1715

Cortés-Avizanda A, Blanco G, DeVault TL, Markandya A, Virani MZ, Brandt J, Donázar JA (2016) Supplementary feeding and endangered species: benefits, caveats and controversies. Front Ecol Environ 14:191–199

Creel S (2001) Social dominance and stress hormones. Trends Ecol Evol 16:491–497

Davies NB, Hartley IR (1996) Food patchiness, territory overlap and social systems: an experiment with dunnocks *Prunella modularis*. J Anim Ecol 65:837–846

Del Hoyo J, Elliott A, Sargatal J (1994) Handbook of the birds of the world, vol 2. New world vultures to guineafowl. Lynx Edicions, Barcelona

Dhondt AA, Schillemans J (1983) Reproductive success of the great tit in relation to its territorial status. Anim Behav 31:902–912

Donázar JA, Hiraldo F, Bustamante J (1993) Factors influencing nest site selection, breeding density and breeding success in the bearded vulture (*Gypaetus barbatus*). J Appl Ecol 30:504–514

Donázar JA, Margalida A, Campión D (2009a) Vultures, feeding stations and sanitary legislation: a conflict and its consequences from the perspective of conservation biology. Munibe 29, Sociedad de Ciencias Aranzadi, San Sebastián

Donázar JA, Margalida A, Carrete M, Sánchez-Zapata JA (2009b) Too sanitary for vultures. Science 326:664

Donázar JA, Cortés-Avizanda A, Carrete M (2010) Dietary shifts in two vultures after the demise of supplementary feeding stations: consequences of the EU sanitary legislation. Eur J Wildl Res 56:613–621

Donázar JA, Cortés-Avizanda A, Fargallo JA, Margalida A, Moleón M, Morales-Reyes Z, Moreno-Opo R et al (2016) Roles of raptors in a changing world: from flagships to providers of key ecosystem services. Ardeola 63:181–235

Emlen ST (1991) The evolution of helping I. An ecological constraints model. Am Nat 119:29–39

Emlen ST, Reeve HK, Sherman PW, Wrege PH, Ratnieks FLW, Shelman-Reeve J (1991) Adaptive vs nonadaptive explanations of behavior: the case of alloparental helping. Am Nat 138:259–270

Fielding AH, Haworth PF (1995) Testing the generality of bird-habitat models. Conserv Biol 9:1466–1481

Forslund P, Pärt T (1995) Age and reproduction in birds: hypotheses and tests. Trends Ecol Evol 10:374–378
Frey H, Knotzinger O, Llopis A (1995) The breeding network. An analysis of the period 1978 to 1995. In: Frey H, Kurzweil J, Bijleveld M (eds) Bearded vulture: reintroduction into the Alps. Annual report 1995 Foundation for the conservation of the bearded vulture, Wassenaar, The Netherlands, pp 13–38
Gandon S, Michalakis Y (2001) Multiple causes of the evolution of dispersal. In: Clobert J, Danchin E, Dhondt AA, Nichols JD (eds) Dispersal Oxford University Press, New York, USA, pp 155-167
Gavashelishvili A, McGrady MJ (2006) Breeding site selection by bearded vulture (*Gypaetus barbatus*) and Eurasian griffon (*Gyps fulvus*) in the Caucasus. Anim Conserv 9:159–170
Gilbert M, Watson RT, Ahmed S, Asim M, Johnson JA (2007) Vulture restaurants and their role in reducing diclofenac exposure in Asian vultures. Bird Conserv Int 17:63–77
Godoy JA, Negro JJ, Hiraldo F, Donázar JA (2004) Phylogeography, genetic structure, and diversity in the endangered bearded vulture (*Gypaetus barbatus*, L) as revealed by mitochondrial DNA. Mol Ecol 13:371–390
González LM, Margalida A, Sánchez R, Oria J (2006) Supplementary feeding as an effective tool for improving breeding success in the Spanish imperial eagle (*Aquila adalberti*). Biol Conserv 129:477–486
Gordon DM (1997) The population consequences of territorial behaviour. Trends Ecol Evol 12:63–66
Green RE, Newton I, Shultz S, Cunningham AA, Gilbert M, Pain DJ, Prakash V (2004) Diclofenac poisoning as a cause of vulture population declines across the Indian subcontinent. J Appl Ecol 41:793–800
Green RE, Taggart MA, Devojit D, Pain DJ, Kumar CS, Cunningham AA, Cuthbert R (2006) Collapse of Asian vulture populations: risk of mortality from residues of the veterinary drug diclofenac in carcasses of treated cattle. J Appl Ecol 43:949–956
Green RE, Donázar JA, Sánchez-Zapata JA, Margalida A (2016) The threat to Eurasian griffon vultures *Gyps fulvus* in Spain from veterinary use of the drug diclofenac. J Appl Ecol 53:993–1003
Guisan A, Thuiller W (2005) Predicting species distributions: offering more that simple habitat models. Ecol Lett 8:993–1009
Guisan A, Zimmermmann NE (2000) Predictive habitat distribution models in ecology. Ecol Model 135:147–186
Guisan A, Edwards TC Jr, Hastie T (2002) Generalized linear and generalized additive models in studies of species distributions: setting the scene. Ecol Model 157:89–100
Guisan A, Tingley R, Baumgartner JB, Naoujokaitis-Lewis I, Sutcliffe PR et al (2013) Predicting species distributions for conservation decisions. Ecol Lett 16:1424–1435
Hamilton IM, Taborsky M (2005) Unrelated helpers will not fully compensate for costs imposed on breeders when they pay to stay. Proc R Soc B 272:445–454
Hartley IR, Davies NB (1994) Limits to cooperative polyandry in birds. Proc R Soc B 257:67–73
Hatchwell BJ, Komdeur J (2000) Ecological constraints, life history traits and the evolution of cooperative breeding. Anim Behav 59:1079–1086
Heg D, van Treuren R (1998) Female–female cooperation in polygynous Oystercatchers. Nature 391:687–691
Heredia R, Donázar JA (1990) High frequency of polyandrous trios in an endangered population of lammergeiers *Gypaetus barbatus* in Northern Spain. Biol Conserv 53:163–171
Heredia R, Heredia B (1990) El quebrantahuesos (*Gypaetus barbatus*) en los Pirineos. Colección Técnica Ministerio de Agricultura, Pesca y Alimentación, Madrid
Hernández M, Margalida A (2008) Pesticide abuse in Europe: effects on the Cinereous vulture (*Aegypius monachus*) population in Spain. Ecotoxicology 17:264–272
Hernández M, Margalida A (2009) Poison-related mortality effects in the endangered Egyptian Vulture (*Neophron percnopterus*) population in Spain: conservation measures. Eur J Wildl Res 55:415–423

Hernández-Matías A, Real J, Moleón M, Palma L, Sánchez-Zapata JA et al (2013) From local monitoring to a broad-scale viability assessment: a case study for the Bonelli's eagle in Western Europe. Ecol Monog 83:239–261

Hernández-Matías A, Real J, Parés F, Pradel J (2015) Electrocution threatens the viability of populations of the endangered Bonelli's eagle (*Aquila fasciata*) in Southern Europe. Biol Conserv 191:110–116

Hiraldo F, Delibes M, Calderón J (1979) El Quebrantahuesos *Gypaetus barbatus* (L.) Monografías 22 Madrid: Instituto para la Conservación de la Naturaleza

Hirzel AH, Posse B, Oggier P-A, Crettenand Y, Glenz C, Arlettaz R (2004) Ecological requirements of reintroduced species and the implications for release policy: the case of the bearded vulture. J Appl Ecol 41:1103–1116

Houston DC, Copsey JA (1994) Bone digestión and intestinal morphology of the bearded vulture. J Raptor Res 28:73–78

Inchausti P, Halley J (2003) On the relation between temporal variability and persistence time in animal populations. J Anim Ecol 72:899–908

Jenouvrier S, Barbraud C, Cazelles B, Weimerskirch H (2005) Modelling population dynamics of seabirds: importance of the effects of climate fluctuations on breeding proportions. Oikos 108:511–522

Jerina K, Debeljak M, Dzeroski S, Kobler A, Adamic M (2003) Modeling the brown bear population in Slovenia. A tool in the conservation management of a threatened species. Ecol Model 170:453–469

Kokko H, Sutherland WJ (1998) Optimal floating and queuing strategies: consequences for density dependence and habitat loss. Am Nat 152:354–366

Korpimaki E, Wiehn J (1998) Clutch size of kestrels: seasonal decline and experimental evidence for food limitation under fluctuating food conditions. Oikos 83:259–272

Krüger O, Lindstrom J (2001) Habitat heterogeneity affects population growth in goshawk *Accipiter gentilis*. J Anim Ecol 70:173–181

Lack D (1968) Ecological adaptations for breeding in birds. Methuen and Co, London

Lambin X, Aars J, Piertney SB (2001) Dispersal, intraspecific competition, kin competition and kin facilitation: a review of the empirical evidence. In: Clobert J, Danchin E, Dhondt AA, Nichols JD (eds) Dispersal Oxford University Press, New York, USA, pp 110–122

Lambrechts MM, Caro S, Charmantier A, Gross N, Galan MJ, Perret P, Cartan-Son M et al (2004) Habitat quality as a predictor of spatial variation in blue tit reproductive performance: a multi-plot analysis in a heterogeneous landscape. Oecologia 141:555–561

Lande R, Enger S, Saether BE (2003) Stochastic population dynamics in ecology and conservation. Oxford University Press, Oxford

Lindenmayer DB, Ritman K, Cunningham RB, Smith JDB, Horvath D (1995) A method for predicting the spatial distribution of arboreal marsupials. Wildl Res 22:445–456

Linklater WL (2003) Science and management in a conservation crisis: a case study with rhinoceros. Conserv Biol 17:968–975

López-Sepulcre A, Kokko H (2005) Territorial defense, territory size and population regulation. Am Nat 166:317–329

Mañosa S, Real J, Codina J (1998) Selection of settlement areas by juvenile Bonelli's eagle in Catalonia. J Raptor Res 32:208–214

Margalida A (2008a) Bearded vultures (*Gypaetus barbatus*) prefer fatty bones. Behav Ecol Sociobiol 63:187–193

Margalida A (2008b) Presence of bone remains in the ossuaries of bearded vultures *Gypaetus barbatus*: storage or nutritive rejection? Auk 125:560–564

Margalida A (2010a) Conservation biology of the last and largest natural population of the European bearded vulture *Gypaetus barbatus* (Linnaeus, 1758). PhD thesis, University of Bern, Bern

Margalida A (2010b) Supplementary feeding during the chick-rearing period is ineffective in increasing the breeding success in the bearded vulture (*Gypaetus barbatus*). Eur J Wildl Res 56:673–678

Margalida A (2012) Baits, budget cuts: a deadly mix. Science 338:192
Margalida A, Bertran J (2000) Breeding behaviour of the bearded vulture (*Gypaetus barbatus*): minimal sexual differences in parental activities. Ibis 142:225–234
Margalida A, Bertran J (2005) Territorial defence behaviour of bearded vulture *Gypaetus barbatus* against conspecifics and heterospecifics. Ethol Ecol Evol 17:51–63
Margalida A, Colomer MA (2012) Modelling the effects of sanitary policies on European vulture conservation. Sci Rep 2:753
Margalida A, Heredia R (2005) Biología de la conservación del quebrantahuesos (*Gypaetus barbatus*) en España Organismo Autonómico Parques Nacionales, Madrid
Margalida A, García D, Bertran J (1997) A possible case of a polyandrous quartet in the bearded vulture (*Gypaetus barbatus*). Ardeola 44:109–111
Margalida A, Bertran J, Boudet J, Heredia R (2004) Hatching asynchrony, sibling aggression and cannibalism in the bearded vulture (*Gypaetus barbatus*). Ibis 146:386–393
Margalida A, García D, Cortés-Avizanda A (2007a) Factors influencing breeding density of bearded vultures, Egyptian vultures and Eurasian griffon vultures in Catalonia (NE Spain): management implications. Anim Biodiv Conserv 30:189–200
Margalida A, Heredia R, Razin M, Hernández M (2007b) Sources of variation in mortality of the bearded vulture *Gypaetus barbatus* in Europe. Bird Conserv Int 17:1–10
Margalida A, Donázar JA, Bustamante J, Hernández F, Romero-Pujante M (2008a) Application of a predictive model to detect long-term changes in nest-site selection in the bearded vultures: conservation in relation to territory shrinkage. Ibis 150:242–249
Margalida A, Mañosa S, González LM, Ortega E, Oria J, Sánchez R (2008b) Breeding non-adults and age effects on productivity in the Spanish imperial eagles *Aquila adalberti*. Ardea 96:173–180
Margalida A, Donázar JA, Carrete M, Sánchez-Zapata JA (2010) Sanitary versus environmental policies: fitting together two pieces of the puzzle of European vulture conservation. J Appl Ecol 47:931–935
Margalida A, Colomer MA, Sanuy D (2011a) Can wild ungulate carcasses provide enough biomass to maintain avian scavenger populations? An empirical assessment using a bio-inspired computational model. PLoS ONE 6:e20248
Margalida A, Moreno-Opo R, Arroyo BE, Arredondo A (2011b) Reconciling the conservation of an endangered species with an economically important anthropogenic activity: interactions between cork exploitation and the cinereous vulture *Aegypius monachus* in Spain. Anim Conserv 14:167–174
Margalida A, Oro D, Cortés-Avizanda A, Heredia R, Donázar JA (2011c) Misleading population estimates: biases and consistency of visual surveys and matrix modelling in the endangered bearded vulture. PLoS ONE 6:e26784
Margalida A, Campión D, Donázar JA (2011d) European vultures' altered behaviour. Nature 480:457
Margalida A, Carrete M, Sánchez-Zapata JA, Donázar JA (2012) Good news for European vultures. Science 335:284
Margalida A, Carrete M, Hegglin D, Serrano D, Arenas R, Donázar JA (2013) Uneven large-scale movement patterns in wild and reintroduced pre-adult bearded vultures: conservation implications. PLoS ONE 8:e65857
Margalida A, Colomer MA, Oro D (2014a) Man-induced activities modify demographic parameters in a long-lived species: effects of poisoning and health policies. Ecol Appl 24:436–444
Margalida A, Campión D, Donázar JA (2014b) Vultures vs livestock: conservation relationships in an emergent human-wildlife conflict. Oryx 48:172–176
Margalida A, Bogliani G, Bowden C, Donázar JA, Genero F, Gilbert M, Karesh et al (2014c) One health approach to the use of veterinary pharmaceuticals. Science 346:1296–1298
Margalida A, Colomer MA, Oro D, Arlettaz R, Donázar JA (2015) Assessing the impact of removal scenarios on population viability of a threatened, long-lived avian scavenger. Sci Rep 5:16962

Margalida A, Pérez-García JM, Afonso I, Moreno-Opo R (2016) Spatial and temporal movements in Pyrenean bearded vultures (*Gypaetus barbatus*): integrating movement ecology into conservation practice. Sci Rep 6:35746

Margalida A, Martínez JM, Gómez de Segura A, Colomer MA, Arlettaz R et al (2017) Supplementary feeding and young extraction from the wild are not a sensible alternative to captive breeding for reintroducing bearded vultures *Gypaetus barbatus*. J Appl Ecol 54:334–340

Martínez-Abraín A, Tavecchia G, Regan H, Jiménez J, Surroca M, Oro D (2012) The effects of wind farms and food scarcity on a large scavenging bird species following an epidemic of bovine spongiform encephalopathy. J Appl Ecol 49:109–117

Mateo-Tomás P, Olea PP, Sánchez-Barbudo IS, Mateo R (2012) Alleviating human-wildlife conflicts: identifying the causes and mapping the risk of illegal poisoning of wild fauna. J Appl Ecol 49:376–385

Mee A, Hall LS (2007) California condors in the 21st century. Series in ornithology Nº 2. Nuttall Ornithological Club and the American Ornithologists' Union, Lancaster

Meretsky VJ, Snyder NFR, Beissinger SR, Clendenen DA, Wiley JM (2000) Demography of the California condor: implications for reestablishment. Conserv Biol 14:957–967

Mills JA, Teplitsky C, Arroyo B, Charmantier A, Becker PH, Birkhead TR, Bize P et al (2015) Archiving primary data: solutions for long-term studies. Trends Ecol Evol 30:581–589

Mills JA, Teplitsky C, Arroyo B, Charmantier A, Becker PH, Birkhead TR, Bize P et al (2016) Solutions for archiving data in long-term studies—a reply to Whitlock et al. Trends Ecol Evol 31:85–87

Mladenoff DJ, Sickley TA, Haight RG, Wydeven AP (1995) A regional landscape analysis and prediction of favorable gray wolf habitat in Northern Great lakes region. Conserv Biol 9:279–294

Mock DW, Drummond H, Stinson H (1990) Avian siblicide. Am Sci 78:438–449

Moleón M, Sánchez-Zapata JA, Margalida A, Carrete M, Owen-Smith N, Donázar JA (2014) Humans and scavengers: evolution of interactions and ecosystem services. Bioscience 64:394–403

Morales-Reyes Z, Pérez-García JM, Moleón M, Botella F, Carrete M, Lazcano C, Moreno-Opo R et al (2015) Supplanting ecosystem services provided by scavengers raises greenhouse gas emissions. Sci Rep 5:7811

Morales-Reyes Z, Pérez-García JM, Moleón M, Botella F, Carrete M, Donázar JA et al (2017) Evaluation of the network of protection areas for the feeding of scavengers (PAFs) in Spain: from biodiversity conservation to greenhouse gas emission savings. J Appl Ecol. doi:10.1111/1365-2664.12833

Moreno-Opo R, Trujillano A, Arredondo A, González LM, Margalida A (2015) Manipulating size, amount and appearance of food inputs to optimize supplementary feeding programs for European vultures. Biol Conserv 181:27–35

Mougeot F, Redpath SM, Moss R, Matthiopoulos J, Hudson PJ (2003) Territorial behaviour and population dynamics in red grouse *Lagopus lagopus scoticus* I. Population experiments. J Anim Ecol 72:1073–1082

Mundy P, Butchart D, Ledger J, Piper S (1992) The vultures of Africa. Academic Press, London

Newton I (1979) Population ecology of raptors. T and AD Poyser, Berkhamsted, UK

Newton I (1992) Experiments on the limitation of bird numbers by territorial behaviour. Biol Rev 67:129–173

Newton I (1998) Population limitation in birds. Academic Press, San Diego

Oaks JL, Gilbert M, Virani MZ, Watson RT, Meteyer CU, Rideout HL et al (2004) Diclofenac residues as the cause of population decline of vultures in Pakistan. Nature 427:630–633

Oppel S, Meirinho A, Ramírez I, Gardner B, O'Connell AF, Miller PI, Louzao M (2012) Comparison of five modelling techniques to predict the spatial distribution and abundance of seabirds. Biol Conserv 156:94–104

Oppel S, Dobrev V, Arkumarev V, Saravia V, Bounas A et al (2016) Assessing the effectiveness of intensive conservation actions: does guarding and feeding increase productivity and survival of Egyptian vultures in the Balkans? Biol Conserv 198:157–164

Oro D, Pradel R, Lebreton JD (1999) Food availability and nest predation influence life history traits in Audouin's gull, *Larus audouinii*. Oecologia 118:438–445

Oro D, Cam E, Pradel R, Martínez-Abraín A (2004) Influence of food availability on demography and local population dynamics in a long-lived seabird. Proc R Soc Lond Ser B 271:387–396
Oro D, Margalida A, Carrete M, Heredia R, Donázar JA (2008) Testing the goodness of supplementary feeding to enhance population viability in an endangered vulture. PLoS ONE 3: e4084
Oro D, Jiménez J, Curcó A (2012) Some clouds have a silver lining: paradoxes of anthropogenic perturbations from study cases on long-lived social birds. PLoS ONE 7:e42753
Ortega E, Mañosa S, Margalida A, Sánchez R, Oria J, González LM (2009) A demographic description of the recovery of the vulnerable Spanish imperial eagle *Aquila adalberti*. Oryx 43:113–121
Owens IPP, Bennett PM (2000) Ecological basis of extinction risk in birds: habitat loss versus human persecution and introduced predators. Proc Natl Acad Sci USA 97:12144–12148
Pen I, Weissing FJ (2000) Towards a unified theory of cooperative breeding: the role of ecology and life history re-examined. Proc R Soc B 267:2411–2418
Perrin N, Mazalov V (1999) Dispersal and inbreeding avoidance. Am Nat 154:282–292
Piper SE, Boshoff AF, Scott HA (1999) Modelling survival rates in the cape griffon *Gyps coprotheres*, with emphasis on the effects of supplementary feeding. Bird Study 46:230S–238S
Razin M, Bretagnolle V (2003) Dynamique spatiotemporelle de la population nord Pyrénéenne de Gypaète barbu. In: Thiollay JM, Sarrazin F (eds) Actes du Colloque International sur la conservation du Gypaète barbu en Europe LPO (Ligue pour la Protection des Oiseaux) Mission FIR. Tende, France, pp 88–94
Real J, Mañosa S (1997) Demography and conservation of western European Bonelli's eagle *Hieraaetus fasciatus* populations. Biol Conserv 79:59–66
Ridley J, Komdeur J, Sutherland WJ (2004) Incorporating territory compression into population models. Oikos 105:101–108
Robb GN, McDonald RA, Chamberlain DE, Bearhop S (2008) Food for thought: supplementary feeding as a driver of ecological change in avian populations. Front Ecol Environ 6:476–484
Robertson BC, Elliott GP, Eason DK, Clout MN, Gemmell NJ (2006) Sex allocation theory aids species conservation. Biol Lett 2:229–231
Rodenhouse NL, Sherry TW, Holmes RT (1997) Site-dependent regulation of population size: a new synthesis. Ecology 78:2025–2042
Rodriguez A, Andrén H (1999) A comparison of Eurasian red squirrel distribution in different fragmented landscapes. J Appl Ecol 36:649–662
Romero LM (2004) Physiological stress in ecology: lessons from biomedical research. Trends Ecol Evol 19:259–266
Rotllant J, Pavlidis M, Kentouri M, Abad ME, Tort L (1998) Non-specific immune responses in the red porgy *Pagrus pagrus* after crowding stress. Aquaculture 156:279–290
Rushton SP, Ormerod SJ, Kerby G (2004) New paradigms for modelling species distributions? J Appl Ecol 41:193–200
Sanz A, Tavecchia G, Pradel R, Minguez E, Oro D (2008) The cost of reproduction and experience-dependent vital rates in a small petrel. Ecology 89:3195–3203
Sarrazin F, Bagnolini C, Pinna JL, Danchin E (1996) Breeding biology during establishment of a reintroduced griffon vulture *Gyps fulvus* population. Ibis 138:315–325
Sæther BE, Bakke Ø (2000) Avian life history variation and contribution of demographic traits to the population growth rate. Ecology 81:642–653
Schaub M, Zink R, Beissmann H, Sarrazin F, Arlettaz R (2009) When to end releases in reintroduction programmes: demographic rates and population viability analysis of bearded vultures in the Alps. J Appl Ecol 46:92–100
Sergio F, Newton I (2003) Occupancy as a measure of territory quality. J Anim Ecol 72:857–865
Sergio F, Tavecchia G, Blas J, López L, Tanferna A, Hiraldo F (2011) Age-structured vital rates in a long-lived raptor: implications for population growth. Basic Appl Ecol 12:107–115
Serrano D, Forero MG, Donázar JA, Tella JL (2004) Dispersal and social attraction affect colony selection and dynamics of lesser kestrels. Ecology 85:3438–3447

Sesé JA, Antor RJ, Alcántara M, Ascaso JC, Gil JA (2005) La alimentación suplementaria en el quebrantahuesos: estudio de un comedero del Pirineo occidental aragonés. In: Heredia R (ed) Margalida A. Biología de la conservación del Quebrantahuesos Gypaetus barbatus en España Madrid, Organismo Autónomo Parques Nacionales, pp 279–304
Sillett TS, Rodenhouse NL, Holmes RT (2004) Experimentally reducing neighbour density affects reproduction and behavior of a migratory songbird. Ecology 85:2467–2477
Simmons RE (1988) Offspring quality and the evolution of cainism. Ibis 130:339–357
Stamps JA (1990) The effect of contender pressure on territory size and overlap in seasonally territorial species. Am Nat 135:614–632
Stinson CH (1979) On the selective advantage of fratricide in raptors. Evolution 33:1219–1225
Tavecchia G, Pradel R, Boy V, Johnson AR, Cézilly F (2001) Sex- and age-related variation in survival and cost of first reproduction in greater flamingos. Ecology 82:165–174
Tavecchia G, Pradel R, Genovart M, Oro D (2007) Density-dependent parameters and demographic equilibrium in open populations. Oikos 116:1481–1492
Virani MZ, Kendall C, Njoroge P, Thomsett S (2011) Major declines in the abundance of vultures and other scavenging raptors in and around the Masai Mara ecosystem. Biol Conserv 144:746–752
Western D, Ssemakula J (1982) Life history patterns in birds and mammals and their evolutionary interpretation. Oecologia 54:281–290
Wilbur SR, Carrier WD, Borneman JC (1974) Supplemental feeding program for California condors. J Wildl Manage 38:343–346
Winkler DW, Dunn PO, McCulloch CE (2002) Predicting the effects of climate change on avian life-history traits. Proc Nat Acad Sci USA 66:13595–13599
Xirouchakis S, Nikolakakis M (2002) Conservation implications of the temporal and spatial distribution of bearded vulture *Gypaetus barbatus* in Crete. Bird Conserv Int 12:269–280
Zuberogoitia I, Zabala J, Martínez JA, Martínez JE, Azkona A (2008) Effect of human activities on Egyptian vulture breeding success. Anim Conserv 11:313–320
Zuberogoitia I, Martínez JE, Margalida A, Gómez I, Azkona A, Martínez JA (2010) Reduced food availability induces behavioural changes in griffon vulture. Orn Fenn 87:52–60

Chapter 16
Monitoring Global Change in High Mountains

Regino Zamora, Antonio J. Pérez-Luque and Francisco J. Bonet

Abstract Long-term ecological research provides essential information to understand the complex dynamics of natural systems. In a global change scenario, high mountains represent an exceptional ecology field lab for long-term research and monitoring, offering an enormous mosaic of ecological conditions existing along mountain slopes. Mountains ecosystems also constitute invaluable observatories of the atmosphere and all the aspects related to climate, atmospheric particle deposition, pollutants, greenhouse gases, or the transport of resistant biological forms. Mountains are sensors for early detection of change. In the Sierra Nevada LTER site (southern Spain), we have been implementing a long-term monitoring programme taking advantage of the high altitude and geographical position of this Mediterranean mountain. We have identified the main expected impacts in the context of global change and analysed the biophysical and socioeconomic data available to assess exposure, sensitivity, and adaptive capacity of ecosystems to future scenarios. The study incorporates a retrospective of past human management of land use, to understand the current state of conservation of the ecosystems and make plausible forecasts on its response to future scenarios. The results show the following: (1) an ancestral human footprint on the ecosystems of Sierra Nevada, particularly evident during the 20th century; (2) a moderate climate warming, with reduction and increased variability in precipitation, as well as a consequent reduction in snow-cover duration during the last few decades; (3) significant changes in biophysical characteristics of rivers and mountain lakes; and (4) shifts in the distribution and phenology of many species of plants and animals along elevation gradients.

Keywords Long-term ecological research LTER · Sierra Nevada · Global change monitoring · Mediterranean mountains · Land-use changes

R. Zamora (✉) · A.J. Pérez-Luque · F.J. Bonet
Departamento de Ecología, Facultad de Ciencias & Instituto Interuniversitario Sistema Tierra en Andalucía (IISTA), Universidad de Granada,
Avda. del Mediterráneo s/n, 18006 Granada, Spain
e-mail: rzamora@ugr.es

J. Catalan et al. (eds.), *High Mountain Conservation in a Changing World*,
Advances in Global Change Research 62, DOI 10.1007/978-3-319-55982-7_16

16.1 Introduction

Mountains are the crown jewels of our natural heritage around the world and yet are locally unique. They provide key ecosystems services particularly sensitive to human impact. The challenge for mountain conservation is to determine ecosystem exposure and sensitivity to environmental changes as well as their ability to adapt to global change in the present and the future (Hansen et al. 2014; Zamora et al. 2016). Despite uncertainties concerning the response of biophysical processes to human impact, there is enough scientific capability to envisage the foreseeable trajectory of major ecological processes and to manage the adaptation of ecosystems to possible future scenarios. In this context, long-term research and monitoring are essential for understanding the dynamics of populations, communities, and ecosystems (Carpenter 1998; Lovett et al. 2007). One of the most prominent examples of a long-term study is the work begun in 1958 in Mauna Loa, Hawaii, which has demonstrated a slow but steady increase in the concentration of atmospheric carbon dioxide (Keeling et al. 1995, 1996)—a trend that continues today and is now investigated through a global network of stations (Battin et al. 2009). Moreover, developing long-time data series is critical for the proper parameterization and validation of environmental models, such as the general circulation models to predict future climate, or related potential changes in species distribution (Burgman et al. 1993; Canham et al. 2003; Berdanier and Clark 2016). Thus, high-quality ecological information collected over extended periods of time can yield valuable insights into changes in ecosystem structure, key ecological processes, and the services provided by ecosystems (e.g. Daily 1997; Lindenmayer et al. 2012).

A full comprehension of many ecological processes requires long-term monitoring because their rate of change is very slow, and/or because extreme events that can have substantial impacts are usually rare (Fahey et al. 2015). Many ecosystems are likely to undergo abrupt changes due to global drivers such as climate change (Barnosky et al. 2012). Designing and implementing large-scale ecosystem management programmes are needed to confront these problems and to provide positive ecological and economic solutions (Pace et al. 2015). Additionally, the use of information from long-term data series enables managers to evaluate and mitigate threats to ecosystem function and services while operating more effectively in the legal and political arenas. This point is important because resource managers, policy-makers, and the general public may be unaware of these values and the critical role of long-term ecological studies in tackling emerging problems of major social concern (Lindenmayer et al. 2012).

16.2 Monitoring Global Change in High Mountains: The Case of Sierra Nevada

To understand the consequences of human impact on the planet, we need systems of reference. Mountain ecosystems may represent the best-preserved reference systems in a given region, providing us the opportunity to compare their dynamic behaviour with anthropic ecosystems surrounding the mountains under a global change scenario (Becker and Bugmann 2001; Huber et al. 2005; Hansen et al. 2014). In this respect, mountain ecosystems are equivalent to controls in the classic experimental designs against which to compare "treatments", in this case, caused by human activities (land-use changes, climatic change, pollution, overexploitation of resources, etc.) in the surrounding anthropic matrix. For example, we can compare the effect of tree species diversity on forest productivity (e.g. Liang et al. 2016) in native mountain forest vs. plantations and/or strongly managed forest in the humanized landscape surrounding the most preserved mountain forest ecosystems.

In a scenario of global change, a high mountain such as Sierra Nevada (Box 16.1) represents an exceptional ecology field lab, offering the advantage of the enormous mosaic of ecological conditions existing along mountain slopes. For instance, the topographic variability of steep mountain slopes creates a multitude of fine-scale environmental conditions that is mirrored in plant species distribution and abundance (Scherrer and Körner 2010, 2011; Scherrer et al. 2011). As a result of this fine-scale spatial distribution, within a short distance, on the same elevation, we can find "Mediterranean", "montane", and "alpine" species depending on the micro-environmental conditions of their habitats (Scherrer and Körner 2010, 2011; Scherrer et al. 2011).

Box 16.1 Sierra Nevada

Sierra Nevada (Andalusia, SE Spain) is a mountainous region covering more than 2000 km^2 with an elevation range of between 860 m and 3482 m a.s.l. The regional climate is Mediterranean, characterized by cool winters and hot summers, with a pronounced summer drought (July–August). The annual average temperature decreases in altitude from 12–16 °C below 1500 m to 0 °C above 3000 m a.s.l. Annual precipitation ranges from less than 250 mm in the lowest parts to more than 700 mm in the summit areas. Additionally, the complex orography of the mountains causes sharp climatic contrasts between the sunny, dry south-facing slopes and the shaded, less-exposed north-facing slopes. This mountain area harbours 27 habitat types of the EU Habitat Directive (92/43/EEC), and it is considered one of the most important biodiversity *hotspots* in the Mediterranean region (Blanca et al. 1998; Cañadas et al. 2014). Sierra Nevada receives legal protection in multiple ways: it is a MAB Biosphere Reserve, Special Area of Conservation (Natura 2000 network); Natural Park and National Park; and Important Bird Area. Sierra Nevada is included at the World Green List of Protected Areas (IUCN), and is part of the Spanish Long-Term Ecological Research (Zamora et al. 2016). The main economic activities in this mountain region are agriculture, tourism, livestock raising, beekeeping, mining, and skiing (Bonet et al. 2010).

Mountains are far more than just lands at high elevations. High mountains such as Sierra Nevada act as sensors for early detection of signs of change, due to the high elevation and geographical position of this Mediterranean mountain.

This includes processes considered more genuinely global and is precisely the processes that can be observed better from the mountains, as exceptional lookouts. In this sense, mountain ecosystems are key observatories of the atmosphere and all the aspects related to climate such as energy balance, UV radiation, atmospheric particle deposition, pollutants, greenhouse gases, or the transport of resistant biological forms and microorganisms (Beniston 2003; Huber et al. 2005; Zamora et al. 2016).

16.3 Conceptual and Method-Related Issues

Given the complexity of natural systems and the variety of factors that influence natural processes, monitoring programmes clearly need conceptual frameworks that help organize the information gathered. These programmes to be effective should be based on well-defined specific questions within a clearly stated framework providing scientific hypotheses and predictable scenarios of change. Otherwise, it could turn into an exercise in "fishing", with the risk of the monitoring programme ending up with a collection (a hodgepodge as broad as the kinds of specialists involved) of pseudo-indicators in which all taxonomic and/or thematic variable entities capable of being measured are jumbled together. In fact, many monitoring initiatives have a narrow thematic scope. In our case, the goal is to identify the best biophysical and socioeconomic indicators of the health of our ecosystems under a global change scenario, and to determine the exposure and sensitivity of ecosystems to environmental changes, and their ability to adapt to global change (Hansen et al. 2014; Zamora et al. 2016). Besides the rigour necessary within a specific scientific framework, a monitoring programme should be guided by human needs, identifying key environmental and economic services that the public receives from ecosystems. In doing so, monitoring programmes are also valuable for society in general as well as to scientists and managers in particular (Lindenmayer and Likens 2009).

The design of the Global Change Monitoring programme in Sierra Nevada has been inspired by the conceptual framework and the thematic areas proposed by the GLOCHAMORE (Global Change in Mountain Regions http://mri.scnatweb.ch/projects/glochamore sponsored initiative UNESCO) (Grabherr et al. 2005) and related initiatives. Our monitoring programme is based on evaluating, using standardized protocols, the composition, structure and functioning of the ecosystems of the Sierra Nevada, and its dynamics over the middle to long term. This programme is designed to identify the effects of global change on key biophysical processes in mountain ecosystems, providing a vision of the trends of change that allow the development of an adaptive capacity. The long-term evaluation of ecosystems function and services in a context of global change is a primary aim of our approach. Long-term series collected in our programme will help us to forecast the evolution of our ecosystems under new scenarios, with the use of modelling tools.

Within the monitoring programme, the way to foster cooperation between research groups of the University of Granada (Spain) and other research institutions and managers and technical teams of the Sierra Nevada National Park is through joint work in the thematic areas we have defined as systems of indicators (Box 16.2). For each of the thematic areas of our research strategy, methods have been defined to evaluate both the state of key ecological functions, as well as the possible impacts of global change on ecosystems of Sierra Nevada.

Box 16.2 Thematic Areas

Climatology
Temporal change in the cryosphere
Palaeo perspective
Land-use and land-cover changes
Atmospheric deposition
Population trends and community changes
Biodiversity changes
Phenological changes
Invasive species
Emerging diseases
Biogeochemical changes in aquatic systems
Primary productivity and carbon fluxes in terrestrial ecosystems
Ecosystem services and socio-economy
Extreme events (wildland fire, floods, landslides)
Assessment of ecosystem management activities

The thematic areas of our research strategy are organized according to our understanding of the causes and consequences of global change, their ecological consequences and the corresponding biotic and socioeconomic responses to changes. For practical organization, each thematic area follows the same logical flow: (1) starting with well-formulated and tractable questions that were posed at the outset of the work, (2) designing high-quality data-collection protocol, with careful attention to field data and field sample storage, and (3) developing collaborative partnerships among scientists and environmental managers.

As far as possible, each thematic area is linked to projects of adaptive management and the development of decision support tools.

16.4 Data Collection Within the Monitoring Protocol

For each of the thematic areas outlined above, a set of monitoring methods has been defined to assess the status of key ecological functions, such as primary productivity, ecosystems functions and services, biogeochemical cycles, etc. (Aspizua et al. 2014). These methods allow us to cover most of the aspects considered to be necessary for evaluating the effects of global change in this mountain region.

Our approach takes into account the high spatial heterogeneity and ecological diversity of the Sierra Nevada mountain range. Data are collected over a hierarchy of spatial scales: fine scale (point and transect data); a somewhat coarser scale but covering the entire space (e.g. pixels of satellite images, polygons of a vegetation map); and administrative boundary scale (i.e. catchment basin, municipality). Also, many of the sampling points that take more detail (points and transects) are spatially aggregated in places with a high density of monitoring protocols.

Our monitoring programme incorporates the temporal dimension from two different perspectives: (1) Historical information (including when possible the palaeo perspective) on the structure and dynamics of the Sierra Nevada ecosystems; and (2) Recent information, focused more on the frequency of data collection. The purpose of the historical reconstruction is to use information about the past in interpreting and understanding the present, and by doing so, try to predict possible future trends. In this regard, it is important to consider the length of the series available for each feature monitored. As with the frequency of the data collection from recent information, we use methods that collect information according to the processes of interest (e.g. periodicities of less than a day (weather stations) to seasonal inventories, or at longer time scales, annually or every several years).

In short, our monitoring programme is composed of a set of scientifically validated protocols that can be described based on a number of attributes, thematic (according to GLOCHAMORE approach), spatial (data-collection scale and the extent of data application), and temporal (length of time series and data-collection periodicity) (Fig. 16.1).

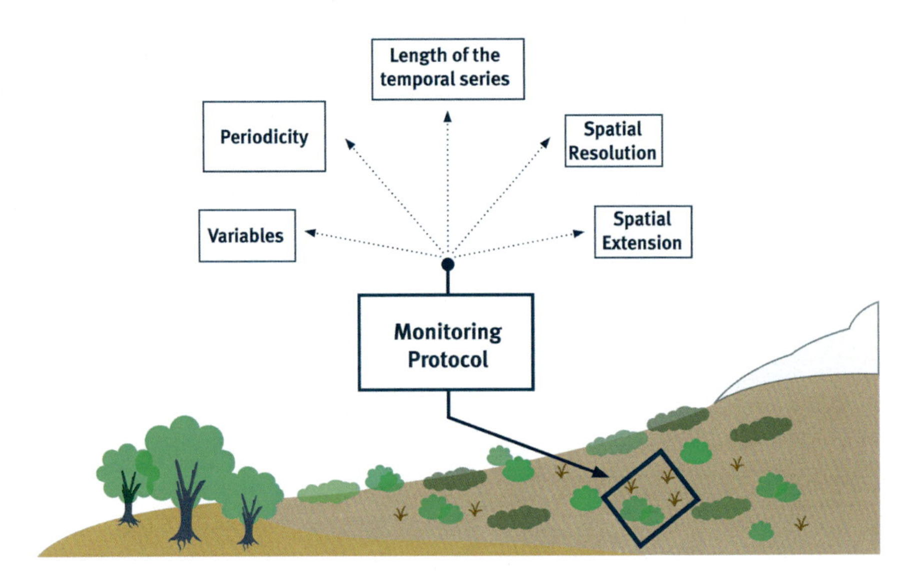

Fig. 16.1 Thematic representation of the five main attributes used to characterize the monitoring protocols. Each attribute is defined using either continuous ranges of values (number of variables or series length) or discrete lists (period of data collection, resolution and spatial extension). Modified from Aspizua et al. (2014)

16.5 Spatial Organization of the Monitoring Programme

To study the potential effects of global change in a mountain range, we needed the sampling units to be spatially distributed according to the specific objectives of the programme and thus optimize and streamline the gathering of environmental information. First, we identified major environmental gradients over the entire mountain (e.g. elevation and exposure gradients). Then, the main ecosystem types along such gradients were identified, such as natural forests (Holm oak and Pyrenean oak forests [*Quercus ilex* and *Q. pyrenaica*], Scot pine forests [*Pinus sylvestris* var. *nevadensis*]), high-mountain shrublands (*Juniperus communis*), pine plantations, etc. These reference ecosystems, arranged over broad environmental gradients of elevation, exposure, and orientation, became the focus of ongoing sampling efforts. Then we installed some multiparametric meteo-stations to measure abiotic variables spatially associated with those reference ecosystems, together with ecological data. To increase the monitoring spatial resolution we installed wireless sensor networks around the meteo-station. These sensors collect abiotic data taking into account topographically controlled features (Scherrer and Körner 2010). In addition, several ecological protocols take data (demography, phenology, abundance of species, etc.) in each reference ecosystem (Aspizua et al. 2014). Finally, data from remote sensing (e.g. snow-cover-related indicators, vegetation indices, productivity) are gathered for the same reference ecosystem. Thus, we have developed the concept of Intensive Monitoring Stations (IMS) that denotes areas with a high density of ecological monitoring protocols coupled with abiotic measurements at several scales (Figs. 16.2 and 16.3).

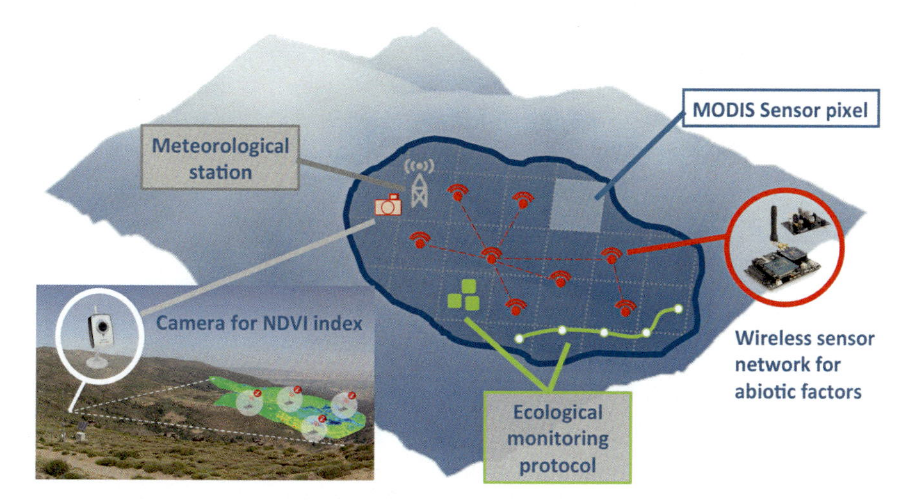

Fig. 16.2 Scheme of the Intensive Monitoring Stations. These are areas with high densities of ecological monitoring protocols located around a meteo-station. They also include a wireless sensor network to measure abiotic factors in selected microhabitats around the station (temperature, moisture, irradiance, etc.), as well as a phenocam. Modified from Zamora et al. (2016)

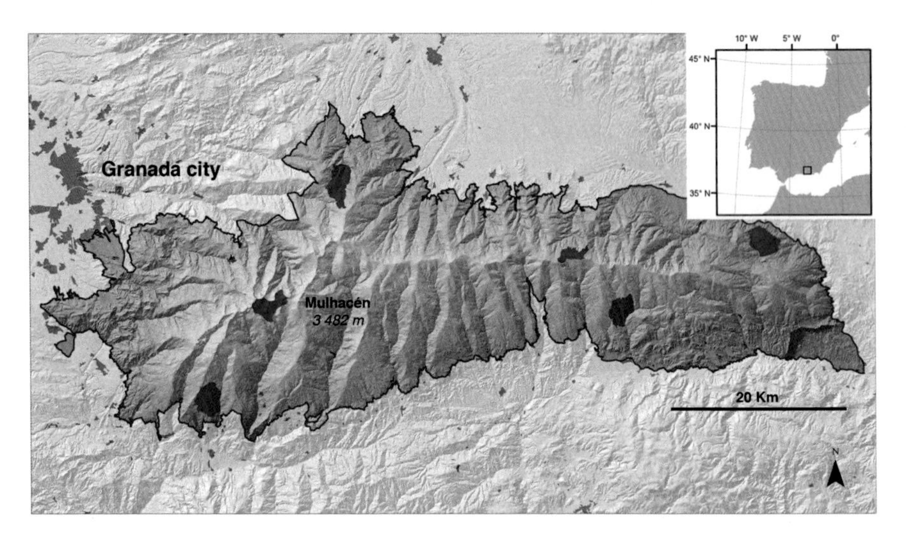

Fig. 16.3 Spatial distribution of the Intensive Monitoring Stations (*dark-blue polygons*). *Blue line* corresponds to boundaries of Sierra Nevada Protected Area. Reprinted with the permission of Zamora et al. (2016)

Practical examples of use for this concentration of layers of ecological information taken at the same site and time are the following cases:

1. To establish robust spatiotemporal associations between variables that are closely related and that may depend each other, for example: (i) consumer/resource relations, such as fleshy fruit abundance and the corresponding abundance of frugivorous birds in the same plot, or flower and pollinator abundance; (ii) measurements of temperature and phenological responses of plants and animals associated on a per-plot basis; (iii) the relation between availability (dry and wet deposition) of aerosols collected by the sensors of the multiparametric tower and the in situ evaluation of processes of eutrophication in nearby aquatic and terrestrial systems; (iv) ground-based collection of photographs acquired from the same fixed location with phenocams for monitor phenological changes in vegetation status and environmental changes over long periods at the same site (Brown et al. 2016).
2. Comparison of data from sensors of a meteo-station with sensors placed in different microhabitats: on steep mountains slopes, the interplay between exposure and vegetation is leading to mosaics of life conditions. For example, the temperature experienced by an organism in a particular microhabitat can be totally different from the conditions measured by the nearest conventional meteo-station, depending on their aerodynamic coupling to the atmosphere. The more strongly an ecosystem is decoupled from atmospheric conditions by topography and vegetation structure, the more thermal microhabitat variation is observed (Scherrer and Körner 2010). Our Intensive Monitoring Station allows to analyse in real time the average atmospheric meteo-value (low spatial resolution), and the

corresponding value for a number of microhabitats (high spatial resolution) found next to the meteo-station.

3. To validate reality-terrain: the great concentration of biophysical information gathered at the same plot of a reference ecosystem makes an IMS the ideal site for the necessary field validation required to interpret the spectral information acquired by remote sensing.

In short, we can consider an IMS as a monitoring hotspot, where heterogeneous data are collected at the same spatial location. The concentration of sampling points can be used to generate raster maps for IMS and thus layers of information can be spatially superimposed and analysed in GIS (Geographical Information System). Perhaps the most practical advantage of such an integrated approach is the cost savings associated with routine measurements to provide data for a wide variety of research purposes as well as for management. Our IMSs are a good example of multidisciplinary research to generate and apply scientific knowledge (Fahey et al. 2015).

16.6 Collection of Long-Term Data Series

The success of any global change programme depends on the quality of the data it collects, manages, and disseminates (Cook and Lineback 2008). Our monitoring programme aims to use information compiled in long-term research conducted in Sierra Nevada by naturalists and scientists belonging to different disciplines over the last several decades. We have collected information from several sources, such as reviews of historical literature published by naturalist scientists (up to 1960, see Titos-Martínez 2002; Ruano and Tinaut 2003; Garzón-Gutiérrez 2012); review of recent literature (after 1970); and information from the grey literature (project authorizations and research reports) related to the Sierra Nevada Protected Area. The aims of our recent literature review are (i) to gather all the scientific publications carried out in Sierra Nevada in the last 50 years; (ii) to identify the main research areas; and (iii) to geolocate the field sampling points. Our review in Web of Science (WoS) has resulted in a total of 1,038 publications related to the Sierra Nevada massif during the 1970–2015 period. The articles were distributed among 66 Web of Science categories. The five top-ranked research areas were "Ecology", "Plant Sciences", "Geosciences", "Environmental Sciences", and "Zoology". Using the information provided in each publication, we created a spatial database with the locations where the research was conducted. The preliminary results reveal that forests and aquatic systems (particularly alpine lakes) were the ecosystem types where more research was done.

We are also interested in compiling information on the current research being done in Sierra Nevada. Thus, we compiled information from research authorization documents for the Sierra Nevada Protected Area, from 2002 to 2015. Authorization documents contain basic information concerning the research activity, date, and location of the research area. After categorizing the raw information according to several attributes (e.g. disciplines, spatial and temporal range, fauna/flora studied, etc.), we integrated it into a normalized and spatial database. Furthermore, we compiled basic metadata on the research projects conducted in Sierra Nevada over the last 15 years from the research projects database of the Spanish Network of National Parks. This approach has enabled us to characterize the main research topics as well as to identify and locate areas with a high concentration of research

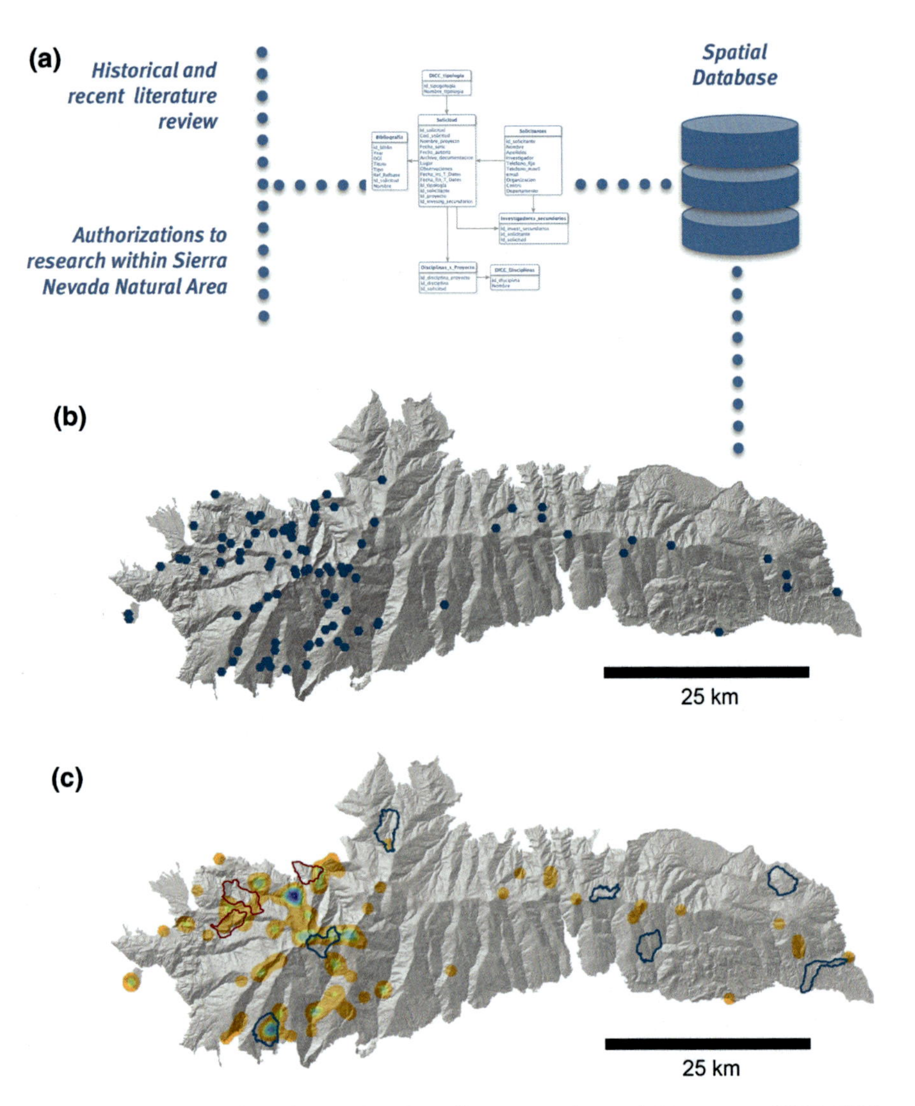

Fig. 16.4 Integration of the information from literature review and from recent (2002–2016) research activities conducted in Sierra Nevada into a spatial database (**a**). Spatial localization of the research activities in the past 15 years in Sierra Nevada (**b**). Map of density of research activities in Sierra Nevada (**c**) with an overlay of the location of the Intensive Monitoring Stations (*blue* and *red polygons*)

projects within Sierra Nevada (Fig. 16.4). Preliminary results have shown a spatial pattern with a high density of research activities at the western side of the mountain range (Fig. 16.4b). In particular, the areas with high densities of research projects correspond to high-mountain terrestrial and aquatic ecosystems, mountain rivers, and forests and shrublands. We also identified a lack of research activities in the central region of the range. The distribution of the research projects shows that the most arid ecosystems of Sierra Nevada (located on the eastern side) have scarcely been studied (Fig. 16.4c).

16.7 Data Management

An increasing demand for more detailed, high-quality data and information about natural resources and ecosystems functions requires trained personnel (resource specialist, data manager, researchers) working in collaboration to steward data and information assets (Cook and Lineback 2008). Information systems are fundamental for managing a large amount of data (Rüegg et al. 2014). Here, we distinguish two types of information: first, raw data collected directly in the field by scientific methods; second, structured information found in scientific papers and reports, slide presentations, videos, etc.

Raw data collected from the field is stored in relational spatial databases. To store processed information, we use relational databases that allow the fuzzy classification of the information into categories using the facilities of the web 2.0. For large amounts of information, it becomes critical to have a catalogue that highlights the main features of this information. This data about data is known as metadata or documentation. Metadata should answer prime questions about data, such as who created them, where they were collected, what method was used, and what organizational principle was used (Michener 2006). Metadata are very useful to share information between different information systems (Schildhauer et al. 2001).

Once an information system is established for the storage of documented data, the next step is to process and analyse raw data to provide information. From a functional and structural standpoint, an information system must be able to document and execute the algorithms that we use to process the data. Ideally, these tasks are run automatically using a scientific workflow software (Barseghian et al. 2010; McPhillips et al. 2009; Bonet et al. 2014). These tools allow the creation and execution of complex workflows by linking several computational steps (algorithms) in order to produce a given final product. Scientific workflow software can be used to run a spatial distribution model or even a simple query to a relational database. The results are also documented using the same standards described above.

We have developed an information system for the Sierra Nevada Global Change Observatory (Fig. 16.5). This system, called *Linaria* (https://linaria.obsnev.es—free access upon registration), acts as a repository storing raw data gathered by the monitoring programme as well as information generated through the processing of

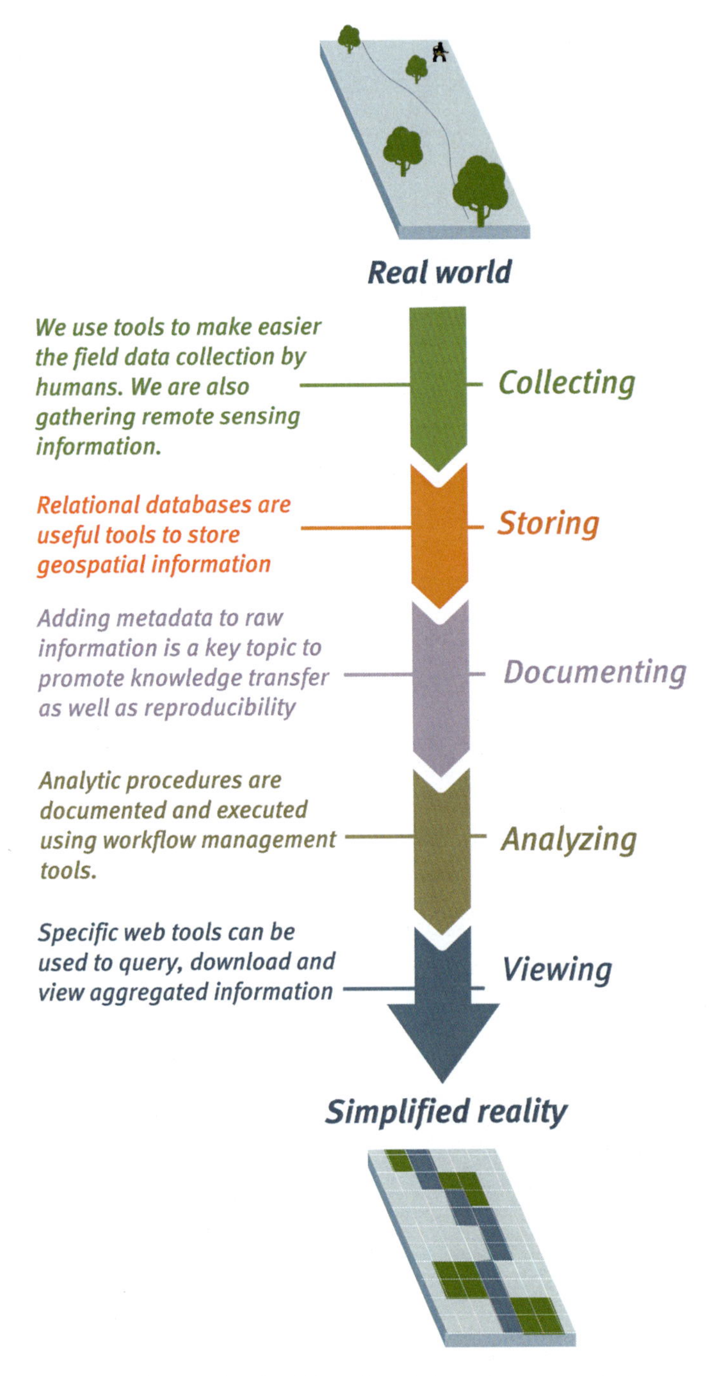

Fig. 16.5 Structure and main functionalities of the information system designed to store and analyse all the information gathered in the Sierra Nevada monitoring programme

such data. Data are stored in a standardized and documented way to facilitate its integration and analysis. Information generated is used in decision-aid tools and decision-making knowledge.

16.8 Distillation of Information and Outreach

Effective science outreach and communication is critical if we are to improve environmental decision-making (Lubchenco 1998; Bennett et al. 2005). Data analysis, elaboration of scientific publications and distillation of the information is the crucial final stage of the information-management processes. A pillar of our monitoring programme is the collaboration with research experts trained for the work of analysing and interpreting the information gathered in the scientific context as well as the development of simulation models. As a result of the analysis and synthesis came scientific publications in peer-reviewed journals, followed by reports that distilled the results of the peer-reviewed publications for decision-makers (Driscoll et al. 2011; Zamora et al. 2016).

16.9 Detecting Changing Signals

In this section, we present a summary of the main results found at the site LTER of Sierra Nevada (Zamora et al. 2016). Specifically, we present a temporal evaluation of the two main drivers of global change: climate and land-use changes as well as the biotic and socioeconomic responses to these changes.

Apart from land-use and climate changes, other drivers of global change (Sala et al. 2000) appear to have a comparatively minor impact on Sierra Nevada. For instance, this massif is relatively far from centres of industrial activity and therefore anthropic pollution detected is low and comes mainly from the city of Granada (c. 235,000 inhabitants). On the contrary, Sierra Nevada is an invaluable watchtower to detect atmospheric depositions of remote origin. For example, for its geographic position in the extreme south of continental Europe, this massif receives aerosol depositions from the Sahara Desert (Morales-Baquero et al. 2006). Also, in Sierra Nevada National Park, no serious problems of invasive species have been detected yet, but other ecosystems situated at lower elevations are affected, especially aquatic ecosystems.

16.9.1 Land-Use Change

Land-use drive changes in the vegetation cover, affecting ultimately the biogeochemical and waters cycles, the stability of mountain slopes, and household economies. Centuries of human activity have changed the structure of the landscape in Sierra Nevada. Multiple pieces of evidence indicate that from about 3,000 years ago to the present, human activity intensified in Sierra Nevada (Anderson et al. 2011; Jiménez-Moreno and Anderson 2012; Jiménez-Moreno 2016). After 3000 BP the frequency of fires increased, and there are signs of grazing as well as mining; more recently, olive cultivation on a large scale at the lowest elevations together with pine reforestation have become characteristic.

The mountain landscapes of Sierra Nevada have undergone massive land-use changes from the past century. The sharp decline in agro-pastoral pressure due to rural depopulation raises major environmental and societal issues. For instance, there is evidence from paintings, forestry maps, and the land registry of the last 100 years (Jiménez-Olivencia et al. 2016; Moreno-Llorca et al. 2016) of a clear land-use shift over the last 50–60 years, due to both the abandonment of agricultural and livestock activities, as well as a surge in active reforestation (pine plantations; Fig. 16.6). Land-use change in the last 50 years has affected over half of the 170,000 ha of Sierra Nevada Protected Area.

We hypothesized that the land-use changes in Sierra Nevada should facilitate the native forest regeneration, and a process of colonization of marginal habitat (abandoned cropland, pine plantations) will occur. The distribution of the oak forests in Sierra Nevada and their degree of conservation are determined both by biophysical variables that are expressed at present (slope, climate, water availability, etc.), and by the dynamics of land-use changes in recent decades (intensive use and subsequent abandonment). We have quantified the ecological functions "seed production and dispersal" and "sapling survival and growth" both at the upper elevational limit as well in marginal habitats (i.e. abandoned croplands and pine plantations). The preliminary results showed that oak forests regenerated more readily on the north-eastern than on the southern slopes, whereas the colonization of abandoned crops seems to occur more intensely in the south than in the north-east (possibly due to the greater incidence of herbivory in the latter slope) (Pérez-Luque et al. 2015; Bonet et al. 2016).

Moreover, the ecosystem transformation associated with land-use changes drastically diminishes the biological legacies, including remnant native woody plants and their propagules. Consequently, the recuperation of community diversity within human-created habitats (such as abandoned cropland, pine plantations) depends heavily on both internal, in situ biological legacies, and external, well-conserved nearby areas, as a source of propagules. The effect of these historical and current ecological factors on current *Quercus* species regeneration under pine plantations has been analysed in Sierra Nevada massif (southern Spain). Our results indicate that native oak forest regeneration ability under pine plantations depends largely on land-use legacies, although nearby, well-conserved areas also provide propagules for colonization from outside the plantation (Gómez-Aparicio et al. 2009; Navarro-González et al. 2013; Pérez-Luque et al. 2016a). The higher the land-use intensity in the past, the weaker biological legacies are and, therefore, the weaker the current native forest regeneration ability, and vice versa.

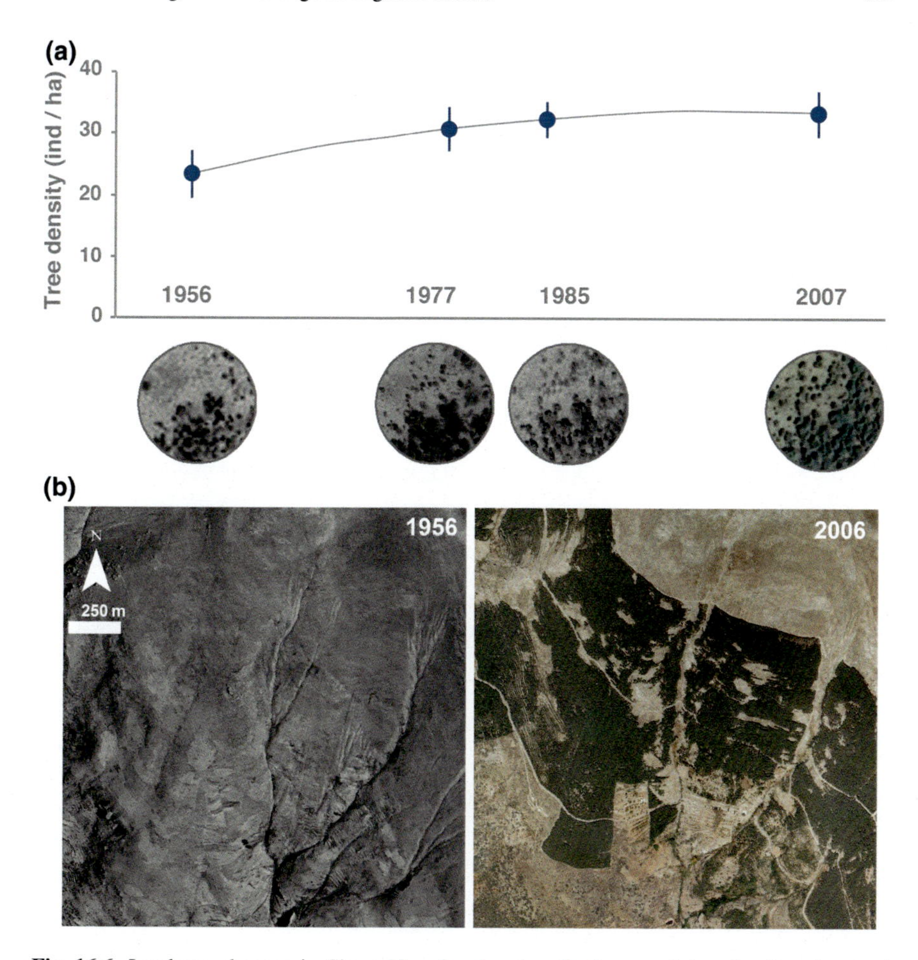

Fig. 16.6 Land-use changes in Sierra Nevada, showing the increased tree density of natural Pyrenean oak forests (*Quercus pyrenaica*) in Sierra Nevada in the past 50 years (**a**). Comparative orthophotography showing the changes at the landscape level due to reforestation activities (pine plantations) (**b**)

Overall, our results suggest that *Quercus* forest regeneration responded more to changes in land use than to the rise in temperature (Pérez-Luque et al. 2015). Separating the respective effects of land use, climatic variability, and their interaction constitutes a major challenge but achievable only by experimental long-term approaches, and by repeated observations of more pristine areas (such as high mountains) from human impacts in order to establish a climate signal on the ecological processes. An understanding of the past and the present changes and subsequent impacts on mountain ecosystems of those major drivers and their interaction is a prerequisite for any management or adaptation strategy.

16.9.2 Climate Change

Rising temperatures and growing aridity become evident in the palaeolimnological indicators of the lakes of the Sierra Nevada (Anderson et al. 2011; Jiménez-Moreno and Anderson 2012). In addition, an increase in the chlorophyll-a content of these lakes over the last 120 years correlates well with temperature data for the corresponding period (Pérez-Martínez 2016). Overall, there has been progressive aridification over the last 7,000 years, with signs of a gradual increase in human activity for the last 3,000 years (Jiménez-Moreno 2016).

An analysis of the temporal pattern of the main climatic variables in Sierra Nevada for the last 50 years has revealed changes in annual temperatures (maximum and minimum) and rainfall (Fig. 16.7). Maximum and minimum temperatures have risen, especially after 1980s (Pérez-Luque et al. 2016b). This agrees with the general pattern of rising temperatures in the second half of the twentieth century

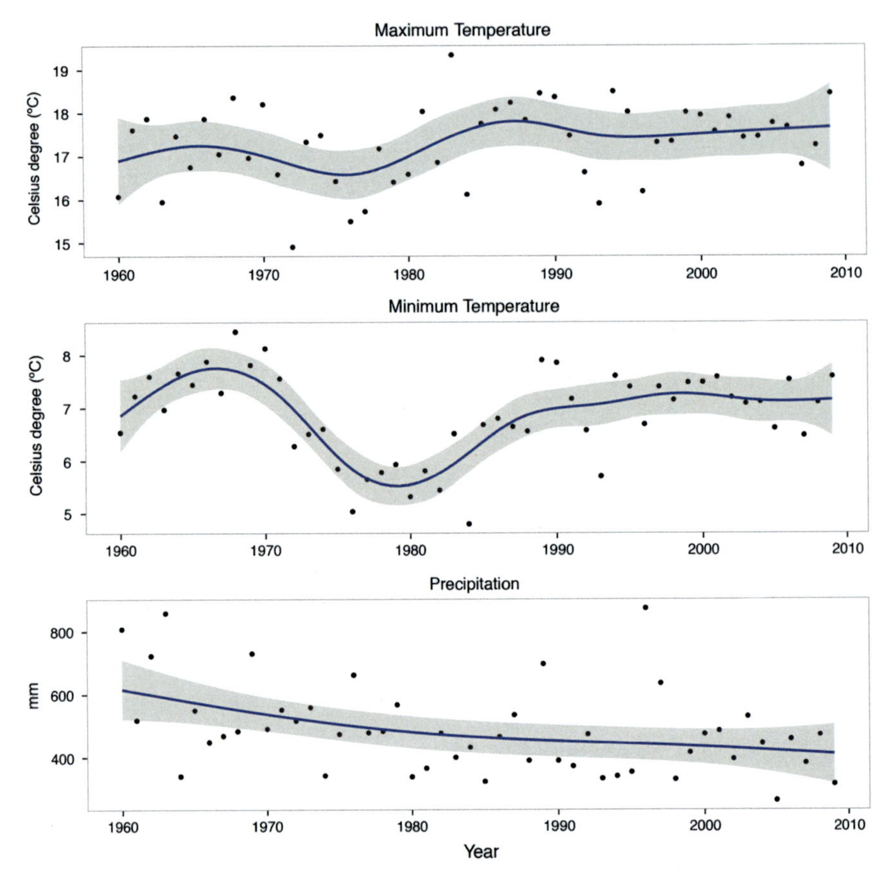

Fig. 16.7 Time course of the annual maximum (*upper panel*) and minimum (*middle panel*) temperatures and annual rainfall (*lower panel*). Average values for all pixels (100 m pixel size) of Sierra Nevada are shown

observed at different scales on the Iberian Peninsula (Galán et al. 2001; De Castro et al. 2005). The precipitation shows a general decreasing pattern, being more pronounced in the western part than in the east. This pattern is consistent with the overall declining trend in the south of the Iberian Peninsula (Rodrigo et al. 1999; Ruiz-Sinoga et al. 2010).

All these changes in climate have implications for the dynamics of snow cover. Satellite information from MODIS (*Moderate Resolution Imaging Spectroradiometer*) for the period 2000–2014 have indicated a decrease in snow-cover duration over 79% of Sierra Nevada. There was a trend towards a delay in the date of the first snowfall and a trend towards earlier melting dates. The change has been more pronounced in magnitude at the summits than in the lowland areas (Bonet et al. 2016). For instance, above 3000 m a.s.l., the duration of snow cover decreased by an average of three days in the last 14 years. A significant recent decline in snow duration has also been reported from other mountain regions of Europe (Scherrer et al. 2004; Moreno-Rodríguez 2005; Marty 2008; Nikolova et al. 2013).

16.9.3 Biotic Responses to Land-Use and Climate Change

Mountain lakes as integrated ecological sensors

The combination of old studies and resampling of the same localities can help to integrate short-term data into long-term datasets (Müller et al. 2010). These long-term datasets can be used to evaluate spatiotemporal changes and trends in biological communities and their relation to drivers of global change, such as land use or climate in mountain regions. For example, an exhaustive review of the research on alpine lakes of Sierra Nevada from 1975 to the present has identified the role of these ecosystems as sentinels of change (Villar-Argaiz and Bullejos 2016; Medina-Sánchez et al. 2016). Unlike other European mountain lakes, the geographic location and geological history of Sierra Nevada cause high-mountain lakes to be exposed simultaneously to several environmental stressors: climatic anomalies (temperature and precipitation), UV radiation, aerosol deposition, and allochthonous nutrient input. High transparency, low nutrient content, and narrow temperature ranges found in high-mountain lakes qualifies them as sentinels of global change (Medina-Sánchez et al. 2016; Villar-Argaiz and Bullejos 2016). However, especially the great simplicity of their biological communities helps us to assess their impact on ecosystem functioning, as well as the evolution of organisms that inhabit them. Long-term monitoring of the population dynamics of pelagic plankton in the Laguna de la Caldera, the largest alpine lake of Sierra Nevada, indicates that phytoplankton has increased in parallel with the increase in the intensity and frequency of atmospheric aerosols (Villar-Argaiz and Bullejos 2016). These results suggest that allochthonous nutrients associated with Saharan intrusions have a fertilizing effect that stimulates the growth of algal biomass. These patterns are consistent with previous studies indicating that the Saharan depositions are important sources of nutrients, especially Phosphorus (Morales-Baquero et al. 2006), encouraging primary productivity in the oligotrophic waters of mountain lakes of Sierra Nevada (Villar-Argaiz et al. 2001).

Elevation range extension

The consequences of climate and land-use changes are typically most evident in mountain habitats, where expansions or contractions in species' distribution ranges along elevation gradients may occur via the migration of species to new areas (Jump and Peñuelas 2005; Lenoir et al. 2008). Species may expand upwards into new areas that become favourable, and retract from those that turn unfavourable. For instance, a comparative study on macroinvertebrate communities in the rivers of Sierra Nevada in the 1980s and today reveals substantial changes (Sáinz-Bariáin et al. 2015). Some species in the 1980s that were associated with the middle elevation stretch of the rivers are today found in the higher reaches. As a result, the diversity of species in the upper reaches of rivers is far greater than 30 years ago. In places where the water temperature has increased more, there has also been a major increase in the diversity of caddisflies. In the case of Plecoptera, it has been found that the lower limits of the distribution of some species have contracted, while the distribution at the upper limits has remain unchanged.

In terrestrial ecosystems a similar pattern has been found. Faunistic inventories conducted 20–40 years ago have been repeated in recent years. Significant changes in the spatial distribution and/or abundance of several groups of species were found. For insects, an elevational migration has been confirmed for dung beetles (Menéndez et al. 2014). The results show that within 25 years, 89% of the species increased their average elevation and upper distribution limits while more than 84% increased their lower distribution limits. The average elevational ascent of dung beetles in Sierra Nevada over 25 years was 400 m, coinciding with a rise of the same magnitude in the butterfly Apollo (Barea-Azcón 2016). Meanwhile, ants (Formicidae, Hymenoptera) have also risen in elevation, at least in the case of two species (*Proformica longiseta* and *Formica fusca/lemani*), which have expanded their upper distribution limits by about 200 m on the southern slopes of Sierra Nevada (González-Megías et al. 2016).

Among vertebrates, equivalent responses have also been found. For instance, passerines have shown marked temporal dynamics over the last 30 years which have been strongly influenced by global change (Zamora and Barea-Azcón 2015). Some generalist mountain species are now more abundant than before in high-mountain areas, while the most typical alpine species have become progressively scarcer. Overall, there has been a decline in montane species, which have been replaced by more thermophilic Mediterranean lowland species. We conclude that, within the global change context, protected mountain areas play a vital role in maintaining biodiversity, as populations can adapt to shifting conditions, moving along elevation gradients, according to their ecological requirements.

Ongoing changes in climate are altering ecological conditions for many plant species, and are most evident at the edge of the geographical distribution of a species, where range expansions or contractions may occur. For example, the demographic structure of both *Pinus sylvestris* and *Juniperus communis* populations differs across the elevational distribution in Sierra Nevada, where low-elevation populations of *P. sylvestris* and *J. communis* are susceptible to decline through reduced growth, seed production, and population regeneration (García et al. 2000; Matías and Jump 2015). Both species presented a significantly reduced proportion of young individuals at the lowermost populations, and a clear dominance of older age classes (Matías and Jump 2015; Rabasa et al. 2013; Benavides et al. 2013; 2015). This contrasts sharply with the pattern found at the treeline, where a higher proportion of saplings appeared and an even distribution was found across age classes. These results indicate an ongoing elevational displacement for both species at the southernmost edge of the geographical distribution area (Fig. 16.8a). Still, evidence for lowland contractions in these woody species is scarce due to generally great individual longevity and relatively slow decline until survival thresholds are exceeded, especially for slow-growing species such as *J. communis* (García and Zamora 2003).

However, factors other than climate affect species growth and reproduction at the same time such as land-use change and herbivory pressure (Zamora et al. 2001; Zamora and Matías 2014; Herrero et al. 2012, 2016), and the result of their interactions are strongly heterogeneous across areas and species. Therefore, research across altitudinal gradients taking into account other factors in addition to climate, such as biotic interactions (mostly herbivory), land use, or local adaptations, are strongly recommended for an accurate forecasting of climate-change consequences on long-lived woody plant species.

Summit vegetation also showed signals of change, with an expansion of some species, and greater vegetation cover. Many alpine plants have become rarer in the period 2001–2008 on the high mountains of the European continent, whereas plants from low elevations become more common (Gottfried et al. 2012). In Sierra Nevada, over the last 11 years, 13 alpine species have disappeared from monitoring plots and, at the same time, five new taxa have appeared (Gottfried et al. 2012; Sánchez-Rojas and Molero Mesa 2016).

Overall, these results for plant and animal species show that a high mountain, such as Sierra Nevada, plays a vital role in maintaining biodiversity in the context of global change. Species populations can adapt to climatic changes elevationally by moving according to their ecological necessities, although plants and animals do not necessarily need to climb several hundred metres in elevation to find suitable new habitats in case of warming but may find conditions matching their "thermal niche" over very short distances, taking advantage of the mosaics of microhabitats characterizing mountains slopes (Scherrer and Koerner 2011). These results indicate that montane ecological communities that are considered rather stable are, in fact, undergoing strong spatial and temporal dynamics because of climate warming (and more so after land-use change). As a result, there is a convergence in biotic responses to these two major drivers of global change (land-use and climatic changes), threatening high-mountain species while at the same time opening opportunities for species from lower elevations. All of this seems to be restricting the distribution range or leading to the extinction of rare, endemic, and/or specialist species, parallel to expansions in the distribution range of generalist species. This may be increasing

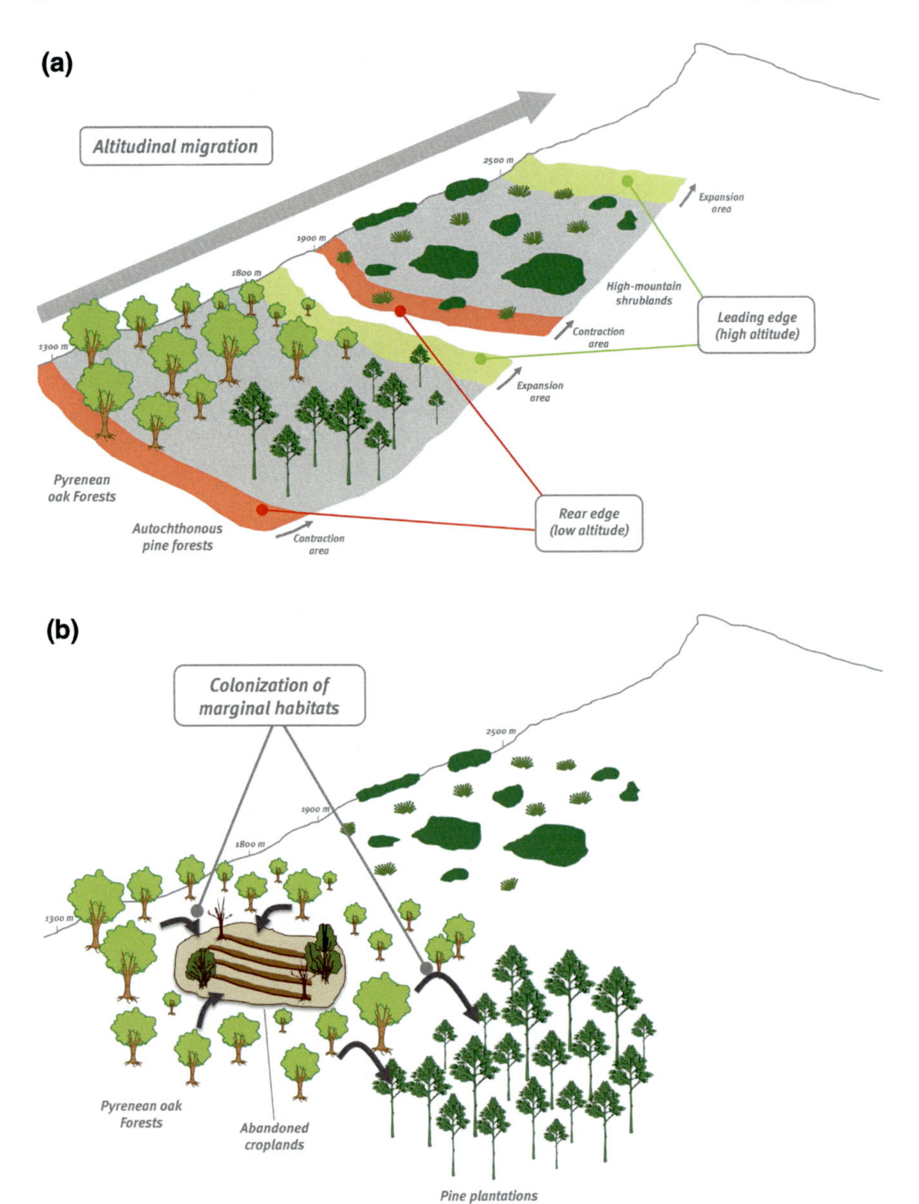

Fig. 16.8 An elevational displacement has been observed for major woody plant species (Pyrenean oak, autochthonous pine and high-mountain shrublands) (**a**). We also found an outstanding woodland expansion towards marginal areas (abandoned croplands and pine plantations) (**b**). Modified from Pérez-Luque et al. (2015)

the similarity of taxonomic composition among communities along elevation gradients. This process can be envisaged as the first stage of a taxonomic homogenization processes (McKinney and Lockwood 1999; Olden and Rooney 2006) or, in other words, a temporal decline in beta diversity. To determine to what degree the mosaic of ecological conditions favoured by the topography of mountains delays (or even halts), this widespread process of biotic homogenization (i.e. favouring beta diversity) is a highly relevant topic for scientific investigation of the future.

The maintenance of the monitoring programme and associated research is crucial to test whether these changes in the composition, abundance and altitudinal distribution, will have consequences in the organization of food webs and in ecosystem functioning, as has already been demonstrated for high-mountain lakes (Medina-Sanchez et al. 2004).

16.9.4 Socioeconomic Responses to Global Change

Changes in land use and climate have direct consequences for human populations and socioeconomic activities. The organization of the mountain economy in Sierra Nevada has drastically changed during the last 60 years. The rural exodus and the decline in traditional land-use practices, mainly agriculture and livestock husbandry, together with re-afforestation, have contributed to a general expansion of forested areas. Today, a socioeconomic analysis must consider different perspectives, ecosystem services, and their multiple uses.

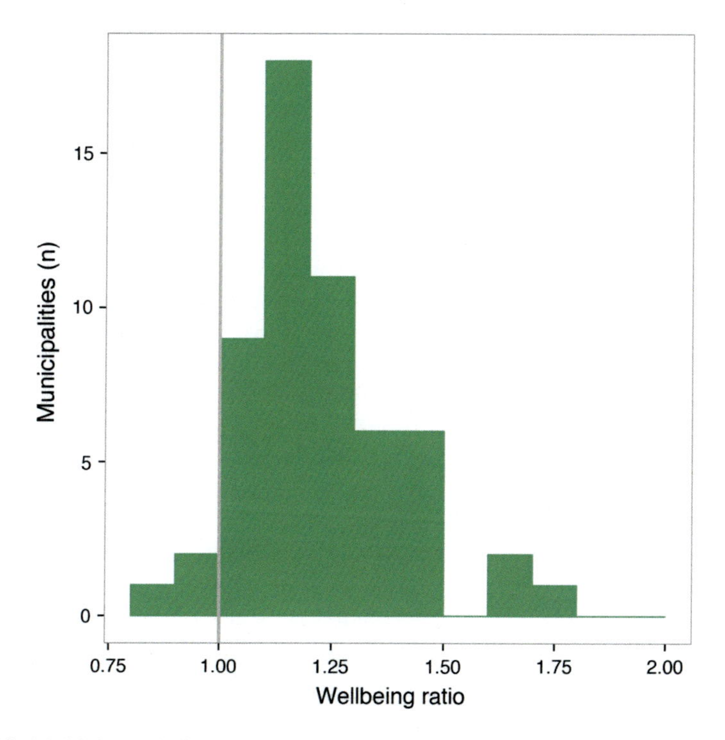

Fig. 16.9 Well-being ratio for the municipalities of Sierra Nevada. Municipalities to the *right* of the *vertical* line represent those that have improved their welfare in the past 20 years

Mountains are sources of multiple ecosystem services, benefiting the welfare of human populations. We performed a temporal analysis of the well-being in the municipalities of the Sierra Nevada mountain range, comparing the welfare at two points in time: before the declaration of the protected area (20 years ago) and at the present. For this, we used data from 22 socioeconomic indicators belonging to eight dimensions of well-being according to Millennium Ecosystem Assessment. For each municipality we computed a synthetic indicator of well-being, using an approach based on compound weighted indices (Bonet et al. 2015). The index shows a significant increase in well-being for 94% of the municipalities analysed (Fig. 16.9). These results agree with those found for an analysis performed at the regional scale, where the increment in well-being was higher for municipalities in protected areas than outside of them (Bonet-García et al. 2015). This indicates that protected areas, in addition to acting as instruments for the conservation of biodiversity and ecological processes, generate ecosystem services valuables for human populations.

16.10 From Monitoring to Action in a Changing World

16.10.1 Cross-Scale Interaction: Think Globally, Act Locally

Long-term monitoring, data analysis, and policy-making can be put into practice at different spatial scales. Cross-scale interactions arise when the driver and response variables in cause–effect relationships operate at different spatial and temporal scales (Soranno et al. 2014). Understanding these spatial and temporal scales requires the creation of a "network of networks" (Peters et al. 2008) for observation, experimentation, modelling, and policy-making. Thus, we have devoted effort to establishing links between our monitoring programme and other similar initiatives. First, Sierra Nevada is part of the Spanish LTER (Long-Term Ecological Research) network (http://www.lter-spain.net). This network provides access to analytical tools, harmonized protocols, ecoinformatics procedures to manage the information and it is also a network of networks working on ecological monitoring. Moreover, our site is connected to regional (Andalusian Network of Global Change Observatories) as well as national (Monitoring Global Change in National Parks) initiatives that share this long-term philosophy. Finally, our site has participated in several EU projects such as eLTER (H2020 project aiming to vertebrate the European LTER network http://www.lter-europe.net/projects/eLTER), ECOPOTENTIAL (H2020 project aiming to quantify ecosystem services using remote sensing http://www.ecopotential-project.eu/), and ADAPTAMED (Life project) aiming to implement real managerial activities to improve resilience in Sierra Nevada ecosystems.

16.10.2 The Application of Monitoring Information to Adaptive Management

Long-term observations are critical for making informed decisions in many environmental management contexts. For example, long-term studies can be important in quantifying the effectiveness of conservation management activities such as ecosystem restoration, where prolonged periods might be required for systems to recover following major human disturbance. As opposed to the traditional management characterized by a lack of monitoring, we have opted for "adaptive management" based on follow-up of actions in order to evaluate the effects of a treatment submitted to testing. We have put into practice this philosophy in Sierra Nevada, proposing key questions from the outset, defining the goals to be pursued with the actions undertaken, and specifying the methodological and analytical details necessary to address these efforts.

In this sense, several projects are already under way, such as actions for the naturalization of pine reforestation, monitoring of the effectiveness of post-fire restoration actions, and analysis of the population dynamics of pine processionary moth *Thaumetopoea pityocampa* (see Chap. 8 in Zamora et al. 2016). We are also collaborating in a LIFE project called ADAPTAMED (protection of key ecosystem services threatened by climate change through adaptive management of Mediterranean socioecological systems), led by the Environmental Ministry of the Regional Government of Andalusia. The project focuses on implementing, monitoring, evaluating, and disseminating adaptive management measures, with an ecosystem approach. The project objectives aim to reduce the negative impact of climatic change in the area of influence of Natural Protected Areas. As a result, an increase in the resilience of the socio-ecosystems concerned is expected, in such a way that their future provision of services will also be improved, in comparison to the scenario of no intervention.

16.10.3 Last, But not the Least: Maintaining Monitoring Over the Long Term

Global change monitoring programmes arise from the visionary effort of some people, but must survive those founding individuals with the challenge to survive the year-to-year political vagaries through consolidation of monitoring programmes into the existing research institutions (Universities, National Research Agencies) and environmental management institutions (public authorities responsible for environmental management). Research institutions have to vertebrate short-term research ordinarily engaged by individual researchers into long-term research and monitoring programme to ensure the continuity for collecting, documenting, and analysing key biophysical and socioeconomic processes. For understanding, modelling, and managing nature, nothing has more value than a long series of field data. To amass this long series of data and maintain it live and operative over time, long-term needs financial and personal resources that are not always available for planning failures. Meanwhile, the use of monitoring information will allow managers to confront and mitigate global change threats to ecosystem function and services by designing the appropriate management measures. If the project is a conscientious, forthright initiative, the continuity of the monitoring programme must necessarily be guaranteed by the above-mentioned institutions, and must be coordinated with the other nodes forming a network of observatories in order to make the diagnosis and forecast of the health of the planet.

Acknowledgements Authors want to thank L. Nagy, an anonymous reviewer, and to the Editors for helpful suggestions. The Sierra Nevada Global-Change Observatory is a joint effort funded by the Andalusian department of environment and land planning (Junta de Andalucía), the Spanish National Park Service (Spanish Government), the European Regional Development Fund, and the Biodiversity Foundation. Our scientific work carried out in Sierra Nevada has been funded mainly by the Andalusian Regional Government (Junta de Andalucía), Spanish ministry of economy and competitiveness, as well as by several EU projects, such as eLTER (H2020. Grant agreement Nº654359), ECOPOTENTIAL (H2020. Grant agreement Nº641762), and ADAPTAMED (Life. 2014. CCA/ES/000612).

References

Anderson RS, Jiménez-Moreno G, Carrión JS, Pérez-Martínez C (2011) Postglacial history of alpine vegetation, fire, and climate from Laguna de Río Seco, Sierra Nevada, Southern Spain. Quaternary Sci Rev 30:1615–1629

Aspizua R, Barea-Azcón JM, Bonet FJ, Pérez-Luque AJ, Zamora R (eds) (2014) Sierra Nevada global-change observatory. Monitoring methodologies. Consejería de Medio Ambiente, Junta de Andalucía

Barea-Azcón JM (2016) The phenology of butterflies in Sierra Nevada. In: Zamora R, Pérez-Luque AJ, Bonet FJ, Barea-Azcón JM, Aspizua R (eds) Change Impacts in Sierra Nevada: challenges for conservation. Consejería de Medio Ambiente y Ordenación del Territorio, Junta de Andalucía, pp 133–137

Barnosky AD, Hadly EA, Bascompte J, Berlow EL, Brown JH, Fortelius M et al (2012) Approaching a state shift in earth's biosphere. Nature 486:52–58

Barseghian D, Altintas I, Jones MB, Crawl D, Potter N, Gallagher J et al (2010) Workflows and extensions to the Kepler scientific workflow system to support environmental sensor data access and analysis. Ecol Inform 5:42–50

Battin TJ, Luyssaert S, Kaplan LA, Aufdenkampe AK, Richter A, Tranvik LJ (2009) The boundless carbon cycle. Nat Geosci 2:598–600

Becker A, Bugmann H (eds) (2001) Global change and mountain regions: the mountain research initiative. IGBP Report 49

Benavides R, Rabasa SG, Granda E, Escudero A, Hódar JA, Martinez-Vilalta J et al (2013) Direct and indirect effects of climate on demography and early growth of pinus sylvestris at the rear edge: changing roles of biotic and abiotic factors. PLoS ONE 8:e59824

Benavides R, Escudero A, Coll L, Ferrandis P, Gouriveau F, Hódar JA et al (2015) Survival vs. growth trade-off in early recruitment challenges global warming impacts on Mediterranean mountain trees. Perspect Plant Ecol Evol Syst 17:369–378

Beniston M (2003) Climatic change in mountain regions: a review of possible impacts. Clim Change 59:5–31

Bennett EM, Peterson GD, Levitt EA (2005) Looking to the future of ecosystem services. Ecosystems 8:125–132

Berdanier AB, Clark JS (2016) Multiyear drought-induced morbidity preceding tree death in Southeastern U.S. forests. Ecol Appl 26:17–23

Blanca G, Cueto M, Martínez-Lirola MJ, Molero-Mesa J (1998) Threatened vascular flora of Sierra Nevada (Southern Spain). Biol Conserv 85:269–285

Bonet FJ, Pérez-Luque AJ, Moreno-Llorca R, Zamora R (2010) Sierra Nevada global change observatory. Structure and basic data. Environment department (Andalusian Regional Government)—University of Granada

Bonet FJ, Perez-Perez R, Benito BM, de Albuquerque FS, Zamora R (2014) Documenting, storing, and executing models in ecology: a conceptual framework and real implementation in a global change monitoring program. Environ Modell Softw 52:192–199

Bonet FJ, Pérez-Luque AJ, Perez-Perez R (2016) Trend analysis (2000–2014) of the snow cover by satellite (MODIS sensor). In: Zamora R, Pérez-Luque AJ, Bonet FJ, Barea-Azcón JM, Aspizua R (eds) Change impacts in Sierra Nevada: challenges for conservation. Consejería de Medio Ambiente y Ordenación del Territorio, Junta de Andalucía, pp 43–46

Bonet-García FJ, Pérez-Luque AJ, Moreno-Llorca RA, Perez-Perez R, Puerta-Piñero C, Zamora R (2015) Protected areas as elicitors of human well-being in a developed region: a new synthetic (socioeconomic) approach. Biol Conserv 187:221–229

Brown TB, Hultine KR, Steltzer H, Denny EG, Denslow MW, Granados J et al (2016) Using phenocams to monitor our changing earth: toward a global phenocam network. Front Ecol Environ 14:84–93

Burgman MA, Ferson S, Akçakaya HR (1993) Risk assessment in conservation biology. Chapman and Hall, New York and London

Canham CD, Cole JJ, Lauenroth WH (2003) Models in ecosystem science. Princeton University Press, Princeton

Cañadas EM, Fenu G, Peñas J, Lorite J, Mattana E, Bacchetta G (2014) Hotspots within hotspots: Endemic plant richness, environmental drivers, and implications for conservation. Biol Conserv 170:282–291

Carpenter SR (1998) The need for large-scale experiments to assess and predict the response of ecosystems to perturbation. In: Successes, limitations, and frontiers in ecosystem science. Springer, New York, pp 287–312

Cook RR, Lineback P (2008) Sierra Nevada network data management plan. National Park Service, Fort Collins, Colorado

Daily GC (ed) (1997) Nature's services: societal dependence on natural ecosystems. Island Press, Washington

de Castro M, Martí-Vide J, Alonso S (2005) El clima de España: pasado, presente y escenarios de clima para el siglo XXI. In: Moreno-Rodríguez J (ed) Evaluación preliminar de los impactos en España por efecto del cambio climático, pp 1–64

Driscoll CT, Lambert KF, Weathers KC (2011) Integrating science and policy: a case study of the hubbard brook research foundation science links program. Bioscience 61(10):791–801

Fahey TJ, Templer PH, Anderson BT, Battles JJ, Campbell JL, Driscoll CT Jr et al (2015) The promise and peril of intensive-site-based ecological research: insights from the Hubbard Brook ecosystem study. Ecology 96:885–901

Galán A, Cañada R, Fernández F, Cervera B (2001) Annual temperature evolution in the southern plateau of Spain from the construction of regional climatic time series. In: Brunet-Inda M, López-Bonillo D (eds) Detecting and modelling regional climate change, pp 99–132

García D, Zamora R (2003) Persistence, multiple demographic strategies and conservation in long-lived Mediterranean plants. J Veg Sci 14:921–926

García D, Zamora R, Gómez JM, Jordano P, Hódar JA (2000) Geographical variation in seed production, predation and abortion in Juniperus communis throughout its range in Europe. J Ecol 88:435–446
Garzón-Gutiérrez J (2012) Revisión histórica de la ornitología en Sierra Nevada. In: Garzón-Gutiérrez J, Henares-Civantos I (eds) Las aves de Sierra Nevada, pp 41–49
Gómez-Aparicio L, Zavala MA, Bonet FJ, Zamora R (2009) Are pine plantations valid tools for restoring Mediterranean forests? An assessment along abiotic and biotic gradients. Ecol Appl 19:2124–2141
González-Megías A, Menéndez R, Tinaut A (2016) Shifts in the elevational ranges of insects in Sierra Nevada: evidence of climate change. In: Zamora R, Pérez-Luque AJ, Bonet FJ, Barea-Azcón JM, Aspizua R (eds) Change impacts in Sierra Nevada: challenges for conservation. Consejería de Medio Ambiente y Ordenación del Territorio, Junta de Andalucía, pp 120–122
Gottfried M, Pauli H, Futschik A, Akhalkatsi M, Barančok P, Benito Alonso JL et al (2012) Continent-wide response of mountain vegetation to climate change. Nat Clim Change 2:111–115
Grabherr G, Björnsen Gurung A, Dedieu J-P, Haeberli W, Hohenwallner D, Lotter AF et al (2005) Long-term environmental observations in mountain biosphere reserves: recommendations from the EU GLOCHAMORE project. Mt Res Dev 25:376–382
Hansen AJ, Piekielek N, Davis C, Haas J, Theobald DM, Gross JE et al (2014) Exposure of U.S. National Parks to land use and climate change 1900–2100. Ecol Appl 24:484–502
Herrero A, Zamora R, Castro J, Hódar JA (2012) Limits of pine forest distribution at the treeline: herbivory matters. Plant Ecol 213:459–469
Herrero A, Almaraz P, Zamora R, Castro J, Hódar JA (2016) From the individual to the landscape and back: time-varying effects of climate and herbivory on tree sapling growth at distribution limits. J Ecol 104:430–442
Huber UM, Bugmann H, Reasoner MA (eds) (2005) Global change and mountain regions: an overview of current knowledge. Springer, New York
Jiménez-Moreno G (2016) Reconstruction of the vegetation from palynological analysis. In: Zamora R, Pérez-Luque AJ, Bonet FJ, Barea-Azcón JM, Aspizua R (eds) Change Impacts in Sierra Nevada: challenges for conservation. Consejería de Medio Ambiente y Ordenación del Territorio, Junta de Andalucía, pp 50–52
Jiménez-Moreno G, Anderson RS (2012) Holocene vegetation and climate change recorded in alpine bog sediments from the Borreguiles de la Virgen, Sierra Nevada, Southern Spain. Quatern Res 77:44–53
Jiménez-Olivencia Y, Porcel-Rodríguez L, Caballero-Calvo A, Bonet FJ (2016) Land-use changes in Sierra Nevada over the last 50 years. In: Zamora R, Pérez-Luque AJ, Bonet FJ, Barea-Azcón JM, Aspizua R (eds) Change impacts in Sierra Nevada: challenges for conservation. Consejería de Medio Ambiente y Ordenación del Territorio, Junta de Andalucía, pp 56–58
Jump AS, Peñuelas J (2005) Running to stand still: adaptation and the response of plants to rapid climate change. Ecol Lett 8:1010–1020
Keeling CD, Whorf TP, Wahlen M, van der Plichtt J (1995) Interannual extremes in the rate of rise of atmospheric carbon dioxide since 1980. Nature 375:666–670
Keeling CD, Chin JFS, Whorf TP (1996) Increased activity of northern vegetation inferred from atmospheric CO_2 measurements. Nature 382:146–149
Lenoir J, Gégout JC, Marquet PA, de Ruffray P, Brisse H (2008) A significant upward shift in plant species optimum elevation during the 20th century. Science 320:1768–1771
Liang J, Crowther TW, Picard N, Wiser S, Zhou M, Alberti G, et al. (2016) Positive biodiversity-productivity relationship predominant in global forests. Science 354:aaf8957
Lindenmayer DB, Likens GE (2009) Adaptive monitoring: a new paradigm for long-term research and monitoring. Trends Ecol Evol 24:482–486
Lindenmayer DB, Likens GE, Andersen A, Bowman D, Bull CM, Burns E et al (2012) Value of long-term ecological studies. Austral Ecol 37:745–757

Lovett GM, Burns DA, Driscoll CT, Jenkins JC, Mitchell MJ, Rustad L et al (2007) Who needs environmental monitoring? Front Ecol Environ 5:253–260
Lubchenco J (1998) Entering the century of the environment: a new social contract for science. Science 279:491–497
Marty C (2008) Regime shift of snow days in Switzerland. Geophys Res Lett 35:L12501
Matías L, Jump AS (2015) Asymmetric changes of growth and reproductive investment herald altitudinal and latitudinal range shifts of two woody species. Global Change Biol 21:882–896
McKinney ML, Lockwood JL (1999) Biotic homogenization: a few winners replacing many losers in the next mass extinction. Trends Ecol Evol 14:450–453
McPhillips T, Bowers S, Zinn D, Ludäscher B (2009) Scientific workflow design for mere mortals. Future Gener Comput Syst 25:541–551
Medina-Sánchez JM, Villar-Argaiz M, Carrillo P (2004) Neither with nor without you: a complex algal control on bacterioplankton in a high mountain lake. Limnol Oceanogr 49:1722–1733
Medina-Sánchez JM, Delgado-Molina JA, Carrillo P (2016) Sentinels of global change (I): mixotrophic algae in La Caldera Lake. In: Zamora R, Pérez-Luque AJ, Bonet FJ, Barea-Azcón JM, Aspizua R (eds) Change impacts in Sierra Nevada: challenges for conservation. Consejería de Medio Ambiente y Ordenación del Territorio, Junta de Andalucía, pp 83–85
Menéndez R, González-Megías A, Jay-Robert P, Marquéz-Ferrando R (2014) Climate change and elevational range shifts: evidence from dung beetles in two European mountain ranges. Global Ecol Biogeogr 23:646–657
Michener WK (2006) Meta-information concepts for ecological data management. Ecol Inform 1:3–7
Morales-Baquero R, Pulido-Villena E, Reche I (2006) Atmospheric inputs of phosphorus and nitrogen to the southwest Mediterranean region: biogeochemical responses of high mountain lakes. Limnol Oceanogr 51:830–837
Moreno-Llorca R, Pérez-Luque AJ, Bonet FJ, Zamora R (2016) Historical analysis of socio-ecological changes in the municipality of Cáñar (Alpujarra, Sierra Nevada) over the last 5 centuries. In: Zamora R, Pérez-Luque AJ, Bonet FJ, Barea-Azcón JM, Aspizua R(eds) Change impacts in Sierra Nevada: challenges for conservation Consejería de Medio Ambiente y Ordenación del Territorio, Junta de Andalucía, pp 59–62
Moreno-Rodríguez J (ed) (2005) Evaluación preliminar de los impactos en España por efecto del cambio climático. Ministerio de Medio Ambiente, Madrid
Müller F, Baessler C, Schubert H, Klotz S (eds) (2010) Long-term ecological research. Between theory and application. Springer, Nueva York
Navarro-González I, Pérez-Luque AJ, Bonet FJ, Zamora R (2013) The weight of the past: land-use legacies and recolonization of pine plantations by oak trees. Ecol Appl 23:1267–1276
Nikolova N, Faško P, Lapin M, Švec M (2013) Changes in snowfall/precipitation-day ratio in Slovakia and their linkages with air temperature and precipitation. Contrib Geophys Geodes 43:141
Olden JD, Rooney TP (2006) On defining and quantifying biotic homogenization. Global Ecol Biogeogr 15:113–120
Pace ML, Carpenter SR, Cole JJ (2015) With and without warning: managing ecosystems in a changing world. Front Ecol Environ 13:460–467
Pérez-Luque AJ, Zamora R, Bonet FJ, Perez-Perez R (2015) Dataset of MIGRAME project (Global change, altitudinal range shift and colonization of degraded habitats in Mediterranean mountains). PhytoKeys 56:61–81
Pérez-Luque AJ, Navarro I, Bonet FJ, Zamora R (2016a) The importance of past land uses in the natural regeneration of Holm oak woodlands under pine plantations. In: Zamora R, Pérez-Luque AJ, Bonet FJ, Barea-Azcón JM, Aspizua R (eds) Change impacts in Sierra Nevada: challenges for conservation. Consejería de Medio Ambiente y Ordenación del Territorio, Junta de Andalucía, pp 66–67
Pérez-Luque AJ, Pérez-Pérez R, Aspizua R, Muñoz JM, Bonet FJ (2016b) Climate in Sierra Nevada: present and future. In: Zamora R, Pérez-Luque AJ, Bonet FJ, Barea-Azcón JM,

Aspizua R (eds) Change impacts in Sierra Nevada: challenges for conservation. Consejería de Medio Ambiente y Ordenación del Territorio, Junta de Andalucía, pp 27–31

Pérez-Martínez C (2016) Analysis of the palaeolimnological indicators in the lakes of Sierra Nevada. In: Zamora R, Pérez-Luque AJ, Bonet FJ, Barea-Azcón JM, Aspizua R (eds) Change impacts in Sierra Nevada: challenges for conservation. Consejería de Medio Ambiente y Ordenación del Territorio, Junta de Andalucía, pp 53–55

Peters DP, Groffman PM, Nadelhoffer KJ, Grimm NB, Collins SL, Michener WK, Huston MA (2008) Living in an increasingly connected world: a framework for continental-scale environmental science. Front Ecol Environ 6:229–237

Rabasa SG, Granda E, Benavides R, Kunstler G, Espelta JM, Ogaya R et al (2013) Disparity in elevational shifts of European trees in response to recent climate warming. Global Change Biol 19:2490–2499

Rodrigo FS, Esteban-Parra MJ, Pozo-Vázquez D, Castro-Díez Y (1999) A 500-year precipitation record in Southern Spain. Int J Climatol 19:1233–1253

Ruano F, Tinaut A (2003) Historia de la Entomología en Sierra Nevada (Sur de España) de 1813 a 1994 (1). Boletín Asociación Española De Entomología 109–126

Ruiz-Sinoga JD, Garcia Marin R, Martinez Murillo JF, Gabarron Galeote MA (2010) Precipitation dynamics in southern Spain: trends and cycles. Int J Climatol 31:2281–2289

Rüegg J, Gries C, Bond-Lamberty B, Bowen GJ, Felzer BS, McIntyre NE et al (2014) Completing the data life cycle: using information management in macrosystems ecology research. Front Ecol Environ 12:24–30

Sáinz-Bariáin M, Zamora-Muñoz C, Soler JJ, Bonada N, Sáinz-Cantero CE, Alba-Tercedor J (2015) Changes in Mediterranean high mountain Trichoptera communities after a 20-year period. Aquat Sci 78:669–682

Sala OE (2000) Global biodiversity scenarios for the year 2100. Science 287:1770–1774

Sánchez-Rojas CP, Molero-Mesa J (2016) High-mountain plant communities: GLORIA. In: Zamora R, Pérez-Luque AJ, Bonet FJ, Barea-Azcón JM, Aspizua R (eds) Change impacts in Sierra Nevada: Challenges for conservation. Junta de Andalucía, Consejería de Medio Ambiente y Ordenación del Territorio, pp 96–98

Scherrer D, Körner C (2010) Infra-red thermometry of alpine landscapes challenges climatic warming projections. Global Change Biol 16:2602–2613

Scherrer D, Körner C (2011) Topographically controlled thermal-habitat differentiation buffers alpine plant diversity against climate warming. J Biogeogr 38:406–416

Scherrer D, Schmid S, Körner C (2011) Elevational species shifts in a warmer climate are overestimated when based on weather station data. Int J Biometeorol 55:645–654

Scherrer SC, Appenzeller C, Laternser M (2004) Trends in Swiss Alpine snow days: the role of local- and large-scale climate variability. Geophys Res Lett 31:L13215

Schildhauer M, Bojilova J, Berkley C, Jones Matthew B (2001) Managing scientific metadata. IEEE Internet Comput 5:59–68

Soranno PA, Cheruvelil KS, Bissell EG, Bremigan MT, Downing JA, Fergus CE et al (2014) Cross-scale interactions: quantifying multi-scaled cause–effect relationships in macrosystems. Front Ecol Environ 12:65–73

Titos-Martínez M (2002) La investigación naturalista sobre Sierra Nevada (1732–1936). Acta Granatense 1:161–169

Villar-Argaiz M, Bullejos FJ (2016) Sentinels of global change (II): herbivorous consumers in La Caldera Lake. In: Zamora R, Pérez-Luque AJ, Bonet FJ, Barea-Azcón JM, Aspizua R (eds) Change impacts in Sierra Nevada: challenges for conservation. Consejería de Medio Ambiente y Ordenación del Territorio, Junta de Andalucía, pp 86–91

Villar-Argaiz M, Medina-Sánchez JM, Cruz-Pizarro L, Carrillo P (2001) Inter- and intra-annual variability in the phytoplankton community of a high mountain lake: the influence of external (atmospheric) and internal (recycled) sources of phosphorus. Freshw Biol 46:1017–1034

Zamora R, Barea-Azcón JM (2015) Long-term changes in mountain Passerine Bird communities in Sierra Nevada (Southern Spain): a 30-Year Case Study. Ardeola 62:3–18

Zamora R, Matías L (2014) Seed dispersers, seed predators, and browsers act synergistically as biotic filters in a mosaic landscape. PLoS ONE 9:e107385

Zamora R, Gómez JM, Hódar JA, Castro J, García D (2001) Effect of browsing by ungulates on sapling growth of Scots pine in a Mediterranean environment: consequences for forest regeneration. Forest Ecol Manag 144:33–42

Zamora R, Pérez-Luque AJ, Bonet FJ, Barea-Azcón JM, Aspizua R (eds) (2016) Global change impacts in Sierra Nevada: challenges for conservation. Consejería de Medio Ambiente y Ordenación del Territorio, Junta de Andalucía

Printed in the United States
By Bookmasters